Molecular Diagnostics: Promises and Possibilities

Mousumi Debnath · Godavarthi B.K.S. Prasad ·
Prakash S. Bisen

Molecular Diagnostics: Promises and Possibilities

Springer

Dr. Mousumi Debnath
Jaipur Engineering College &
Research Centre
Dept. Biotechnology
Jaipur-302022
India

Dr. Godavarthi B.K.S. Prasad
Jiwaji University
School of Studies in Biochemistry
Gwalior-474011
India

Dr. Prakash S. Bisen
Jiwaji University
School of Studies in Biotechnology
Gwalior-474011
India

ISBN 978-90-481-3260-7 e-ISBN 978-90-481-3261-4
DOI 10.1007/978-90-481-3261-4
Springer Dordrecht Heidelberg London New York

Library of Congress Control Number: 2009941372

Printed on acid-free paper

Springer is part of Springer Science+Business Media (www.springer.com)

Preface

Molecular diagnostics represents a revolution postponed, perhaps, but one that is now clearly underway. Early and accurate diagnosis in medicine goes big way in prevention and treatment of communicable as well as non-communicable disorders. Traditional tests typically screen for a biochemical constituent or specific antibody or antigen derived from infectious agent to a particular disease. Molecular diagnostic tests typically analyze key DNA, RNA, or protein biomarkers (analytes) to identify a disease, determine its course, evaluate response to therapy, or predict individual predisposition to a disease. The techniques applied involve analysis of DNA sequences, DNA methylation patterns, gene expression profiles, proteins, protein expression, or combinations of these biomarkers. Such biomarkers provide direct information about genotypic and/or phenotypic changes associated with specific diseases or responses to treatment. Biomarker analysis has also become an important tool in drug discovery, preclinical drug development, and patient monitoring during clinical trials. The first few chapters of the book deals with the concepts of genomics and molecular biology.

Specificity and sensitivity of DNA based tests are much higher, and the test is quicker than identification by traditional laboratory methods. In the case of human immunodeficiency virus (HIV), the initial detection limit in the late 1980s was 22 days post infection. However, the advent of nucleic acid testing for HIV RNA shortened this time to 11 days post infection. This highlights another distinct advantage offered by molecular diagnostics over immunoassays – amplification. The polymerase chain reaction (PCR) allows for the billion-fold amplification of available genetic material, thereby allowing the detection of minute amounts of genetic material and greatly increasing the sensitivity of the assay. No such amplification technique exists for proteins.

A series of technologies developed over the years to address the challenges of molecular diagnostics include eight major areas: amplification technologies (gene and signal), blotting technologies, electrophoretic technologies, microarray technologies, probe technologies, restriction fragment-length polymorphism (RFLP) analysis, RNA inhibition analysis, and single nucleotide polymorphism (SNP) analysis and software. These developments have set the stage for excellent growth potential in molecular diagnostics.

Current molecular diagnostics are primarily single-analyte tests involving the detection of a single gene or protein. However, many disease-related processes are multifactorial, involving the abnormal expression of multiple genes or proteins. Second-generation molecular diagnostics are anticipated to utilize novel detection technologies and multiplexing platforms to allow the measurement of a large number of analytes simultaneously. These innovations will increasingly utilize multiplexing platforms such as DNA microarrays that perform parallel biomarker analyses.

Molecular diagnostics also has an additional ace up its sleeve – the capability to multiplex. A single assay can be used for the detection of multiple genetic targets. An example of this would be the simultaneous detection of HIV, hepatitis C and hepatitis B. This multiplexing capability helps save time, costs and represents a more effective diagnostic tool than an immunoassay for a single analyte. The next few chapters is focused on development of various nucleic acid technologies starting from PCR to microarrays.

Molecular diagnostics got its big break following the successful sequencing of the human genome in 2003. DNA Testing has emerged fully from research into clinical practice and becoming a dominant platform in clinical medicine. This allowed the identification of portions of host or infectious agent genome that are responsible for disorder, as well as those that could be used as a marker to predict adverse reaction, dosing and efficacy to a particular drug. Applications in molecular diagnostics currently range from screening for infectious diseases and genetic disorders. It could test for genetic mutations and genome changes associated with cancers, or for the sequence or biomarker associated with a type of cancer. It has been particularly successful in diagnosing colorectal cancer, Methicillin-resistant *Staphylococcus aureus* (MRSA), breast cancer, respiratory viruses, prostate cancer and human papillomavirus (HPV). It could also be used to sequence individual genomes to identify a predisposition to developing disorders such as Alzheimer's disease, Parkinson's disease, diabetes and alcoholism. The molecular diagnostics is going to play an important role in pharmacogenomics, or personalized medicine. These methods allows to test an individual's genetic disposition to a specific drug for most effective dosage and also helps to detect if the patient would show side effects. Molecular diagnosis aids in determining the right dosage of the right drug and at the right time for each patient. Molecular diagnostics is considered to work faster than traditional microbiology and more sensitive than traditional immunoassay testing. The capabilities of molecular diagnostics are increasingly being recognized due to global trends in emerging infectious diseases. Molecular assays have the capability to differentiate viral subtypes, detect genetic predispositions to a disease and monitor the course of a disease. The severe acute respiratory syndrome (SARS) corona virus and avian influenza outbreak in the recent past created an increasing awareness for this technology. The next few chapters of the book deals with the applications of molecular diagnostics in infectious, non-infectious and metabolic disorders.

Semi and fully automated molecular testing platforms are being developed that address the issues of pre-analytical sample preparation and cross contamination. This will eventually allow molecular diagnostics to capture a larger segment of the infectious disease diagnostics market and achieve wider penetration in clinical diagnostics. The last chapter of the book deals with molecular diagnostics, its segments, market potential and current and future trends. The major shortcoming of molecular diagnostics lies in its high cost and the high level of technical skill needed to run complex assays. However, there is no match for the specificity and sensitivity of molecular assays. The adoption of the technology is increasing globally and the costs of molecular assays are expected to drop following the expiration of various PCR patents. Despite these obstacles, the speed and sensitivity of molecular diagnostics would gain more market share from traditional immunoassay testing, especially as costs come down.

We thank Mr. Devendra Singh, Mr. Gulab Singh Thakur, Mr. Bhagwan S. Sanodiya, Mr. Vikas Dubey for their valuable assistance in preparation of this book. We are particularly indebted to Mr. Vikas Dubey, Web Designer, Vikrant Institute of Technology and Management, Gwalior for front cover image design. Finally we owe an unquantifiable debt to our families, who has seen the book built sentence by sentence and whose encouragement has never flagged.

Jaipur, India — Mousumi Debnath
Gwalior, India — Godavarthi B.K.S. Prasad
Gwalior, India — Prakash S. Bisen

Contents

1 Introduction to Molecular Diagnostics 1
1.1 Prologue 1
1.2 Clinical Diagnostics Entering a New Phase 2
1.3 Concept of Molecular Diagnostics 3
1.4 Molecular Diagnostic Technology and Health Care Industries 4
1.5 Molecular Diagnostic Approaches 6
1.5.1 Nucleic Acid Test Systems 7
1.5.2 Gene-Based Diagnostics 8
1.6 Conclusion 10
References 10

2 Omics Technology 11
2.1 Prologue 11
2.2 Concept of Omics 12
2.3 Genome and Genomics 13
2.4 Transcriptome and Transcriptomics 15
2.5 Proteome and Proteomics 17
2.6 Metabolome and Metabolomics 21
2.7 Integration of Omics 23
2.8 Conclusion 27
References 29

3 Recombinant DNA Pharmaceuticals 33
3.1 Prologue 34
3.2 Concept 34
3.3 Tools of Recombinant DNA Technology 35
3.3.1 Common Enzymes Used in Recombinant DNA Technology 36
3.3.2 Vectors 38
3.3.3 Hosts 38
3.3.4 DNA Libraries 42
3.3.5 Library Screening 43
3.3.6 Polymerase Chain Reaction 44

3.3.7 Analysis of Cloned DNA Sequences 46
3.3.8 Expression of Cloned Genes 46
3.4 Therapeutic Applications of Recombinant DNA Derived Pharmaceuticals 47
3.4.1 Recombinant Therapeutic Proteins 47
3.4.2 Recombinant DNA Technology and Medicine 49
3.5 Conclusion 51
References 52

4 Aptamers: In Vitro DNA Selection 55
4.1 Prologue 55
4.2 Concept 56
4.3 Selection of Aptamers 57
4.4 Applications of DNA and RNA Aptamers 60
4.4.1 Aptamers for Detection of Pathogens 61
4.4.2 Aptamers for Therapy 62
4.4.3 Aptamers to Detect Change in Gene Expression . . . 63
4.4.4 Aptamers as Diagnostics and Therapeutics 64
4.4.5 Aptamers as Radiopharmaceutical Tools 64
4.5 Conclusion 65
References 66

5 Reporter Gene 71
5.1 Prologue 71
5.2 Concept 71
5.3 Reporter Gene Assay 74
5.3.1 *Luc* Reporter Gene System 75
5.3.2 *Ruc* Reporter Gene System 76
5.3.3 Green Fluorescent Protein (GFP) Reporter Gene System 76
5.3.4 Human Growth Hormone (hGH) Reporter Gene System 77
5.3.5 Beta-Galactosidase Reporter Gene System 77
5.3.6 CAT Reporter Gene System 78
5.3.7 *LacZ* Reporter Gene System 78
5.3.8 GUS Reporter Gene System 79
5.4 Transformation and Transfection Assays 79
5.5 Gene Expression Assays 80
5.6 Promoter Assays 80
5.7 Role of Reporter Gene Assay in Molecular Diagnostics 80
5.8 Conclusion 82
References 82

6 Impact of HGP on Molecular Diagnostics 85
6.1 Prologue 85
6.2 Origin of the Human Genome Project 85

6.3 Goals and Issues of HGP 86
6.4 The Human Genome and Genetic Variation 88
6.5 Human Genome Project and Health Services 89
6.6 Single Gene Disease 89
6.7 New and Individualised Drug Treatments 91
6.8 The Achievements and Challenges of the HGP 91
6.9 Information Transfer 92
6.10 Impact of the Human Genome Project on Epidemiologic Research 93
6.11 Impact of the Human Genome Project on Our Genomic Makeup 94
6.12 Conclusion 95
References 96

7 Molecular Diagnosis in the Post Genomic and Proteomic Era 97
7.1 Prologue 97
7.2 The Genomic Era 97
7.3 Concept 98
7.4 The Post Genomic Era 101
7.5 Advantages of Combining Multiple Types of Data 102
7.6 Biomedical Research in the Postgenomic Era 104
7.7 Conclusion 106
References 107

8 Ethics, Patents and Regulations 111
8.1 Prologue 111
8.2 Concept of Ethical, Legal, and Social Issues 111
8.3 Ethical and Social Issues in Diagnostic Molecular Genetics . . 112
8.4 Confronting the Ethical, Legal and Social Issues 114
8.5 Genes and Disease 115
8.6 The Existence of Genetic Information 117
8.6.1 Genetic Treatment of Disease 119
8.6.2 Genetic Testing 120
8.6.3 Discrimination 122
8.7 Impact of the Human Genome Project at the Interface Between Patent and FDA Law 123
8.8 Regulatory Process 125
8.9 Conclusion 126
References 127

9 Polymerase Chain Reaction 129
9.1 Prologue 129
9.2 The Concept of Polymerase Chain Reaction 130
9.3 PCR Optimisation 132
9.3.1 Magnesium Concentration 132
9.3.2 Buffer Concentration 132

9.3.3 Enzyme Concentration 133
9.3.4 PCR Primer Design 133
9.3.5 Template Quality 133
9.3.6 Template Quantity 134
9.3.7 Cycling Parameter 134
9.3.8 PCR Enhancers and Additives 135
9.3.9 Nucleic Acid Cross-Contamination 135
9.4 Advances in the PCR Technique 136
9.4.1 RT-PCR 137
9.4.2 "Hot Start" PCR 137
9.4.3 Long Range PCR 138
9.4.4 Inverse PCR 138
9.4.5 Anchored PCR 138
9.4.6 Nested Primer PCR 139
9.4.7 Colony PCR 139
9.4.8 Quantitative PCR 139
9.4.9 Real-Time PCR 140
9.4.10 Rapid Amplification of cDNA Ends (RACE) 141
9.4.11 AFLP 141
9.4.12 In Situ PCR 142
9.5 Cloning PCR Products 142
9.5.1 T-A Cloning Strategy 142
9.5.2 Incorporation of Restriction Sites in Primers 143
9.6 PCR as a Diagnostic Tool 143
9.7 Conclusion 147
References 148

10 In Situ Hybridization 153
10.1 Prologue 153
10.2 The Concept of In Situ Hybridization 154
10.2.1 The Process 154
10.3 Disadavantages of Radioactive Probes 156
10.4 Solving the Problem: Advent of FISH Technique 156
10.4.1 Fluorescence In Situ Hybridization 157
10.4.2 FISH in Action 159
10.5 Applications of FISH as a Diagnostic Tool for Research 160
10.5.1 To Analyze the Onset of Specific Gene Expression . . 161
10.5.2 Analysis of the Chromosome Structure 161
10.5.3 Localisation of RNA Transcripts 162
10.5.4 FISH as a Molecular Cytogenetic Technique to Understand Diseases 162
10.5.5 FISH for Detection of Pathogens 165
10.6 Recent Advances of In Situ Hybridisation Technology 165
10.7 Conclusion 166
References 167

11 Immunoassay 171
11.1 Prologue 171
11.2 Concept 172
11.3 Types of Immunoassay 172
11.4 Recent Advances in the Field of Immunodiagnostics 174
11.5 Clinical Applications of Immunoassay 175
11.6 Utilization and Interpretation of Immunological Tests 177
11.7 Conclusion 178
References 178

12 Phage Display 181
12.1 Prologue 181
12.2 Concept 181
12.3 Phage-Display Libraries as Populations of Replicable, Mutable Chemicals 182
12.4 Practical Applications of Phage Display 184
12.4.1 Target Receptors Used in Affinity Selection 184
12.4.2 Epitope Mapping and Mimicking 185
12.4.3 Identifying New Receptors and Natural Ligands . . . 186
12.4.4 Drug Discovery 186
12.4.5 Epitope Discovery – A New Route to Vaccines and Diagnostics 187
12.4.6 Selection of DNA-Binding Proteins 188
12.4.7 Landscape Libraries as a Source of New Materials . . 189
12.4.8 Phage Display-Combinatorial Chemistry on the Cheap 190
12.4.9 Cloning Allergens by Phage Display 190
12.5 Conclusion 191
References 192

13 Microarray 193
13.1 Prologue 193
13.2 The Concept of Microarray 195
13.3 Current Challenges of Microarrays 198
13.3.1 Gene Discovery 199
13.3.2 Gene Expression Profiling 200
13.3.3 Pharmacogenomics and Microarray 202
13.3.4 Molecular Diagnostic Research 204
13.4 Conclusion 205
References 206

14 DNA Biosensors 209
14.1 Prologue 209
14.2 The Concept of DNA Biosensor 210
14.3 Applications of Biosensors 211
14.4 Advantages of Biosensors 212

14.5 Development of DNA Hybridization Biosensor 213
14.6 DNA Biosensor for Molecular Detection of Pathogens 217
14.7 Biosensors as Analytical Tools in the Food and Drink Industries 218
14.8 Potential of Biosensor for Environmental Monitoring 220
14.9 Conclusions & Future Challenges 221
References 223

15 Molecular Microbiological Testing 227
15.1 Prologue 227
15.2 Concept 228
15.3 Advent of Improved Diagnostics 229
15.4 Traditional Microbial Typing 230
15.4.1 Biotyping 230
15.4.2 Antibiograms, Resistograms, and Bacteriocin Typing 231
15.4.3 Protein Analysis 231
15.4.4 Phage Analysis 231
15.4.5 Chromatographic Analysis 231
15.5 Nucleic Acid-Based Typing Systems 232
15.5.1 Plasmid Analysis 232
15.5.2 Restriction Enzyme Pattern 232
15.5.3 Ribotyping 232
15.5.4 Random Amplified Polymorphic DNA (RAPD) . . . 232
15.5.5 Nucleic Acid Probes 233
15.5.6 Polymerase Chain Reaction 233
15.6 Current Application of Molecular Diagnostics 234
15.6.1 Clinical Microbiology 234
15.6.2 Clinical Epidemiology and Infection Control 237
15.7 Promise of Molecular Testing 238
15.8 Assay Validation-Analytic Sensitivity and Specificity 239
15.9 Conclusion 241
References 241

16 Proteomic Technology 245
16.1 Prologue 245
16.2 Concept 246
16.3 Methods to Perform Proteomic Analysis 247
16.4 Functional Proteomics 249
16.5 Implications of the Human Genome Project 249
16.6 Measurement Using a Proteomic Approach 251
16.7 Use of Mass Spectrometry in Proteomics 252
16.8 Proteomics: From Basic Research to Diagnostic Application 253
16.9 Role of In Vitro Diagnostics (IVD) in Health Care 254
16.10 Goals of Proteomics 257
16.11 Conclusion 258
References 259

17 Nanomedicine . . . 261
17.1 Prologue . . . 261
17.2 Concept . . . 262
17.3 Nanotechnology in Medicine . . . 264
17.3.1 Use of Quantum Dots . . . 265
17.3.2 Cell Targeting . . . 266
17.3.3 New Nanocarriers for Drug Delivery . . . 267
17.3.4 Gene Therapy . . . 270
17.3.5 Molecular Medicine . . . 272
17.3.6 Construction of Nanostructure Template . . . 273
17.3.7 Targeting Drugs to Cells . . . 274
17.4 Nanodiagnostics . . . 276
17.5 Molecular Electronics . . . 278
17.6 Nanotechnology Biosensor for Pathogen Detection . . . 280
17.7 Nanotechnology Composite Materials for Next Generation Biomedical Applications . . . 280
17.8 A Carbon Nanotube Biosensor . . . 281
17.9 Conclusion & Future Directions . . . 282
References . . . 283

18 Biomarkers . . . 287
18.1 Prologue . . . 287
18.2 Concept of Biomarkers . . . 288
18.3 Biomarkers on the Horizon . . . 289
18.3.1 DNA Biomarkers . . . 290
18.3.2 RNA Biomarkers . . . 290
18.3.3 Protein Biomarkers . . . 291
18.4 New Applications for Established Biomarkers . . . 294
18.4.1 Development Pathways . . . 295
18.4.2 Identification of Biomarkers . . . 295
18.4.3 Chronology of Biomarker Development . . . 296
18.4.4 Role of Biomarkers in Cancer Detection, Diagnosis, and Prognosis . . . 297
18.4.5 Biomarkers in Cancer Therapy . . . 299
18.4.6 Biomarkers in Early Detection . . . 302
18.4.7 Biomarkers in Risk Assessment . . . 302
18.5 Organization of a Program in Protein Biomarker Discovery . . . 303
18.5.1 Cancer Site Teams . . . 304
18.5.2 Biomarker Mines . . . 304
18.5.3 Informatics Platform . . . 305
18.5.4 Reagents Core . . . 305
18.5.5 Incorporating Technology Improvements . . . 306
18.6 Conclusion . . . 307
References . . . 307

19 Diagnosis and Monitoring of Infections 309
19.1 Prologue 309
19.2 Concept of Diagnosis and Monitoring of Infections 310
19.2.1 Detection and Identification of Pathogens Without Target Amplification 311
19.2.2 Nucleic Acid Amplification 312
19.2.3 Detecting Antimicrobial-Drug Resistance 314
19.3 Molecular Diagnostics of Infections 316
19.3.1 Sexually Transmitted Infections 316
19.3.2 Vector Borne Disease 318
19.3.3 Viral Infections 318
19.3.4 Bacterial Infections 321
19.3.5 Fungal Infections 323
19.3.6 Practical Applications of Molecular Methods in the Clinical Microbiology Laboratory 323
19.4 Conclusion 327
References 328

20 Diagnosis of Mutation and Genetic Disorders 331
20.1 Prologue 331
20.2 Concept 332
20.3 Factors Regulating a Genetic Disease 333
20.4 Genetic Testing 334
20.4.1 Cytogenetic Testing 334
20.4.2 Biochemical Testing 335
20.4.3 Molecular Testing 336
20.5 Current Status of Molecular Diagnosis of Some Common Genetic Diseases 336
20.5.1 Cystic Fibrosis 337
20.5.2 Duchenne Muscular Dystrophy 338
20.5.3 Haemophilia A 338
20.5.4 Haemophilia B 338
20.5.5 Phenylketonuria 339
20.5.6 β Thalassaemia 339
20.5.7 Wilson's Disease (Hepatolenticular Degeneration) . . 339
20.5.8 α Antitrypsin Deficiency 340
20.5.9 Familial Hypercholesterolaemia and Other Lipoprotein Disorders 340
20.5.10 Huntington's Disease Gene 340
20.5.11 Multiple Endocrine Neoplasia 341
20.5.12 Factor V (Leiden) Mutation 342
20.5.13 Hemochromatosis Gene, HFE 342
20.5.14 Colon Cancer Gene, APC 343
20.6 Microarray Analysis for Detection of Complex Pattern of Genes 343

20.7 Conclusion 344
References 344

21 Diagnosis of Complex Diseases 347
21.1 Prologue 347
21.2 Concept 348
21.3 Background Information 349
21.4 Detecting New Metabolic Disease Pathways 350
21.5 Phenotypic Variations in Simplex Diseases 350
21.6 The Genetic Background of Complex Disorders 351
21.7 Dissecting Interactions Between Genes and Environment . . . 351
21.7.1 Alzheimer Disease 353
21.7.2 Cancer 356
21.7.3 Multiple Sclerosis 360
21.7.4 Diabetes 363
21.7.5 Hemophilia 364
21.7.6 Obesity 367
21.7.7 Schizophrenia 370
21.8 Conclusion 374
References 375

22 Biochips 383
22.1 Prologue 383
22.2 Concept 383
22.3 Design of Biochip 385
22.4 Applications of Biochips 385
22.4.1 Biochip and Clinical Medicine 386
22.4.2 Biochips and Biosensors 387
22.4.3 Biochip and Pharmacogenetics 387
22.4.4 Biochip and Biowarfare 388
22.4.5 Biochip and Diagnostics 388
22.4.6 Biochip and Cancer 389
22.4.7 Biochip Department of Defense 389
22.5 Biochip Market 390
22.6 Future of Biochips 390
References 391

23 Personalised Medicine 393
23.1 Prologue 393
23.2 Concept 394
23.3 Practice of Medicine in the Twenty-First Century 396
23.4 Role of Personalized Medicine in Drug Discovery 396
23.5 Molecular Diagnosis Will Determine Prognosis and Therapy 399
23.6 Personalized Medicine and Genetic Markers 400
23.7 Challenges of Realizing the Promise of Personalized Medicine 401

23.8 Personalised Medicine and Pharmacogenomics 403
23.9 Personalized Medicine and Diseases 406
23.10 Personalized Medicine and Diagnostics Industry 406
23.11 Impact of the US Patent System on the Promise of Personalized Medicine 408
23.12 Ethical Legal and Social Issues of Personalised Medicine . . . 409
23.13 Conclusion 410
References 410

24 Biopharmaceutical Industry and Health Care 413
24.1 Prologue 413
24.2 Concept 414
24.3 Biopharmaceutical Research 415
24.4 Opportunities in Healthcare 416
24.5 Molecular Diagnostics and Health Care 417
24.6 Global Context 418
24.7 Emerging Biopharmaceuticals 420
24.8 Challenges of the Biopharmaceutical Industry 423
References 424

25 Forensic Medicine 425
25.1 Prologue 425
25.2 Concept 426
25.3 Forensic Medicine and DNA Fingerprinting 428
25.4 Applications of DNA Methylation Markers in Forensic Medicine 431
25.5 Forensic Medicine and Anthropometry 431
25.6 Clinical Forensic Medicine 432
25.7 Conclusion 433
References 433

26 Pharmacogenomics 435
26.1 Prologue 435
26.2 Concept 436
26.3 Predicting Drug Response on Gene Variation 437
26.4 Drug Development and Testing Benefit from Pharmacogenomics 437
26.5 Applications and Benefits of Pharmacogenomics 438
26.5.1 Better, Safer Drugs the First Time 438
26.5.2 More Accurate Methods of Determining Appropriate Drug Dosages 438
26.6 Benefits of Pharmacogenomic Testing 439
26.6.1 Patient's Ability to Metabolize Drugs 439
26.6.2 Age-Related Genetic Variations 440
26.7 Recent Reports of Pharmacogenomics Use 441
26.8 Interpreting Pharmacogenomic Tests 442
26.9 Barriers to Pharmacogenomics Progress 443

26.10 The Promise of Pharmacogenomics 444
26.11 Impact of Drug Discovery, Development, and Marketing Affected by Pharmacogenomics 446
26.12 Pharmacogenomics Initiatives 448
26.13 Pharmacogenomics and the Pharmaceutical Industry 448
26.14 Applied Pharmacogenetics 449
26.15 Ethical Concerns 450
26.16 Molecular Diagnostic Methods for Optimizing Drug Therapy . 451
26.17 Challenges for the Future 452
References 453

27 Gene Technology in Forensic Sciences 455
27.1 Prologue 455
27.2 Concept 456
27.3 DNA Typing and Genetic Markers 456
27.4 DNA Based Methods for Identification of Individuals 459
27.4.1 Non-PCR Based Methods 459
27.4.2 PCR Based Methods 460
27.5 Ethical, Legal, and Social Concerns About DNA Databanking 463
27.6 Future of DNA Typing Systems 463
References 464

28 Food Diagnostics 467
28.1 Prologue 467
28.2 Concept 468
28.3 Tools of Molecular Diagnostics to Assess Food Quality 469
28.4 Food Allergy 470
28.5 Food-Borne Disease 470
28.6 Commercial Rapid Food-Safety Diagnostics 471
28.7 Methods for Analysis of Food 472
28.7.1 Rapid Methods 473
28.8 Future Technologies 476
28.8.1 Microarrays 476
28.8.2 Sensor Technology 477
28.8.3 Flow Cytometry 478
28.8.4 Bacteriophage-Based Techniques 478
28.8.5 Riboprinting and Pulse-Field Gel Electrophoresis . . 479
28.8.6 Applications and Limitations of Rapid Methods . . . 479
28.9 Business Outlook of Food Diagnostics 479
28.10 Conclusion 480
References 481

29 Rapid Diagnostic Methods for Biowarfare 483
29.1 Prologue 483
29.2 Concept 484
29.3 Characteristics of Biological Weapons 485

29.3.1 Anthrax 487
29.3.2 Plague 487
29.3.3 Tularemia 487
29.3.4 Melioidosis 488
29.3.5 Viral Hemorrhagic Fevers 488
29.3.6 Other Viral Fevers 488
29.3.7 Trichothecene Mycotoxins 488
29.3.8 Aflatoxin 489
29.4 Biodefense 489
29.4.1 Viral Agents – Poxviridae 490
29.4.2 Botulinal Toxins 490
29.4.3 Mycotoxins 491
29.5 Combating Detection of Biowarfare Agents 491
29.5.1 Prophylatic & Therapeutic Approaches 491
29.5.2 Detection Methods 495
29.6 Impact of Biological Weapons 495
29.7 Tools for Self Defence Against Bioweapons 496
29.8 Genetic Engineering and Biological Warfare 497
29.9 Impact of Genomics and Genetic Technology 499
29.10 Conclusion 500
References 500

30 Segments of Molecular Diagnostics – Market Place 503
30.1 Prologue 504
30.2 Concept 504
30.3 Market Drivers 506
30.4 Trends in Infectious Diseases Testing Market 507
30.5 Trends in Cancer Diagnostic Testing World Markets 507
30.6 Trends in Cardiac Marker Diagnostic Testing Markets 508
30.7 Point of Care Diagnostic Testing World Markets 508
30.8 Next Generation Molecular Diagnostics 509
30.9 The Rise of Companion Diagnostics 510
30.10 Market Considerations and Forecasts 510
References 513

Index 515

Chapter 1
Introduction to Molecular Diagnostics

Abstract Molecular diagnostics has become a growing part of the clinical laboratory. It includes all tests and methods to identify a disease and understand the predisposition for a disease analyzing DNA or RNA of an organism. Rapid advances in molecular diagnostics enable basic research and results in practical diagnostic tests. The basic application is to determine changes in sequence or expression levels in crucial genes involved in disease. The use of molecular diagnostics, such as pre-implantation diagnostics or predictive genetic testing, still has technical problems as well as novel, and to date unclear, social, ethical and legal implications. The scope of molecular diagnostics in molecular medicine could be expanded well beyond current nucleic acid testing. It plays an important role in practice of medicine, public health, pharmaceutical industry, forensics and biological warfare and drug discovery. The molecular diagnostic marketplace offers a growth opportunity given the interest in utilizing molecular tools to precisely target therapeutics.

Keywords Clinical diagnostics · Molecular biology · Health care · Molecular diagnostics · Genetics · Diagnostic medicine · Prognostic · Molecular diagnostic industry · Clinical testing · Therapeutics · PCR · Microarray

1.1 Prologue

The past few years has seen enormous progress in molecular genetics. Techniques such as Southern blotting, Northern blotting, polymerase chain reaction, nucleotide sequencing, chromosome walking and genetic transfer allow the specific isolation, characterization and modification of genetic information (Boerman et al., 2001). Molecular studies allow the laboratory to become predictive. Now statements can be made about events that occur to a patient in the future. This new technology returns results that give an indication that the patient may be at risk for the disease long before it becomes symptomatic disease.

Molecular pathology is in a state of rapid evolution featuring continuous technology development and new clinical opportunities for drug selection predicting efficiency, toxicity and monitoring disease outcome. Thus major advances are being

M. Debnath et al., *Molecular Diagnostics: Promises and Possibilities*,
DOI 10.1007/978-90-481-3261-4_1, © Springer Science+Business Media B.V. 2010

made in the science of genetics, resulting in increased use of molecular technology in clinical laboratory.

1.2 Clinical Diagnostics Entering a New Phase

Biotechnology, in all its forms, is the fastest-growing discipline in the modern laboratory, whether be it clinical, research or forensic (Jain, 2002). The volume of research from the past is translating to a growing list of new diagnostic tests for the patient oriented clinical laboratory. Optimists have characterized these developments as "mankind about to determine its destiny," whereas others are afraid that scientists are about to "start playing God."

Molecular biology has revolutionized biological and biomedical research and has become an indispensable tool in clinical diagnostics (Fig. 1.1). It has developed more than any other science in the last 10 years. Until this time, the laboratory had been descriptive in nature. It could measure events that were currently going on by evaluating the chemistry, hematology or anatomical pathology.

Major advances are being made in the science of genetics, resulting in the increased use of molecular technology in the clinical laboratory. A wide variety of drugs in late preclinical and clinical development are being targeted to disease specific gene and protein defects that will require co-approval of diagnostics and therapeutics products by regulating agencies. An increasingly educated public will demand more information about their predisposition for serious diseases, how these potential illnesses can be detected in an early stage, when they can be arrested or cured with new therapies custom-designed for their individual clinical status. To respond to this demand, major pharmaceutical companies will form partnerships with diagnostics companies or develop their own in-house capabilities that will permit efficient production of more effective and less toxic integrated personalized medicine drug and test products. For clinical laboratories and pathologists, this integration of diagnostics and therapeutics represents a major new opportunity to emerge as leaders of the new medicine, guiding the selection, dosage, route of administration, and multidrug combinations and producing increased efficacy and reduced toxicity of pharmaceutical products.

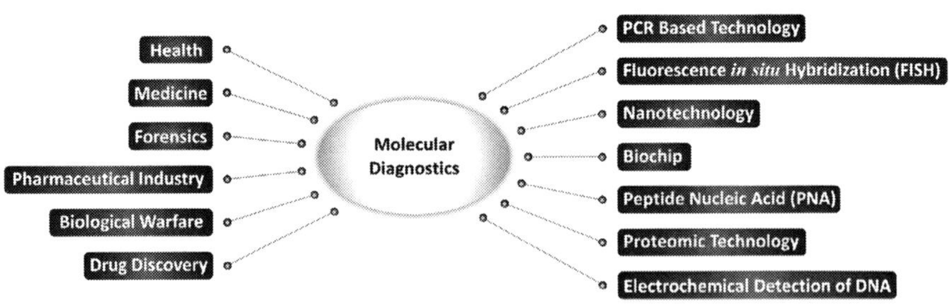

Fig. 1.1 Applications of molecular diagnostics in the field of clinical diagnostics

The clinical laboratory has a long history of developing, managing, and providing assays and test results for diagnosing disease. A good example of this is the development of immunoassays and the enhancements made to this technology with the introduction of monoclonal antibodies. Clinical molecular genetics is now part of the mainstream of medical care in the United States (Amnos and Patnaik, 2002). All commercial clinical reference laboratories now have a molecular genetic diagnostic unit, many of which are in contractual agreement with third party payers to provide services. Gene discovery provides valuable insight into the mechanisms of disease processes and gene-based markers will enable clinicians to study disease predisposition, as well as improved methods for diagnoses, prognosis, and monitoring of therapy. The broad range of mutation spectrum and type performed in the clinical laboratory requires the use of multiple technologies rather than a single typing platform. Platform choice depends on such diverse factors as local expertise, test volume, economies of scale, R&D budget, and royalties. Test validation is a major hurdle and positive control samples are often not readily available. Oversight and the regulatory environment for clinical molecular genetics laboratories in the United States are evolving rapidly. Several government agencies and private organizations are currently involved in revision of specific laboratory standards, including the Secretary's Advisory Committee on Genetic Testing (SACGT), Food and Drug Administration (FDA), Center for Disease Control (CDC), College of American Pathologists (CAP), American College of Medical Genetics (ACMG).

Over the past 50 years, the clinical laboratory has evolved into a complex, technology-driven enterprise with the principal tasks of diagnosing and screening for disease, monitoring health and therapeutic response, and gauging deviations from normal physiology in humans and animals. Advances in diagnostic medicine, on the other hand, have come through the application of science and technology as a result of a coordinated effort among academia, industry, government, and private institutions. We are now entering the era of Molecular Diagnostics and Pathology, which is bringing forth the newest and most powerful science and technology available for the modern-day practice of diagnostic laboratory medicine. Among the numerous important areas to consider with molecular diagnostics are the emerging issues concerning the development of genetic assays and their use for testing individual patient responses or suitability for pharmaceutical drugs. The rapidly growing area of molecular diagnostics is ideally suited to clinical laboratories (Hall et al., 2003).

1.3 Concept of Molecular Diagnostics

Molecular diagnosis is the detection of pathogenic mutations in DNA and RNA samples to aid in detection, diagnosis, subclassification, prognosis, and monitoring response to therapy. It covers current molecular biological techniques used to identify the underlying molecular defects in inherited disease. Principles underlying nucleic acid based diagnosis originate from localization, identification, and characterization of genes responsible for human disease. It also determines how these genes and proteins are interacting in a cell. It focuses upon patterns, gene and protein activity patterns in different types of cells. Molecular diagnostics uncovers these sets of changes and captures this information as expression patterns. Also called "molecular signatures," these expression patterns are improving the clinicians' ability to diagnose diseases. Molecular Diagnostics is constantly translating new discoveries and novel technologies into useful clinical tests that provide a molecular fingerprint of tumors and they are predictive of the response to specific therapies. Molecular diagnostic technologies offer the potential for moving from diagnostics to prognostics.

More than 80% of molecular tests performed today address infectious disease detection and management. The test menu for infectious diseases includes human immunodeficiency virus (HIV), hepatitis B virus (HBV), hepatitis C virus (HCV), human papilloma virus (HPV), cytomegalovirus (CMV), *Chlamydia trachomatis, Neisseria gonorrhoeae*, and *Mycobacterium tuberculosis*. Other molecular tests are applied in forensic medicine, paternity testing, tissue typing, oncology, and food and beverage testing. The molecular diagnostic testing can be performed on very small amounts of tissue obtained from small biopsies and fine-needle aspirates. Testing can even be performed on archival material (material that has been collected in the past and fixed in formalin). It is usually not necessary to take extra tissue for these tests, since most of them can be performed on tissue that is left over after routine diagnostic analysis.

1.4 Molecular Diagnostic Technology and Health Care Industries

Advancing knowledge and technologies are pushing the diagnostics industry to the front, and the molecular diagnostics industry in particular is emerging as a powerful health care player with tremendous potential. Molecular diagnostics is the fastest growing segment of clinical testing today, comprising 3–4 times the growth of the core business, depending on the country. The molecular diagnostic market is expected to grow 15–17% annually in the short term. Anticipation of such steady growth is based on expectations that maturation of the technology, the development of diagnostic applications, and expansion of the market into developing countries will lead to increased acceptance and market penetration of the tests. Continuing simplification of the technology and instrumentation used in molecular diagnostic tests holds the key to this phase of development. In the US, this sector is predicted to expand to 30% of the total service diagnostics revenues by 2015. Traditionally, it has only included testing for infectious disease (such as HIV and HPV); genetics (such as cystic fibrosis, and BRCA1 and 2); and, to a lesser extent, oncology (such as leukemia and lymphoma molecular classification). However, trends are emerging based on new molecular tests for the diagnosis and management of several diseases beyond cancer, infectious disease and genetics. This includes testing for immunity, metabolism, the central nervous system and cardiovascular function. In addition, new technologies which analyze multiple genes (10–10,000), single cells or new metabolites are now available, providing rapid information to better classify or manage disease.

Recent progress in the development of molecular diagnostics in medicine has been rapid and the hope has been expressed by some people that it will soon be possible to have a detailed "genetic readout" to assist in the diagnosis of treatment of a variety of diseases (Hall et al., 2003). The genomic discovery will fuel the diagnostic marketplace because every gene-based therapy might need an accompanying (molecular) diagnostic test. At approximately 30,000 genes for the human genome and with an estimated diagnostic significance of about 5%, the commercialization of 1,500 gene-based tests can be expected.

The way is paved by infectious disease testing and blood banking applications, but pharmacogenetic, predisposition diagnostics and molecular cancer diagnostics applications will follow soon and post strong numbers in the years to come. Cancer for instance, represents a set of diseases with vast unmet clinical need for improved diagnostics and therapeutics. Cancer in the developed western countries with prevalence between 1.0 and 1.5% depicts a large market with a high potential for R&D and fields like therapy specific diagnostics (theranostics). The function of such technology is to determine changes in sequence or expression levels in crucial genes involved in cancer.

Molecular diagnostics are seen as being a major breakthrough in medical science. Rapid progress in this field is expected to fundamentally change health care (Jain, 2002). While the technology itself has broad applications to such areas as agricultural bioterrorism and environmental science, medical and medical research applications have provided the primary focus. As a result, great things have been predicted for the development and use of new therapeutics based on molecular testing approaches.

Gene-based molecular diagnostics is changing the practice of medicine and will continue to do so for the foreseeable future. The primary molecular diagnostics approaches are based on two major database sciences (genomics and proteomics) and their relationship to disease and metabolic processes (functional genomics and functional proteomics). The genome or proteome that forms the basis for these tests is very often of human origins, but may also be from a pathogen. As a result, the breadth of application for molecular diagnostics is virtually unlimited. With all this opportunity, however, progress has been slow in the development, introduction and commercialization of molecular diagnostics. This has been related to the difficulty and expense of performing such assays, and the utility of the test procedures in the absence of substantive therapeutic options based on the additional data provided by these assays.

The main objectives of this analysis are:

(1) To identify viable technology drivers through a comprehensive look at platform technologies for molecular diagnostics, including probe-based nucleic acid assays, microarrays and sequencing;

(2) To obtain a complete understanding of the chief molecular diagnostics tests i.e., predictive, screening, prognostic, monitoring, pharmacogenomic and theranostic from their basic principles to their applications;
(3) To discover feasible market opportunities by identifying high-growth applications in different clinical diagnostic areas and by focusing on expanding markets, such as communicable diseases, cardiology and oncology; and
(4) To focus on global industry development through an in-depth analysis of the major world markets for molecular diagnostics, including growth forecasts.

The molecular diagnostic industry is characterized by a diverse, constantly changing technology base that continuously produces new opportunities and applications which has become a growing part of the clinical laboratory in recent years. The trend will not only continue but will show significant growth in the future. Similarly, a growing understanding of the molecular basis of cancer and other chronic diseases is opening up new realms of medicine to the possibilities of molecular diagnostic testing. Molecular signatures can also be important for the detection of minimal residual or recurrent disease and can help distinguish a metastasis from a second primary tumor. The tests are highly accurate and sensitive. When interpreted within the appropriate clinical situation and combined with other diagnostic procedures, these tests can help health care providers diagnose lymphomas and other tumors earlier so that appropriate treatment can be started.

Genetic screening tests, despite some restrictions are a promising area for future expansion of in vitro diagnostic market. Molecular diagnostics is being combined with therapeutics and forms an important component of integrated healthcare. Molecular diagnostic technologies are also involved in development of personalized medicine based on pharmacogenetics and pharmacogenomics. Currently, there has been a considerable interest in developing rapid diagnostic methods for point-of-care and biowarfare agents such as anthrax.

To date, the technical elements of molecular therapy have not yet fulfilled their expectations. In the broad spectrum of obstetrics and gynecology, new molecular discoveries are influenced not only by technical but also by socioeconomic and political considerations. These include, for example, free access to genetic testing, patents for genes and the financial monopoly over molecular medication (Beckmann et al., 2003).

1.5 Molecular Diagnostic Approaches

Molecular diagnostics focuses primarily on nucleic acids. Rapid advances in molecular diagnostics both enable basic research and result in practical diagnostic tests (Dennis and Wittwer, 2009). These types of tests include the analysis of DNA, RNA (nucleic acids) micro RNAs, and complex proteomic or metabolomic pattern array based tests. In the foreseeable future, proteomic/metabolomic tests may form their own sub-category (Fig. 1.2). However, they are presently included among the general group of molecular diagnostics.

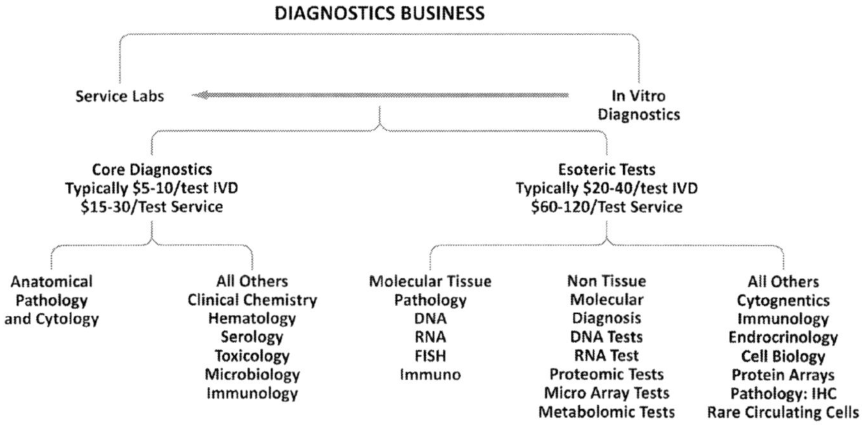

Fig. 1.2 The diagnostic business includes all service laboratories and in vitro diagnostics

1.5.1 Nucleic Acid Test Systems

The field of molecular diagnostics is represented primarily by the nucleic acid test (NAT), which is used to detect genomic DNA or RNA sequences and to characterize genomic aberration or sequence polymorphisms. For the past 20 years, NATs have been mainly hybridization-based assays, with detection a function of label molecules coupled to the probe sequence. Some commercial NATs are based on direct hybridization detection of ribosomal RNA. The high sensitivity of the assay is attributable partly to the fact that there are thousands of ribosomal RNA molecules per cell. To be able to detect less-abundant RNA or DNA molecules at increased sensitivity usually requires some form of amplification to generate a detectable signal.

Amplification strategies vary, but techniques can be categorized as target, probe, or signal amplifications. Target amplification employs various sequences of enzymatic reactions to amplify the targeted nucleic acid fragment, that is, to produce multiple copies of this target sequence. Most notable of the target amplification tests is polymerase chain reaction (PCR), developed by Roche Diagnostics Corp. (Indianapolis). Also worth mentioning are strand-displacement amplification (SDA) from BD (Franklin Lakes, NJ), transcription-mediated amplification (TMA) from Gen-Probe Inc. (San Diego), and nucleic acid sequence-based amplification (NASBA) from Organon Teknika Corp. (Durham, NC). All of the hybridization-based assays may use chromogenic, fluorescent, or optical reporter labels to produce the measurable light units detected with an optical readout device. Today, polymerase chain reaction (PCR) is the gold standard for molecular assays, owing to its intrinsic versatility and flexibility. Advances in PCR, multiplexing, sequencing, and other technologies are propelling both new and old companies forward with novel capabilities. Roche's PCR-based product lines, which have application strictly in the infectious diseases area, have captured 47% of the molecular diagnostics market.

This circumstance will not change until molecular assays for cancer and heritable diseases become part of the routine test menu.

The objective of molecular test manufacturers is to develop homogeneous closed-tube molecular test systems that are suited for diagnostic applications. A number of labeling technologies for homogeneous molecular assays have been developed that employ various configurations of probe and fluorescence label, and incorporate different versions of fluorescence resonance energy transfer (FRET). In a FRET-based assay, a fluorescent-tagged molecular probe is excited by a laser. The fluorophore responds by emitting energy that is transferred to a second, quenching, fluorophore. If these two fluorophore molecules are very close to each other, the fluorescence is quenched.

The desire to find better ways to design and manufacture these labels and probes, so as to achieve better performance and a lower cost per test, drives the development of new labeling methods in this area. Now, homogeneous-label methods are widely used by molecular laboratories in in-house PCR tests for infectious diseases, heritable diseases, and cancer. Several IVD manufacturers are incorporating these methods in their own molecular assays. Although homogeneous assays often provide the best sensitivity in detection, the limited number of different fluorophores available for simultaneous detection of multiple analytes from a single sample makes this assay platform less attractive for use in multiplex assays.

1.5.2 Gene-Based Diagnostics

Molecular biology has revolutionized biological and biomedical research and has become an indispensable tool in clinical diagnostics. Gene expression profiling, and modifications at the protein levels, will be critical to cancer classification and staging. Lately, demand for multiplexing platforms that perform parallel nucleic acid or protein analyses in diagnostic applications has been steadily increasing.

There has been a dramatic rise in the use of DNA microarrays for medical research over the last few years. Although generally perceived as a method for testing thousands of gene fragments with low sensitivity and reproducibility, the GeneChip from Affymetrix Inc. (Santa Clara, CA) is an example and advanced microarray platforms are being devised for diagnostic ends. New platforms, like the Flow-Thru Chip developed by MetriGenix Inc. (Gaithersburg, MD), promise high assay performance, amenability to automation, and the flexibility to handle low- to high-throughput diagnostic applications. These microarrays will be targeting tens to hundreds of analytes rather than thousands. More important, they will combine bioinformatic and microfluidic technologies with the microarray in order to miniaturize and simplify the tests. These new tools allow us to analyze thousands of genes simultaneously leading to a better understanding of gene function and dysfunction.

The first commercially available microarray, the AmpliOnc I from Abbott, was developed from a subsidiary's Vysis GenoSensor microarray system. It is designed to detect abnormal increases in gene copies for 58 different genes. A number of biochip makers are pursuing collaborative arrangements with IVD manufacturers and drug manufacturers to exploit the possibility of their microarray technologies being used in clinical product development. In addition to the microarray platform, bead-based technologies are also gaining recognition for their potential utility in clinical multiplex testing. Each bead incorporates a different label dye and thus has a different spectral address. One can be distinguished from another by its unique wavelength. If each bead is associated with only one specific analyte, then multiplex assays of nucleic acid or protein can be performed. The flexibility of the multiplex format of bead-based assays is the biggest advantage these assays have over the microarray platform. The LabMAP system from Luminex Corp. (Austin, TX) is an example. However, this system carries a disadvantage in that it is limited to fewer than 100 elements per test. Other technologies such as mass spectroscopy and denaturing high-pressure liquid chromatography have also recently been making their way into the molecular diagnostics field.

The involvement of thousands and millions of genetic markers in specific diseases and physiological pathways has just begun to be biologically validated. Development of biological assay content in this field will take much longer than technology development. But anticipation is high that, with the large-scale gene-discovery efforts being undertaken by many government and private institutions, molecular diagnostics will come to maturity in the next 5–10 years.

The market potential for oncological and genetic-disease molecular testing is smaller in the near term than that for infectious-disease testing. For the foreseeable future, in-house molecular tests, offered as analyte specific reagents (ASRS), will remain an integral part of the testing in this field. Only an estimated 25% of the 4,000 known hereditary diseases can be analyzed by means of molecular testing techniques today. Additional factors, including technical, patent, and ethical issues, may further restrict the development and market penetration of genetic diagnostic tests.

In this molecular medicine era, it has become widely accepted that almost every disease process results from the interaction of the environment and individual lifestyle with a specific genetic predisposition. There will be no more "one drug fits all," and pharmaceutical medicines will be prescribed to targeted patient populations case by case on the basis of the individual's genomic profile. Furthermore, with early detection and intervention available, most of the historically life-threatening diseases will become merely chronic disease conditions that require long-term management. Pharmacogenomic testing for patient drug susceptibility, dose response and drug resistance thus holds exciting potential for molecular diagnostics in years to come. According to a presentation at a recent industry conference, 370 biopharmaceutical drugs are in late stages of clinical development. Management of several medical conditions, including diabetes, asthma, autoimmune disease, and cardiac disease, will greatly benefit from these gene-based diagnostics and therapeutic interventions.

1.6 Conclusion

Molecular diagnostic technologies will play an important role in practice of medicine, public health, pharmaceutical industry, forensics and biological warfare in the twenty-first century. This includes several polymerase chain reactions (PCR) based technologies, fluorescence in situ hybridization (FISH), peptide nucleic acids (PNA), electrochemical detection of DNA, biochips, nanotechnology (Jain, 2003) and proteomic technologies.

Molecular diagnostics has evolved rapidly during the past decade and has an impact on the practice of medicine as well as many other applications including drug discovery. Advances in biotechnology have been incorporated into molecular diagnostics. There has been a distinct trend in miniaturization with development of biochips and micro fluidics. This trend has continued with the development of nanotechnology. Nanotechnologies are now being applied to molecular diagnostics to refine and extend the limits of detection (Jain, 2003). The molecular diagnostics offers a growth opportunity in utilizing molecular tools to precisely target therapeutics. The impact of single nucleotide polymorphisms (SNPs) and other molecular diagnostic markers is now being felt. This will grow as more molecular markers are validated and launched for their diagnostic and prognostic value. Ideally, therapy will be based on this information, aiming at the most effective mechanism based treatment with the least toxicity (Zlotogorski et al., 2002). Molecular diagnostics, in the future, will be expected to provide gene profile based personalized therapeutic approaches.

References

Amos, J., Patnaik, M., 2002, Commercial molecular diagnostics in the U.S.: The Human Genome Project to the clinical laboratory. *Hum Mutat* 19:324–333.

Beckmann, M.W., Strick, R., Strissel, P.L., Fasching, P.A., Oppelt, P., Pöhls, U.D., Malur, S.U., Ackermann, S., 2003, Aspects of molecular diagnostics and therapy in obstetrics and gynecology. *Expert Rev Mol Diagn* 3:279–287.

Boerman, O.C., Oyen, W.J.C., Corstens, F.H.M., 2001, Progress in gene therapy: seeing is believing. *J Nucl Med* 42(8):1235–1237.

Dennis Lo, Y.M., Wittwer, C.T., 2009, Molecular diagnostics: at the cutting edge of translational research. *Clin Chem* 55:601.

Hall, A.G., Coulthard, S.A., Irving, J.A., 2003, Molecular diagnostics: a healthcare perspective. *Expert Rev Mol Diagn* 3:13–16.

Jain, K.K., 2002, Current trends in molecular diagnostics. *Med Device Technol* 13:14–18.

Jain, K.K., 2003, Nanodiagnostics: application of nanotechnology in molecular diagnostics. *Expert Rev Mol Diagn* 3:153–161.

Zlotogorski, A., Panteleyev, A.A., Aita, V.M., Christiano, A.M., 2002, Clinical and molecular diagnostic criteria of congenital atrichia with papular lesions. *J Invest Dermatol* 118:887–890.

Chapter 2
Omics Technology

Abstract Omics has become the new mantra in molecular research. "Omics" technologies include genomics, transcriptomics, proteomics and metabolomics. Genomics had revealed the static sequences of genes and proteins and focus has now been shifted to their dynamic functions and interactions. Transcriptomics, proteomics and metabolomics reveal the biological function of the gene product. The "-omic-" technologies are high-throughput technologies and they increase substantially the number of proteins/genes that can be detected simultaneously to relate complex mixtures to complex effects in the form of gene/protein expression profiles. The primary aim of omic technologies is the nontargeted identification of all gene products (transcripts, proteins, and metabolites) present in a specific biological sample. The powerful "omics" technologies have opened new avenues towards biomarker discovery, identification of signaling molecules associated with cell growth, cell death, cellular metabolism and early detection of cancer. Omics will not only have an impact on our understanding of biological processes, but the prospect of more accurately diagnosing and treating disease will soon become a reality.

Keywords Omics · Genome · Proteomics · Transcriptomics · Metabolomics · Systems biology · High throughput technology · Genomics · Human genome project · Gene sequence · Gene transcript · Herceptin · Gleevec · Molecular signature · Expression profile · Expression signature · Computational tools · Oncology · Proteome · Biomarkers · Yeast two hybrid analysis · Clinical diagnostics · Protein–protein interaction · Mass spectroscopy · Protein microarray · Metabolome · Metabolic profiling · Toxicogenomics · Nutrigenomics

2.1 Prologue

Over the last decade, we have witnessed a fundamental change in how biomedical research is carried out and we can now assess the impact of the Human Genome Project on drug discovery and development (Bilello, 2005). With the completion of a rough draft of the human genome, many researchers are looking at how genes and proteins interact to form other proteins. The human genome sequence has revealed that sequence variations are very common. Genetic approaches have already linked

M. Debnath et al., *Molecular Diagnostics: Promises and Possibilities*,

DOI 10.1007/978-90-481-3261-4_2,

a large number of genetic variations (polymorphisms) with human diseases and adverse reactions from exposure to drugs or toxicants, suggesting sensitivity of genome to drugs and environmental agents, disease susceptibilities, and therapeutic responses. The application of the "-omic-" technologies may lead to a change of paradigms and open the new vistas in medicine.

2.2 Concept of Omics

New technologies that permit simultaneous monitoring of many hundreds or thousands of macro- and small molecules promise to allow functional monitoring of multiple (or perhaps all) key cellular pathways simultaneously. The new "global" methods of measuring families of cellular molecules, such as RNA, proteins, and intermediary metabolites have been termed "-omic" technologies, based on their ability to characterize all, or most, members of a family of molecules in a single analysis (Fig. 2.1). With these new tools, we can now obtain complete assessments of the functional activity of biochemical pathways, and of the structural genetic (sequence) differences among individuals and species, that were previously unattainable. The terms "Ome" and "Omics" are derivations of the suffix *-ome*, which has been appended to a variety of previously existing biological terms to create names for fields of endeavor like genome, proteome, transcriptome and metabolome that are either speculative or have some tangible meaning in particular contexts.

Omics has become the new mantra of molecular biology. "Omic" technologies include genomics, transcriptomics (gene expression profiling), proteomics and metabolomics. The recent availability of masses of omic data is responsible for the major growth spurt of systems biology. Pharmaceutical companies and others need to make sense out of all this omic information in order to take the next step in overcoming their innovation deficits. Systems biology provides the methods, computational capabilities, and inter-disciplinary expertise to facilitate this jump. The technology platform of genomics, transcriptomics, proteomics and metabolomics provide high-throughput technologies. They increase substantially the number of proteins/genes that can be detected simultaneously and have the potential to relate complex mixtures to complex effects in the form of gene/protein expression profiles.

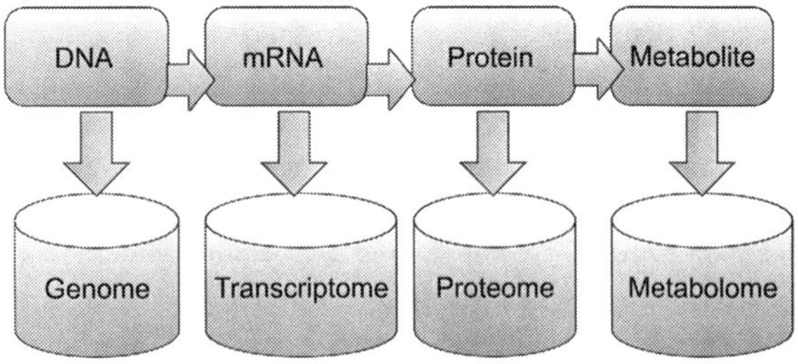

Fig. 2.1 The Central Dogma and the interacting "ome" includes the study of genome, proteome, transcriptome and metabolome

Commercial applications are emerging at an accelerating pace as pharmaceutical and biotechnology research organizations start to combine various forms of omic data into more comprehensive computer models, and bioinformatics companies increasingly turn their attention toward offering systems biology solutions to drug developers and diagnostics companies.

2.3 Genome and Genomics

In biology the genome of an organism is its whole hereditary information and is encoded in the DNA (or, for some viruses, RNA). This includes both the genes and the non-coding sequences of the DNA (Min et al., 1972). More precisely, the genome of an organism is a complete DNA sequence of one set of chromosomes; for example, one of the two sets that a diploid individual carries in every somatic cell. The term genome can be applied specifically to mean the complete set of *nuclear DNA* (i.e., the "nuclear genome") but can also be applied to organelles that contain their own DNA, as with the mitochondrial genome or the chloroplast genome. The genome contains the coded instructions necessary for the organism to build and maintain itself. Genomics is the study of an organism's genome, or genetic material. Genomics emerged as a new field of molecular biology where novel technologies were exploited in order to understand the complex, biological function of the genome. It revealed the static sequences of genes.

The molecular biology revolution and the advent of genomic technologies are facilitating rapid advances in our understanding of the molecular details of cell and tissue function. These advances have the potential to transform toxicological and clinical practice, and are likely to lead to the supplementation or replacement of traditional biomarkers of cellular integrity, cell and tissue homeostasis, and morphological alterations that result from cell damage or death. The increasing amount of genomic and molecular information is the basis for understanding higher-order biological systems, such as the cell and the organism, and their interactions with the environment, as well as for medical, industrial and other practical applications (Kanehisa et al., 2006). The knowledge of full genomes has created the possibility for the field of functional genomics, mainly concerned with patterns of gene expression during various conditions. Microarray and bioinformatics are the most important tools of genomics.

A major branch of genomics concerns with sequencing genomes of various organisms. In 1972, Walter Fiers and his team at the Laboratory of Molecular Biology of the University of Ghent (Ghent, Belgium) were the first to determine the sequence of a gene: the gene for bacteriophage MS2 coat protein. In 1976, the team determined the complete nucleotide-sequence of bacteriophage MS2-RNA. The first DNA-based genome to be sequenced in its entirety was that of bacteriophage Φ-X174; (5,368 bp), sequenced by Frederick Sanger (Sanger et al., 1977). The first free-living organism to be sequenced was that of *Haemophilus influenzae* (1.8 Mb) in 1995, and since then genomes are being sequenced at a rapid pace.

A rough draft of the human genome was completed by the Human Genome Project in early 2001, creating much fanfare.

As of January 2005, the complete sequence was known of about 1,000 viruses, 220 bacterial species and roughly 20 eukaryote organisms, of which about half are fungi. Most of the bacteria whose genomes have been completely sequenced are problematic disease-causing agents, such as *Haemophilus influenzae*. Of the other sequenced species, most were chosen because they were well-studied model organisms or promised to become good models. Yeast (*Saccharomyces cerevisiae*) has long been an important model organism for the eukaryotic cell, while the fruit fly *Drosophila melanogaster* has been a very important tool (notably in early pre-molecular genetics). The worm *Caenorhabditis elegans* is an often used simple model for multicellular organisms. The zebrafish *Brachydanio rerio* is used for many developmental studies on the molecular level and the flower *Arabidopsis thaliana* is a model organism for flowering plants. The Japanese pufferfish (*Takifugu rubripes*) and the spotted green pufferfish (*Tetraodon nigroviridis*) are interesting because of their small and compact genomes, containing very little non-coding DNA compared to most species. The mammals dog (*Canis familiaris*), brown rat (*Rattus norvegicus*), mouse (*Mus musculus*; Lein et al., 2007) and chimpanzee (*Pan troglodytes*) are all important model animals in medical research.

Genomics has the potential to revolutionize the practice of medicine, but despite significant scientific advances, very few genomics-based tests or treatments have reached consumers. The application of this genomics revealed that gene expression profiles induced by single drugs and the ones induced by the combination of the same drugs can be entirely different. The increasing amount of genomic and molecular information is the basis for understanding higher-order biological systems, such as the cell and the organism, and their interactions with the environment, as well as for medical, industrial and other practical applications. (Kanehisa et al., 2006).

The combination of genomic technologies with a knowledge of gene sequence and sequence conservation has made available markers that facilitate the correlation of genetic variation with biological outcomes, and "-omic" technologies allow efficient biochemical characterization of functional pathways – providing new markers of the susceptibility of individuals to different diseases and response to specific therapies. The powerful -omic technologies allow efficient monitoring of gene transcripts, proteins, and intermediary metabolites, making it possible to monitor a large number of key cellular pathways simultaneously. This has enabled the identification of key biomarkers and signaling molecules associated with cell growth, cell death, and cellular metabolism. These new markers facilitate monitoring of functional disturbance, molecular and cellular damage, and damage-response (MacGregor, 2004).

Over the past two decades, advances in genomic technology have allowed laboratories to generate vast amounts of biological data. These data include gene sequences, protein structures, information on gene expression and metabolic pathways. Automated instrumentation has enabled large volumes of data to be generated and automatically stored in computer databases, and this data has as many different formats as there are instruments. In addition to the new information gathered from genomic technologies, pharmaceutical and biotechnology companies have large amounts of "legacy data" – data inherited from their own and other sources on chemical structures and properties of compounds, and clinical, phenotypic and toxicological information. Most of this is stored in older types of databases designed for the particular type of data, and a major computational challenge is to integrate the new genomic information with current database systems in order to facilitate decision-making. Molecular biology and genetics has excelled in the creation of molecular diagnostics to identify and characterize disease processes. In the case of many diseases it has markedly altered clinical practice. In acute care settings it has led to increasingly accurate diagnosis and has impacted clinical treatment and outcomes. In HIV disease management, viral load and resistance genotyping has altered patient treatment decisions. Molecular diagnostics has potential role in multifactorial disease treatment such as breast cancer and this is an example of targeting a therapy to a specific genetic profile (Pegram and Slamon, 2000), however, Slamon and his co-workers demonstrated that amplification of the HER-2/neu gene was a significant predictor of both overall survival and time to relapse in patients with breast cancer. Similarly it took almost 20 years from the molecular understanding of the role of the *bcr-abl* fusion protein to the development of the therapeutic Gleevec. Nonetheless gene based testing technologies clearly offer a great advantage in identifying patient populations which can benefit from therapy. More importantly, the clear demonstration that one can proceed from a genetics-based disease understanding to therapeutic efficacy is a critical learning. While the impact early in the discovery phase is most significant, the application of genomics can contribute throughout the drug development process from initial discovery all the way through life cycle management.

2.4 Transcriptome and Transcriptomics

Transcriptomics refers to the comprehensive scanning of the nearly fifty thousand currently known genes that are transcribed into RNA molecules from the three-billion-letter human genome. Each cell utilizes (expresses) different genes at different times in its development and under different physiological conditions. In general, tissues express similar sets of genes that can be used to identify those tissues in the absence of any other information. For example, the brain expresses about 30% of all of the known genes; those specific transcripts are different from the transcribed genome in the heart. We can therefore define molecular signatures based on expression profiles, and these profiles can then be used to automatically separate normal cells or tissues into their correct category.

This method of expression profiling can also be done with disease states. In Alzheimer's disease, neurons show altered gene expression patterns when compared to normal neurons. This information can be used as molecular diagnostic criteria in the absence of histopathology for tangles because it is, in fact, a surrogate marker of that cellular condition. Transcriptomics builds upon the more classical study of gene function where Northern blots are used for transcriptional analysis of individual genes. However, instead of examining one gene at a time this new technology analyses in one single go the entire transcriptome, which is the full set of all messenger RNA molecules present at any given time in a defined population of cells. This reveals an instant picture of all genes being actively transcribed, and since the technique is a least partly quantitative it also tells about the expression level of each gene. For example, by comparing such transcriptional analyses from plants exposed to different environmental conditions it is possible to quickly identify all genes involved in the adaptive process. All genes and most proteins can be regarded as instruments for making up the biochemical composition and there by the physiological identity of an organism. The same can be done for clinical states that may be less obvious or for which there does not exist a diagnostic marker (Example Autism). This approach will also facilitate testing in a pre-symptomatic scenario to encourage early interventions that may result in better outcomes.

Transcriptomics approach facilitates sub-classification of disorders that on the surface appear to be similar. This strategy has been shown elegantly for a number of disorders, primarily in oncology, where one can define outcome and survival rates and drug response rates based on an expression profile. The development of microarray technologies and other high-throughput strategies are becoming increasingly important in biology because they permit to monitor expression levels of thousands of genes in only one experiment. Availability of computational tools to compare and analyse these expression profiles in a suitable way for biological interpretation is need of the day (Subramanian et al., 2005).

Microarray technologies at the transcriptomic level can be high throughput and excellent for diagnostic purposes, but often their dynamic range, informational and experimental complexities limit their utility as diagnostic tools as compared with other assessments of mRNA abundance e.g. real time RT-PCR. One of the early applications of microarray was in mode of action (MOA) studies for isoniazid (INH) an antibiotic effective in the control of tuberculosis. Studies of *M. tuberculosis* exposed to INH indicated up regulation of a number of genes relevant to the MOA. In addition to possibly increasing the number of targets by identifying genes within pathways, the generation of "signature transcriptome analysis (TA) profiles" may be useful in predicting the MOA of new chemical entities. This approach has been shown to be quite useful in the search for tumor-specific markers and in distinguishing potential therapeutic targets in neoplastic cells. Gene expression profiles of tumors, for example, facilitate identification of subsets of genes that function as prognostic disease markers or biologic predictors of therapeutic response, thus permitting clinicians to distinguish indolent from aggressive tumors which require immediate therapeutic intervention.

Although microarray studies can reveal the relative amounts of different mRNAs in the cell, levels of mRNA are not directly proportional to the expression level of the proteins they code for. The number of protein molecules synthesized using a given mRNA molecule as a template is highly dependent on transcription-initiation features of the mRNA sequence; in particular, the ability of the promoter region is a key determinant in the recruiting of ribosomes for protein translation. A study of 158,807 mouse transcripts revealed that 4,520 of these transcripts form antisense partners that are base pair complementary to the exons of genes (Katayama et al., 2005). These results raise the possibility that significant numbers of "antisense RNA-coding genes" might participate in the regulation of the levels of expression of protein-coding mRNAs.

2.5 Proteome and Proteomics

With the avalanche of genomic information and improvements in analytical technology, proteomics is becoming increasingly important for the study of many different aspects. Since proteins serve as important components of major signaling and biochemical pathways, studies at protein levels are essential to reveal molecular mechanisms underlying disease processes.

Proteomics is the application of evolving technologies to analyze gene products, *proteins*, on a large scale. This concerns protein expression profiles, protein modifications and protein networks in relation to cell function and biological processes e.g. development, health and disease (Macaulay et al., 2005). With the mapping of the human genome, proteomics has rapidly emerged as an exciting new field of research, one that complements rather than replaces genomics. The genome holds all of an organism's genes, which carry the codes for the proteins that it needs to function. Proteomics is the integrated study of proteins and their biological functions and processes. It allows the study of protein structure and protein–protein interactions, and today, this branch of science is becoming a major tool in biomedical and drug development research.

Since proteins play a central role in the life of an organism, proteomics is instrumental in discovery of biomarkers that indicate a particular disease. A surprising finding of the Human Genome Project is that there are far fewer protein-coding genes in the human genome than proteins in the human proteome (20,000–25,000 genes vs. about 1,000,000 proteins). The human body may contain more than 2 million proteins, each having different functions (http://en.wikipedia.org/wiki/Proteomics). The protein diversity is thought to be due to alternative splicing and post-translational modification of proteins. The discrepancy implies that protein diversity cannot be fully characterized by gene expression analysis, thus proteomics is useful for characterizing normal and abnormal cellular processes.

To catalog all human proteins, their functions and interactions is a great challenge for scientists and an international collaboration with these goals is co-ordinated by the Human Proteome Organization (HUPO). Most proteins function in collaboration with other proteins, and one goal of proteomics is to identify which proteins interact. This often gives important clues about the functions of newly discovered proteins. Several methods are available to probe protein–protein interactions. The traditional method is yeast two-hybrid analysis. New methods include protein microarrays, immunoaffinity chromatography followed by mass spectrometry, and combinations of experimental methods such as phage display and computational methods

(Wu et al., 2005). Proteomics uses various technologies such as (a) One- and two-dimensional gel electrophoresis to identify the relative mass of a protein and its isoelectric point (b) X-ray crystallography (c) Nuclear magnetic resonance to characterize the three-dimensional structure of peptides and proteins (d) low-resolution techniques such as circular dichroism (e) Fourier transform infrared spectroscopy and small angle X-ray scattering (SAXS) to study the secondary structure of proteins and (f) Mass spectrometry (no-tandem), often MALDI-TOF is used to identify proteins by peptide mass finger printing. Protein profiling with MALDI-TOF MS can be of high use in clinical diagnostics.

Affinity chromatography, yeast two hybrid techniques, fluorescence resonance energy transfer (FRET), and Surface Plasma Resonance (SPR) are used to identify protein–protein and protein-DNA binding reactions. X-ray Tomography is used to determine the location of labeled proteins or protein complexes in an intact cell. Soft ware based image analysis is utilized to automate the quantification and detection of spots within and among gels samples. While this technology is widely utilized, the intelligence has not been perfected yet. Because most potential molecular markers and targets are proteins, proteomic profiling is expected to yield more direct answers to functional and pharmacological questions than does transcriptional profiling (Nishizuka et al., 2003).

The molecular biology revolution and the advent of genomic and proteomic technologies are facilitating rapid advances in our understanding of the molecular details of cell and tissue function. These advances have the potential to transform toxicological and clinical practice and are likely to lead to the supplementation or replacement of traditional biomarkers of cellular integrity, cell and tissue homeostasis, and morphological alterations that result from cell damage or death.

Elucidation of cellular responses to molecular damage, including evolutionarily conserved inducible molecular defense systems, suggests the possibility of new biomarkers based on molecular responses to functional perturbations and cellular damage. Our improved understanding of the molecular basis of various pathologies suggests that monitoring specific molecular responses may provide improved prediction of human outcomes. Responses that can be monitored directly in the human should provide "bridging biomarkers" that may eliminate much of the current uncertainty in extrapolating from laboratory models to human outcome. More importantly the evaluation of one or a few biomarkers is severely limited by the fact that many disease syndromes share inflammatory components or involve the remodelling of extracellular matrix proteins common to many disease processes (e.g. fibrosis). Some of the diseases have characteristic biomarkers. These biomarkers are used by the physicians for diagnosis. Glycogen synthesis kinase has been a leader in advocating asystematic, highly parallel, combinatorial approach to assemble "disease specific signatures", as well as resolving patterns in response to therapeutic agents. In Alzheimer's disease, elevations in beta secretase creates amyloid/beta-protein, that causes plaque to build up in the patient's brain, which causes dementia (Hye et al., 2006). Targeting this enzyme decreases the amyloid/beta-protein and so slows the progression of the disease. A procedure to test for the increase in amyloid/beta-protein is immunohistochemical staining, in which antibodies bind to specific antigens or biological tissue of amyloid/beta-protein. Heart disease is commonly assessed using several key protein based biomarkers (Vasan, 2006). Standard protein biomarkers for CVD include interleukin-6, interleukin-8, serum amyloid A protein, fibrinogen, and troponins. cTnI cardiac troponin I increases in concentration within 3–12 h of initial cardiac injury and can be found elevated days after an acute myocardial infarction. A number of commercial antibody based assays as well as other methods are used in hospitals as primary tests for acute myocardal infarction. Proteomic analysis of kidney cells and cancerous kidney cells is producing promising leads for biomarkers for renal cell carcinoma and developing assays to test for this disease (Perroud et al., 2006). In kidney-related diseases, urine is a potential source for such biomarkers. Identification of urinary polypeptides as biomarkers of kidney-related diseases allows to diagnose the severity of the disease several months before the appearance of the pathology (Rogers et al., 2003).

Proteomic technologies rely on the ability to separate a complex mixture so that individual proteins are more easily processed with other techniques. Well-known methods include low-throughput sequencing through Edman degradation. Higher-throughput proteomic techniques are based on mass spectrometry, commonly peptide mass fingerprinting on simpler instruments, or De novo repeat detection sequencing on instruments capable of more than one round of mass spectrometry. Recent successes illustrate the role of mass spectrometry-based proteomics as an indispensable tool for molecular and cellular biology and for the emerging field of systems biology. The ability of mass spectrometry to identify and, increasingly, to precisely quantify thousands of proteins from complex samples can be expected to impact on biology and medicine (Aebersold and Mann, 2003).

Identifying all protein–protein interactions in an organism is a major objective of proteomics. A related goal is to know which protein pairs are present in the same protein complex. High-throughput methods such as yeast two-hybrid (Y2H) and affinity purification coupled with mass spectrometry (APMS) have been used to detect interacting proteins on a genomic scale. However, both Y2H and APMS methods have substantial false-positive rates. Aside from high-throughput interaction screens, other gene- or protein-pair characteristics may also be informative of physical interaction. Therefore it is desirable to integrate multiple datasets and utilize their different predictive value for more accurate prediction of co-complexed relationship. Many proteins participate in cellular processes as members of protein complexes of varying size. It is believed that combinatorial interactions among proteins serve as an important basis for the biological complexity of higher organisms. Therefore, increased knowledge about protein–protein interactions and protein complexes will greatly aid our understanding of protein function (Blagoev et al., 2003; Zhang et al., 2004).

Advances in mass spectrometry (MS), such as surface-enhanced laser desorption ionization (SELDI), hold great promise for early ovarian cancer detection through proteomic profiling of patient serum. Because thousands of proteins and peptides can be characterized and quantified at the same time, large amount of valuable data are obtained for identifying characteristic and effective biomarkers for ovarian cancer detection. Several advanced data mining algorithms have been reported to be promising for diagnosis of early-stage ovarian cancer (Petricoin et al., 2002; Wulfkuhle et al., 2003; Tirumalai et al., 2003; Zhu et al., 2003; Yanagisawa et al., 2003; Pan et al., 2005). However, considerable controversy has been generated, and there remain some critical issues such as reproducibility and robustness of these methods, which make the proteomic profiling approach less congenial for achieving definite diagnosis (Diamandis, 2004; Baggerly et al., 2005; Ransohoff, 2005). Kristiansen et al. (2004) have carried out a comprehensive characterization of human bile to define the bile proteome by one-dimensional gel electrophoresis and lectin affinity chromatography followed by liquid chromatography tandem mass spectrometry. The strategy can be applicable for a detailed proteomic analysis of most body fluids. In combination with "tagging" approaches for differential proteomics, the proteomics approach could serve to identify cancer biomarkers from any body fluid (Baggerly et al., 2005).

In terms of the drug development process emphasis has been placed upon staging and progression rather than diagnosis. Recent development of highly multiplexed plate, microbead and protein microarray technologies (Liotta et al., 2005) with larger numbers of analytes (with clinical relevance) gave rise to applications such as protein profiling of plasma from complex disease states. Protein microarray and assay detection systems based on this technology were assumed to emerge as the correlates of DNA microarrays in the identification, quantitation and functional analysis of proteins. In concept, protein microarray is a fairly simple process, antibodies or other affinity reagents are attached to a surface and a cell lysate or serum, is added. After a period of time to allow the antigens present to bind to their cognate antibodies, bound antigen is detected using fluorescently tagged secondary antibodies. Alternatively lysates containing fluorescently tagged or radioactively labelled proteins can be added directly to the array to permit direct detection of the analyte. Currently, "cytokine-chemokine array" strategies appear to be more robust, since a number of well characterized reagents and commercial arrays are available and many disease syndromes display characteristic patterns of cytokine and chemokine expression. Other functional protein arrays are being used to study protein–protein, protein-nucleic acid, protein-allergen, protein-carbohydrate, protein-lipid, enzyme-substrate, and protein-drug interactions. While antibody reagents are not available for most proteins, a number of initiatives have been undertaken to expand the number of antibodies, aptamers and defined proteins for use in quantitative and function based assays.

One of the most promising developments to come from the study of human genes and proteins has been the identification of potential new drugs for the treatment of disease. This relies on genome and proteome information to identify proteins associated with a disease. When a particular protein is implicated in a disease, its 3D structure provides the information to design drugs to interfere with the action of the protein. A molecule that fits the active site of an enzyme, but cannot be released by the enzyme, will inactivate the enzyme. This is the basis of new drug-discovery tools, which aim to find new drugs to inactivate proteins involved in disease. The genetic differences among individuals can be exploited to develop personalized drugs that are more effective for the individual (Arora et al., 2005).

A computer technique which attempts to fit millions of small molecules to the three-dimensional structure of a protein is called "virtual ligand screening". The computer rates the quality of the fit to various sites in the protein, with the goal of either enhancing or disabling the function of the protein, depending on its function in the cell. A good example of this is the identification of new drugs to target and inactivate the HIV-1 protease. The HIV-1 protease is an enzyme that cleaves a very large HIV protein into smaller, functional proteins. The virus cannot survive without this enzyme; therefore, it is one of the most effective protein targets for killing HIV.

An important use of proteomics is using specific protein biomarkers to diagnose disease. Understanding the proteome, the structure and function of each protein and the complexities of protein–protein interactions will be critical for developing the most effective diagnostic techniques and disease treatments in the future .The long-term challenge of proteomics is enormous: to define the identities, quantities, structures and functions of complete complements of proteins, and to characterize how these properties vary in different cellular contexts. One critical step in tackling this goal is the generation of sets of clones that express a representative of each protein of a proteome in a useful format, followed by the analysis of these sets on a genome-wide basis. Such studies enable genetic, biochemical and cell biological technologies to be applied on a systematic level, leading to the assignment of biochemical activities, the construction of protein arrays, the identification of interactions, and the localization of proteins within cellular compartments (Phizicky et al., 2003).

2.6 Metabolome and Metabolomics

Since the emergence of genomics, several other "omic" techniques have come to the front and promising one is "metabolomics" (Schmidt, 2004). Metabolome refers to the complete set of small-molecule metabolites (such as metabolic intermediates, hormones and other signalling molecules, and secondary metabolites) to be found within a biological sample, such as a single organism (Oliver et al., 1998). Metabolomics (also referred to as metabonomics) is the comprehensive study of the metabolome, the repertoire of biochemicals (or small molecules) present in cells, tissue, and body fluids. One of the major challenges of this approach, as with other "omic" technologies, is that the metabolome is context-dependent and will vary with pathology, developmental stage and environmental factors. Following the rapid developments in genome, transcriptome and proteome technologies, there is a growing interest in metabolome research. The study of metabolism at the global or "-omics" level is a new but rapidly growing field that has the potential to have an impact upon medical practice as the metabolic state reflects what has been encoded by the genome and modified by environmental factors (Kaddurah-Daouk, 2006).

In January 2007, scientists at the University of Alberta and the University of Calgary completed the first draft of the human metabolome (Wishart et al., 2007).

They have catalogued and characterized 2,500 metabolites, 1,200 drugs and 3,500 food components that can be found in the human body (Harrigan and Goodacre, 2003). In animals and humans, metabolic profiling of blood and urine components to characterize toxicity and disease states such as inborn errors of metabolism has been ongoing since the introduction of gas-chromatography mass-spectrometry (GC-MS) in the mid-1960s. Techniques such as NMR spectroscopy can be used for a very wide range of components of blood or urine, and advanced data analysis techniques locate the key descriptors (biomarkers) of effect. In addition to NMR, other spectroscopic methods such as MS, GC/MS can produce metabolic profiles or signatures of toxicity, disease or drug efficacy.

Thus, the possibility of globally profiling the metabolome of an organism is a genuine analytical challenge, as by definition this must also take into consideration all relevant factors that influence metabolism. Despite these challenges, the approach has already been applied to understand the metabolism in a range of animal models, and has more recently started to be projected into the clinical situation (Brindle et al., 2002). The use of metabolomic data to predict the health trajectories of individuals will require bioinformatic tools and quantitative reference databases. These databases containing metabolite profiles from the population must be built, stored and indexed according to metabolic and health status. Building and annotating these databases with the knowledge to predict how a specific metabolic pattern from an individual can be adjusted with diet, drugs and lifestyle to improve health represents a logical application of the biochemistry knowledge that the life sciences have produced over the past 100 years (German et al., 2005). Using information from the completely sequenced genomes and known metabolic pathways of various microorganisms, there are ongoing attempts to completely model the metabolism of a cell or microorganism.

In metabolomics research, there are several steps between the sampling of the biological condition under study and the biological interpretation of the results of the data analysis (German et al., 2002). First, the biological samples are extracted and prepared for analysis. Subsequently, different data pre-processing steps (Fiehn, 2002) are applied in order to generate "clean" data in the form of normalized peak areas that reflect the (intracellular) metabolite concentrations. These clean data can be used as the input for data analysis. However, it is important to use an appropriate data pretreatment method before starting data analysis. Data pretreatment methods convert the clean data to a different scale (for instance, relative or logarithmic scale). Hereby, they aim to focus on the relevant (biological) information and to reduce the influence of disturbing factors such as measurement noise. Procedures that can be used for data pretreatment are scaling, centering and transformations. The ability to use metabolomics approaches for classification and mechanistic studies may influence and augment our ability to study and address the aging process scientifically and clinically (Kristal and Shurubor, 2005).

Metabolic profiling (especially of urine or blood plasma samples) can be used to detect the physiological changes caused by toxic insult of a chemical (or mixture of chemicals). In many cases, the observed changes can be related to specific syndromes, e.g. a specific lesion in liver or kidney. This is of particular relevance to pharmaceutical companies wanting to test the toxicity of potential drug candidates: if a compound can be eliminated before it reaches clinical trials on the grounds of adverse toxicity, it saves the enormous expense of the trials (Lindon et al., 2004). Metabolites are the key regulators of systems homeostasis. As such, concentration changes of specific groups of metabolites may reflect systemic responses to environmental, therapeutic or genetic interventions. Thus, the study of metabolites is a powerful tool for the characterization of complex phenotypes as well as for the development of biomarkers for specific physiological responses. Metabolic signatures provide prognostic, diagnostic, and surrogate markers for a disease state; the ability to subclassify disease; biomarkers for drug response phenotypes (pharmacometabolomics); and information about mechanisms of disease. Indeed, sophisticated metabolomic analytical platforms and informatics tools have recently been developed that make it possible to define initial metabolic signatures for several diseases (Kaddurah-Daouk, 2006). Therefore, metabolomics is a valuable platform for studies of complex diseases and the development of new therapies, both in non-clinical disease model characterization and clinical settings.

Applications to preclinical drug safety studies are illustrated by the Consortium for Metabolomic Toxicology, a collaboration involving several major pharmaceutical companies (Brindle et al., 2002). The use of metabolomics in disease diagnosis and therapy monitoring and the concept of pharmaco-metabonomics as a way of predicting an individual's response to treatment can be studied (Ellis and Goodacre, 2006). Combination drug therapies with individualized optimization are likely to become a major focus. Metabolomics incorporates the most advanced approaches to molecular phenotype system readout and provides the ideal theranostic technology platform for the discovery of biomarker patterns associated with healthy and disease states, for use in personalized health monitoring programs and for the design of individualized interventions (German et al., 2003). Today, clinicians capture only a very small part of the information contained in the metabolome, as revealed by a defined set of blood chemistry analyses to define health and disease states. Examples include measuring glucose to monitor diabetes and measuring cholesterol for cardiovascular health. Such a narrow chemical analysis could potentially be replaced in the future with a metabolic signature that captures global biochemical changes in disease and upon treatment.

2.7 Integration of Omics

Developing a new drug is a tedious and expensive undertaking. The recently developed high-throughput experimental technologies, encompassing genomics, transcriptomics, proteomics and metabolomics provide for the first time ever the means to comprehensively monitor the disease processes at the molecular level. They increase substantially the number of proteins/genes that can be detected simultaneously and have the potential to relate complex mixtures to complex effects in the form of gene/protein expression profiles. The primary aim of "omic" technologies is the nontargeted identification of all gene products (transcripts, proteins, and metabolites) present in a specific biological sample. By their nature, these technologies reveal unexpected properties of biological systems. A second and more challenging aspect of omic technologies is the refined analysis of quantitative dynamics in biological systems. The "-omics" technologies facilitate the systematic characterization of a drug target's physiology, thereby helping to reduce the typically high attrition rates in discovery projects, and improving the overall efficiency of pharmaceutical research processes. Currently, the bottlenecks for taking full advantage of these experimental technologies are the rapidly growing volumes of automatically produced biological data and lack of scalable database systems and computational tools (Fischer, 2005).

High-throughput analysis is essential considering data at the "omic" level. There are four major types of high-throughput measurements that are commonly performed: genomic SNP analysis (i.e., the large-scale genotyping of single nucleotide polymorphisms), transcriptomic measurements (i.e., the measurement of all gene expression values in a cell or tissue type simultaneously), proteomic measurements (i.e., the identification of all proteins present in a cell or tissue type), and metabolomic measurements (i.e., the identification and quantification of all metabolites present in a cell or tissue type). Each of these four is distinct and offers a different perspective on the processes underlying disease initiation and progression as well as on ways of predicting, preventing, or treating disease (Venkatesh and Harry, 2002).

Genomic SNP genotyping measures a person's genotypes for several hundred thousand single nucleotide polymorphisms spread throughout the genome. These SNPs tend to be quite common (with typically at least 5% of the population having at least one copy of the less frequent allele), and not strictly causal of the disease. Rather, SNPs can act in unison with other SNPs and with environmental variables to increase or decrease a person's risk of a disease. This makes identifying important SNPs difficult; the variation in a trait that can be accounted for by a single SNP is fairly small relative to the total variation in the trait. Even so, because genotypes remain constant (barring mutations to individual cells) throughout life, SNPs are potentially among the most useful measurements for predicting risk.

Transcriptomic measurements (often referred to as gene expression microarrays or "gene chips" are the oldest and most established of the high-throughput methodologies). Gene expression levels influence traits more directly than SNPs, and so significant associations are easier to detect. While transcriptomic measures are not as useful for pre-disease prediction (because a person's gene expression levels very far in advance of disease initiation are not likely to be informative because they have the potential to change so significantly), they are very well-suited for either early identification of a disease (i.e., finding people who have gene expression levels characteristic of a disease but who have not yet manifested other symptoms) or classifying patients with a disease into subgroups (by identifying gene expression levels that are associated with either better or worse outcomes or with higher or lower values of some disease phenotype).

Proteomics is similar in character to transcriptomics. Like transcriptomic measures, though, proteomic measures are excellent for early identification of disease or classifying people into subgroups. As with proteomics, the metabolites are measured in a very fast serial process. NMR is typically used to both identify and quantify metabolites. This technology is newer and less frequently used than the other technologies, but similar caveats apply. Measurements of metabolites are dynamic as are gene expression levels and proteins, and so are best suited for either early disease detection or disease subclass identification.

The new omics technologies seem set to fulfill their huge expectations and in combination they might prove extremely valuable in functional gene analyses. Genomic and transcriptomic studies are mostly conducted by DNA microarray technologies. Proteomics and metabolomics have no standardized procedures yet, but usually, proteome analysis is done by two-dimensional gel electrophoresis, Liquid chromatography-mass spectrometry, while metabolome analysis is conducted through Gas chromatography-mass spectrometry, Liquid chromatography-mass spectrometry and Liquid chromatography – Nuclear magnetic resonance and MALDI-TOF. Transcriptomics provides the tool for deciphering gene expression networks, and proteomics links these networks to protein products. The third crucial partner is metabolomics, which defines the metabolic network(s) linked to gene expression. NMR and mass spectrometry enable the broad screen analysis of the metabolome and its transformation pathways, transcending classical targeted metabolic studies. Omics will not only have an impact on our understanding of biological processes, but the prospect of more accurately diagnosing and treating disease will soon become a reality (Loughlin, 2007).

Development of microarrays has permitted global measurement of gene expression at the transcript level and provided a glimpse into the coordinated control and interactions between genes (Mutch et al., 2002). Additional "transcriptomic" and other "omic" analyses will further advance medical research. Protein profiling (proteome), more comprehensive analysis of gene function by knocking down or out of candidate genes (phenome), protein–protein interaction (interactome), and protein cellular and subcellular localization (localizome) are all needed for a deeper understanding of the development, physiology, and pathology of any organ or tissue. Each technique will collectively contribute to the diagnosis, prevention, and therapy of the great number of heart diseases (Coen et al., 2004).

The unprecedented advances in molecular biology during the last two decades have resulted in a dramatic increase in knowledge about gene structure and function, an immense database of genetic sequence information, and an impressive set of efficient new technologies for monitoring genetic sequences, genetic variation, and global functional gene expression. These advances have led to a new sub-discipline of toxicology: "toxicogenomics". We define toxicogenomics as"the study of the relationship between the structure and activity of the genome (the cellular complement of genes) and the adverse biological effects of exogenous agents". Toxicogenomics help to monitor the expression of multiple genes, proteins, and metabolites simultaneously. It combines new technologies in genomics, proteomics, and metabolomics (Fig. 2.2) with traditional tools of pathology and toxicology to study biological response to drugs and other environmental xenobiotics. The biological response to environmental exposure is so complex and involves so many interactive factors that the use of a systems biology analytical approach is required.

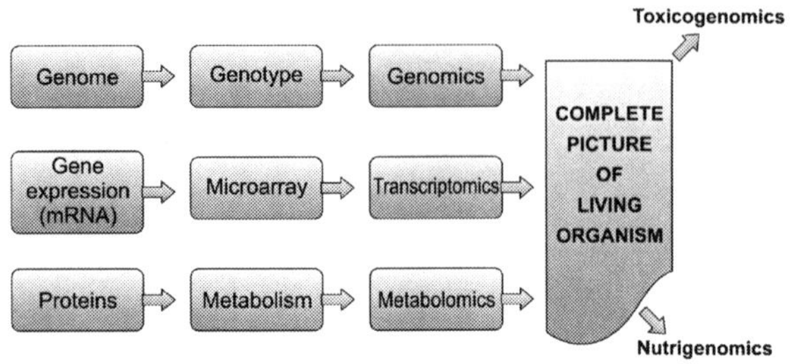

Fig. 2.2 Integrating omics technology: One of the challenges of systems biology is to integrate genomics, proteomic, transcriptomic, and metabolomic information to give a more complete picture of living organisms

These new technologies will lead to new families of biomarkers that permit characterization and efficient monitoring of cellular perturbations, provide an increased understanding of the influence of genetic variation on toxicological outcomes, and allow definition of environmental causes of genetic alterations and their relationship to human disease. The broad application of these new approaches will likely erase the current distinctions among the fields of toxicology, pathology, genetic toxicology, and molecular genetics. Instead, a new integrated approach will likely emerge that involves a comprehensive understanding of genetic control of cellular functions, and of cellular responses to alterations in normal molecular structure and function (Aardema and MacGregor, 2002).

It was realized however, that many, possibly thousands of other gene polymorphisms might result in minor deviations in nutritional biochemistry, where only marginal or additive effects would result from these deviations. The tools to study the physiological impact were not available at the time and are only now becoming available. "Nutrigenomics" is the application of the sciences of genomics, transcriptomics, proteomics and metabolomics to human nutrition, especially the relationship between nutrition and health. Nutrigenomics is associated with the issue of personalized nutrition, since claims are being made that differences in genotype should result in differences in the diet and health relationship. The recent advances in nutrigenomics studies are owed to the completion of human genome project and the new biomics technologies that provide means for the simultaneous determination of the expression of many thousands of genes at the mRNA (transcriptomics), metabolites (metabolomics) and protein (proteomics) levels. The role of metabolomics in nutrigenomics (German et al., 2003) requires understanding, and ultimately regulating, a multitude of nutrient-related interactions at the gene, protein and metabolic levels. These new disciplines and their attendant technologies are changing the paradigms of health research. A number of genetic variations have been shown to increase the susceptibility to diet-related diseases. These include variants that have been associated with Type 2 diabetes mellitus, obesity, cardiovascular diseases, some autoimmune diseases and cancers. Nutrigenetics aims to study these susceptible genes and provide dietary interventions for individuals at risk of such diseases.

Hyperlipidemia is usually associated with atherosclerosis and coronary heart disease (Kaput, 2004). Therapy includes lifestyle changes as alterations in the patient's diet, physical activity and treatment with pharmaceuticals as statins. However, individuals respond differently to the treatment. This was attributed to genetic variations within the population. Genetic variations in genes encoding for apolipoproteins, some enzymes and hormones can alter individual sensitivity to developing cardiovascular diseases (Ordovas and Mooser, 2004). Nutrients can contribute to the development of cancers especially colon, gastric and breast cancer. Several gene variants have been identified as susceptibility genes. One example is the N-Acetyltransferase (NAT) gene. Studies have shown that the NAT2 fast acetylator genotype had a higher risk of developing colon cancer in people who consumed relatively large quantities of red meat (Mutch et al., 2005). The advent to access comprehensive sets of information (i.e., genome, transcriptome, proteome, interactome, phenome, and localizome) has brought a new way of global thinking to biological questions, and analysis using these sets is increasingly the choice for many investigations. Although these approaches may appear as a substitute for the more traditional "reductionist" approach that tackles one or few genes or gene products at one time, they are, in reality, complementary and have the potential to greatly enhance the traditional approach (Abu-Issa and Kirby, 2004).

Less than a decade ago, describing the complexity of chemical behavior in biological systems was severely limited because realistic models presented combinatorial and other problems beyond the capabilities of most computers. In the field of bioinformatics, for example, major advances were made not from faster statistical analysis of data after the acquisition, but from the integration of computational and data acquisition technologies. It is now possible to consider how to evaluate the vast amounts of information generated by "omic" technologies using data-mining tools made possible by rapid advances in computational storage capacity and speed. Within the systems biology framework, functional analyses at the level of gene expression (transcriptomics), protein translation (proteomics), and, more recently, the metabolite network (metabolomics) have become increasingly popular. Metabolomics experiments aim to quantify all metabolites in a cellular system (cell or tissue) under defined states and at different time points so that the dynamics of any biotic, abiotic, or genetic perturbation can be accurately assessed.

With the progress in genetics, biochemical disorders with high nutritional relevance were linked to a genetic origin. The next generation omics will include genome resequencing, methylation analysis, mRNA tag profiling, small RNA identification, transcriptome sequencing and also the functional elements (Fig. 2.3). The "omic," or comprehensive, approach can yield mountains of new information in a relatively short time. Progress in applying the "omic" approach is still in its early exponential phase and does not look like it will plateau any time soon.

2.8 Conclusion

The technology platform of genomics, proteomics and metabolomics ("-omic-" technologies) are high-throughput technologies. They increase substantially the number of proteins/genes that can be detected simultaneously and have the potential

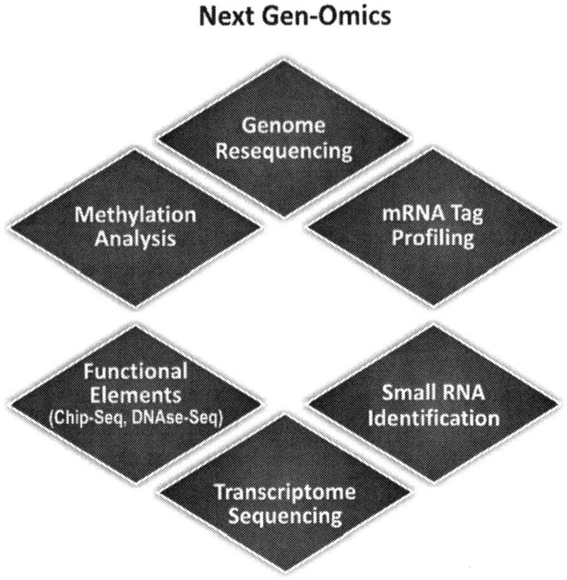

Fig. 2.3 The next generation omics will include all the outcomes after integrating omics

to relate complex mixtures to complex effects in the form of gene/protein expression profiles. By their nature, these technologies reveal unexpected properties of biological systems. A second and more challenging aspect of omic technologies is the refined analysis of quantitative dynamics in biological systems. Several recent technological advances now make it possible to develop molecular profiles using genomic, proteomic, and metabolomic methods in order to identify the effects that chemicals may have on living organisms or the environment. Although the technology continues to change and improve, conducting these types of analyses is no longer a question of capability. In summary, the new omics technologies seem set to fulfill huge expectations and in combination they might prove extremely valuable in functional gene analyses. It is, however, necessary to further standardize and automate the methods of especially proteomics and metabolomics in order to make efficient and reproducible high-throughput analyses. Omics will not only have an impact on our understanding of biological processes, but the prospect of more accurately diagnosing and treating disease will soon become a reality.

Advances in "omics" technologies have the potential to revolutionize our approach to disease diagnosis, prognostication and development of novel therapeutics. However, the promise of rapid advances in medicine "from the lab bench to the bedside" has not manifested as of yet. Indeed it appears that the translational applications of genomic-based research have preceded the development of both (i) a conceptual framework for disease understanding and (ii) effective tools that can exploit the vast amounts of data derived from these efforts. In reality great progress has been made, however understanding processes such as disease progression (or drug response) requires systematic insight into dynamic (and temporal) differences in gene regulation, interaction and function (Bilello, 2005).

Omics is a progressive and useful concept in biology. It can revolutionize the way biology is done and how we see life in the future. However, omics fields itself is undergoing change. Some omics do not have livable niche. As time goes by, practically useful omics will survive and go into the mainstream biology while some others will die out until revived in an unexpected way. The omic methods will be powerful aids in understanding biology and advancing clinical practice, but it was never going to be as easy as some led us to believe.

References

Aardema, M.J., MacGregor, J.T., 2002, Toxicology and genetic toxicology in the new era of "toxicogenomics": impact of "-omics" technologies. *Mutat Res* 499:13–25.

Abu-Issa, R., Kirby, M.L., 2004, Take heart in the age of "Omics". *Circ Res* 95:335.

Aebersold, R., Mann, M., 2003, Mass spectrometry-based proteomics. *Nature* 422:198–207.

Arora, P.S., Yamagiwa, H., Srivastava, A., Bolander, M.E., Sarkar, G., 2005, Comparative evaluation of two two-dimensional gel electrophoresis image analysis software applications using synovial fluids from patients with joint disease. *J Orthop Sci* 10:160–166.

Baggerly, K.A., Morris, J.S., Edmonson, S.R., Coombes, K.R., 2005, Signal in noise: evaluating reported reproducibility of serum proteomic tests for ovarian cancer. *J Natl Cancer Inst* 97: 307–309.

Bilello, J.A., 2005, The agony and ecstasy of "OMIC" technologies in drug development. *Curr Mol* 5:39–52.

Blagoev, B., Kratchmarova, I., Ong, S.E., Nielsen, M., Foster, L.J., Mann, M., 2003, A proteomics strategy to elucidate functional protein–protein interactions applied to EGF signaling. *Nat Biotechnol* 21:315–318.

Brindle, J.T., Antti, H., Holmes, E., Tranter, G., Nicholson, J.K., Bethell, H.W.L., Clarke, S., Schofield, P.M., McKilligin, E., Mosedale, D.E., Grainger, D.J., 2002, Rapid and non-invasive diagnosis of the presence and severity of coronary heart disease using 1H NMR-based metabonomics. *Nature Med* 8:1439–1444.

Coen, M., Ruepp, S.U., Lindon, J.C., Nicholson, J.K., Pognan, F., Lenz, E.M., Wilson, I.D., 2004, Integrated application of transcriptomics and metabonomics yields new insight into the toxicity due to paracetamol in the mouse. *J Pharm Biomed Anal* 35:93–105.

Diamandis, E.P., 2004, Analysis of serum proteomic patterns for early cancer diagnosis: drawing attention to potential problems. *J Natl Cancer Inst* 96:353–356.

Ellis, D.I., Goodacre, R., 2006, Metabolic fingerprinting in disease diagnosis: biomedical applications of infrared and Raman spectroscopy. *Analyst* 131:875–885.

Fiehn, O., 2002, Metabolomics the link between genotypes and phenotypes. *Plant Mol Biol* 48:155–171.

Fiers, W., Contreras, R., Duerinck, F., Haegeman, G., Iserentant, D., Merregaert, J., Min Jou, W., Molemans, F., Raeymaekers, A., Van den Berghe, A., Volckaert, G., Ysebaert, M., 1976, Complete nucleotide-sequence of bacteriophage MS2-RNA – primary and secondary structure of replicase gene. *Nature* 260:500–507.

Fischer, H.P., 2005, Towards quantitative biology: integration of biological information to elucidate disease pathways and to guide drug discovery. *Biotechnol Annu Rev* 11:1–68.

German, J.B., Hammock, B.D., Watkins, S.M., 2005, Metabolomics: building on a century of biochemistry to guide human health. *Metabolomics* 1:3–9.

German, J.B., Roberts, M.A., Fay, L., Watkins, S.M., 2002, Metabolomics and individual metabolic assessment: the next great challenge for nutrition. *J Nutr* 132:2486–2487.

German, J.B., Roberts, M.A., Watkins, S.M., 2003, Genomics and metabolomics as markers for the interaction of diet and health: lessons from lipids. *J Nutr* 133:2078S–2083S.
Harrigan, G.G., Goodacre, R., 2003, Metabolic Profiling: Its Role in Biomarker Discovery and Gene Function Analysis. Kluwer Academic Publishers, Boston, ISBN 1-4020-7370-4.
Hye, A., Lynham, S., Thambisetty, M., Causevic, M., Campbell, J., Byers, H.L., Hooper, C., Rijsdijk, F., Tabrizi, S.J., Banner, S., Shaw, C.E., Foy, C., Poppe, M., Archer, N., Hamilton, G., Powell, J., Brown, R.G., Sham, P., Ward, M., Lovestone, S., 2006, Proteome-based plasma biomarkers for Alzheimer's disease. *Brain* 129:3042–3050.
Kaddurah-Daouk, R., 2006, Metabolic profiling of patients with schizophrenia. *PLoS Med* 3:363.
Kanehisa, M., Goto, S., Hattori, M., Aoki-Kinoshita, K.F., Itoh, M., Kawashima, S., Katayama, T., Araki, M., Hirakawa, M., 2006, From genomics to chemical genomics: new developments in KEGG. *Nucleic Acids Res* 34:D354–D357.
Kaput, J., 2004, Diet-disease gene interactions. *Nutrition* 20:26–31.
Katayama, S., Tomaru, Y., Kasukawa, T., Waki, K., Nakanishi, M., Nakamura, M., Nishida, H., Yap, C.C., Suzuki, M., Kawai, J., et al., 2005, "Antisense transcription in the mammalian transcriptome" by the RIKEN Genome Exploration Research Group and Genome Science Group (Genome Network Project Core Group) and the FANTOM Consortium. *Science* 309:1564–1566.
Kristal, B.S., Shurubor, Y.I., 2005, Metabolomics: opening another window into aging. *Sci Aging Knowl Environ*, (26), pe 1.
Kristiansen, T.Z., Bunkenborg, J., Gronborg, M., Molina, H., Thuluvath, P.J., Argani, P., Goggins, M.G., Maitra, A., Pandey, A., 2004, A proteomic analysis of human bile. *Mol Cell Proteomics* 3:715–728.
Lein, E.S., Hawrylycz, M.J., Ao, N., Ayres, M., Bensinger, A., Bernard, A., Boe, A.F., Boguski, M.S., Brockway, K.S., Byrnes, E.J., et al., 2007, Genome-wide atlas of gene expression in the adult mouse brain. *Nature* 445:168–176.
Lindon, J.C., Holmes, E., Nicholson, J.K., 2004, Metabonomics technologies and their applications in physiological monitoring, drug safety assessment and disease diagnosis. *Biomarkers* 9(1): 1–31.
Liotta, L.A., Lowenthal, M., Mehta, A., Conrads, T.P., Veenstra, T.D., Fishman, D.A., Petricoin, E.F., III, 2005, Importance of communication between producers and consumers of publicly available experimental data. *J Natl Cancer Inst* 97:315–319.
Loughlin, M.F., 2007, Using 'omic' technology to target *Helicobacter pylori. Drug Discov* 2: 1041–1051.
Macaulay, I.C., Carr, P., Gusnanto, A., Ouwehand, W.H., Fitzgerald, D., Watkins, N.A., 2005, Platelet genomics and proteomics in human health and disease. *J Clin Invest* 115: 3370–3377.
MacGregor, J.T., 2004, Biomarkers of cancer risk and therapeutic benefit: new technologies, new opportunities, and some challenges. *Toxicol Pathol* 1:99–105.
Min Jou, W., Haegeman, G., Ysebaert, M., Fiers, W., (1972) Nucleotide sequence of the gene coding for the bacteriophage MS2 coat protein. *Nature* 237:82–88.
Mutch, D.M., Berger, A., Mansourian, R., Rytz, A., Roberts, M.A., 2002, The limit fold change model: a practical approach for selecting differentially expressed genes from microarray data. *BMC Bioinformatics* 3:17.
Mutch, D.M., Wahli, W., Williamson, G., 2005, Nutrigenomics and nutrigenetics: the emerging faces of nutrition. *FASEB J* 19:1602–1616.
Nishizuka, S., Charboneau, L., Young, L., Major, S., Reinhold, W.C., Waltham, M., Kouros-Mehr, H., Bussey, K.J., Lee, J.K., Espina, V., Munson, P.J., Petricoin, E., 3rd, Liotta, L.A., Weinstein, J.N., 2003, Proteomic profiling of the NCI-60 cancer cell lines using new high-density reverse-phase lysate microarrays. *Proc Natl Acad Sci USA* 100:14229–14234.
Oliver, S.G., Winson, M.K., Kell, D.B., Baganz, F., 1998, Systematic functional analysis of the yeast genome. *Trends Biotechnol* 16:373–378.
Ordovas, J.M., Mooser, V., 2004, Nutrigenomics and nutrigenetics. *Curr Opin Lipidol* 15:101–108.

Pan, S., Zhang, H., Rush, J., Eng, J., Zhang, N., Patterson, D., Comb, M.J., Aebersold, R., 2005, High throughput proteome screening for biomarker detection. *Mol Cell Proteom* 4: 182–190.

Pegram, M., Slamon, D., 2000, Biological rationale for HER2/neu (c-erbB2) as a target for monoclonal antibody therapy. *Semin Oncol* 27:13–19.

Perroud, B., Lee, J., Valkova, N., Dhirapong, A., Lin, P.Y., Fiehn, O., Kültz, D., Weiss, R.H., 2006, Pathway analysis of kidney cancer using proteomics and metabolic profiling. Mol Cancer 5:64–82.

Petricoin, E.F., Ardekani, A.M., Hitt, B.A., Levine, P.J., Fusaro, V.A., Steinberg, S.M., Mills, G.B., Simone, C., Fishman, D.A., Kohn, E.C., Liotta, L.A., 2002, Use of proteomic patterns in serum to identify ovarian cancer. *Lancet* 359:572–577.

Phizicky, E., Bastiaens, P.I., Zhu, H., Snyder, M., Fields, S., 2003, Protein analysis on a proteomic scale. *Nature* 422:208–215.

Ransohoff, D.F., 2005, Lessons from controversy: ovarian cancer screening and serum proteomics. *J Natl Cancer Inst* 97:315–319.

Rogers, M.A., Clarke, P., Noble, J., Munro, N.P., Paul, A., Selby, P.J., Banks, R.E., 2003, Proteomic profiling of urinary proteins in renal cancer by surface enhanced laser desorption ionization, and neural-network analysis: identification of key issues affecting clinical potential utility. *Cancer Res* 63:6971–6983.

Sanger, F., Air, G.M., Barrell, B.G., Brown, N.L., Coulson, A.R., Fiddes, C.A., Hutchison, C.A., Slocombe, P.M., Smith, M., 1977, Nucleotide sequence of bacteriophage phi X174 DNA. *Nature* 265:687–695.

Schmidt, C., 2004, Metabolomics takes its place as latest up-and-coming "omic"science. *J Natl Cancer Inst* 96:732–734.

Subramanian, A., Tamayo, P., Mootha, V.K., Mukherjee, S., Ebert, B.L., Gillette, M.A., Paulovich, A., Pomeroy, S.L., Golub, T.R., Lander, E.S., Mesirov, J.P., 2005, Gene set enrichment analysis: a knowledge-based approach for interpreting genome-wide expression profiles. *Proc Natl Acad Sci USA* 102:15545–15550.

Tirumalai, R.S., Chan, K.C., Prieto, D.A., Issaq, H.J., Conrads, T.P., Veenstra, T.D., 2003, Charaterization of the low molecular weight human serum proteome. *Mol Cell Proteom* 10:1096–1103.

Vasan, R.S., 2006, Biomarkers of cardiovascular disease: molecular basis and practical considerations. *Circulation* 113:2335–2362.

Venkatesh, T.V., Harry, B., 2002, Harlow integromics: challenges in data integration. *Genome Biol* 3:reports 4027.1–reports 4027.3.

Wishart, D.S., Tzur, D., Knox, C., Eisner, R., Guo, A.C., Young, N., Cheng, D., Jewell, K., Arndt, D., Sawhney, S., et al., 2007, HMDB: the human metabolome database. *Nucleic Acids Res* 35:521–526.

Wu, S.L., Kim, J., Hancock, W.S., Karger, B., 2005, Extended Range Proteomic Analysis (ERPA): a new and sensitive LC-MS platform for high sequence coverage of complex proteins with extensive post-translational modifications-comprehensive analysis of beta-casein and epidermal growth factor receptor (EGFR). *J Proteome Res* 4:1155–1170.

Wulfkuhle, J.D., Liotta, L.A., Petricoin, E.F., 2003, Proteomic applications for the early detection of cancer. *Nat Rev Cancer* 3(4):267–275.

Yanagisawa, K., Shyr, Y., Xu, B.J., Massion, P.P., Larsen, P.H., White, B.C., Roberts, J.R., Edgerton, M., Gonzalez, A., Nadaf, S., Moore, J.H., Caprioli, R.M., Carbone, D.P., 2003, Proteomic patterns of tumour subsets in non-small-cell lung cancer. *Lancet* 362:433–439.

Zhang, Z., Bast, R.C., Jr., Yu, Y., Li, J., Sokoll, L.J., Rai, A.J., Rosenzweig, J.M., Cameron, B., Wang, Y.Y., Meng, XY., et al., 2004a, Three biomarkers identified from serum proteomic analysis for the detection of early stage ovarian cancer. *Cancer Res* 64:5882–5890.

Zhu, W., Wang, X., Ma, Y., Rao, M., Glimm, J., Kovach, J.S., 2003, Detection of cancer-specific markers amid massive mass spectral data. *Proc Nat Acad Sci USA* 100:14666–14671.

Chapter 3
Recombinant DNA Pharmaceuticals

Abstract Modern molecular medicine encompasses the utilization of many molecular biological techniques in the analysis of disease, disease genes and disease gene function. The study of disease genes and their function in an unaffected individual has been possible by the development of recombinant DNA and cloning techniques. The term gene cloning covers a range of techniques that makes it possible to manipulate DNA in a test tube and also to return it to living organism where it function normally. The tools of gene cloning includes vectors, genes and the enzymes. Overcoming the biological and methodological obstacles posed by cell factories to the production or rDNA pharmaceuticals is a main challenge in the further development of protein-based molecular medicine. Recombinant DNA technologies might have exhausted conventional cell factories and new production systems need to be deeply explored and incorporated into the production pipeline. On the other hand, a more profound comprehension of host cell physiology and stress responses to protein production would necessary offer improved tools (either at genetic, metabolic or system levels) to favour high yield and high quality protein production. Apart from the expected incorporation of unusual mammalian hosts such as transgenic animals or plants, microbial cells appear as extremely robust and convenient hosts, and gaining knowledge about the biological aspects of protein production would hopefully enhance the performance of such hosts beyond the current apparent limitations. In this regard, not only commonly used bacteria and yeasts but unconventional strains or species are observed as promising cell factories for forthcoming recombinant drugs. Their incorporation into productive processes for human pharmaceuticals would hopefully push the trend of marketed products and fulfil the increasing demands of the pharmacological industry.

Keywords Gene cloning · Recombinant DNA technology · Vectors · Genes · Molecular medicine · Pharmaceuticals · Humulin · Post translational modifications · Clone · cDNA cloning · cDNA sequences · Genomic DNA libraries · cDNA library · Genomic cloning · Polymerase chain reaction · Recombinant DNA · Recombinant proteins · Glycosylation · Insect cell lines · Hamster cell lines · Hybridoma cell lines · Human cell lines · Recombinant DNA pharmaceuticals · Biopharming · Screening · Gene transfer · Gene therapy

M. Debnath et al., *Molecular Diagnostics: Promises and Possibilities*,
DOI 10.1007/978-90-481-3261-4_3,

3.1 Prologue

Recombinant DNA technology has now made it possible to produce proteins for pharmaceutical applications (Manning et al., 1989). This has been accomplished successfully in microbial cells such as bacteria and yeasts. In early 1980s, the FDA approved the clinical use of recombinant human insulin from recombinant *Escherichia coli* (Humulin-US/Humuline-EU) for the treatment of diabetes, being the first recombinant pharmaceutical to enter the market. One hundred and fifty one recombinant pharmaceuticals have so far been approved for human use by the Food and Drug Administration (FDA) and/or by the European Medicines Agency (EMEA) (Ferrer-Miralles et al., 2009). The versatility and scaling-up possibilities of the recombinant protein production opened up new commercial opportunities for pharmaceutical companies. Since the approval of recombinant insulin, other recombinant DNA drugs have been marketed in parallel with the development and improvement of several heterologous protein production systems. This has generated specific strains of many microbial species adapted to protein production, and has allowed the progressive incorporation of yeasts and eukaryotic systems for this purpose. Among the approved protein-based recombinant pharmaceuticals licensed up to January 2009 by the FDA and EMEA, 45 (29.8%) are obtained in *E. coli*, 28 (18.5%) in *Saccharomyces cerevisiae*, 17 (11.2%) in hybridoma cells, 1 in transgenic goat milk, 1 in insect cells and 59 (39%) in mammalian cells.

Recombinant DNA (rDNA) technologies now offer a very potent set of technical platforms for the controlled and scalable production of polypeptides of interest by relatively inexpensive procedures. When massively expressed in host like bacteria, recombinant proteins often tend to misfold and accumulate as soluable and insoluable aggregates (de Marco et al., 2005. This fact indicates that despite the diverse bottlenecks and obstacles that microbial systems pose to the efficient production of functional mammalian proteins, namely lack or unconventional post-translational modifications, proteolytic instability, poor solubility and activation of cell stress responses, among others, they represent convenient and powerful tools for recombinant protein production. The entering into the market of a progressively increasing number of protein drugs produced in non-microbial systems has not impaired the development of products obtained in microbial cells, proving the robustness of the microbial set of cellular systems developed for protein drug production.

3.2 Concept

Genetic engineering and biomedical research have experienced such a revolutionary change since the past 30 years with the development of gene manipulation. Most biomolecules exist in low concentrations & as complex, mixed populations. This problem was solved in 1970 using the molecular biologist's favourite bug, *E. coli*, a normally innocuous commensal occupant of the human gut. The basis of the term recombinant DNA refers to the recombining of different segments of DNA (Gluck et al., 1992). Constructing a recombinant DNA molecule involves cutting the DNA into fragments with restriction endonucleases and rejoining the fragments in novel arrangements with ligase. Propagating the molecule in a microorganism, or cloning, is necessary to increase the number of gene copies to facilitate detection of the gene of interest and to produce the protein it encodes (Black, 1989) in a desirable host. The purpose of these approaches is used to identify genes, to isolate the gene of interest, to amplify the gene if necessary, and to clone genes (Carroll, 1993).

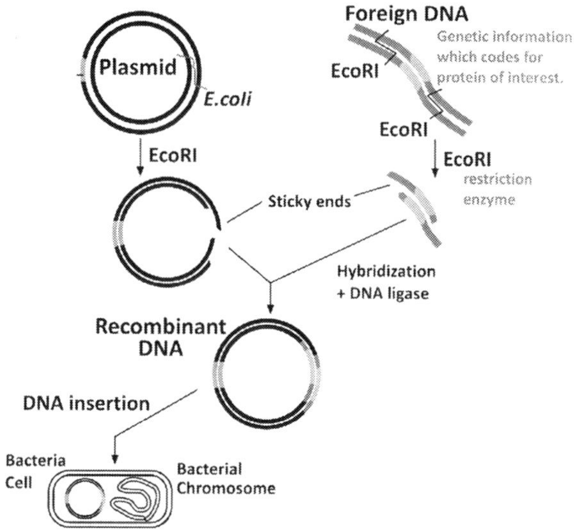

Fig. 3.1 General concept of recombinant DNA technology

3.3 Tools of Recombinant DNA Technology

Recombinant DNA is a DNA molecule constructed outside of living cells by joining natural or synthetic DNA segments to DNA molecules that can replicate in a living cell. Most of the techniques of molecular genetics are either directly or indirectly dependent on molecules discovered in and isolated from bacteria. These molecules have functions in the everyday life of bacteria, but scientists have learned to exploit them as tools used for the manipulation and study of DNA. Any fragment of DNA can be cloned once it is introduced into a suitable vector for transforming a bacterial host. Cloning refers to the process of preparing multiple copies. cDNA cloning is a central technology in molecular biology. cDNA sequences are used to determine mRNA transcript structures, including splice junctions, open reading frames (ORFs) and 5′- and 3′-untranslated regions (UTRs). cDNA clones are valuable reagents for functional studies of genes and proteins. "Genomic cloning" refers to the production of a library of cloned DNAs representing the entire genome of a particular organism. From either of these types of libraries one can isolate a single cDNA or gene clone. In order to clone either cDNAs or copies of genes a *vector* is required to *carry* the cloned DNA. Once cloned, DNA fragments can be characterized by restriction enzyme mapping, by DNA sequencing, or by gene expression studies.

3.3.1 Common Enzymes Used in Recombinant DNA Technology

The ability to manipulate DNA in vitro (outside the cell) depends entirely on the availability of purified enzymes that can cleave, modify and join the DNA molecule in specific ways. At present, no purely chemical method can achieve the ability to manipulate the DNA in vitro in a predictable way. Only enzymes are able to carry out the function of manipulating the DNA. Each enzyme has a vital role to play in the process of genetic engineering. The technique basically involves the use of several recently discovered enzymes (Table 3.1) that facilitate cutting and joining DNA to construct recombinant molecules, which can be transferred from one organism to another (Siddiqui, 1982).

Table 3.1 Enzymes used in gene cloning

Enzyme(s)	Activity	Comments
Restriction endonucleases	Recognize specific nucleotide sequences and cleaves the DNA within or near to the recognition sequences	Used to specifically recognize and cleave the double stranded DNA
Reverse transcriptase (RT)	Retrovirally encoded RNA-dependent DNA polymerase	Used to convert mRNA in to a complimentary DNA (cDNA) copy for the purpose of cloning cDNAs
RNase H	Recognizes RNA-DNA duplexes and randomly cleaves the phosphodiester backbone of the RNA	Used primarily to cleave the mRNA strand that is annealed to the first strand of cDNA generated by reverse transcription
DNA polymerase	Synthesis of DNA	Used during most procedures where DNA synthesis is required, also used in in vitro mutagenesis
Klenow DNA polymerase	Proteolytic fragment of DNA polymerase that lacks the $5' \rightarrow 3'$ exonuclease activity	Used to incorporate radioactive nucleotides into restriction enzyme generated ends of DNA, also can be used in place of DNA polymerase
DNA ligase	Covalently attaches a free $5'$ phosphate to a $3'$ hydroxyl	Used in all procedures where to molecules of DNA need to be covalently attached
Alkaline phosphatase	Removes phosphates from $5'$ ends of DNA molecules	Used to allow $5'$ ends to be subsequently radiolabeled with the phosphate of ATP in the presence of polynucleotide kinase, also used to prevent self-ligation of restriction enzyme digested plasmids and lambda vectors.

Table 3.1 (continued)

Enzyme(s)	Activity	Comments
Polynucleotide kinase	Introduces phosphate of ATP to 5′ ends of DNA	It is a bifunctional 5′-kinase/3′-phosphatase that aids in the repair of broken termini in RNA by converting 3′-PO_4/5′-OH ends into 3′-OH/5′-PO_4 ends, which are then sealed by RNA ligase
DNase I	Randomly hydrolyzes the phosphodiester bonds of double-stranded DNA	Is used in the identification of regions of DNA that are bound by protein and thereby protected from DNase I digestion, also used to identify transcriptionally active regions of chromatin since they are more susceptible to DNase I digestion
S_1 nuclease	Exonuclease that recognizes single stranded regions of DNA	Used to remove regions of single strandedness in DNA or RNA-DNA duplexes
Exonuclease III	Exonuclease that removes nucleotides form the 3′ end of DNAs	Used to generate deletions in DNA for sequencing, or to map functional domains of DNA duplexes
Terminal transferase	DNA polymerase that requires only a 3′-OH, lengthens 3′ ends with any dNTP	Used to introduce homopolymeric (same dNTP) *tails* onto the 3′ ends of DNA duplexes, also used to introduce radiolabeled nucleotides on the 3′ ends of DNA
T3, T17 and SP6 RNA polymerase	Bacterial virus encoded RNA polymerase, recognize specific nucleotide sequences for initiation of transcription	Used to synthesize RNA in vitro
Taq and Vent DNA polymerase	Thermostable DNA polymerases	Used in PCR

One vital type of molecule discovered in bacteria is the restriction endonuclease. These enzymes recognize specific base sequences within DNA molecules, and cleave specific phosphodiester bonds within that sequence. The key to the in vitro utilization of restriction endonucleases is their strict nucleotide sequence specificity. The different enzymes are identified by being given a name indicating the bacteria from which they were isolated, e.g. the enzymes EcoRI which recognizes the sequences, 5′-GAATTC-3′, was isolated from *E. coli*. One unique feature of restriction enzymes is that the nucleotide sequences they recognize are palindromic, i.e. they are the same sequences in the 5′ → 3′ direction of both strands. Some restriction endonucleases make staggered symmetrical cuts away from the center of their recognition site within the DNA duplex, some make symmetrical cuts in the middle of their recognition site while still others cleave the DNA at a distance from the recognition sequence (Fig. 3.1). Enzymes that make staggered cuts leave the resultant DNA with cohesive or sticky ends. Enzymes that cleave the DNA at the center of the recognition sequence leave blunt-ended fragments of DNA. It was also shown that an enormous variety of different restriction-modification systems, each with their own characteristic pair of nuclease and methylase for certain DNA recognition sequences, exist in different bacterial species.

Any two pieces of DNA containing the same sequences within their sticky ends can anneal together and be covalently ligated together in the presence of DNA ligase. Any two blunt-ended fragments of DNA can be ligated together irrespective of the sequences at the ends of the duplexes. Other modifying enzymes (Table 3.1) includes phosphatases, kinases, single-strand specific nucleases, etc. that allow precise modifications to pieces of DNA to be made in vitro in order to add, remove or alter the structure of DNA. The RNA modifying enzymes – e.g. exonucleases, RNA ligase, reverse transcriptase. RNA is much more difficult to work with in vitro because the enzymes available are generally not as sophisticated as the set available which modify DNA. DNA Polymerases allow the synthesis of DNA from a pre-existing template in vitro, which can then be used for a variety of purposes, including sequence analysis by chain-termination methods. Recently, thermostable polymerases have become important, e.g. Taq DNA polymerase from *Thermus aquaticus*. This bacterium has evolved to grow in hot springs at temperatures which kill most other species. These enzymes allow the amplification of as little as one molecule of DNA into a large amount by means of repeated cycles of melting, primer annealing & extension by the enzyme which is not destroyed by the high temperatures used in this process. This is known as the polymerase chain reaction.

3.3.2 Vectors

Vectors are segments of DNA replicated in living cells and engineered to allow the insertion of other DNA fragments. There are many types of vectors, including virus DNA, plasmid/virus hybrids, and artificial eukaryotic chromosomes (a shuttle vector is one that can replicate in 2 or more types of cells (e.g., *E. coli* and yeast)), but plasmids are the most common and easiest to understand. The insertion of DNA into a vector is done by cutting both the vector and DNA of interest with the same restriction enzyme. The sticky ends on each piece of DNA will be complementary to each other, due to the palindromic nature of the restriction enzyme recognition sites. Such DNA molecules in solution will transiently interact with each other, and join together via hydrogen bonds formed between sticky ends, forming a recombinant DNA molecule. If DNA ligase is present in the mixture with the two different fragments of DNA, any transiently-forming recombinant DNA molecules can be joined permanently to each other.

3.3.3 Hosts

Once created, these recombinant DNA molecules are inserted into various hosts. The commonly used host is bacteria where by a process known as transformation these bacteria are successfully transformed with the DNA. They can be selected using the selectable marker. These transformed bacteria can then be grown up in great quantities, so that large amounts of the DNA can be recovered. This process, known as DNA cloning, is used for storage and amplification of recombinant DNA molecules.

1. *Escherichia coli (E. coli)*

The enterobacterium *E. coli* is the first-choice microorganism for the production of recombinant proteins, and widely used for primarily cloning, genetic modification and small-scale production for research purposes (Sorensen and Mortensen, 2005). Recombinant proteins obtained in *E. coli* lack the post-translational modifications (PTMs) which are present in most of eukaryotic proteins (Marston, 1996). PTMs play a crucial role in protein folding, processing, stability, final biological activity, tissue targeting, serum half-life and immunogenicity of the protein; therefore PMT deficient version might be insoluble, unstable or inactive.

On the other hand, the frequencies with which the different codons appear in *E. coli* genes are different from those occurring in human genes, and this is directly related to the abundance of specific tRNAs. Therefore, genes that contain codons rare for *E. coli* may be inefficiently expressed by this organism and cause premature termination of protein synthesis or amino acid misincorporation, thus reducing the yield of expected protein versions. This problem can be solved either by site-directed replacement of rare codons in the target gene by codons that are more frequently used in *E. coli*, or, alternatively, by the co-expression of the rare tRNAs (*E. coli* strains BL21 codon plus and Rosetta were designed for this purpose). In addition, initial methionine removal depends on the side chain of the penultimate amino acid of N-terminal in final recombinant proteins produced in *E. coli* although it can be efficiently removed using recombinant methionine aminopeptidase. Around 10% of full-length eukaryotic proteins tested in this system have been successfully produced in soluble form in *E. coli*.

2. *Saccharomyces cerevisiae*

Production in yeast is usually approached when the target protein is not produced in a soluble form in the prokaryotic system or a specific PTM, essential for its biological activity, cannot be produced artificially on the purified product. Yeasts are as cost effective, fast and technically feasible as bacteria and high density cell cultures can also be reached in bioreactors (Branduardi et al., 2008). Even more, mutant strains that produce high amounts of heterologous protein are already available. Even though yeasts are able to perform many PTMs as O-linked glycosylation, phosphorylation, acetylation and acylation, the main pitfall of this expression system is related to N-linked glycosylation patterns which differ from higher eukaryotes, in which sugar side chains of high mannose content affect the serum half-life and immunogenicity of the final product. Although less studied than in bacteria, the production of recombinant proteins also triggers conformational stress responses and produced proteins fail sometimes to reach their native conformation.

3. Insect Cell Lines

Cultured insect cells are used as hosts for recombinant baculovirus infections. The production of a recombinant viral vector for gene expression is time-consuming, the cell growth is slow when compared with former expression systems, the cost of growth medium is high and each protein batch preparation has to be obtained from fresh cells since viral infection is lethal. PTMs are also an important limitation of this expression system because of the simple non-syalated N-linked glycosylation which is translated in a rapid clearance from human sera.

4. Hybridoma Cell Lines

Hybridomas are fusion cells of murine origin (B-cells and myeloma tumour cells) that are able to express specific monoclonal antibodies against a determined antigen, thus possessing therapeutic potential. Clone selection may account for the progressive enrichment of cells displaying a glycosylation profile with reduced potency and undesirable immunogenic reaction since the human immune system recognizes mouse antibodies as foreign.

5. Hamster Cell Lines

Most of the therapeutic proteins approved so far have been obtained using transgenic hamster cell lines, namely 49 in Chinese hamster ovary cells (CHO) and 1 in baby hamster kidney cells (BHK). The main advantage of this expression system is that cells can be adapted to grow in suspension in serum free media (SFM), protein-free and chemically defined media. This fact increases the biosafety of final products reducing risk of introducing prions of bovine spongiform encephalopathy (BSE) from bovine serum albumin and of infectious variant Creutzfeldt – Jakob disease (vCJD) from human serum albumin. In addition, recombinant products can be secreted into the chemical defined media, which simplifies both upstream and downstream purification process. PTMs in this expression system are almost the same as in human cell lines, although some concerns about comparability in the glycosylation pattern have arisen when comparing different batches of the same manufacturer product. Further development of chemically defined media and fine description of growth conditions would help to overcome this issue.

6. Human Cell Lines

In the recent years, three therapeutic proteins produced in human cell lines have been approved, namely Dynepo-erithropoietin, Elaprase-irudonate-2-sulfatase and Replagal-alfa-galactosidase A. These products are fully glycosylated human proteins, so this expression system should be addressed when heavily glycosylation is needed. In general, recombinant biopharmaceuticals obtained from mammalian cells cover a wider spectrum of pathological conditions than those obtained from microbes, and the distribution of applications is less biased than when observing products from *E. coli* or *S. cerevisae* (Fig. 3.2).

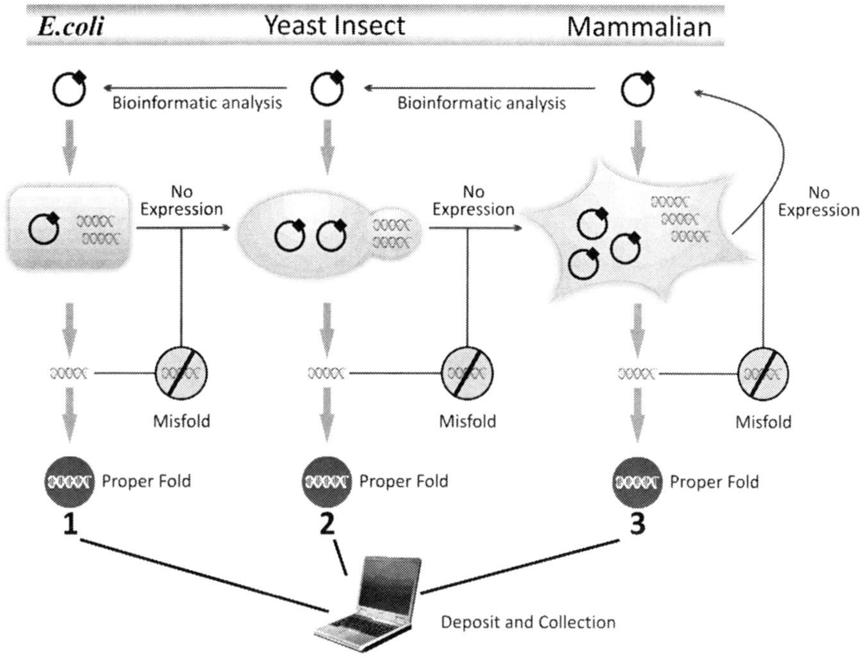

Fig. 3.2 Importance of post translational modifications for recombinant proteins

7. Transgenic Animals

Transgenic animals (avian and mammals) have been successfully used for the production of recombinant proteins secreted into egg white and milk respectively. Protein production using transgenic farm animals supposes a great biotechnological challenge in terms of safety concerns such as transmission of infectious diseases (including viral and prion infections) or adverse allergenic, immunogenic and autoimmune responses. In 2006, ATryn was the first and so far single approved rDNA biopharmaceutical using transgenic animals and validated manufacturer technology platform. It contains human antithrombin (432 amino acids) with 15% glycosylated moieties and is secreted into the milk of transgenic goats. Another product obtained from the milk of transgenic rabbits (Rhucin) has been recently denied for its approval by the EMEA although more tests of repeated treatment are underway to try again its approval. Despite such limited progress, if pharmacovigilance after patient treatment does not reveal any adverse side effects, we might envisage, in the next years, an increase in the approval rate of recombinant protein products from transgenic animal origin.

Recombinant DNA biopharmaceuticals obtained from bacterial, yeast or mammalian cell culture bioreactors are quite effective as therapeutic agents although production costs are relatively high. One way to address the economic-cost benefit hurdle is through the use of transgenic organisms to manufacture biopharmaceuticals. Biopharming would dramatically reduce the cost of recombinant therapeutic proteins not only in the initial construction of production facilities but also the scale-up process and the final recombinant protein yield. Nonetheless, the fact that regulatory guidelines are being developed as the same time that the establishment of protein production processes is creating uncertainty within biotechnological companies to fulfil drug administration requirements.

8. Transgenic Plants

Transgenic plants have been used as recombinant protein producers for research and diagnostic uses due to the advantageous low cost of cultivation, high mass production, flexible scale-up, lack of human pathogens and addition of eukaryotic PTMs. The first recombinant protein product obtained from transgenic tobacco (Staub et al., 2000; Travis et al., 1985) was human growth hormone and since then, many other products have been obtained (including antibodies, the surface antigen of the Hepatitis-B-Virus, industrial enzymes and milk proteins). Again, the main disadvantage is related to the plant specific PMTs introduced in recombinant proteins which produce adverse immune responses. Moreover, the possibility to spread the proteins in open fields and the negative public perception of the transgenic plants precludes the use of plants as an attractive expression system of therapeutic proteins.

3.3.4 DNA Libraries

DNA libraries are mixtures of different restriction enzyme-digested DNA molecules ligated into vectors. Libraries are constructed using the same general approach outlined above, except that instead of inserting one type of DNA fragment into the vector, there are thousands or even millions of different DNA fragments inserted into vector molecules. There are essentially two types of DNA libraries, based on the source of the DNA:

Genomic DNA Libraries are made from genomic DNA. Genomic DNA molecules are very large, so they must be fragmented into small enough pieces to insert into vectors. This is typically done through digestion with one or more appropriate restriction endonucleases, mechanical shearing, or a combination of the two processes. The DNA is then ligated into the vector, which could be a plasmid, but is more often a cosmid or a viral chromosome.

cDNA Libraries are made from cDNA, which are DNA copies of mRNA molecules. To make cDNA, mRNA is isolated from a tissue or whole organism, and DNA is copied from the mRNA template using an enzyme called reverse transcriptase. This enzyme works like a DNA polymerase, except that it uses RNA as a template instead of DNA. The resulting cDNA molecules are then engineered so that they have restriction enzyme recognition sites at each end of every molecule, which allows them to be digested and inserted into a vector as outlined previously. However, EST sequencing samples a cDNA library at random, and it recovers transcripts with low expression levels inefficiently.

The difference between these two libraries is the nature of the DNA found in the library. A cDNA library, because it is derived from mRNA molecules, will only contain DNA representing transcribed genes. A genomic library will also contain gene sequences, but will also contain a lot of DNA that is not genes, because genes typically make up only about 1–5% of the total genomic DNA. However, this type of library will contain regulatory DNA sequences (such as enhancers and promoter sequences) that would not be present in a cDNA library. Therefore, the choice of library type depends on what type of DNA sequence a researcher wishes to study.

3.3.5 Library Screening

Plating out the phage is the first step of the screening process. Replicas are produced by placing absorbent paper, nitrocellulose or nylon membranes, over the agar plates in such a way that the pattern of plaques on the original plate is maintained. Each replica is then incubated with an appropriate nucleic acid probe that is labelled in some way (e.g. radiolabelling) so as to facilitate selection. If the probe is a radio-labelled fragment of DNA, the location of the positive clones may be detected by autoradiography.

To find a specific fragment of DNA in a library, screening cDNA libraries can be done using antibodies against the protein product encoded by the gene of interest. Some vectors used to create libraries allow the inserted DNA to be transcribed after transformation into bacteria. These vectors have promoters built in, and since cDNA is copied from mRNA, it is capable of encoding protein. The transcripts would then be translated into protein in the bacteria. Bacteria transformed with the library can be grown in duplicate copies on plates and on filters in a way that individual bacterial colonies are produced, with each colony originating from a single bacterial cell, so all bacteria in a colony would contain the same recombinant DNA molecule, and therefore produce the same protein. Many thousands of colonies can be grown on a series of filters, and the bacteria lysed (broken open), liberating the protein, which would then adhere to the filter. Filters can be exposed to antibodies that are labeled in some way (usually with a fluorescent or radioactive tag); the antibody to the protein of interest will bind only to the colony that contained the recombinant DNA molecule encoding that protein. Once the colony is identified, the duplicate colony from a bacterial plate can be picked, and grown up to produce more of that DNA. This method doesn′t work well with genomic libraries, because most recombinant clones in a genomic library do not encode protein.

Another way to screen a library is with a piece of DNA similar to the DNA of interest, such as a similar gene from a different organism, for example. This strategy takes advantage of the ability of DNA to denature and renature. In this case, transformed bacterial colonies are grown in duplicate on plates and filters. The bacteria on the filters are lysed, and the DNA is denatured by treating with an alkaline solution. The filters are then exposed to a solution containing a single-stranded DNA "probe", labeled with a radioactive or chemical tag, under conditions that promote renaturation. If the probe DNA is in excess, sequences strongly related to the probe will renature with the probe, and can be detected by exposure to X-ray film. As with the antibody approach, the corresponding colony on the plates can be picked and isolated. All of the bacteria in this colony descended from a single bacterial cell, and therefore they all contain the same genetic complement. This means that the plasmid (or whichever vector was used) in every bacterial cell in the colony will contain the same insert.

Once isolated, a bacterial colony can be inoculated into bacterial growth medium, which provides the bacteria with the necessary nutrients for fast reproduction. In a matter of hours, the total number of bacterial cells can be increased over one billion-fold. The plasmid (or other vector) DNA can then be isolated from the bacteria, producing a sizable yield. In this way, it is possible to amplify the amount of a specific sequence of DNA. Bacteria from the growth culture can also be stored frozen at –70°C, and will remain viable for many years. Therefore, once a particular clone from a library is selected, it can be stored for years without having to rescreen the library for that sequence.

3.3.6 Polymerase Chain Reaction

In the late 1980s, an alternative technique was developed by which DNA can be amplified many times over. This technique does not require the use of living bacteria or other cells, and is essentially a DNA replication reaction done in a test tube. We will discuss this technique in the chapter on PCR.

PCR can be used in the analysis of disease genes (Table 3.2) by being able to amplify detectable amounts of specific fragments of DNA. The amplified fragments from disease genes may be larger, due to insertions, or smaller, due to deletions. The dramatic amplification of DNA by PCR allows the analysis of disease genes in extremely small samples of DNA. For example, only a small number of fetal cells need be extracted from amniotic fluid in order to analyze for the presence of specific disease genes. Additionally, single point mutations can be detected by modified PCR techniques such as the ligase chain reaction (LCR) and PCR-single-strand conformational polymorphisms (PCR-SSCP) analysis. The PCR technique also can be used to identify the level of expression of genes in extremely small samples of material, e.g. tissues or cells from the body. This technique is termed reverse transcription-PCR (RT-PCR). Many inherited disorders are due to single nucleotide changes within critical regions of the affected gene (e.g. sickle cell anemia).

Many inherited disorders (Table 3.2) are due to single nucleotide changes within critical regions of the affected gene (e.g. sickle cell anemia). The PCR-SSCP technique can detect single mutations in genes due to the altered conformation mobility of the single strands of DNA within an electrophoresis gel harboring the mutation relative to the wild-type strands that do not. Specific PCR primers are made that span the sequences of a given disease gene where a mutation is known to exist and the region amplified by PCR. The same region of the wild-type gene is PCR amplified. The two strands of wild-type PCR product will migrate differently than the two strands of mutant PCR product. Even single point mutations lead to the strands of amplified DNA existing in different conformations which alter their mobility when subjected to electrophoresis in non-denaturing gels.

The most common techniques utilized in gene therapy studies is the introduction of the corrected gene into bone marrow cells, skin fibroblasts or hepatocytes. The vectors most commonly utilized are derived from retroviruses and utilize only the transcriptional promoter regions of these viruses (the LTRs) to drive expression of the gene of interest. The advantage of retroviral-based vector systems is that expression occurs in most cell types.

A number of human inherited disorders have been corrected in cultured cells and several diseases (e.g. malignant melanoma and severe combined immunodeficiency disease, SCID) are currently being treated by gene therapy techniques indicating that gene therapy is likely to be a powerful therapeutic technique against a host of diseases in coming years. The technique of PCR has also revolutionized forensic science, because it allows amplification of DNA from minute samples such as blood droplets, which then allows DNA fingerprints to be performed.

Table 3.2 Diseases and disorders detected by PCR and their related gene

Disease	Affected gene
Adenosine deaminase deficiency	Adenosine deaminase (ADA)
Lesch-Nyhan syndrome	Hypoxanthine-guanine phosphoribosyltransferase (HGPRT)
α-1-Antitrypsin deficiency	α-1-Antitrypsin
Cystic fibrosis	Cystic fibrosis transmembrane conductance (CFTR) protein
Fabry disease	β-Galactosidase
Gaucher's disease	Glucocerebrosidase
Sandhoff-Jatzkewitz disease	Hexosaminidase A and B
Tay-Sachs disease	Hexosaminidase A
Familial hypercholesterolemia (FH)	LDL receptor
Glucose-6-phosphate dehydrogenase deficiency	Glucose-6-phosphate dehydrogenase
Maple syrup urine disease	α-keto acid decarboxylase
Phenulketonuria (PKU)	Phenylalanine
Ornithine transcarbamylase deficiency	Ornithine transcarbamylase
Retinoblastoma (Rb)	Rb gene product
Sickle-cell anemia	Point mutation in globin gene resulting in improper folding of protein
β-Thalassemia	Mutations in β-globin gene that result in loss of synthesis of protein
Hemophilia A	Factor VIII
Hemophilia B	Factor IX
von Wilebrand disease	von Wilebrand factor (vWF)

3.3.7 Analysis of Cloned DNA Sequences

Once DNA has been cloned and amplified, it needs to be analyzed and characterized. Two common ways to do this are restriction enzyme mapping and DNA sequencing, both of which rely upon a technique known as gel electrophoresis.

Restriction enzyme mapping entails digesting the DNA of interest with various restriction enzymes, singly and in pairs. By running the various reactions on agarose gels, it is possible to determine the sizes of the digested fragments (by comparing to DNA standard fragments of known size). This in turn allows mapping of the different restriction enzyme recognition sites relative to each other. Sometimes, it is necessary to visualize only a few restriction fragments out of a large set of fragments; a technique known as Southern blotting is used.

The analysis of cloned cDNAs and genes involves a number of techniques. The initial characterization usually involves mapping of the number and location of different restriction enzyme sites. This information is useful for DNA sequencing since it provides a means to digest the clone into specific fragments for sub-cloning, a process involving the cloning of fragments of a particular cloned DNA. Once the DNA is fully characterized cDNA clones can be used to produce RNA in vitro and the RNA translated in vitro to characterize the protein. Clones of cDNAs also can be used as probes to analyze the structure of a gene by Southern blotting or to analyze the size of the RNA and pattern of its expression by Northern blotting. Northern blotting is also a useful tool in the analysis of the exon-intron organization of gene clones since only fragments of a gene that contain exons will hybridize to the RNA on the blot. Western blotting involves the analysis of proteins following attachment to a solid support. The proteins are separated by size SDS-PAGE and electrophoretically transferred to nitrocellulose or nylon filters. The filter is then probed with antibodies raised against a particular protein.

Ultimately, to properly study and utilize a piece of DNA, we need to know its sequence of bases. This can be done using a technique called DNA sequencing, which utilizes a modified DNA replication reaction. Sequencing of DNA can be accomplished by either chemical or enzymatic means. The original technique for sequencing, Maxam and Gilbert sequencing, relies on the nucleotide-specific chemical cleavage of DNA and is not routinely used any more. The enzymatic technique, Sanger sequencing, involves the use of dideoxynucleotides (2′,3′-dideoxy) that terminate DNA synthesis and is, therefore, also called dideoxy chain termination sequencing.

3.3.8 Expression of Cloned Genes

The main reason we clone and characterize genes is to produce the recombinant protein. This is done for a variety of reasons. Sometimes, the protein is a product that will be harvested for commercial sale by a biotechnology or pharmaceutical company. Production of the protein can also be used in an indirect way to understand its function. By expressing the gene in cells or an organism (for example, expressing a gene at a time or in a place that it wouldn't normally be expressed), it may be possible to observe the effect that expression has on the organism, and infer the function from that. Expression of a gene employs a class of vector known as an expression vector. These vectors have promoters (often from viral genes) engineered into the vector near the multiple cloning site. Any gene inserted into the multiple cloning site therefore comes under the control of the promoter, allowing expression of the gene under appropriate conditions. Expression may be in vivo (in a living cell or organism) or in vitro (in a test tube using cell-free transcription and translation systems that are available commercially, allowing production of the protein relatively purely).

3.4 Therapeutic Applications of Recombinant DNA Derived Pharmaceuticals

Successful development of recombinant DNA-derived pharmaceuticals, a new class of therapeutic agents, is determined by a variety of factors affecting the selection and positioning of the compound under development. Technical development of recombinant DNA technology puts genetics in the mainstream of medicine. Techniques for gene transfer have been improved to introduce synthetic sequences into the cellular genomes. High quality of recombinant products covering monoclonal antibodies, vaccines thrombolytics have been developed by this techniques. Candidate cells and culture systems have been screened for their capacity to synthesise the proper recombinant molecules.

3.4.1 Recombinant Therapeutic Proteins

Recombinant DNA technology has now made it possible to produce proteins for pharmaceutical application (Manning et al., 1989). Consequently, proteins produced via biotechnology now comprise a significant portion of the drugs currently under development. Substitution therapy with recombinant DNA derived human proteins is in therapeutic application to replace their counterparts from native source by human pharmacologically active proteins which cannot be isolated from their natural source. For recombinant DNA derived proteins where the mode of action is known short development times frames can be expected allowing an early return on investment. Isolation, purification, formulation, and delivery of proteins represent significant challenges to pharmaceutical scientists, as proteins possess unique chemical and physical properties. These properties pose difficult stability problems.

After more than half a century of treating diabetics with animal insulin, recombinant DNA technologies and advanced protein chemistry made human insulin preparations available in the early 1980s (Vajo et al., 2001). Human insulin produced by recombinant DNA technology is the first commercial health care product derived from this technology (Johnson, 1983). Large quantities of biosynthetic human proinsulin have recently become available through recombinant DNA technology (Revers et al., 1984). A chimeric bovine GH (amino acids Met-Asp-Gln-greater than 1-23) and human GH (hGH) (amino acids 24-191) plasmid was constructed and expressed in *E. coli* (Binder et al., 1989). Human follicle-stimulating hormone (FSH) (Loumaye et al., 1995) and tissue-type plasminogen activator (Tiefenbrunn et al., 1985) are now produced in vitro by recombinant DNA technology.

Approved therapeutic protein-based products from *E. coli* include hormones (human insulin and insulin analogues, calcitonin, parathyroid hormone, human growth hormone, glucagons, somatropin and insulin growth factor 1), interferons (alfa-1, alfa 2a, alfa-2b and gamma-1b), interleukins 11 and 2, light and heavy chains raised against vascular endothelial growth factor-A, tumor necrosis factor alpha, cholera B subunit protein, B-type natriuretic peptide, granulocyte colony stimulating factor and plasminogen activator. Noteworthy, most of the recombinant pharmaceuticals produced in *E. coli* are addressed for the treatment of infectious diseases or endocrine, nutritional and metabolic disorder disease groups (Werner, 1990; Redwan et al., 2008). The approved protein products produced in yeast are obtained exclusively in *S. cerevisiae* and correspond to hormones (insulin, insulin analogues, non glycosylated human growth hormone somatotropin, glucagon), vaccines (hepatitis B virus surface antigen) and virus-like particles (VLPs) of the major capsid protein L1 of human papillomavirus type 6, 11, 16 and 18, urate oxidase from *Aspergillus flavus*, granulocyte-macrophage colony stimulating factor, albumin, hirudin of *Hirudo medicinalis* and human platelets derived growth factor. As in the case of *E. coli*, most of the recombinant pharmaceuticals from yeast are addressed to either infectious diseases or endocrine, nutritional and metabolic disorders, being these therapeutic areas the most covered by microbial products. Interestingly, several yeast species other than *S. cerevisiae* are being explored as sources of biopharmaceuticals and other proteins of biomedical interest (Porro et al., 2005). In addition, current metabolic engineering approaches and optimization of process procedures (Graf et al., 2008; Mattanovich et al., 2004) are dramatically expanding the potential of yeast species for improved production of recombinant proteins.

There is only one approved biopharmaceutical product containing recombinant proteins from infected insect cell line Hi Five, Cervarix, consisting of recombinant papillomavirus C-terminal truncated major capsid protein L1 types 16 and 18. Nonetheless, this expression system has been extensively used in structural studies since correctly folded eukaryotic proteins can be obtained in a secreted form in serum free media which enormously simplifies protein capture in purification protocols.

Genetic engineering has been applied to obtain humanized monoclonal antibodies using either recombinant mammalian cells producing chimeric antibodies or genetically modified mice to produce human-like antibodies. One such product, Remicade, which binds tumour necrosis factor-alpha, is a pharmaceutical blockbuster used in the treatment of Crohn's disease. Most of the therapeutic proteins approved so far have been obtained using transgenic hamster cell lines, namely 49 in Chinese hamster ovary cells (CHO) and 1 in baby hamster kidney cells (BHK). The main advantage of this expression system is that cells can be adapted to grow in suspension in serum free media (SFM), protein-free and chemically defined media.

This fact increases the biosafety of final products reducing risk of introducing prions of bovine spongiform encephalopathy (BSE) from bovine serum albumin and of infectious variant Creutzfeldt – Jakob disease (vCJD) from human serum albumin. In addition, recombinant products can be secreted into the chemical defined media, which simplifies both upstream and downstream purification process. PTMs in this expression system are almost the same as in human cell lines, although some concerns about comparability in the glycosylation pattern have arisen when comparing different batches of the same manufacturer product. Further development of chemically defined media and fine description of growth conditions would help to overcome this issue.

Recombinant structural proteins include hepatitis B virus vaccine and CD4 protein, and recombinant modifier proteins include tissue plasminogen activator and superoxide dismutase (agents that split or splice organic molecules). In the future, gene defects associated with genetic diseases will be unraveled, leading to the production of new therapeutic agents designed to counteract or actually reverse those defects. Recombinant protein drugs will be further tailored to enhance their activity and specificity. These advances are so novel and momentous that patent protection has been extended not only to recombinant drugs but to the recombinant microorganisms in which they are manufactured. In cloning genes, investigators directly use the protein designs that occur naturally. To use recombinant DNA technology to functionally analyze mutations introduced into cloned eukaryotic genes, a rapid procedure is necessary to assay the steps along the gene expression pathway. If one assays shortly after its introduction into mammalian cells, it can be shown that this recombinant plasmid programs the synthesis of correctly spliced and polyadenylylated insulin mRNA that functions in the synthesis and secretion of rat proinsulin. This system permits rapid analysis of cloned in vitro – engineered mutations and the programming of eukaryotic cells to manufacture proteins that they normally do not synthesize. Basic research will soon lead to the engineering of novel proteins with specified functions.

3.4.2 Recombinant DNA Technology and Medicine

Recombinant DNA technology has already had a major impact on our understanding of microbiology, cell biology, developmental biology (Dawid, and Wahli, 1979) and genetic diseases and it will certainly have extensive applications in laboratory medicine. In the nearest future, high biotechnologies should provide a great quantity of recombinant products, covering a high variety of diseases (clotting factors, cytokines, hormones, monoclonal antibodies, vaccines, thrombolytics). Besides, genetic engineering offers new therapeutic outlets: transgenic animals, genic therapy, antisense technology (Descamps, 1994). In case of a single gene defect a number of appropriate gene probes are available for prenatal diagnosis. Knowledge of the genetic disorders enables in some cases early onset of therapy or the option for abortion. However, gene technology which enables the diagnosis should not be viewed from an ethical point of view but rather the action taken when diagnostic results are available.

Gene therapy for a single gene defect still is at the early stage of development. Only few patients have been treated in various indications. Difficult to be overcome are the low frequency and unspecific integration of inserted DNA into the chromosome, lack of sufficient transcription control and short half-lives of the integrated gene. From an ethical perspective gene therapy complies with the therapeutic concept of medicine. Antisense oligonucleotides are under clinical development for blockage of the synthesis of oncogenes and viral proteins. Stability of oligonucleotides as well as selectivity for specific cells will have to be overcome for broader application. Its therapeutic application is in accordance with the ethical principles of medicine.

Recombinant DNA procedures have now been applied to the problem of the identification of molecular defects in man that account for heritable diseases, somatic mutations associated with neoplasia, and acquired infectious diseases. Genetic tests will predict common diseases as well as many rare ones and touch the lives of most people (Holtzman, 1988). Thus recombinant DNA technology has rapidly expanded our ability to diagnose disease (Berg et al., 1974). Substantial advances in the simplification of procedures for diagnostic purposes have been made, and the informed physician has gained in diagnostic accuracy as a consequence of these developments. The wide application of recombinant DNA diagnostics will depend on simplicity, speed of results, and cost containment.

As yet recombinant DNA technology does not appear to have widespread diagnostic application in pathology. However, it does have a useful role to play in specific circumstances in at least three main areas:

1. It can provide precise diagnostic information about genetic diseases, allowing appropriate counselling, and indicating future directions for research on therapeutic intervention, e.g. gene therapy
2. Micro-organisms can be identified more sensitively and specifically, in fresh or fixed tissue samples, and their genomes can be analysed in fine detail, providing information relevant to the aetiology, epidemiology and pathogenesis of many diseases
3. In tumour pathology the main application so far has been to resolve diagnostic problems associated with leukaemias and lymphomas, when other diagnostic procedures have been inconclusive. Specific chromosomal translocations, involving recognized genes, are particularly amenable to diagnosis by these means. Diagnostic applications to solid tumours are yet to be identified, although significant insights into tumorigenesis have been obtained, and these may ultimately lead to the development of useful markers for prognostic and therapeutic purposes.

The techniques of restriction endonuclease analysis of DNA, nucleic acid hybridization after electrophoretic separation of nucleic acid fragments, and molecular cloning of bacterial, viral, and human genes are already being used in epidemiologic studies and the prenatal diagnosis of certain genetic diseases, such as sickle cell anemia and the thalassemias. New insights into genes that may be involved in human cancer are being developed and may lead to improved methods for diagnosis and classification of tumors. Techniques for gene transfer have been improved to introduce synthetic sequences into the cellular genomes. Candidate cells and culture systems have been screened for their capacity to synthesise the proper recombinant molecules. In the nearest future, high biotechnologies should provide a great quantity of recombinant products, covering a high variety of diseases (clotting factors, cytokines, hormones, monoclonal antibodies, vaccines, thrombolytics). This tendency has been largely initiated, and is accompanied by the quick development of basic sciences such as glycobiology and molecular biology, together with the development of a specific regulatory environment. Besides, genetic engineering offers new therapeutic outlets: transgenic animals, gene therapy, antisense technology (Descamps, 1994).

Recombinant DNA drug products have been developed that represent the communicator, structural, and modifier classes of proteins. Recombinant communicator proteins include interferons alfa-2a and alfa-2b and granulocyte-macrophage colony-stimulating factor (immune system modulators) epidermal growth factor and erythropoietin (tissue repair promoters) and human insulin, growth hormone, and atrial peptide (metabolism modulators). Regulation of expression of specific genes has impacted on understanding of cancer development and immunopathologic diseases. An even greater impact has been the production of new therapeutic molecules. Cytokines have been produced by recombinant DNA technology and are established as effective therapeutic modalities for cancer and infectious diseases. High quality pharmaceutical molecules have been produced which can substantially regulate cell function. Interferons typify this progress and are now licensed for both viral and neoplastic disease in more than 43 countries around the world.

Parallel development of monoclonal antibody (MAb) and recombinant DNA technologies has enabled the design and production of several potent biologics that can specifically block the deleterious effects of TNFα. Two MAbs (infliximab and adalimumab), as well as a receptor fusion protein (etanercept), have thus far proven safe and efficacious in pivotal rheumatoid arthritis clinical trials. In a recent study recombinant proteins were reported to be used as specific molecular markers to differentiate between acute and chronic infections. Holec-Gasior and his colleagues (2009) reported GRA2 and ROP1 recombinant antigens as potential markers for detection of *Toxoplasma gondii*-specific immunoglobulin G in humans with acute Toxoplasmosis. These evolving technologies provide the foundation from which future biotherapies can be derived as targets are identified and validated in diverse disease states.

3.5 Conclusion

Overcoming the biological and methodological obstacles posed by cell factories to the production or rDNA pharmaceuticals is a main challenge in the further development of protein-based molecular medicine. Recombinant DNA technologies might have exhausted conventional cell factories and new production systems need to be deeply explored and incorporated into the production pipeline. Interestingly, a plateau in the rate of rDNA drug approval during the last 2–3 years is becoming perceivable, irrespective of the production system. Although it might be observed as a transient event, this fact seems instead to indicate that the current production systems could be near to the exhaustion regarding their ability to hold the production of complex proteins, protein complexes or the so-called difficult-to-express heterologous proteins (Jana and Deb, 2005). Desirably, recent insights about system's biology of recombinant cells and hosts, and specially, arising novel concepts on recombinant protein quality (Gonzalez-Montalban, 2007; de Marco, 2007) and host stress responses would enlarge the possibilities for metabolic and process engineering aiming to the economically feasible production of new, more complex drugs. Indeed, pushed by fast advances in molecular medicine the pharmaceutical industry is urgently demanding improved production systems and novel and cheaper drugs. Their incorporation into productive processes for human pharmaceuticals would hopefully push the trend of marketed products and fulfil the increasing demands of the pharmacological industry.

References

Berg, P., Baltimore, D., Boyer, H.W., 1974, Potential biohazards of recombinant DNA molecules. *Science* 185:303.

Binder, L., Vogel, T., Hadary, D., 1989, Chimeric bovine-human growth hormone prepared by recombinant DNA technology: binding properties and biological activity. *Mol Endocrinol* 3:923–930.

Black, W.J., 1989, Drug products of recombinant DNA technology American. *J HospPharm* 46:1834–1844.

Branduardi, P., Smeraldi, C., Porro, D., 2008, Metabolically engineered yeasts: 'potential' industrial applications. *J Mol Microbiol Biotechnol* 15:31–40.

Carroll, W.L., 1993, Introduction to recombinant-DNA technology American. *J Clin Nutr* 58:249S–258S.

Dawid, I.B., Wahli, W., 1979, Application of recombinant DNA technology to questions of developmental biology: a review. *Dev Biol* 69:305–328.

de Marco, A., 2007, Protocol for preparing proteins with improved solubility by co-expressing with molecular chaperones in *Escherichia coli*. *Nat Protoc* 2:2632–2639.

de Marco, A., Schroedel, A., 2005, Characterization of the aggregates formed during recombinant protein expression in bacteria. *BMC Biochem* 6:10. doi: 10.1186/1471-2091-6-10.

Descamps, J.L., 1994, Therapeutic applications of genetic engineering. *Ann Pharm Fr* 52: 294–302.

Ferrer-Miralles, N., Domingo-Espín, J., Corchero, J.L., Vázquez, E., Villaverde, A., 2009, Microbial factories for recombinant pharmaceuticals. *Microb Cell Fact* 8:17, doi:10.1186/1475-2859-8–17.

Gluck, S.L., Nelson, R.D., Lee, B.S., 1992, Introduction to the principles and methods of recombinant DNA technology. *Semin Nephrol* 12:467–494.

Gonzalez-Montalban, N., 2007, Recombinant protein solubility-does more mean better. *Nat Biotechnol* 25:718–720.

Graf, A., Gasser, B., Dragosits, M., Sauer, M., Leparc, G.G., Tüchler, T., Kreil, D.P., Mattanovich, D., 2008, Novel insights into the unfolded protein response using *Pichia pastoris* specific DNA microarrays. *BMC Genomics* 9:390.

Holec-Gasior, L., Kur, J., Hiszczyńska-Sawicka, E., 2009, GRA2 and ROP1 recombinant antigens as potential markers for detection of *Toxoplasma gondii*-specific immunoglobulin G in humans with acute toxoplasmosis. *Clin Vaccine Immunol* 16:510–514.

Holtzman, N.A., 1988, Recombinant DNA technology, genetic tests, and public policy. *Am J Hum Genet* 42:624–632.

Jana, S., Deb, J.K., 2005, Strategies for efficient production of heterologous proteins in *Escherichia coli*. *Appl Microbiol Biotechnol* 67:289–298.

Johnson, I.S., 1983, Human insulin from recombinant DNA technology. *Science* 219:632–637.

Lomedico, P.T., 1982, Use of recombinant DNA technology to program eukaryotic cells to synthesize rat proinsulin: a rapid expression assay for cloned genes. *PNAS* 79:5798–5802.

Loumaye, E., Campbell, R., Salat-Baroux, J., 1995, Human follicle-stimulating hormone produced by recombinant DNA technology: a review for clinicians. *Hum Reprod Update* 1:188–199.

Manning, M.C., Patel, K., Borchardt, R.T., 1989, Stability of protein pharmaceuticals. *Pharm Res* 6:903–918.

Marston, F.A., 1986, The purification of eukaryotic polypeptides synthesized in *Escherichia coli*. *Biochem J* 240:1–12.

Mattanovich, D., Gasser, B., Hohenblum, H., Sauer, M., 2004, Stress in recombinant protein producing yeasts. *J Biotechnol* 113:121–135.

Porro, D., Sauer, M., Branduardi, P., Mattanovich, D., 2005, Recombinant protein production in yeasts. *Mol Biotechnol* 31:245–259.

Redwan, e., Matar, S.M., El-Aziz, G.A., Serour, E.A., 2008, Synthesis of the human insulin gene: protein expression, scaling up and bioactivity. *Prep Biochem Biotechnol* 38:24–39.

Revers, R.R., Henry, R., Schmeiser, L., Kolterman, O., Cohen, R., Bergenstal, R., Polonsky, K., Jaspan, J., Rubenstein, A., Frank, B., 1984, The effects of biosynthetic human proinsulin on carbohydrate metabolism. *Diabetes* 33:762–770.

Siddiqui, M.A., 1982, Recombinant DNA technology and its application to developmental biology. *J Craniofac Genet Dev Biol* 2:75–92.

Sorensen, H.P., Mortensen, K.K., 2005, Soluble expression of recombinant proteins in the cytoplasm of *Escherichia coli*. *Microb Cell Fact* 4:1.

Staub, J.M., Garcia, B., Graves, J., Hajdukiewicz, P.T., Hunter, P., Nehra, N., Paradkar, V., Schlittler, M., Carroll, J.A., Spatola, L., Ward, D., Ye, G., Russell, D.A., 2000, High-yield production of a human therapeutic protein in tobacco chloroplasts. *Nat Biotechnol* 18:333–338.

Tiefenbrunn, A.J., Robinson, A.K., Kurnik, P.B., Ludbrook, P.A., Sobel, B.E., 1985, Clinical pharmacology in patients with evolving myocardial infarction of tissue-type plasminogen activator produced by recombinant DNA technology. *Circulation* 71:110–116.

Travis, J., Owen, M., George, P., 1985, Isolation and properties of recombinant DNA produced variants of human alpha 1-proteinase inhibitor. *J Biol Chem* 260:4384–4389.

Vajo, Z., Fawcett, J., Duckworth, W.C., 2001, Recombinant DNA technology in the treatment of diabetes: insulin analogs. *Endocr Rev* 22:706–717.

Werner, R.G., Pommer, C.H., 1990, Successful development of recombinant DNA-derived pharmaceuticals. *Arzneimittelforschung* 40:1274–1283.

Chapter 4
Aptamers: In Vitro DNA Selection

Abstract Aptamers are a new class of therapeutic and diagnostic reagents identified as binding molecules to numerous small compounds, proteins and rarely even to complete pathogen particles. Aptamers are typically selected from libraries of random DNA (or RNA) sequences through systematic evolution of ligands by exponential enrichment (SELEX). This method is an in vitro method of selection of DNA or RNA sequences, involving several rounds of alternating steps of partitioning of candidate oligonucleotides and their PCR amplification. These range from inhibition of receptors and enzymes to the identification of small molecules in sensor applications, and from the development of targeted therapeutic to the design of novel diagnostic and imaging agents. Furthermore, aptamers have been designed for targets that cover a wide range of diseases, from HIV to tropical diseases, cancer and inflammation. Their easy development and flexibility of use and manipulation, offers further potential as effective diagnostic tools.

Keywords Aptamers · Pathogens · Therapeutics · Amplification · DNA selection · RNA aptamers · Oligomers · SELEX · Oligonucleotide library · MonoLEX · Combinatorial library · DNA aptamers

4.1 Prologue

Effective catalysis involves the specific recognition of a target molecule with high affinity. Although many catalytic functions have evolved naturally, mostly in proteins, the need for new catalytic functions is growing; for example to convert industrial byproducts to useful products. To develop chemicals that recognize new targets such as industrial byproducts, combinatorial molecular biology are now-a-days being used. Aptamers are such new products that form unique structures to meet the requirements for more selective binding and sensitive recognition modules for increasingly sensitive detection methods that are required in natural sciences and medicine for the detection and quantization of molecules.

M. Debnath et al., *Molecular Diagnostics: Promises and Possibilities*,
DOI 10.1007/978-90-481-3261-4_4,

4.2 Concept

Aptamers are short, single-stranded nucleic acid (DNA or RNA) oligomers that have been selected in vitro to bind with high affinity and specifically to a certain molecular target (Lou, 2009; Ellington and Szostok, 1990). They are 50–100 base pair in length that bind to proteins bind with Kds (equilibrium constant) in the range of 1 pM–1 nM similar to monoclonal antibodies. The molecular target may be amino acids, drugs, proteins, nucleic acids small organic compounds, and even entire organisms (Joshi et al., 2009; Mairal, 2008). The term "aptamer" is derived from Latin *aptus* (= fitting) and Greek *meros* (= part) and was chosen to illustrate "lock and key model" between aptamers and their binding partners. Due to the multitude of structures which can be formed by aptamers, they are able to recognize virtually all classes of substrates and bind them perfectly fitting in analogy to antigen-antibody interactions. In such a way, aptamers were developed for atoms and small molecules, for amino acids, peptides, polysaccharides, and proteins, but also for complex targets like viruses and protozoa. The selected sequences have the ability to recognize specific ligands by forming binding pockets and can bind to nucleic acids (Le Tinevez et al., 1998; Pileur et al., 2003), proteins (Bock et al., 1992; Lupold et al., 2002) or small organic compounds (Berens et al., 2001; Burgstaller and Famulok, 1996).The aptamer-target interaction takes place via structural compatibilities of the two binding partners, via electrostatic interactions, like Van der Waals interactions, ionic or dipole forces, hydrogen bonds, or stacking interactions. These nucleic acid ligands bind to nucleic acid, proteins. Aptamer recognition affinity and specificity is comparable to those of monoclonal antibodies and aptamers can be selected to recognize and bind a wide range of targets, including toxic compounds and inherently non-immunogenic molecules that antibodies cannot be raised against. Furthermore, as aptamers are 10–100 times smaller than antibodies, they are expected to achieve higher tumour penetration than their counterparts (Jayasena, 1999).

Aptamers have many potential uses in intracellular processes studies, medicine and technology. In addition to the genetic information encoded by nucleic acids they also function as highly specific affinity ligands by molecular interaction based on the three dimensional folding pattern. The three dimensional complex shape of a single stranded oligonucleotide is primarily due to the base composition led intramolecular hybridization that initiates folding to a particular molecular shape. This molecular shape assists in binding through shape specific recognition to it targets leading to considerable three dimensional structure stability and thus the high degree of affinity. Natural examples of molecular shape recognition interactions of nucleic acids with proteins are tRNA, ribozymes, DNA binding proteins and DNAzymes.

DNA aptamers are heat and protease resistant without stabilizing modifications. Due to chemical synthesis, aptamer production can easily be scaled up (Jellinck et al., 1993). Once identified, aptamers can be reliably synthesized by automated methods. During synthesis, aptamers can be easily modified to facilitate further modifications such as attachment points for enzymes or a variety of other reagents, and the introduction of modified bases for nuclease resistance (Brody and Gold,

2000; Eaton et al., 1997; Gewirtz, 1999). Furthermore, aptamers can be stored until needed and are resistant to denaturation and degradation when lyophilised.

During the selection process aptamers with the highest affinity to a target structure are isolated from a pool of oligonucleotides with random positions for nucleic acids, which, depending on the length of the aptamers, contain up to 10^{15} different sequence variants, representing a significantly larger pool of variants than antibodies do. In addition to these evident benefits, aptamers exhibit some drawbacks, including their reduced resistance to nucleases and the problem of delivery and clearance in therapeutic application. First attempts to stabilize aptamers were promising (Ruckman et al., 1998; Morris et al., 1998).

4.3 Selection of Aptamers

Theoretically it is possible to select aptamers virtually against any molecular target. Aptamers have been selected for small molecules, peptides, proteins as well as viruses and bacteria. The aptamers are selected by incubating the target molecule in a large pool of oligonucleotide (usually 40–60 mers); the pool size of the oligonucleotide is from 1,010 to 1,020. The large pool size of the oligonucleotide ensures the selection and isolation of the specific aptamer. The structural and informational complexity of the oligonucleotide pool and its functional activity is an interesting and active area to develop an algorithm based development of nucleic acid ligands. Aptamers can distinguish between closely related but non-identical members of a protein family, or between different functional or conformational states of the same protein. In a striking example of specificity, an aptamer to the small molecule theophylline (1,3-dimethylxanthine) binds with 10,000-fold lower affinity to caffeine (1,3,7-trimethylxanthine) that differs from theophylline by a single methyl group.

The SELEX method (systematic evolution of ligands by exponential enrichment) (Ellington and Szotok, 1990) is an oligonucleotide-based combinatorial library approach that has been extensively used to isolate high-affinity ligands (called aptamers) for a wide variety of proteins and small molecules (Jellinek et al., 1993). This is a method in which single stranded oligonucleotides are selected from a wide variety of sequences, based on their interaction with a target molecule. Using this process, it is possible to develop new aptamers in as little as 2 weeks.

Starting point of each SELEX process is a synthetic random DNA oligonucleotide library (RNA, ssDNA or modified oligonucleotide molecules). Normally, 10^{15} different molecules are employed. Regarding this multitude of possible structures, it is very likely to select at least one binding partner (aptamer) for virtually each target. In order to amplify these target binding structures, the oligonucleotide library is subjected to an adequate selection process in the presence of the target molecules. During this process, relevant nucleic acid molecules are separated from non-relevant ones. Target binding aptamers subsequently are amplified and repeatedly subjected to interactive selection processes until relevant nucleic acid molecules dominate the oligonucleotide pool. The enriched aptamer pool is cloned and several individual aptamers can be sequenced and characterized. As a result, the

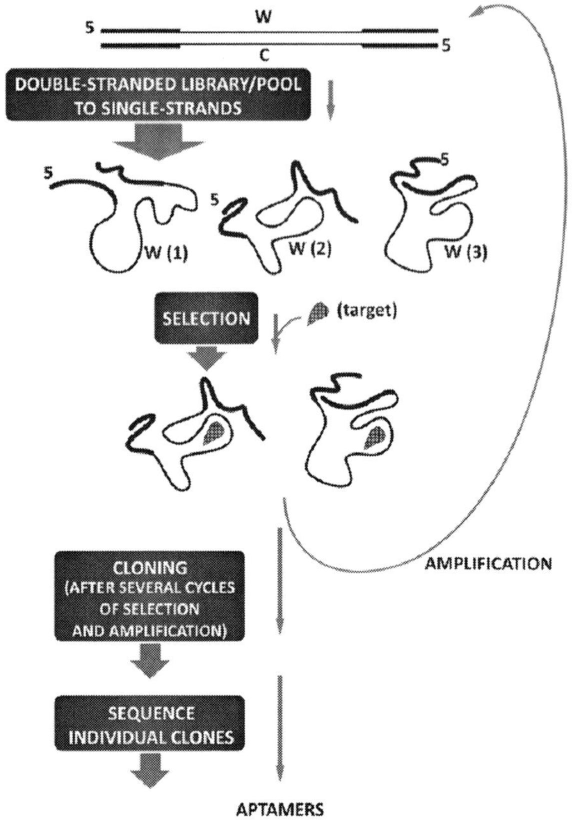

Fig. 4.1 The SELEX method is an oligonucleotide-based combinatorial library approach to isolate high-affinity ligands called aptamers for a wide variety of proteins and small molecules

selected aptamers are ready to be ultilized according to the given task (Fig. 4.1). However, this procedure requires repetitive cycles and is therefore very time-consuming. The in vitro selection of RNA and DNA ligands against specific targets obeys the same rules as natural selection. For this purpose a partial randomized synthetic DNA template is constructed containing a random inner region that is flanked on both sides by constant sequences. The random sequence classically consists of 15–75 random positions where all four bases are incorporated with equal probabilities. This pool containing 10^{12}–10^{15} different sequences can be either directly used for selection or first transcribed to RNA using T7 RNA polymerase. In this case a T7 promoter site needs to be placed on the 5′ site of the DNA template. The random DNA/RNA pool is exposed to the protein target and the best fitting molecules in the selection pool are culled and amplified.

The procedure will be repeated with increasing stringency until the previous random pool is purified to a few molecules with the desired binding properties. The final pool is cloned into a bacterial vector and individual colonies are sequenced (Fig. 4.2). The previous random regions are aligned and searched for consensus

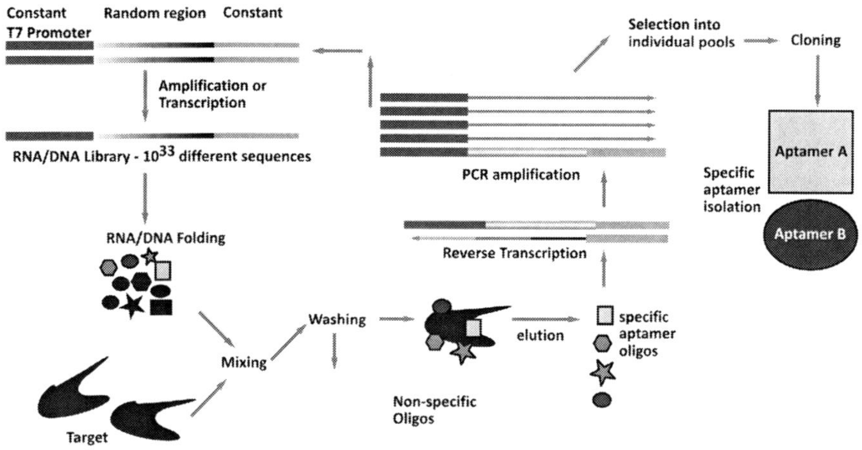

Fig. 4.2 The SELEX method can be used to select specific regions and specific aptamers

motifs. Consensus motifs often located in stem-loop structures are thought to mediate binding specificity.

The first SELEX experiment on single stranded RNA oligonucleotides was published by Tuerk and Gold, when investigating the bacteriophage T4 DNA polymerase (Tuerk and Gold, 1990). For RNA aptamers, another selection process has been facilitated and accelerated by Cox and co-workers, who established an automated selection procedure based on magnetic beads. This process has been adjusted and optimized by several others (Giordano et al., 1999; Katzin and Colli, 1983). An automated selection procedure for RNA and DNA aptamers without magnetic beads has also been reported. Due to its 2′-hydroxyl group, however, RNA is easily subjected to hydrolysis by RNases, which can diminish its applicability if it is not stabilized by chemical modifications. An alternative is the employment of Spiegelmers (L-RNAs), which represent the mirror images of the naturally occurring D-RNAs. Since the L-forms of nucleic acids do not occur in nature, there are no enzymes to degrade them. DNA aptamers, lacking the 2′-hydroxyl group, are also more stable than unmodified RNAs in biological samples.

To select DNA aptamers specific for Vaccinia virus (VACV), aptamer selection can be performed in two steps. An initial chromatography was performed on proteins from virus-free cell culture supernatant to eliminate aptamers binding to cell debris or media components. Subsequently, a second chromatography was carried out on resin-bound heat-inactivated VACV particles. For direct recovery of high-affinity-binding aptamers, the resin of the affinity column was physically segmented into slices and the amount of retained binding aptamers was estimated by quantitative real-time PCR. High concentrations of aptamers not eluted during the extended washing procedure were identified on several affinity column segments.

As an alternative to the SELEX process with 7 to more than 30 selection and amplification cycles requiring large amounts of the target molecule, a new selection process was established. Based on reports about selecting functional oligonucleotides and the potential of oligonucleotides as non-Watson-Crick-type binders to peptides and proteins, a new one-step aptamer isolation protocol called MonoLEX is used. This is used to retrieve DNA aptamers that have the potential to bind specifically to virus particles. The MonoLEX approach combined a single affinity chromatography step with subsequent physical segmentation of the affinity resin and one single final exponential amplification step of bound aptamers. A schematic representation of SELEX is given in Fig. 4.1. Specific aptamers were selected from a combinatory library of oligonucleotides characterized by two flanking primers of known sequence and an internal region of 20 random nucleotide positions (N20 DNA library). Under adequate chromatographic conditions, like flow laminarity, sufficient capacity and homogeneity of the resin, high-affinity-binding aptamers stuck to the target, whereas weakly binding oligonucleotides could be removed from the resin by ample washing. Since the selection is an in vitro process, depending on the final application of the aptamer, the selection conditions can be adapted accordingly.

This procedure exploits the advantages of automation (i.e., it allows for a fast, reproducible, and parallelized selection of DNA aptamers). Only little manual handling is required. Merely the preparation of the solutions, precipitation, and amplification steps have to be performed manually. All other selection steps, like coupling of the target to the solid surface, interaction of DNA with the target molecule, wash and elution steps, as well as the single-stranded DNA preparation can be carried out automatically (Wochner et al., 2008).

In literature we can now identify a large number of aptamers selected for a very broad spectrum of targets. Furthermore, the first aptamer has recently successfully reached the pharmaceutical market. This is an anti-VEGF aptamer for the treatment of macular degenerative disease that exerts its action by attacking the neo-vascularization characteristic of this disease, under the name of Macugen (Eyetech/Pfizer) (Ng et al., 2006; Kourlas and Schiller, 2006).

4.4 Applications of DNA and RNA Aptamers

Aptamers offer the utility for biotechnological and therapeutic applications (Fig. 4.3) as they offer molecular recognition properties that reveal the commonly used biomolecule, antibodies. Aptamers are functional molecules, usually DNA or RNA oligonucleotides, with the appropriate sequence and structure to form a complex with a target molecule. They are able to bind tightly and selectively to disease markers and can greatly benefit disease diagnosis and therapy. In addition to their discriminate recognition, aptamers offer advantages over antibodies as they can be engineered completely in a test tube. They are readily produced by chemical synthesis, possess desirable storage properties, and elicit little or no immunogenicity in therapeutic applications. When compared to antibodies, aptamers have advantages and disadvantages as therapeutic and biological reagents (Jayasena, 1999). Aptamers present faster tissue penetration and wider applicability

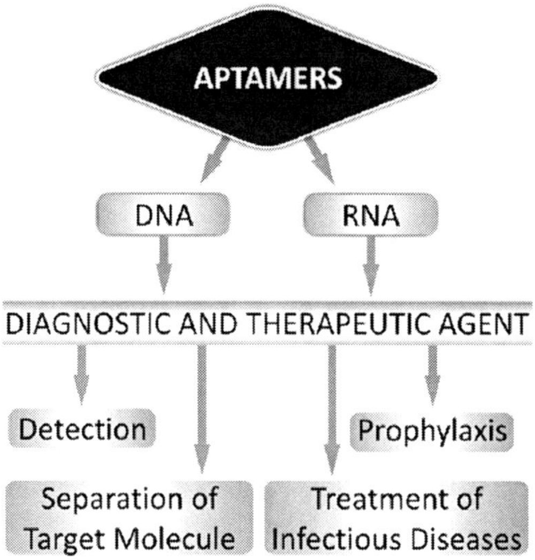

Fig. 4.3 Aptamers have diverse applications in the field of diagnostics and therapeutics

and present the opportunity for simple base modifications to improve functionality by comparison. Furthermore their small size (molecular weights between 3,000 and 20,000) may reduce steric hindrance, increasing surface coverage during immobilization (Deng et al., 2001). The in vitro selection process (SELEX) can be more precisely monitored than can organismal immunization, and the affinities and specificities of aptamers can thus be better tailored than can those of antibodies. Aptamers can be more readily engineered than antibodies for biological/medicinal use (Jayasena, 1999).

Their application is very diverse. Aptamers have been used as molecular recognition elements in analytical systems for detection, separation, or purification of target molecules. They also play an important role in medical therapy (Pan et al., 1995) and environmental analysis (Homann and Göringer, 2001). Aptamers have been selected for various purposes, making use of their high specificity, versatility and affinity in target recognition. Examples include aptamers that recognize proteins, peptides, dyes, amino acids, nucleotides, and drugs among others, to act as biosensors (Potyrailo et al., 1998), probes (Pavski and Le, 2001) and anti-clotting agents (Bock et al., 1992), whilst the first aptamer has made it to the market gaining FDA approval.

4.4.1 Aptamers for Detection of Pathogens

Aptamers promise to be such additional candidates for prophylaxis and treatment of infectious diseases, as they can be directed against a wide variety of target molecules like toxins or even complete microorganisms (Xu and Ellington, 1996; Hamm,

1996). Like the antibodies, aptamers bind to their target by their three-dimensional structure with high affinity and specificity. Compared to antibodies, aptamers have several advantages, as they are selected in vitro which also enables selection against toxic or weakly immunogenic targets. Especially in scenarios where infections with new, emerging pathogens have to be detected or treated, a fast technique providing specific detection tools may be helpful. The spectrum for different applications of aptamers like A38, either on a sensor for virus detection or for treatment of infections, requires an extremely high stability of the aptamer. The use of DNA aptamers facilitates such a high stability without modifications, even in body fluids.

Many different aptamers have been selected by various SELEX-based protocols for a wide variety of targets ranging from small molecules to whole cells and bacteria. Some of these aptamers were shown to possess inhibitory activity with certain HIV strains (McEver, 1995), to block cell binding of human cytomegalovirus (CMV) (Bevilacqua et al., 1994) or influenza virus hemagglutinin (Hicke et al., 1996) or were selected as tools for differentiation of closely related influenza strains (Almeida-de-Faria et al., 1999). Some other examples include aptamers to HIV related targets (Andreola et al., 2001; de Soultrait et al., 2002; Duzgunes, 2001), hepatitis C (Biroccio et al., 2002; Hwang et al., 2000; Vo et al., 2003), *Trypanosoma cruzi* (Homann and Göringer, 1999; Homann and Göringer, 2001; Ulrich et al., 2002), prion proteins (Sayer et al., 2004; Weiss et al., 1997), thrombin (Bock et al., 1992; Griffin et al., 1993; Holland et al., 2000), anti-angiogenesis vascular endothelial growth factor (Blank et al., 2001; White et al., 2003), prostate specific membrane antigen (Lupold et al., 2002), Botulinum neurotoxin type A (Lou, 2009).

4.4.2 Aptamers for Therapy

The most important challenge in fighting almost any disease is to be able to detect and treat it in the early stages. For most diseases there are changes in gene expression and subsequent protein products that could be used for detection. However, disease-initiated changes often occur in the depths of our tissues and are not visible from the surface. Therefore another challenge for the developing new technology to fight disease is to find ways of non-invasive imaging (e.g. no biopsy or surgery) of the body's status and to find ways to target small sites inside the body where the disease process is initiating (Szpechcinski and Grzanka, 2006).

Due to the high degree of specificity and targeted nature of aptamer behaviour towards the target molecule, it is expected that the dosage of therapeutic aptamer to be used can be the smaller, this strategy may reduce costs and alleviate bystander effects caused by the aptamer-conjugate toxic side effects (Gewirtz, 1999). Furthermore, more pharmokinetic data is readily available that demonstrates aptamer quick clearance. PEGylation is currently been studied to improve aptamer circulation for longer periods of time (Pendergrast et al., 2005).

DNA aptamers against a tumor marker protein, human vascular endothelial growth factor (VEGF165) have been reported recently (Ikebukuro et al., 2007). These specific DNA aptamers against VEGF165 would be a useful sensing element for cancer diagnosis. These small single-stranded nucleic acids are claimed to act as mimics of antibodies in that they can recognise molecular targets with high specificity and are able to carry therapeutic agents, radioisotopes, directly to solid tumour masses or to individual cells that may have metastasised with greater efficiency (Brody and Gold, 2000; Cerchia et al., 2002; Sooter and Ellington, 2002). The anti-VEGF aptamer pegaptanib has already received FDA approval for treatment for human ocular vascular disease (Ulrich et al., 2006).

Another success story is the E2F decoy aptamer that binds the transcription factor E2F, aimed at reducing the proliferation of cardiac and vascular cells in cardiovascular diseases as well as in the development of malignancies (Fabbrizio et al., 1999). This aptamer is a naturally developed aptamer, because it contains a known consensus sequence that is found naturally in the human body. This aptamer is currently in Phase three trials study by Corgenetech, Inc. and being evaluated for its efficacy at restraining coronary and peripheral vascular graft failure.

Aptamers also present the possibility to work as antidotes and more importantly the potential of creating a self-antidote, simply through the creation of a self-aptamer that disrupts the interaction with the target, thus allowing for a controlled therapy. This feature is not usually present in any of other biological therapy.

However, drawbacks such as their cost, synthesis, bioavailability, pharmacokinetics properties and delivery hindered aptamers from immediate application in the pharmaceutical industry and more importantly the fact that antibodies are a proven biological therapeutic and thus aptamers would need to be more flexible, more stable and more efficient than antibodies in order to replace the later. Thus, aptamer development needs to focus on their advantages in relation to other more proven biological therapeutics and the possibilities they present.

4.4.3 Aptamers to Detect Change in Gene Expression

Aptamers can also be used to detect a change in gene expression by binding the protein product of the gene. Aptamers are being developed in two ways to detect changes in gene expression in vivo. The first is to prepare probes in which the aptamer is part of a regulated nucleic acid sequence that is complementary to a particular mRNA. The second is to develop allosteric aptamers that will bind specific cell surface proteins that are the result of altered gene expression. Binding the cell surface protein brings about a structural change in the aptamer that result in its binding an imaging agent or drug to treat the disease.

Many aptamers have been selected to recognize specific proteins. Some of these aptamers are currently in use clinically or are being investigated in clinical trials for treating disease. Allosteric aptamers called "cis-linked aptamers for medical procedures" (CLAMPs) are being developed that will recognize a particular cell surface protein and respond by binding another ligand once it has bound to the surface of a diseased cell. The other ligand could be a radiolabeled ligand that can be used to detect the diseased cell. Or it could be a prodrug that will release a toxic compound when brought near to the diseased cell.

4.4.4 Aptamers as Diagnostics and Therapeutics

The idea of using aptamers as therapeutic molecules have been suggested by several groups. Now aptamers are at the forefront of the list of the most promising agents as potential nucleic acid pharmaceuticals. A whole series of aptamers that can bind protein targets inside and outside the cell can be found in the literature, which illustrates the potential of DNA or RNA aptamers as therapeutic modalities. These aptamers are suitable for applications based on molecular recognition of a target molecule including diagnostics and therapeutics (Ulrich et al., 2006). The use of the SELEX method has been extended to complex targets such as red blood cell membranes (Morris et al., 1998) and the membrane-bound nicotinic acetylcholine receptor, whole virus particles (Pan et al., 1995) and African trypanosomes (Homann and Göringer, 1999). These ligands have dissociation constants in the picomolar to low nanomolar range for their protein targets. The SELEX methodology has also successfully been used to develop RNA antagonists of CD4 epitopes on mononuclear mouse cells (Davis et al., 1998) and selectin cell surface molecules (O'Connel et al., 1996; Jenison et al., 1998). These selected RNA antagonists prevented the P-selectin-dependent neutrophil-platelet adhesion in vitro. Now, RNA or DNA aptamers can be obtained for almost every target whether complex or small.

The potential utility of aptamers as therapeutic agents is considerably enhanced by chemical modifications that lend resistance to nuclease attack (Davis et al., 1998; Ito et al., 1998). This resistance is either achieved by the addition of phosphorothioates (Andreola et al., 2001) or by the substitution of the 2′-OH groups of pyrimidines with 2′-F, 2′-NH_2, or 2′-OMe. Aptamers recognize epitopes with the same specificity as antibodies but in contrast to antibodies they possess low immunogenicity and are not subject to proteolytic degradation.

The selection against complete Vaccinia virus particles resulted in a 64-base DNA aptamer specifically binding to orthopoxviruses as validated by dot blot analysis, Surface Plasmon Resonance, Fluorescence Correlation Spectroscopy and real-time PCR, following an aptamer blotting assay. The same oligonucleotide showed the ability to inhibit in vitro infection of Vaccinia virus and other orthopoxviruses in a concentration-dependent manner (Nitsche et al., 2007).

4.4.5 Aptamers as Radiopharmaceutical Tools

The advent of techniques such as SELEX has allowed the study and manipulation of nucleic acids and has propelled the discovery of molecular tools for the treatment of human disease. Equipped with these new tools, the implementation of simple steps for the formation of a new class of therapeutic agents was one step away. The idea of attaching radionucleotides to a biological molecule is not new (Bernardo-Filho et al., 2005). Long and detailed studies have been carried out in the past with some success with antibodies, peptides or antisense targeting. Aptamer labelling was a natural step in the direction of molecular targeted radiopharmaceutical development, (Ferreira and Missailidis, 2007) propelled both by the understanding of the biological properties of these molecules, specially their chemical malleability and pharmacokinetics, and the simplicity of the chemistry involved in the labelling of aptamers for pharmacoimaging and pharmacotherapy (Britz-Cunningham and Adelstein, 2003; Cerchia et al., 2002; Urbain, 1999).

Two DNA aptamers directed at two separate exosites on the human alpha-thrombin have been evaluated for their potential as thrombus imaging agents. Both aptamers were analysed and ODN2 was found to be suitable for imaging with rapid clearance rates. However, the conclusions of the study also found that the slow mass transfer to the clot worked against the in-vivo thrombin-dependent imaging (Doughan et al., 2003). More recently an aptamer that binds to the extracellular matrix protein Tenascin-C was 99mTc-labeled and analysed for the potential use as radiopharmaceutical in a variety of animal models. In this study rapid uptake and rapid clearance of the radiolabeled aptamer from blood and other tissues provided clear tumour imaging, thus suggesting potentially imaging and therapeutic applications (Hicke et al., 2006).

4.5 Conclusion

When the first aptamers were developed 15 years ago, it was predicted that they had an enormous potential as a feasible alternative to antibodies. Although the application of antibodies for the detection and therapy of infectious agents is well established, aptamers have some potential advantages over antibodies that may fill the gaps that antibody-based applications possess and that expedite the use of aptamers in the virologist's laboratory. These advantages include their small size, eased cell penetration, rapid and cheap synthesis including a variety of chemical modifications and the fact that aptamers are non-immunogenic. Probably their crucial benefit is a selection process performed in vitro. Thus, aptamers can be selected against targets that are either weakly immunogenic or toxic, like toxin proteins. These targets are usually not applicable for in vivo production of monoclonal antibodies in mice; however, high affinity aptamers have been selected against some of these problematic targets. Moreover, the chemical and physical conditions for aptamer selection can be adapted to the real environment in which the aptamer will finally be applied. This includes cross-selection against similar targets that have to be excluded from an aptamer's detection pattern.

With the FDA approval of the first aptamer-based drug Macugen and the recent publication of an aptamer-based cocaine sensor, aptamers are starting to tap their full potential with the increasing need for additional specific detection tools. With the improving automation of the aptamer selection process and the growing knowledge of pathogen-specific targets, the number of aptamers used in diagnostics and therapy of infectious diseases will dramatically increase in the future.

References

Almeida-de-Faria, M., Freymüller, E., Colli, W., Alves, M.J., 1999, Trypanosoma cruzi: characterization of an intracellular epimastigote-like form. *Exp Parasitol* 92:263–274.

Andreola, M.L., Pileur, F., Calmels, C., Ventura, M., Tarrago-Litvak L., Toulme, J.J., Litvak, S., 2001, DNA aptamers selected against the HIV-1 RNase H display in vitro antiviral activity. *Biochemistry* 40:10087–10094.

Berens, C., Thain, A., Schroeder, R., 2001, A tetracycline-binding RNA aptamer. *Bioorg Med Chem* 9:2549–2556.

Bernardo-Filho, M., Santos-Filho, S., Moura, E., Maiworm, A., Orlando, M., Penas, M., Cardoso, V., Bernardo, L., Brito, L., 2005, Drug interaction with radiopharmaceutical: a review. *Braz Arch Biol Technol* 48:13–27.

Bevilacqua, M.P., Nelson, R.M., Mannori, G., Cecconi, O., 1994, Endothelial-leukocyte adhesion molecules in human disease. *Annu Rev Med* 45:361–378.

Biroccio, A., Hamm, J., Incitti, I., De Francesco, R., Tomei, L., 2002, Selection of RNA aptamers that are specific and high-affinity ligands of the hepatitis C virus RNA-dependent RNA polymerase. *J Virol* 76:3688–3696.

Blank, M., Weinschenk, T., Priemer, M., Schluesener, H., 2001, Systematic evolution of a DNA aptamer binding to rat brain tumor microvessels. Selective targeting of endothelial regulatory protein pigpen. *J Biol Chem* 276:16464–16468.

Bock, L.C., Griffin, L.C., Latham, J.A., Vermaas, E.H., Toole, J.J., 1992, Selection of single-stranded DNA molecules that bind and inhibit human thrombin. *Nature* 355:564–566.

Britz-Cunningham, S.H., Adelstein, S.J., 2003, Molecular targeting with radionuclides: state of the science. *J Nucl Med* 44:1945–1961.

Brody, E.N., Gold, L., 2000, Aptamers as therapeutic and diagnostic agents. *Rev Mol Biotech* 74:5–13.

Burgstaller, P., Famulok, M., 1996, Structural characterization of a flavin-specific RNA aptamer by chemical probing. *Bioorg Med Chem Lett* 6:1157–1162.

Cerchia, L., Hamm, J., Libri, D., Tavitian, B., de Franciscis, V., 2002, Nucleic acid aptamers in cancer medicine. *FEBS Lett* 528:12–16.

Davis, K.A., Lin, Y., Abrams, B., Jayasena, S.D., 1998, Staining of cell surface CD4 with 2′-F-pyrimidine-containing RNA aptamers for flow cytometry. *Nucleic Acids Res* 26:3915–3924.

De Soultrait, V.R., Lozach, P.Y., Altmeyer, R., Tarrago-Litvak, L., Litvak, S., Andreola, M.L., 2002, DNA aptamers derived from HIV-1 RNase H inhibitors are strong anti-integrase agents. *J Mol Biol* 324:195–203.

Deng, Q., German, I., Buchanan, D., Kennedy, R.T., 2001, Retention and separation of adenosine and analogues by affinity chromatography with an aptamer stationary phase. *Anal Chem* 73:5415–5421.

Dougan, H.W.J., Stafford, A.R., Gillespie, K.D., Klement, P., Hobbs, J.B., Lyster, D.M., 2003, Evaluation of DNA aptamers directed to thrombin as potential thrombus imaging agents. *Nucl Med Biol* 30:61–72.

Duzgunes, N., Simoes, S., Konopka, K., Rossi, J.J., Pedroso de Lima, M.C., 2001, Delivery of novel macromolecular drugs against HIV-1. *Exp Opin Biol Ther* 1:949–970.

Eaton, B.E., Gold, L., Hicke, B.J., Janjie, N., Jucker, F.M., Sebesta, D.P., Tarasow, T.M., Willis, M.C., Zichi, D.A., 1997, Post-SELEX combinatorial optimization of aptamers. *Bioorg Med Chem* 5:1087–1096.

Ellington, A.D., Szostak, J.W., 1990, In vitro selection of RNA molecules that bind specific ligands. *Nature* 346:818–822.

Fabbrizio, E., Le Cam, L., Polanowska, J., Kaczorek, M., Brent, R., Sardet, C. 1999, Inhibition of mammalian cell proliferation by genetically selected peptide aptamers that functionally antagonize E2F activity. *Oncogene* 18:4357–4363.

Ferreira, C.S.M., Missailidis, S., 2007, Aptamer-based therapeutics and their potential in radiopharmaceutical design. *Braz Arch Biol Technol* 50(no.spe Curitiba):63–76, doi:10.1590/S151689132007000600008.

Gewirtz, A.M., 1999, Oligonucleotide therapeutics: clothing the emperor. *Curr Opin Mol Ther* 1:297–306.
Giordano, R., Fouts, D.L., Tewari, D., Colli, W., Manning, J.E., Alves, M.J., 1999, Cloning of a surface membrane glycoprotein specific for the infective form of *Trypanosoma cruzi* having adhesive properties to laminin. *J Biol Chem* 274:3461–3468.
Griffin, L.C., Toole, J.J., Leung, L.L.K., 1993, The discovery and characterization of a novel nucleotide-based thrombin inhibitor. *Gene* 137:25–31.
Hamm, J., 1996, Characterization of antibody-binding RNAs selected from structurally constrained libraries. *Nucl Acids Res* 24:2220–2227.
Hicke, B.J., Stephens, A.W., Gould, T., Chang, Y.F., Lynott, C.K., Heil, J., Borkowski, S., Hilger, C.S., Cook, G., Warren, S., Schmidt, P.G., 2006, Tumor targeting by an aptamer. *J Nucl Med* 47:668–678.
Hicke, B.J., Watson, S.R., Koenig, A., Lynott, C.K., Bargatze, R.F., Chang, Y.F., Ringquist, S., Moon-McDermott, L., Jennings, S., Fitzwater, T., Han, H.L., Varki, N., Albinana, I., Willis, M.C., Varki, A., Parma, D., 1996, DNA aptamers block L-selectin function in vivo. Inhibition of human lymphocyte trafficking in SCID mice. *J Clin Invest* 98:2688–2692.
Holland, C.A., Henry, A.T., Whinna, H.C., Church, F.C., 2000, Effect of oligodeoxynucleotide thrombin aptamer on thrombin inhibition by heparin cofactor II and antithrombin. *FEBS Lett* 484:87–91.
Homann, M., Göringer, U.H., 1999, Combinatorial selection of high affinity RNA ligands to live African trypanosomes. *Nucl Acids Res* 27:2006–2014.
Homann, M., Göringer, H.U., 2001, Uptake and intracellular transport of RNA aptamers in African trypanosomes suggest therapeutic "Piggy-Back" approach. *Bioorg Med Chem* 9: 2571–2580.
Hwang, J., Fauzi, H., Fukuda, K., Sekiya, S., Kakiuchi, N., Shimotohno, K., Taira, K., Kusakabe, I., Nishikawa, S., 2000, The RNA aptamer-binding site of hepatitis C virus NS3 protease. *Biochem Biophys Res Comm* 279:557–562.
Ikebukuro, K., Hasegawa, H., Sode, K., 2007, Selection and characterization of DNA aptamers against VEGF165 with aptamer blotting method and its application. *Nucleic Acids Symp Ser* 51:399–400, doi:10.1093/nass/nrm200.
Ito Y, Teramoto N., Kawazoe, N., Inada, K., Imanishi, Y., 1998, Modified nucleic acid for systematic evolution of RNA ligands by exponential enrichment. *J Bioact Compat Pol* 13:114–123.
Jayasena, S.D., 1999, Aptamers: an emerging class of molecules that rival antibodies in diagnostics. *Clin Chem* 45:1628–1650.
Jellinek, D., Lynott, C.K., Rifkin, D.B., Janjić, N., 1993, High-affinity RNA ligands to basic fibroblast growth factor inhibit receptor binding. *Proc Nat Acad Sci* 90:11227–11231.
Jenison, R.D., Jennings, S.D., Walker, D.W., 1998, Oligonucleotide inhibitors of P-selectin-dependent neutrophil-platelet adhesion. *Antisense Nucl Acid Drug Dev* 8:265–279.
Joshi, R., Janagama, H., Dwivedi, H.P., Senthil Kumar, T.M., Jaykus, L.A., Schefers, J., Sreevatsan, S., 2009, Selection, characterization, and application of DNA aptamers for the capture and detection of Salmonella enterica serovars. *Mol Cell Probes* 23:20–28.
Katzin, A.M., Colli, W., 1983, Lectin receptors in *Trypanosoma cruzi*: an N-acetyl-D-glucosamine-containing surface glycoprotein specific for the trypomastigote stage. *Bioch Biophys Acta* 727:403–411.
Kourlas, H., Schiller, D.S., 2006, Pegaptanib sodium for the treatment of neovascular age-related macular degeneration: a review. *Clin Ther* 28:36–44.
Le Tinevez, R., Mishra, R.K., Toulme, J.J., 1998, Selective inhibition of cell-free translation by oligonucleotides targeted to a mRNA hairpin structure. *Nucleic Acids Res* 26: 2273–2278.
Lou, X., 2009, Micromagnetic selection of aptamers in microfluidic channels. *Proc Natl Acad Sci* 106:2989–2994.
Lupold, S.E., Hicke, B., Lin, Y., Coffey, D.S., 2002, Identification and characterization of nuclease-stabilized RNA molecules that bind human prostate cancer cells via the prostate-specific membrane antigen. *Cancer Res* 62:4029–4033.

Mairal, T., 2008, Aptamers molecular tools for analytical applications. *Anal Bioanal Chem* 390:989–1007.

McEver, R.P., Moore, K.L., Cummings, R.D., 1995, Leukocyte trafficking mediated by selectin-carbohydrate interactions. *J Biol Chem* 270(19):11025–11028.

Morris, K.N., Jensen, K.B., Julin, C.M., Weil, M., Gold, L., 1998, High affinity ligands from in vitro selection: complex targets. *Proc Natl Acad Sci USA*, 95:2902–2907.

Ng, E.W.M., Shima, D.T., Calias, P., Cunningham, E.T., Jr., Guyer, D.R., Adamis, A.P., 2006, Pegaptanib, a targeted anti-VEGF aptamer for ocular vascular disease. *Nat Rev Drug Dis* 5:123–132.

Nitsche, A., Kurth, A., Dunkhorst, A., Pänke, O., Sielaff, H., Junge, W., Muth D., Scheller, F., Stöcklein, W., Dahmen, C., Pauli, G., Kage, A., 2007, One-step selection of Vaccinia virus-binding DNA aptamers by MonoLEX. *BMC Biotechnol* 7:48.

O'Connell, D., Koenig, A., Jennings, S., Hicke, B., Han, H.L., Fitzwater, T., Chang, Y.F., Varki, N., Parma, D., Varki, A., 1996, Calcium-dependent oligonucleotide antagonists against L-selectin. *Proc Natl Acad Sci USA* 93:5883–5887.

Pan, W., Craven, R.C., Qiu, Q., Wilson, C.B., Wills, J.W., Golovine, S., Wang, J.F., 1995, Isolation of virus-neutralizing RNAs from a large pool of random sequences. *Proc Natl Acad Sci USA* 92:11509–11513.

Pavski, V., Le, X.C., 2001, Detection of human immunodeficiency virus type1 reverse transcriptase using aptamers as probes in affinity capillary electrophoresis. *Anal Chem* 73:6070–6076.

Pendergrast, P.S., Marsh, H.N., Grate, D., Healy, J.M., Stanton, M., 2005, Nucleic acid aptamers for target validation and therapeutic applications. *J Biomol Tech* 16:224–234.

Pileur, F., Andreola, M.L., Dausse, E., Michel, J., Moreau, S., Yamada, H., Gaidamakov, S.A., Crouch, R.J., Toulme, J.J., Cazenave, C., 2003, Selective inhibitory DNA aptamers of the human RNase H1. *Nucleic Acids Res* 31:5776–5788.

Potyrailo, R.A., Conrad, R.C., Ellington, A.D., Hieftje, G.M., 1998, Adapting selected nucleic acid ligands (aptamers) to biosensors. *Anal Chem* 70:3419–3425.

Ruckman, J., Green, L.S., Beeson, J., Waugh, S., Gillette, W.L., Henninger, D.D., Claesson-Welsh, L., Janjić, N., 1998, 2′-Fluoropyrimidine RNA-based aptamers to the 165-amino acid form of vascular endothelial growth factor (VEGF165). Inhibition of receptor binding and VEGF-induced vascular permeability through interactions requiring the exon 7-encoded domain. *J Biol Chem* 273:20556–20567.

Sayer, N.M., Cubin, M., Rhie, A., Bullock, M., Tahiri-Alaoui, A., James, W., 2004, Structural determinants of conformationally selective, prion-binding aptamers. *J Biol Chem* 279: 13102–13109.

Sooter, L.J., Ellington, A.D. (2002), Reflections on a novel therapeutic candidate. *Chem Biol* 9:857–858.

Szpechciñki, A., Grzanka, A., 2006, Aptamers in clinical diagnostics. *Postepy Biochem* 52: 260–270.

Tuerk, C., Gold, L., 1990, Systematic evolution of ligands by exponential enrichment: RNA ligands to bacteriophage T4 DNA polymerase. *Science* 3:505–510.

Ulrich, H., Magdesian, M.H., Alves, M.J.M., Colli, W., 2002, In vitro selection of RNA aptamers that bind to cell adhesion receptors of *Trypanosoma cruzi* and inhibit cell invasion. *J Biol Chem* 277:20756–20762.

Ulrich, H., Trujillo, C.A., Nery, A.A., Alves, J.M., Majumder, P., Resende, R.R., Martins, A.H., 2006, DNA and RNA aptamers: from tools for basic research towards therapeutic applications. *Comb Chem High Throughput Screen* 9:619:632.

Urbain, J.L., (1999), Oncogenes, cancer and imaging. *J Nucl Med* 40:498–504.

Vo, N.V., Oh, J.-W., Lai, M.M.C., 2003, Identification of RNA ligands that bind hepatitis C virus polymerase selectively and inhibit its RNA synthesis from the natural viral RNA templates. *Virology* 307:301–316.

Weiss, S.P.D., Neumann, M., Groschup, M.H., Kretzschar, H.A., Famulok, M., Winnacker, E.L., 1997, RNA aptamers specifically interact with the prion protein PrP. *J Virol* 71: 8790–8797.

White, R.R., Shan, S., Rusconi, C.P., Shetty, G., Dewhirst, M.W., Kontos, C.D., Sullenger, B.A., 2003, Inhibition of rat corneal angiogenesis by a nuclease-resistant RNA aptamer specific for angiopoietin-2. *Proc Natl Acad Sci* USA 100: 5028–5033.

Wochner, A., Menger, M., Orgel, D., Cech, B., Rimmele, M., Erdmann, V.A., Glokler, J.A., 2008, DNA aptamer with high affinity and specificity for therapeutic anthracyclines. *Anal Biochem* 373:34–42.

Xu, W., Ellington, A.D., 1996, Anti-peptide aptamers recognize amino acid sequence and bind a protein epitope. *Biochemistry* 93:7475–7480.

Chapter 5
Reporter Gene

Abstract Reporter genes have become an invaluable tool in studies of gene expression. Reporter gene technology is widely used to monitor the cellular events associated with signal transduction and gene expression. Based upon the splicing of transcriptional control elements to a variety of reporter genes (with easily measurable phenotypes), it "reports" the effects of a cascade of signaling events on gene expression inside cells. The principal advantage of these assays is their high sensitivity, reliability, convenience, and adaptability to large-scale measurements. They are widely used in biomedical and pharmaceutical research and also in molecular biology and biochemistry.

Keywords Reporter · Gene · Signaling · Inducer · Therapeutics · Luc · X-gal · GFP · Lac · Ruc · HGH · CAT · GUS · LacZ · Promoter assay · Promoter · Molecular imaging · Reporter gene construct · Reporter gene assay

5.1 Prologue

Developments in molecular genetics have given us insights, at the molecular level, into vital processes in living organisms, such as embryonic development, growth regulation, differentiation, pathogenesis, and carcinogenesis. Insights into the mechanism of pathologic processes, such as developmental disorders and carcinogenesis, have stimulated efforts to develop therapeutic approaches to prevent or correct these processes. Techniques to directly change the genetic information of a cell have raised high expectations of the therapeutic potential of genetic manipulation. In vitro, genes have been introduced successfully into cells, thus changing the genotype and the phenotype of cells. These developments have raised hopes that diseases appearing to be incurable can soon be cured.

5.2 Concept

Genetic reporter systems represent an extensive toolbox for the study of regulatory promoter and enhancer sequences as well as for the study of transcription factors

M. Debnath et al., *Molecular Diagnostics: Promises and Possibilities*,
DOI 10.1007/978-90-481-3261-4_5,

(Alam and Cook, 1990; Kain and Ganguly, 2001). A gene consists of two functional parts, one is a DNA-sequence that gives the information about the protein that is produced (coding region). The other part is a specific DNA-sequence linked to the coding region; it regulates the transcription of the gene (promoter). The promoter is either activating or suppressing the expression of the gene. A reporter gene is one that codes for a product that is easy to measure. These genes are often used in situations in which the product of some gene of interest is difficult to assay quantitatively.

They are used to replace other coding regions whose protein products are difficult to assay. The most typical use of a reporter gene is to analyze how a gene is regulated. Let's consider a typical gene as being composed of a regulatory region, lying upstream of the transcription start point, and a structural region, including the open reading frame and any 5′ or 3′ untranslated regions (UTR's).

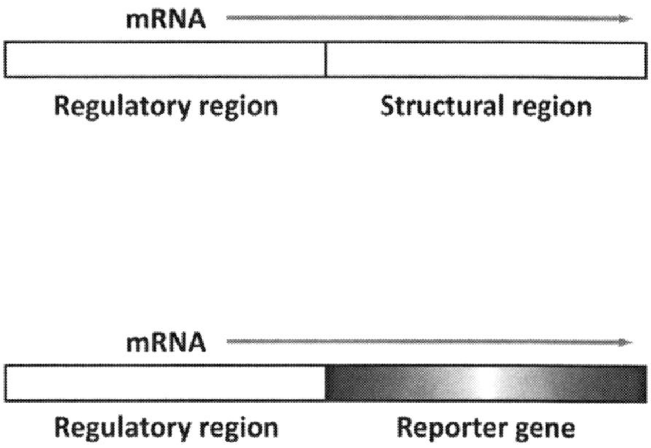

Fig. 5.1 The reporter gene affects the expression of the structural gene

The reporter gene would typically be attached to the regulatory region, using recombinant DNA methods, making an operon fusion like the second one shown in Fig. 5.1. In this arrangement, anything that ordinarily affects the expression of the natural gene would also affect the expression of the reporter gene. Among the more commonly used reporter genes are shown in Table 5.1.

The purpose of the reporter gene assay is to measure the regulatory potential of an unknown DNA-sequence. This can be done by linking a promoter sequence to an easily detectable reporter gene such as that encoding for the firefly luciferase.

Reporter genes have become an invaluable tool in studies of gene expression (Naylor, 1999). They are widely used in biomedical and pharmaceutical research and also in molecular biology and biochemistry. A reporter gene enables researchers to track and study another gene in cell cultures, animals, and plants. Because most gene therapy techniques only work on a small number of individuals, researchers need to use a reporter gene to identify which cells have taken up the gene currently under study, and which have incorporated it into their chromosomes. Regulatory sequences of interest are combined with a reporter construct of choice and are subsequently assayed in conjunction with relevant transcription factors (Phippard and Manning, 2003). Reporter genes can be attached to other sequences so that only the reporter protein is made or so that the reporter protein is fused to another protein (fusion protein). Reporter genes can "report" many different properties and events:

- o the strength of promoters, whether native or modified for reverse genetics studies;
- o the efficiency of gene delivery systems;
- o the intracellular fate of a gene product, a result of protein traffic;
- o interaction of two proteins in the two-hybrid system or of a protein and a nucleic acid in the one-hybrid system;
- o the efficiency of translation initiation signals;
- o and the success of molecular cloning efforts.

Table 5.1 Activity of some reporter genes

Reporter genes and protein	Activity and measurement
GUS (β-glucuronidase)	Transfers radioactive acetyl groups to chloramphenicol; detection by thin layer chromatography and autoradiography
CAT (chloramphenicol acetyltransferase)	Hydrolyzes colorless galactosides to yield colored products
GAL (β-glucuronide)	Hydrolyzes colorless glucuronides to yield colored products
LUC (luciferase)	Oxidizes luciferin, emitting photons
GFP (green fluorescent protein)	Fluoresces on irradiation with UV

If the reporter system is well chosen then the level of reporter gene expression will correlate with the transcriptional activity of the introduced transgenic factors. In order to assure such a correlation, it is important that the reporter gene does not disturb the metabolism of the transformed cells and that the gene is not endogenously expressed by the target cells creating background signals.

Expression of reporter genes can be measured by enzyme activity assay of the expressed enzyme encoded by the reporter gene using chromo-, fluoro- or luminogenic substrates; immunological assay of the expressed protein encoded by the reporter gene (reporter gene ELISA); or by histochemical staining of cells or tissues typically to localize enzymatic activity ectopically expressed from reporter gene constructs in transformed cells.

Reporter genes are widely used to study gene expression and regulation mechanisms in living cells. Not all expressed enzymes are easily detectable, so reporter genes were introduced into cellular DNA to investigate gene function by means of a measurable property, the luminescence. These reporter genes are for example: firefly luciferase, beta-glycosidase, alkaline phosphate, beta-glucuronidase, beta-glucosidase. By far the most popular reporter is the firefly luciferase from the American firefly (*Photinus pyralis*). The high sensitivity, easy handling, short process time and a high quantum yield of the bioluminescence reaction make this method to the "method of choice" to understand gene regulation. Typical commercial assay kits are optimized for extended half life time of more than five minutes. Most reporter genes are placed downstream genetically of the promoter region, but close to the gene under study. This ensures that these genes are expressed together, and are not separated during cell division by crossover events. Sometimes, reporter genes are simply placed in a vector independent of a chromosome, and other techniques used to identify the gene under study (Pardy, 1994).

Reporter genes offer a big advantage, because one doesn't need a separate assay for each regulatory region being studied. Also, for many reporter genes simple indicator plate assays have been developed that allow one to determine levels of expression by the colour of a bacterial or yeast colony, and perhaps to isolate mutants based on changes in the colour. For instance, a bacterial strain making substantial amounts of beta-galactosidase will form a blue colony on a plate with X-gal; this allows one to identify cells with different levels of beta-galactosidase, and to use X-gal plates for any regulatory region driving expression of beta-galactosidase.

5.3 Reporter Gene Assay

Study of gene expression can be facilitated by using a reporter gene assay (Fig. 5.2). Instead of directly measuring the level of target gene mRNA, one can clone the promoter region of the gene of interest in front of a reporter gene and measure the reporter gene expression as a reflection of the expression of the gene of interest (Kain and Ganguly, 2001). A simple *lacZ*-fusion system is to measure the activity of the reporter gene product β-galactosidase (Thibodeau et al., 2004). Different strategies of making the fusion construct and their applications have been reported. This method is particularly useful to dissect the promoter region of the gene of interest and is also used in other experimental protocols such as the yeast two-hybrid analysis.

To introduce a reporter gene into an organism, scientists place the reporter gene and the gene of interest in the same DNA construct to be inserted into the cell or organism. For bacteria or eukaryotic cells in culture, this is usually in the form of a circular DNA molecule called a plasmid. It is important to use a reporter gene that is not natively expressed in the cell or organism under study, since the expression of the reporter is being used as a marker for successful uptake of the gene of interest.

Commonly used reporter genes that induce visually identifiable characteristics usually involve fluorescent proteins; examples include the gene that encodes jellyfish green fluorescent protein (GFP), which causes cells that express it to glow green under UV light, and the enzyme luciferase (Phippard and Manning, 2003; Nordeen, 1988), which catalyzes a reaction with a luciferin to produce light. Another common reporter in bacteria is the *lacZ* gene, which encodes the protein β-galactosidase. This enzyme causes bacteria expressing the gene to appear blue when grown on a medium that contains the substrate analog X-gal (an inducer molecule such as

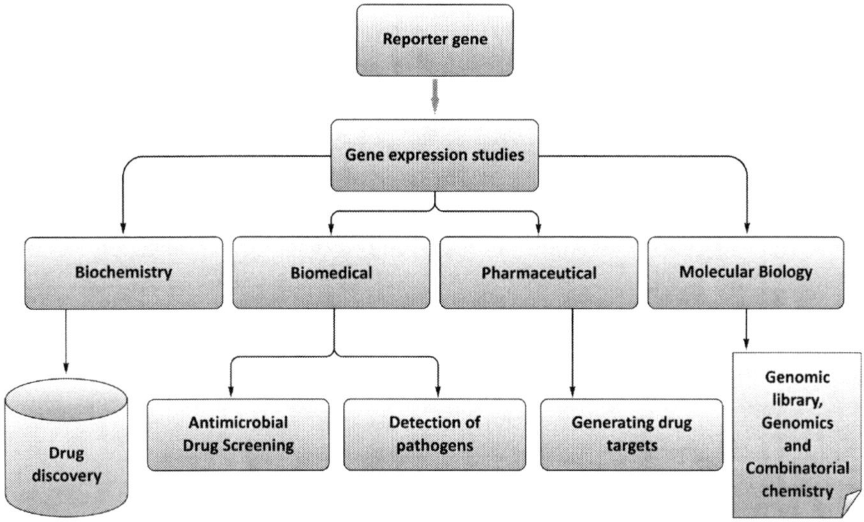

Fig. 5.2 Reporter gene can be used as a molecular diagnostic tool

IPTG is also needed under the native promoter). An example of a selectable-marker reporter in bacteria is the chloramphenicol acetyltransferase (CAT) gene (Altschmied and Duschl, 1997; Pardy, 1994), which confers resistance to the antibiotic chloramphenicol. Various detection methods (Jeffereson, 1989; Gillissen et al., 1998; Nordeen, 1988; Altschmied and Duschl, 1997) are used to measure expressed reporter gene protein. These include luminescence, absorbance and fluorescence.

5.3.1 Luc *Reporter Gene System*

In nature, the luciferase reaction occurs in the peroxisomes of a specialized light organ in fireflies (*Photinus pyralis*). The luciferase reaction emits a yellow-green light (560 nm) and requires the co-factors ATP, Mg^{2+}, O^2 and the substrate luciferin. The glow is widely used as an assay for luciferase activity to monitor regulatory elements that control its expression (Olesen et al., 2002). *Luc* is particularly useful as a reporter gene since it can be introduced into living cells and into whole organisms such as plants, insects, and even mammals. *Luc* expression does not adversely affect the metabolism of transgenic cells or organisms. In addition, the *luc* substrate luciferin is not toxic to mammalian cells, but it is water-soluble and readily transported into cells. Since *luc* is not naturally present in target cells the assay is virtually background-free. Hence, the *luc* reporter gene is ideal for detecting low-level gene expression (Nordeen, 1988). A second reporter system based on luciferase expressed by the *ruc* gene from Renilla (*Renilla reniformis*) has also become available (Wang et al., 2002). The activities of firefly and Renilla luciferase can be combined into a dual reporter gene assay.

Luciferase expression is measured by adding ATP and luciferin to cell lysates and then analyzing bioluminescence by spectroscopy or with a scintillation counter allowing for the detection of even a few hundred thousand enzyme molecules.

The development of camera based imaging systems to visualize *luc* gene expression in vivo has made rapid progress; most impressively, in the generation of transgenic mice carrying the *luc* gene. In some of these mice strains *luc* expression is triggered by intoxication or the inflammation processes. The effects of disease to mice as well as the results of therapy can be studied non-invasively and in real time (www.xenogen.com).

Luciferin was initially isolated by extraction from *Photinus pyralis* which is a tedious process. Despite manufacturing bulk quantities of luciferin chemically, commercial grade luciferin is still specified "synthetic".

5.3.2 Ruc *Reporter Gene System*

Renilla luciferase serves as a marker protein in bioluminescent fusion constructs of the *ruc* gene (Wang et al., 2002). It is similar to the above described *luc* vectors which is useful in low-light imaging of gene expression and of regulation processes in living cells as well as in transgenic plants. Coelenterazine is the best substrate for Renilla luciferase generating 40% more light than native coelenterazine with a considerably higher initial light intensity (Yu and Szalay, 2002).

The firefly luciferase gene (*luc*) has proven to be very useful in reporter gene research and drug discovery assays, when used with luciferin and ATP: firefly luciferase/luciferin emits light at 560 nm, Renilla luciferase/coelenterazine emits light at 475 nm. Because these two reporter systems emit light at quite different wave lengths, it is possible to use fire fly luciferase/luciferin and Renilla luciferase/coelenterazine as a dual reporter system.

In dual reporter gene systems two distinct reporter enzymes can be measured within one sample, whereby the second reporter is often used as an internal control. Dual reporter luminescence assays preferably using firefly luciferase as the first reporter enzyme, have become very popular, e.g. for measuring transcriptional activity in studies of the structural or physiological basis of regulated gene expression (Yu and Szalay, 2002).

5.3.3 Green Fluorescent Protein (GFP) Reporter Gene System

The green fluorescent protein from the *Aequorea victoria* jellyfish emits green fluoresence without the need for any enzyme or co-factors (Phillip, 2001). Gene expression and protein localization is possible in situ and in vivo: no secondary transformation is required. Reporter gene expression can be qualitatively monitored live e.g. for the study of dynamic processes inside the cell (Chalfie et al., 1994). There are a number of GFP proteins in use. Selecting GFP with minimal overlap in light emission allows dual reporter gene studies, for instance, to localize different sites of expression or, the relative effects of different regulatory proteins (Chalfie, 1995).

Most recently, bioluminescence resonance energy transfer (BRET) detection systems have been introduced. These systems are based on excitation energy transfer from a bioluminescent donor molecule, e.g. Renilla luciferase to a fluorescent acceptor molecule, like green fluorescent protein (GFP). It is known that luciferase interacts rapidly with GFP in a highly specific manner to form an equilibrium complex responsible for this very efficient energy-transfer phenomenon.

Renilla luciferase as well as GFP are both attached to the proteins to be examined (Yu and Szalay, 2002), e.g. by chemical synthesis or by use of genetic engineering methods. In the latter case the gene fusion constructs are used to transfect or transform cells and express the fusion proteins. In the presence of coelenterazine, a BRET signal is generated when the fusion proteins are associated. Excellent discrimination of the *ruc* and GFP signals results when coelenterazine is used as the *ruc* substrate (Wang et al., 2002).

As the efficiency of the non-radioactive energy transfer from the Renilla luciferase fusion construct to the GFP conjugate depends on steric effects, for instance, on the distance of the protein molecules and on their relative orientation, the method can be used to study protein–protein interaction as well as to screen for inhibition of such an interaction: in drug discovery applications or receptor-ligand studies. The new spectroscopic techniques and assay methods built on BRET are fascinating tools for investigation in vivo.

5.3.4 Human Growth Hormone (hGH) Reporter Gene System

The human growth hormone (hGH) encoded reporter protein is secreted into the culture medium by transfected cells (Selden et al., 1986). Thus no cell lysis is required for detection. Detection of the secreted hGH requires the use of hGH antibodies. Such antibodies used to be radiolabeled in order to allow detection of the antibody hGH complex. Today hGH is assayed by indirect ELISA. First, the hGH from the supernatant of the culture medium binds to the antibody on the plate. Subsequently, the bound hGH is detected in two steps via a digoxigenin coupled anti-hGH antibody and a peroxidase-coupled anti-digoxigenin antibody. Bound peroxidase is quantified by incubation with a peroxidase substrate such as TMB (3,3′,5,5′-tetramethylbenzidine,T-2100; Selden et al., 1986).

5.3.5 Beta-Galactosidase Reporter Gene System

β-galactosidase is a hydrolase enzyme that catalyzes the hydrolysis of β-galactosides into monosaccharides. Substrates of different β-galactosidases include ganglioside GM1, lactosylceramides, lactose, and various glycoproteins. Alternate or nicknames are "beta-gal" or "β-gal". Lactase is often confused as an alternate name for β-galactosidase, but it is actually simply a sub-class of β-galactosidase (Thibodeau et al., 2004).

The active site of β-galactosidase catalyzes the hydrolysis of its disaccharide substrate via "shallow" and "deep" binding. Monovalent potassium ions (K^+) as well as divalent magnesium ions (Mg^{2+}) are required for the enzyme's optimal activity. The beta-linkage of the substrate is cleaved by a terminal carboxyl group on the side chain of a glutamic acid.

In *E. coli*, Glu-461 was thought to be the nucleophile in the substitution reaction. However, it is now known that Glu-461 is an acid catalyst. Instead, Glu-537 is the actual nucleophile, binding to a galactosyl intermediate. In humans, the nucleophile of the hydrolysis reaction is Glu-268 (McCarter et al., 1997). β-galactosidase is measured by a photometric assay (Pardy, 1994).

5.3.6 CAT Reporter Gene System

Microbial chloramphenicol acetyltransferase (CAT) catalyzes the transfer of acetyl groups from acetyl coenzyme A to chloramphenicol and represents the oldest reporter gene system still in broad use today. The assay is performed by incubating transformed cells with chloramphenicol carrying a radioactive label. CAT activity can be assayed by autoradiography of lysate subjected to thin layer chromatography (TLC). Acetylated and non-acetylated chloramphenicol can readily be separated by TLC of lysate. Hence, the presence of labeled acetylated chloramphenicol correlates to the expression of CAT (Zhang et al., 2003). In an effort to limit the use of radioactivity in the lab the method is rapidly being replaced by immunological detection of the CAT enzyme by ELISA.

5.3.7 LacZ *Reporter Gene System*

Reporter genes have become standard genetic tools used to evaluate either the transcriptional or the translational activity associated with genes of interest, whose products cannot be easily assayed. The lacZ gene from *E. coli* has been used very effectively to quantify such regulated activities in many different organisms. Colonies expressing lacZ appear blue on culture media containing X-gal (B-7150). The prokaryotic beta-galactosidase encoded by lacZ from the *lac* operon catalyzes the hydrolysis of beta-galactosides, most prominently the hydrolysis of lactose. Galactosidase activity in lysates of transfected cells can be assayed via absorption of o-nitrophenolate from o-nitrophenyl-ß-D-galactoside (ONPG,N-4190), fluorescence of 4-methylumbelliferone (from 4-methyl-umbelliferyl-ß-galactopyranoside, M-5550), or via chemiluminescence of 1,2-dioxetan-galactopyranoside derivatives. The lacZ is often co-transfected together with other reporter constructs such as luciferase serving as an internal control.

The lacZ reporter gene is widely used as a convenient tool to identify recombined plasmids during molecular cloning procedures. Since the complete lacZ gene is too large to be carried by a plasmid, cloning vectors contain only the alpha region of the lacZ gene. Expression of the alpha region of lacZ yields an inactive protein. However, the truncated protein encoded by the alpha region can combine into active beta-gal when complemented with a protein from lacZ *lacking* the alpha region. Such "intragenic complementation" can be achieved if a plasmid carrying the alpha region of lacZ is inserted into a bacterial cell containing a mutated alpha region of the lacZ gene. These cells will turn beta-gal positive if the plasmid is present.

DNA cloning into lacZ vectors occurs such that the fragment is likely to insert into the alpha lacZ sequence on the plasmid which renders the alpha-region dysfunctional. In this case intragenic complementation no longer occurs. Hence, cells carrying plasmids containing foreign DNA will remain beta-gal negative. Beta-gal negative cells can readily be isolated from positive ones by incubation of bacteria on chromogenic culture media containing X-gal (5-bromo-4-chloro-3-indoxyl-beta-D-galactopyranoside, B-7150) and IPTG (isopropyl-beta-D-thiogalactopyranoside, I-8000) followed by manual selection of the colorless colonies.

5.3.8 GUS Reporter Gene System

GUS encodes the beta-glucuronidase enzyme from *E. coli* and it has been extensively used as a marker gene (Jefferson, 1989). The main advantage of the GUS reporter gene system is the stable expression of GUS enzyme in plant cells (Gilissen et al., 1998). GUS activity can readily be detected by chromogenic or fluorogenic enzyme substrates. Typically, very little background signal is encountered because the intrinsic expression of endogenous beta-glucuronidase genes is not significant in plants. In addition, plant metabolism remains largely unaffected by the presence of *E. coli* beta-glucuronidase. Various beta-glucuronic acid substrates are available for detection of GUS expression. The most widely used fluorogenic substrate for detection of beta-glucuronidase activity in vitro is 4-methylumbelliferyl beta-D-glucuronide (MUG, 4-methylumbelliferyl-beta-D-glucuronic acid dihydrate, M-5700). For other applications the use of X beta-glucuronide (5-bromo-4-chloro-3-indoxyl-beta-D-glucuronic acid, cyclohexylammonium salt, B-7300), which yields an insoluble blue precipitate in the presence of beta-glucuronidase, has become standard.

5.4 Transformation and Transfection Assays

Transfection and transformation are the basic two ways of expressing a foreign or modified gene in an organism. They are effective in only a small percentage of a population subjected to the techniques. Thus, a method for identifying those few successful gene uptake events is necessary. Reporter genes used in this way are normally expressed under their own promoter independent from that of the introduced gene of interest; the reporter gene can be expressed constitutively (that is, it is "always on") or inducibly with an external intervention such as the introduction of IPTG in the β-galactosidase system. As a result, the reporter gene's expression is independent of the gene of interest's expression, which is an advantage when the gene of interest is only expressed under certain specific conditions or in tissues that are difficult to access.

In the case of selectable-marker reporters such as CAT, the transfected population of bacteria can be grown on a substrate that contains chloramphenicol. Only those cells that have successfully taken up the construct containing the CAT gene will survive and multiply under these conditions (Zhang et al., 2003).

5.5 Gene Expression Assays

Reporter genes can also be used to assay for the expression of the gene of interest, which may produce a protein that has little obvious or immediate effect on the cell culture or organism. In these cases the reporter is directly attached to the gene of interest to create a gene fusion. The two genes are under the same promoter and are transcribed into a single messenger RNA molecule. The mRNA is then translated into protein. In these cases it is important that both proteins be able to properly fold into their active conformations and interact with their substrates despite being fused. In building the DNA construct, a segment of DNA coding for a flexible polypeptide linker region is usually included so that the reporter and the gene product will only minimally interfere with one another.

5.6 Promoter Assays

Reporter genes can be used to assay for the activity of a particular promoter in a cell or organism. In this case there is no separate "gene of interest"; the reporter gene is simply placed under the control of the target promoter and the reporter gene product's activity is quantitatively measured. The results are normally reported relative to the activity under a "consensus" promoter known to induce strong gene expression (Koken et al., 1994). The use of reporter enzyme provides a rapid and sensitive method for the analysis of transgene expression.

5.7 Role of Reporter Gene Assay in Molecular Diagnostics

Reporter gene assays have emerged as a very general strategy for indirectly monitoring various intracellular events. Furthermore, reporter genes are being used to monitor gene/cell therapies, including the location(s), time variation, and magnitude of gene expression. The reporter gene assay has been used for the detection of pathogens (Collins and Franzblau, 1997). Rubella virus was detected using replicon based reporter gene assay (Tzeng et al., 2005). Influenza A virus replication was also detected by virus inducible reporter gene (Lutz et al., 2005).

Some progress was reported in generating a vast number of drug targets through genomics and large compound libraries through combinatorial chemistry have stimulated advancements in drug discovery through the development of new high throughput screening (HTS) methods (Deo and Daunert, 2001). Automation and HTS techniques are also highly desired in fields such as clinical diagnostics. Luminescence-based assays have emerged as an alternative to radiolabel-based assays in HTS as they approach the sensitivity of radioactive detection along with ease of operation, which makes them amenable to miniaturization. Luminescent proteins provide the advantage of reduced reagent and operating costs because they can be produced in unlimited amounts through the use of genetic engineering tools. In that regard, the use of two naturally occurring and recombinantly produced luminescent proteins from the jellyfish *Aequorea victoria*, namely, aequorin and the green fluorescent protein (GFP), has attracted attention in a number of analytical applications in diverse research areas. Aequorin is naturally bioluminescent and has therefore, virtually no associated background signal, which allows its detection down to attomole levels. GFP has become the reporter of choice in a variety of applications given that it is an autofluorescent protein that does not require addition of any co-factors for fluorescence emission (Shaner et al., 2005). Furthermore, the generation of various mutants of GFP with differing luminescent and spectral properties has spurred additional interest in this protein Inouye and Tsuji, 1994.

The green fluorescent protein (GFP) gene offers many advantages as a viability reporter for high-throughput antimicrobial drug screening (Tsien, 1998). However, screening for antituberculosis compounds by using GFP driven by the heat shock promoter, hsp60, has been of limited utility due to the low signal-to-noise ratio. The optimized GFPMA is sufficiently simple, robust, and inexpensive (no reagent costs) to be used for routine high-throughput screening for antituberculosis compounds (Changsen et al., 2003). In another study a throughput screening of G protein coupled receptors were performed using reporter gene assay (Durocher et al., 2000).

Most current molecular genetic imaging strategies are "indirect," coupling a "reporter gene" with a complimentary "reporter probe." Reporter gene constructs can be driven by constitutive promoter elements and used to monitor gene therapy vectors and the efficacy of transgene targeting and transduction, as well as to monitor adoptive cell-based therapies. Inducible promoters can be used as "sensors" to regulate the magnitude of reporter gene expression and can be used to provide information about endogenous cell processes. Reporter systems can also be constructed to monitor mRNA stabilization and specific protein-protein interactions. Promoters can be cell specific and restrict transgene expression to certain tissue and organs (Serganova and Blasberg, 2005).

Molecular imaging is a relatively new discipline, which developed over the past decade, initially driven by in situ reporter imaging technology (Blasberg, 2003). Human reporter genes will play an increasingly more important role in the clinical application of positron-emission-tomography-based reporter gene imaging

(Serganova et al., 2007). Examples of highly expressed cell membrane receptors include specific membrane somatostatin receptors (hSSTrs). The transporter group includes the sodium iodide symporter (hNIS) and the norepinephrine transporter (hNET). The endogenous enzyme classification includes human mitochondrial thymidine kinase 2 (hTK2). Initial applications of reporter gene imaging in patients deal with gene therapy and adoptive cell-based therapies. These studies from the availability of efficient human reporter systems can provide critical monitoring information for adenoviral-based, retroviral-based and lenteviral-based gene therapies, oncolytic bacterial and viral therapies, and adoptive cell-based therapies. Translational applications of noninvasive in vivo reporter gene imaging are likely to include: (a) quantitative monitoring of gene therapy vectors for targeting and transduction efficacy in clinical protocols by imaging the location, extent and duration of transgene expression; (b) monitoring of cell trafficking, targeting, replication and activation in adoptive T-cell and stem/progenitor cell therapies; (c) and assessments of endogenous molecular events using different inducible reporter gene imaging systems.

5.8 Conclusion

Reporter gene technology is widely used to monitor the cellular events associated with signal transduction and gene expression. Based upon the splicing of transcriptional control elements to a variety of reporter genes (with easily measurable phenotypes), it "reports" the effects of a cascade of signalling events on gene expression inside cells. The principal advantage of these assays is their high sensitivity, reliability, convenience, and adaptability to large-scale measurements. With the advances in this technology and in detection methods, it is likely that luciferase and green fluorescent protein will become increasingly popular for the non-invasive monitoring of gene expression in living tissues and cells. Such techniques will be important in defining the molecular events associated with gene transcription, which has implications for our understanding of the molecular basis of disease and will influence our approach to gene therapy and drug development. The future appears to be promising for the continued expansion of the use of reporter genes in the many evolving biomedically related arenas (Min and Gambhir, 2008).

References

Alam, J., Cook, J.L., 1990, Reporter genes: application to the study of mammalian gene transcription. *Anal Biochem* 188:245–254.

Altschmied, J., Duschl, J., 1997, Set of optimized luciferase reporter gene plasmids compatible with widely used CAT vectors. *Biotechniques* 23:436–438.

Blasberg, R.G., 2003, Molecular imaging and cancer. *Mol Cancer Ther* 2:335–343.

Chalfie, M., 1995, Green fluorescent protein. *Photochem Photobiol* 62:651–656.

Chalfie, M., Tu, Y., Euskirchen, G., Ward, W.W., Prasher, D.C., 1994, Green fluorescent protein as a marker for gene expression. *Science* 263:802–805.

Changsen, C., Franzblau, S.G., Palittapongarnpim, P., 2003, Improved green fluorescent protein reporter gene-based microplate screening for antituberculosis compounds by utilizing an acetamidase promoter. *Antimicrob Agents Chemother* 47:3682.

Collins, L., Franzblau, S.G., 1997, Microplate alamar blue assay versus BACTEC 460 system for high-throughput screening of compounds against *Mycobacterium tuberculosis* and *Mycobacterium avium*. *Antimicrob Agents Chemother* 41:1004–1009.

Deo, S.K., Daunert, S., 2001, Luminescent proteins from Aequorea victoria: applications in drug discovery and in high throughput analysis. *Fresen J Anal Chem* 369:258–266.

Durocher, Y., Perret, S., Thibaudeau, E., Gaumond, M.H., Kamen, A., Stocco, R., Abramovitz, M., 2000, A reporter gene assay for high-throughput screening of G-protein-coupled receptors stably or transiently expressed in HEK293 EBNA cells grown in suspension culture. *Anal Biochem* 284:316–326.

Gilissen, L.J., Metz, P.L., Stiekema, W.J., Nap, J.P., 1998, Biosafety of *E. coli* beta-glucuronidase (GUS) in plants. *Transgenic Res* 7:157–163.

Inouye, S., Tsuji, F., 1994, Aequorea green fluorescent protein. Expression of the gene and fluorescence characteristics of the recombinant protein. *FEBS Lett* 341:277–280.

Jefferson, R.A., 1989, The GUS reporter gene system. *Nature* 342:837–838.

Kain, S.R., Ganguly, S., 2001, Overview of genetic reporter systems. *Curr Protoc Mol Biol* Chapter 9:Unit9.6.

Koken, S.E., Greijer, A.E., Verhoef, K., van Wamel, J., Bukrinskaya, A.G., Berkhout B., 1994, Intracellular analysis of in vitro modified HIV Tat protein. *J Biol Chem* 269:8366–8375.

Lutz, A., Dyall, J., Olivo, P.D., Pekosz, A., 2005, Virus-inducible reporter genes as a tool for detecting and quantifying influenza A virus replication. *Virol Methods* 126:13–20.

McCarter, J., Burgoyne, D., Miao, S., Zhang, S., Callahan, J., Withers, S., 1997, Identification of Glu-268 as the catalytic nucleophile of human lysosomal beta-galactosidase precursor by mass spectrometry. *J Biol Chem* 272:396–400, PMID 8995274.

Min, J.J., Gambhir, S.S., 2008, Molecular imaging of PET reporter gene expression. *Handb Exp Pharmacol* 185(Pt 2):277–303.

Naylor, L.H., 1999, Reporter gene technology: the future looks bright. *Biochem Pharmacol* 58(5):749–757.

Nordeen, S.K., 1988, Luciferase reporter gene vectors for analysis of promoters and enhancers. *Biotechniques* 6:454–458.

Olesen, C.E., Yan, Y.X., Liu, B., Martin, D., D'Eon, B., Judware, R., Martin, C., Voyta, J.C., Bronstein, I., 2000, Novel methods for chemiluminescent detection of reporter enzymes. *Methods Enzymol* 326:175–202.

Pardy, K., 1994, Reporter enzymes for the study of promoter activity. *Mol Biotechnol* 2: 23–27.

Phillips, G., 2001, Green fluorescent protein–a bright idea for the study of bacterial protein localization. *FEMS Microbiol Lett* 204 (1):9–18. PMID 11682170.

Phippard, D., Manning, A.M., 2003, Screening for inhibitors of transcription factors using luciferase reporter gene expression in transfected cells. *Methods Mol Biol* 225:19–23.

Selden, R.F., Howie, K.B, Rowe, M.E., Goodman, H.M., Moore, D.D., 1986, Human growth hormone as a reporter gene in regulation studies employing transient gene expression. *Mol Cell Biol* 6:3173–3179.

Serganova, I., Blasberg, R., 2005, Reporter gene imaging: potential impact on therapy. *Nuc Med Biol* 32:763–780.

Serganova, I., Ponomarev, V., Blasberg, R., 2007, Human reporter genes: potential uses in clinical studies. *Nuc Med Biol* 34:791–807.

Shaner, N., Steinbach, P., Tsien, R., 2005, A guide to choosing fluorescent proteins. *Nat Methods* 2:905–909.

Thibodeau, S.A., Fang, R., Joung, J.K., 2004, High-throughput beta-galactosidase assay for bacterial cell-based reporter systems. *Biotechniques* 36:410–415.

Tsien, R., 1998, The green fluorescent protein. *Annu Rev Biochem* 67:509–44. PMID 9759496.

Tzeng, W.-P., Zhou, Y., Icenogle, J., Frey, T.K., 2005, Novel replicon-based reporter gene assay for detection of rubella virus in clinical specimens. *J Clin Microbiol* 43:879–885.

Wang, Y., Yu, Y.A., Shabahang, S., Wang, G., Szalay, A.A., 2002, Renilla luciferase-Aequorea GFP (*Ruc-GFP*) fusion protein, a novel dual reporter for real-time imaging of gene expression in cell cultures and in live animals. *Mol Genet Genomics* 268:160–168.

Yu, Y.A., Szalay, A.A., 2002, A Renilla luciferase-Aequorea GFP (*ruc-gfp*) fusion gene construct permits real-time detection of promoter activation by exogenously administered mifepristone in vivo. *Mol Genet Genomics* 268:169–178.

Zhang, K., Kurachi, S., Kurachi, K., 2003, Limitation in use of heterologous reporter genes for gene promoter analysis: silencer activity associated with the chloramphenicol acetyltransferase reporter gene. *J Biol Chem*, 278:4826–4830.

Chapter 6
Impact of HGP on Molecular Diagnostics

Abstract The Human Genome Project has heralded a whole new era in our understanding of the molecular basis of disease. New opportunities now arise to predict disease by genetic testing, and in some cases to prevent disease through surveillance or other specific interventions. Increasingly it will be possible to test for predisposition to disease, to develop new treatments or to tailor available treatments more specifically to an individual's genetic make-up.

Keywords Single gene disease · Human genome project · HGP · HUGO · Functional genomics · Individual drug treatment · Pharmacogenetics · Drug design · Systems biology · Genetic test · HapMap · Genome medicine

6.1 Prologue

In the past decade, with the availability of technology and knowledge fuelled by the investment and interest in the Human Genome Project, molecular diagnostics has recently enabled laboratories to offer diagnostic and predictive tests for inherited disorders (Ferrari et al., 2003). Large amounts of genetic information can be determined at increasing efficiencies. The Human Genome Project (HGP) will change medicine and medical research irrevocably. The obvious gains in genetic knowledge from the HGP, together with the advances which will flow into bioinformatics, biotechnology and the potential for novel therapeutic agents, will ensure that the financial investment in the HGP is repaid many times over. The HGP's costs in terms of ethical and social issues remain to be determined, but it is to be hoped that these will not detract from the scientific and medical achievements.

6.2 Origin of the Human Genome Project

The Human Genome Project, the mapping of our 30,000–50,000 genes and the sequencing of all our DNA, will have major impact on biomedical research and the whole of therapeutic and preventive health care. The tracing of genetic diseases

M. Debnath et al., *Molecular Diagnostics: Promises and Possibilities*,
DOI 10.1007/978-90-481-3261-4_6, © Springer Science+Business Media B.V. 2010

to their molecular causes is rapidly expanding diagnostic and preventive options. The increased insights into molecular pathways, gained from high-throughput "functional genomics", using DNA-chip and protein-chip approaches and specially designed animal model systems, will open great prospects for pharmacological and genetic therapies. Powerful bioinformatics and biostatistics will further improve our pattern recognition and accelerate progress. A rapidly expanding area of high expectations is that of "pharmacogenomics": the design of more effective drugs with lower toxicity through tailoring of drug treatment to individual, genetically determined differences in drug metabolism. Not only will this decrease the cost of health care through reduction of adverse drug reactions, but a better stratification of populations will also provide more statistical power farther upstream in drug trials. However, the optimal benefits from the current explosion of "data mining" will only be realized when the basic data are made and kept publicly accessible, while at the same time safeguarding the protection of intellectual property arising from downstream inventions. This is one of the goals of HUGO, the international Human Genome Organization, established 13 years ago to assist coördination of data acquisition and exchange and societal implementation of the genome project. Additional points of attention in this historic endeavour are the prevention of stigmatization and discrimination and the safeguarding of a worldwide balance in the contribution by and benefits to different populations, while respecting the diversity in cultures and traditions (van Ommen, 2002).

The Human Genome Project (HGP) is an effort to understand the genetic instructions that make up a human. The human genome is the DNA that resides within every human cell. By sequencing the genome, scientists will identify the location and composition of all of our 100,000 or so genes. The genes carry information for making the proteins that direct the make up the each human. Among other things, proteins control human development, physiology, and resistance to disease. Knowledge of gene sequences is tremendously beneficial to scientists studying all facets of living things. The complete genome sequence of human and many other species provides a new starting point for understanding our basic genetic make-up and how variations in genetic instructions result in human disease or other variants of human phenotypes (Peltonen, 2001). Realization of this goal will include characterization of all the genes that cause or predispose to disease, which will most certainly lead to the development of powerful new tools for diagnosis, prevention, and treatment in all medical fields.

6.3 Goals and Issues of HGP

The HGP was conducted as collaboration between large-scale sequencing centres coordinated by the National Institute for Human Genome Research and the Department of Energy. Following the developmental phase resulting in genetic and physical maps and technology for high throughput sequencing, the US project was organized into 9 large-scale genomic sequencing centres distributed among universities and research institutes.

As new technologies for understanding and working with DNA were developed in the 1980s, the idea arose to sequence the entire human genome systematically, and with this idea arose the HGP. HGP is an attempt to map completely the entire spectrum of genetic materials that can be found in all human beings. It is a research effort initiated by the DOE and jointly managed by the DOE and the National Institutes of Health (NIH). A unifying fact of human genetics is that all humans have genomes – the complete set of our genetic instructions on the 23 pairs of chromosomes within each of us – that are 99.9% identical in sequence. The DNA molecules that carry these genetic instructions are linear, information-containing molecules made up of four simple bases or building blocks – adenine, cytosine, guanine and thymine – that pair up to form the well-known DNA double helix. Although each one of our cells contains about six billion base pairs of DNA, three billion from each parent, we think that our differences are determined by only about one base pair in each thousand (Collins and Patrinos, 1998).

The genome project now embraces three main technical goals: genetic linkage maps to trace the inheritance of chromosome regions through pedigrees; physical maps of large chromosome regions, to enable the direct study of DNA structure in search of genes; and substantial DNA sequence information, enabling the correlation of DNA changes with alterations in biological function. If history were logical, then the genome project would have grown from a discussion of each in turn and how to bring them together into a coherent plan. History is not logical, however, and it was DNA sequencing technology rather than genetic linkage mapping that gave rise to the idea of a human genome project.

One goal of the HGP is to localize all of the estimated 80,000 genes on the human chromosomes and to determine the sequence of all three billion units of DNA that constitute one set of those chromosomes. This information will vastly accelerate studies that will characterize what those genes do and how disease can result from errors in their functioning. An important element of the HGP is to enable technologies that will allow biologists to uncover gene function more efficiently. This is important because while the HGP describe the human genome in molecular detail, its longer term and more profound impact will be to reveal critical mechanisms of human biology and supply the medical context within which investigations on the molecular pathology of human diseases can most efficiently take place. This will lead to a future medicine in which prevention will be firmly rooted in mechanistic knowledge and potential interventions can be more targeted and effective.

The promise of the HGP (Fig. 6.1) goes far beyond medicine to many other areas of science. In addition to its many implications for medicine and human health which includes mutation detection, more accurate risk assessment, more precise disease diagnosis, more rapid characterization of genetic damage and repair processes, and the identification of precision pharmaceuticals based on intimate biological knowledge this international project is creating technologies and resources that will be applied to the characterization of the genomes of other living

Fig. 6.1 Implication of HGP is to reach the mankind for better health and cheaper diagnostic treatment

organisms. This information will, in turn, provide us with important new practical applications in energy, environmental protection, agriculture, and industrial processes.

6.4 The Human Genome and Genetic Variation

The genome contains the genetic information that is present in every cell in the body. The most important sequences in the genome (though only a small percentage of the total) are those that code for the production of proteins: the structural and functional units of all living cells. It is these sequences that correspond to the genes identified by classical genetics. While we all carry the same set of genes, their precise sequences may vary from individual to individual (that is, we carry different alleles), leading to variation in the structure or amounts of the associated proteins and thereby to variation in characteristics such as the colour of the hair or eyes (Risch et al., 2003). Of more importance when considering health and health care, these variations may also lead to differences in susceptibility to disease, or in response to a particular drug treatment.

In general, genetic variation has subtle effects on susceptibility to common diseases such as coronary heart disease or diabetes: many genes are involved, and they interact both with one another and with environmental and lifestyle factors to influence whether and when disease will develop. Some sets of these gene variants may cluster in a particular family, so that a family history of disease may be apparent. In rare cases, however, a change (or "mutation") in a single gene may have a drastic effect, largely over-riding any environmental effects, or effects of other genes, and making disease virtually certain. Examples of such highly "penetrant" single-gene diseases, which show clear inheritance patterns within families, include cystic fibrosis and Huntington's disease. There are also some rare subsets of common diseases such as breast and bowel cancer that are caused by mutations in single genes (Miller and Kumar, 2001).

6.5 Human Genome Project and Health Services

As detailed information emerges about the sequence of the DNA and the expression and properties of the resulting proteins we are building up a greater understanding of biological mechanisms and the ways in which they can go wrong in human disease. Increasingly, diseases may no longer be classified by clinical criteria, but on the basis of their underlying molecular pathology. Together with this new molecular understanding may arise new options for prevention and treatment.

Disease prevention relies on an understanding of the factors that underlie health and disease (Khoury, 2003). The old dichotomy between "genetic" and "environmental" disease has now been replaced by the realisation that both genetic and environmental factors are at work in almost all diseases. Thus, an important theme of health promotion will be to identify the modifiable risk factors for disease that interact with genetic variation and may be used to target preventive action. In terms of what we can currently do about prevention or promotion of public health, however, we can distinguish between the single gene disorders, where there are even now some clinical applications, and those diseases in which the genetic and environmental interaction is much more complex.

We now know that genetic predisposition has an effect in almost all diseases. Diseases with a known genetic component include most of the common diseases such as diabetes, stroke, coronary heart disease, cancers, dementia, depression, and many others – all major contributors to the morbidity and mortality of the population. However, the way in which genetic and environmental factors interact to cause disease is highly complex. Much work will be required to unravel these complexities and determine appropriate strategies for disease prevention. We need to understand the significance of these various factors in populations and sub-populations. The analysis in large populations of the interaction between genetic variation and environmental factors is the relatively new science of genetic epidemiology. Its study will require new techniques for DNA testing, new statistical techniques for analysing hundreds of thousands of people and the utmost care to protect individual privacy.

6.6 Single Gene Disease

With the completion of the human genome project, attention is now rapidly shifting towards the study of individual genetic variation. The most abundant source of genetic variation in the human genome is represented by single nucleotide polymorphisms (SNPs), which can account for heritable inter-individual differences in complex phenotypes. Identification of SNPs that contribute to susceptibility to common diseases will provide highly accurate diagnostic information that will facilitate early diagnosis, prevention, and treatment of human diseases (Suh and Vijg, 2005). Already the genetic cause has been identified for more than 200 single-gene diseases. Many of these are devastating conditions that kill their victims during childhood or early adulthood. For most of these diseases genetics has so far, in general, offered only the option of reproductive choice for affected families. The ability to test for conditions such as Duchenne muscular dystrophy either before birth or even pre-implantation has enabled some individuals to avoid the devastating consequences of severe childhood disability and death for their family.

Ultimately, families affected by single-gene diseases hope that information about the molecular defect involved will enable effective treatments to be developed. The "Holy Grail" for many families is gene therapy, whereby the defective gene in tissue(s) affected by the disease would be replaced by a normal copy. There has been one notable success of gene therapy, to treat a very rare genetic immunodeficiency syndrome, and there is some promising current research on gene therapy for haemophilia, but gene therapy cures for most single-gene diseases still seem to be some way off.

For late-onset single-gene diseases, there are additional considerations. Here, the emphasis falls on predictive testing for adults. If no treatment is available, as in the case of Huntington's disease with its dreadful neurological and psychiatric symptoms, individuals face grave personal dilemmas in considering whether they really want to know their own future.

In some other adult-onset single-gene conditions, there is more incentive for predictive testing. Rare single-gene subsets of two common diseases provide examples where there are real opportunities for prevention. It is important to try to identify individuals carrying these gene mutations because their risk of disease is higher and there are effective interventions to reduce their risk. Strategies for identifying these people usually rely on testing individuals in families that have a strong family history of the disease, or in which one family member is known to carry a causative mutation.

Familial hypercholesterolemia (FH), which is one of the most common single gene disorders in the population, is associated with premature development of arteriosclerosis and early onset of coronary heart disease. However, there is now great potential to prevent or even reverse the arteriosclerosis in patients with this condition due to advances in drug therapy, particularly the development of HMG CoA reductase inhibitors (statins).

Hereditary Non-polyposis Colorectal Cancer (HNPCC) is a rare single-gene subset of colorectal cancer that is caused by a mutation in any one of several different genes that code for DNA repair enzymes. People who have inherited a mutation may have up to an 80% lifetime risk of cancer. Once identified, they can be offered a programme of bowel surveillance by colonoscopy, which can find and remove early growths and almost completely eradicate this risk. The advances in our knowledge of molecular biology and the Human Genome Project have led to greater understandings of genes and gene alterations. These exciting discoveries have led to new ways of thinking about cancers, tumor growth, and how genetic technology may be harnessed to both understand diagnose and alter the course of cancer therapy. Selectively targeting the molecular/genetic abnormalities which give rise to cancers may ultimately result in less toxic treatment approaches which may affect a variety of issues related to quality of life.

6.7 New and Individualised Drug Treatments

Some of the most immediate ways in which the new genetics resulting from the Human Genome Project will make an impact are through developments in drug treatment. Many of the current problems in drug treatment stem from the variable response of patients. This affects efficacy (whether the drug works) and safety (whether or not it produces adverse effects). These differences depend on individual variation in the way in which drugs are absorbed and handled within the body as well as different reactions at target sites. This field of work is known as pharmacogenetics. It is expected that within the next few years there will be many individually tailored prescribing guidelines (Phillips et al., 2001).

Drug design will also be revolutionized as researchers create new classes of drugs based on a reasoned approach to data from gene sequencing and protein structure rather than by the traditional trial and error method. Greater knowledge of the human genome will eventually lead to increasing possibilities to predict future health. However, some of the personal and societal consequences are uncertain. We do not know how people might behave in the face of more precise information about their individual risk. Some may not want the information. Some people may react to the knowledge that they are susceptible to a particular disease by taking every precaution, whilst others may regard disease as pre-ordained and feel that nothing they do will make any difference.

The question of discrimination may also arise. People worry that their genetic information may be made available to others, their insurer or employer, for example. The arguments about this issue are very complex, but it is important that rational policies are in place to avoid the possibility of discrimination and the danger of creating a "genetic underclass". These are examples of the ethical, legal and social issues which arise in associations with the new genetics and which are being explored in detail by such bodies as the Human Genetics Commission.

6.8 The Achievements and Challenges of the HGP

In 1995, for example, the highest resolution physical maps for human chromosomes were completed by the Los Alamos National Laboratory and the Lawrence Livermore National Laboratory for chromosomes 16 and 19, respectively. The chromosome 19 map has already contributed to the characterization of the genetic defect underlying the disease of myotonic dystrophy and to the description of the unusual genetic mechanism of aberrant triplet repeats that is now known to contribute to the onset of at least nine diseases, including Huntington's disease. Genes mapped to chromosome 16 include those involved in Batten's disease, polycystic kidney disease, Crohn's disease, forms of breast and prostate cancer (Miki et al., 1994), and Fanconi's anemia, as well as many others. In addition, the DNA repair genes HHR23A, XRCC1, and ERCC2, as well as genes involved in olfactory receptors, Alzheimer's disease, and one form of migraine headache, have been discovered on chromosome 19.

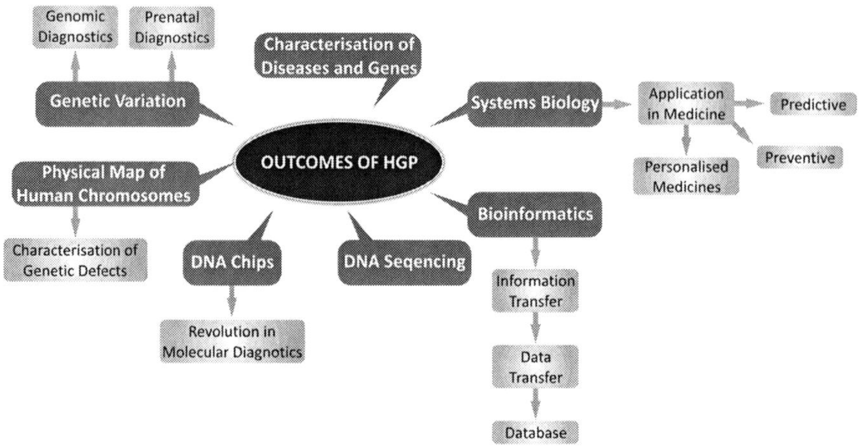

Fig. 6.2 The outcomes of the Human genome project allows us to unravel some of the mysteries of life

Lee Hood at the University of Washington in Seattle has sequenced nearly 700,000 consecutive base pairs of DNA from the human T cell receptor complex, as well as a comparable amount of mouse T cell receptor DNA. These sequences have provided surprising insights into the evolution and function of certain white blood cells important for many immune responses to invading pathogens and are widely thought to be involved in both autoimmune diseases and protection against early tumor development. The Human Genome Project has catalyzed the emergence of a new approach to biology termed systems biology (Fig. 6.2). Systems biology analyzes all the interrelationships of the elements in a biological system, rather than studying them one at a time, as has been the modus operandi in biology for the past 30 years. This systems approach has also emerged in the context of the view biology is an informational science and the development of high-throughput tools for capturing biological information and powerful new computational tools for analyzing it. The application of systems approaches to medicine will lead to the rise of predictive, preventive, and personalized medicine over the next 15–20 years, giving rise to a total transformation in how medicine is practiced.

Additional accomplishments include advances in technologies that are speeding up DNA sequencing, among them the development of novel "vectors" (critical for the manipulation of DNA in fragment sizes that can readily be characterized and studied), particularly Bacterial Artificial Chromosomes or BACs. A DOE-funded BAC library contributed to the discovery of the Breast Cancer-1 gene by supplying the particular BAC containing the appropriate DNA fragment from chromosome 17.

6.9 Information Transfer

To take full advantage of the wealth of information generated by the Human Genome Project, epidemiologists and genetic epidemiologists must have the ability to easily access the data, as well as a working knowledge of the retrieval process to properly query, analyze, and interpret the desired information. A primary component of the genome initiative is the development of comprehensive computer databases to assimilate the tremendous amount of mapping and DNA sequence data and to provide links to the scientific and medical literature. Numerous databases have been established to provide organized storage and efficient dissemination of the genome mapping and sequencing data. The informatics movement has generated global computer networks with on-line access over the internet that permit remote access and retrieval of raw or computed data. The seemingly unlimited potential of this technology is readily available to epidemiologic researchers who may be unfamiliar with genetics or genome informatics through simplified accession programs and database helplines. These databases are invaluable to epidemiologic research and should be familiar to all genetic epidemiologists because they contain vast amounts of information regarding the genes and molecular defects that contribute to human disease, methods for rapid detection of mutations and polymorphisms (if available), comprehensive descriptions of disease phenotypes, and the status of treatment and intervention strategies.

6.10 Impact of the Human Genome Project on Epidemiologic Research

The discipline of genetic epidemiology has greatly expanded the applicability and utility to the public of genetic advancements, including the mapping of genes responsible for Mendelian diseases (such as cystic fibrosis) and the development of models to predict disease (such as the multiple-step mechanism carcinogenesis). The greatest challenges confronting genetic epidemiology, however, are the common chronic diseases with late age-of-onset which exert a tremendous burden on public health as measured by morbidity, mortality, and cost. The greatest impact and benefit to public health from genetic epidemiologic research will likely come from uncovering and better understanding the genetic etiology of the common chronic diseases (such as coronary artery disease and diabetes) and the common forms of cancer (such as breast and colon cancer). Until recently, genetic epidemiology made inferences primarily from statistical analyses of the distribution of disease or other traits among family members. Direct measures of genetic information were rare and, with the exception of the human leukocyte antigen (HLA) complex, were limited to red cell antigens and polymorphic red cell and plasma enzymes.

For the field of genetic epidemiology to achieve its full potential and better characterize the genetic etiology of the common chronic diseases, high quality genetic markers were necessary for gene mapping, and improved methods needed to be developed to detect and quantify functional alleles (Goldstein et al., 1973). The Human Genome Project and other developments in molecular biology are providing the necessary tools for epidemiology and genetic epidemiology to uncover the molecular mechanisms for variation in the distribution of disease among families and populations. Although there have been other conceptual and technical advances (particularly in the area of computational methods) that have advanced the field of genetic epidemiology.

6.11 Impact of the Human Genome Project on Our Genomic Makeup

The Human Genome Project has already fuelled the discovery of more than 1,800 disease genes (Venter et al., 2001). As a result of the Human Genome Project, today's researchers can find a gene suspected of causing an inherited disease in a matter of days, rather than the years it took before the genome sequence was in hand. There are now more than 1,000 genetic tests for human conditions. These tests enable patients to learn their genetic risks for disease and also help healthcare professionals diagnose disease.

The Human Genome Project (HGP) has brought unprecedented insight into our genomic makeup (Venter et al., 2001). Genomic diagnostics makes practical use of the wealth of HGP data by determining correlations between clinical phenotypes and their underlying genotypes, enabling us to give etiological faces to diseases for which none existed before. An area where genomic diagnostics is anticipated to have great impact is prenatal testing. Because so much remains unknown about what causes spontaneous abortions or the development of idiopathic mental retardation, the medical community is eager to apply the latest in comprehensive genomic diagnostics to the prenatal setting.

However, within medicine, few concerns generate as much anxiety in those it affects – namely, expecting parents – as prenatal diagnostics. This anxiety is justified considering full understanding of the clinical relevance of the HGP data often lags behind the technologies that utilize that information. With complete appreciation for every genotype/phenotype consequence yet to be attained, care should be taken when using this information in the prenatal setting. Additionally, sensitivities from sociological, psychological and ethno-religious perspectives further demands that prenatal genomic analysis be approached with exceptional caution and responsibility.

A primary technology that has parlayed the information from the HGP into clinical utility is that of array-based comparative genomic hybridization (aCGH). Because this method makes use of the content of the human genome to screen a clinical sample for genomic copy number changes and hence many disease associations simultaneously in a patient, aCGH has become the most anticipated test in prenatal genetics. Combimatrix Molecular Diagnostics (CMDX) is a pioneer in the conversion of the HGP data into clinical diagnostics. CMDX is leading the vanguard of prenatal testing by offering cutting-edge genomic diagnostics while cognizant of its associated sensitivities and issues, ensuring medical practitioners receive the very best, responsible, and appropriate genomics based prenatal diagnostic testing.

And each step of the way, just like in the HGP, new techniques are developed that are applicable to clinical medicine. Our knowledge of human genetics for identifying new genes and understanding how they work increases in unpredictable ways.

This promises to expand the number and types of clinical laboratory testing that are offered to patients.

Rapid discoveries of novel genes for a variety of human diseases are anticipated as genomic maps are- come more detailed and methods for mapping and characterizing disease genes become more refined (Sachidanandam et al., 2001). Understanding the genetics behind ocular disorders opens a number of opportunities to address disease management, not least of which is the development of molecular diagnostics.

Molecular diagnostics is being revolutionized by the completion of the human genome project and by the development of highly advanced technologies for DNA testing (Ferrari et al., 2003). One of the most important challenges is the introduction of high throughput systems such as DNA chips into diagnostic laboratories. DNA microchips are small devices permitting rapid analysis of genetic information, exploiting miniaturization of all components and automation of operational procedures. The most important biochip applications include gene expression and genetic variation identification and both may improve human molecular diagnostics.

Recent and continuing developments in genome technology and analytic methods provide the tools and raw materials for unravelling the complexities of the common chronic diseases and common forms of cancer.

One major step toward such comprehensive understanding was the development in 2005 of the HapMap, which is a catalogue of common genetic variation, or haplotypes, in the human genome. HapMap data have accelerated the search for genes involved in common human diseases, and have already yielded impressive results in finding genetic factors involved in conditions ranging from age-related blindness to obesity.

6.12 Conclusion

In the future, it is clear that genetic discoveries, including the vast amount of information being made available by the HGP, will lead to opportunities for treatment. The progress in treatment will tend to reduce the dilemmas surrounding genetic testing. In some cases, the neurologist will still be left with the dilemma of a disease that can be accurately diagnosed but for which no effective therapy is yet available. However, the vast array of biological information made available by the genome project will also allow widespread approaches of emerging technologies such as gene therapy, in which defective gene sequences may be replaced. In the case of other disorders, the knowledge gained by genetic dissection may lead to new, more effective drug therapies or immediate approaches to improving health through altering lifestyle (Evans, 1998). Genome medicine in the next century is likely to take on an entirely new role in society based on the beginnings inherent in the HGP (Nathans and Orkins, 2009).

The human genome project was borne of technology, grew into a science bureaucracy in the U.S. and throughout the world and is now being transformed into a hybrid academic and commercial enterprise. The next phase of the project promises to veer more sharply toward commercial application, exploiting the rapidly growing body of knowledge about DNA structure to the pursuit of practical benefits. This Project has catalyzed the emergence of a new approach to biology termed systems biology. The application of systems approaches to medicine will lead to the rise of predictive, preventive, and personalized medicine over the next 15–20 years, giving rise to a total transformation in how medicine is practiced.

References

Collins, F.S., Patrinos, A., 1998, New goals for the U.S. Human Genome Project: 1998–2003 *Science* 282(5389):682–689.

Evans, G.A., 1998, The Human Genome Project: applications in the diagnosis and treatment of neurologic disease. *Arch Neurol* 55(10):1287–1290.

Ferrari, M., Stenirri, S., Bonini, P., Cremonesi, L., 2003, Molecular diagnostics by microelectronic microchips. *Clin Chem Lab Med* 41(4):462–467.

Goldstein, J.L., Schrott, H.G., Hazzard, W.R., Bierman, E.L., Motulsky, A.G., 1973, Hyperlipidemia in coronary heart disease II. Genetic analysis of lipid levels in 176 families and delineation of a new inherited disorder, combined hyperlipidemia. *Clin Invest* 52(7):1544–1568.

Khoury, M.J., 2003, Genetics and genomics in practice: the continuum from genetic disease to genetic information in health and disease. *Gen Med* 5(4):261–268.

Miki, Y., Swensen, J., Shattuck-Eidens, D., Futreal, P.A., Harshman, K., Tavtigian, S., Liu, Q., Cochran, C., Bennett, L.M., Ding, W., et al., 1994, A strong candidate for the breast and ovarian cancer susceptibility gene, BRCA1. *Science* 266(5182):66–71.

Miller, M.P., Kumar, S., 2001, Understanding human disease mutations through the use of interspecific genetic variation. *Hum Mol Genet* 10(21):2319–2328.

Nathans, D.G., Orkins, S.H., 2009, Musing on genome medicine: genome wide association studies. *Genome Med* 1(1):3

Peltonen, L., 2001, The molecular dissection of human diseases after the human genome project. *Pharmacogenome J* 1(1):5–6.

Phillips, K.A., Veenstra, D.L., Oren, E., Lee, J.K., Sadee, W., 2001, Potential role of pharmacogenomics in reducing adverse drug reactions: a systematic review. *JAMA* 286(18):2270–2279.

Risch, N., Burchard, E., Ziv, E., Tang, H., 2003, Categorization of humans in biomedical research: genes, race, and disease. *Genome Biol* 3(7):1–12.

Sachidanandam, R., Weissman, D., Schmidt, S.C., Kakol, J.M., Stein, L.D., Marth, G., Sherry, S., Mullikin, J.C., Mortimore, B.J., Willey, D.L., 2001, A map of the human genome sequence variation containing 1.42 million single nucleotide polymorphisms. *Nature* 409(6822): 928–933.

Suh, Y., Vijg J., 2005, SNP discovery in associating genetic variation with human disease phenotypes. *Mutat Res* 573(1–2):41–53.

Van Ommen, G.-J.B., 2002, The Human Genome Project and the future of diagnostics, treatment and prevention. *J Inher Metab Dis* 25(3):183–188.

Venter, J.C., Adams, M.D., Myers, E.W., Li, P.W., Mural, R.J., Sutton, G.G., Smith, H.O., Yandell, M., Evans, C.A., Holt, R.A., et al., 2001, The sequence of the human genome. *Science* 291(5507):1304–1351.

Chapter 7
Molecular Diagnosis in the Post Genomic and Proteomic Era

Abstract The completion of the human genome project is widely regarded as a turning point in biology and medicine. As we usher in the "post-genomic" era, we are faced with an explosion of new information, which is leading to dramatic changes in the way we are able to study and manipulate life. Over the last few years, bimolecular research has progressed to functional genomics and the application of this knowledge to advance our understanding of health and disease. It is clear that genomic information alone is not sufficient to completely explain disease states. Post-genomic approaches attempt to contribute to our understanding of this interaction, with each approach capturing a different angle of the global picture. Intuitively, the next step forward is to integrate these datasets, an approach that, if successful, could be much more informative and predictive than working exclusively on a single platform.

Keywords Genomics · Transcriptomics · Proteomics · HGP · Biomedical research · Drug discovery · Sequencing · Functional genomics · Microarrays · Environmentome · Multigenetic diseases · Candidate markers · Personalized medicine

7.1 Prologue

The year 2003 marked the completion of the first draft of the human genome sequence. We now find ourselves in the modern genetic era, trying to understand and utilize a previously unimaginable amount of biological information. Scientists have the daunting task of finding new ways to use the human genome sequence to improve drug discovery and development efforts and to find better ways to combat disease for the benefit of humanity (Peltonen and McKusick, 2001).

7.2 The Genomic Era

The genomic era has brought with it a basic change in experimentation, enabling researchers to look more comprehensively at biological systems. The sequencing of the human genome coupled with advances in automation and parallelization

M. Debnath et al., *Molecular Diagnostics: Promises and Possibilities*,
DOI 10.1007/978-90-481-3261-4_7,

technologies have afforded a fundamental transformation in the drug target discovery paradigm, towards systematic whole genome and proteome analyses. In conjunction with novel proteomic techniques, genome-wide annotation of function in cellular models is possible. Overlaying data derived from whole genome sequence, expression and functional analysis will facilitate the identification of causal genes in disease and significantly streamline the target validation process. Moreover, several parallel technological advances in small molecule screening have resulted in the development of expeditious and powerful platforms for elucidating inhibitors of protein or pathway function. Conversely, high-throughput and automated systems are currently being used to identify targets of orphan small molecules. The consolidation of these emerging functional genomics and drug discovery technologies promises to reap the fruits of the genomic revolution.

In 1990, the human genome project was established to sequence the human genome (Collins et al., 2003), with the aim of applying the acquired genomic data to improve disease diagnosis and determine genetic susceptibility. The publication of the first draft sequence of the human genome in 2001(Venter et al., 2001) was thus followed by a rapid growth of different approaches to extract useful information from the genomic sequence. Following the publication of the complete human genomic sequence, the post-genomic era is driven by the need to extract useful information from genomic data (Tang et al., 2009). What is now to be done is to work out the function of the genes. Hence, the post-genome era has been called *functional genomics*, which includes *proteomics*, the technology and strategies required to determine the function of proteins. How this will be accomplished remains to be determined, but new technologies will be needed.

The use of microarrays (a method by which the expression of many thousands of genes can be identified very rapidly with microchips) is an early, promising strategy in functional genomics. Bioinformatics will need to come up with more sophisticated programs by which the function of genes can be predicted. The traditional "wet-lab" approach to research might even give way to a complete "in-silico" (i.e., computer) strategy! The first challenge – sequencing the genome – has been accomplished. Many more even larger challenges await us as we set out to determine the function of all human genes.

The next decade will see the study of the transcriptome representing all of the transcripts or RNA copies of the genes in a cell; proteome, representing all of the proteins in a cell, tissue or individual; and metabolome, representing all of the molecular components of a cell or tissue produced by the proteins (Venter et al., 2008).

7.3 Concept

The molecular diagnostics industry is experiencing explosive growth. Innovation and strategic planning is needed to realize the commercial potential from applied genomics. The number of product offerings for molecular profiling that allow the early detection, prognosis, and treatment of disease are increasing substantially. But the advance of companion diagnostics has been slow to take hold.

Molecular diagnostics is here to stay. Not only are molecular diagnostic techniques being developed for new tests as new disease genes are being discovered, but molecular diagnostic techniques also are being developed that replace traditional tests, especially in the areas of bacteriology and virology. Molecular diagnostics may provide several clinical advantages including the production of a clear unequivocal test, the provision of critical information to inform genetic counseling, the establishment of prenatal screening services and the identification of individuals at a pre-symptomatic stage so that environmental factors, if relevant, may be adjusted. The ultimate goal, of course, would be to design therapeutics based on the genetic information.

Perhaps the greatest utility of the vast genetic information being generated by the human genome initiative is in primary prevention programs. Methods for identifying asymptomatic individuals at risk for genetic disease and the development of more efficacious intervention strategies are becoming paramount as health care costs escalate and medical genetics shifts to early detection and prevention of disease. With these abilities will come the need to fully integrate genetic information into large prospective studies, and intervention trials to accurately predict disease risk and synthesize new approaches to risk reduction. Understanding the role of genes in human disease will improve our understanding of genetic disease etiology as well as our ability to predict disease. Insight into the genetic basis of chronic disease etiology will have immediate impact by suggesting novel therapeutic approaches and aiding new drug discovery. At least 350 biotechnology-based products resulting from the Human Genome Project are currently in clinical trials.

Having the complete sequence of the human genome is similar to having all the pages of a manual needed to make the human body. The challenge now is to determine how to read the contents of these pages and understand how all of these many, complex parts work together in human health and disease. The wait is over, and the post genomic era has arrived. Genomics, transcriptomics, proteomics, metabolomics, epidemiological data and microbial data provide different angles to our understanding of gene-environment interactions and the determinants of disease and health. Our goal and our challenge are to integrate these very different types of data and perspectives of disease into a global model suitable for dissecting the mechanisms of disease and for predicting novel therapeutic strategies. We are in for a veritable revolution, the diagnosis of infections, development of vaccines, new treatments for old bacteria, predictive medicine, and innovative therapies. A truly amazing future awaits us (Fig. 7.1). What is now required is that more training in genetics be imparted and more genetic centers be opened so that the tremendous benefits of this technology are utilized for the benefit of many patients in India now and in future.

A similar transition has occurred in infectious disease following the publication of complete genome sequence for *Mycobacterium tuberculosis* (Young, 2001), *Vibrio cholera*, and at least 30 other microbial species. Genetic material isolated from pathogenic organisms in ancient bone tissue or herbarium samples collected from 1845 through 1847 during the Irish potato famine has been used to characterize the infectious agent using polymerase chain reaction or sequencing of ribosomal

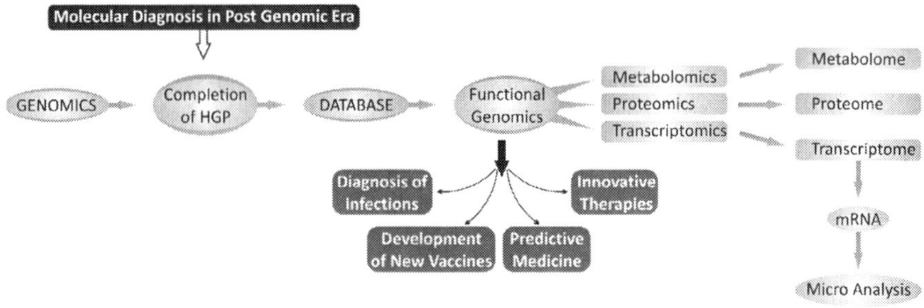

Fig. 7.1 The challenges of the post genomic era

DNA and mitochondrial DNA. It will now be possible to integrate bacterial post genomic data with human post genomic data for historical and current investigation into the pathogenesis of infectious disease, antibiotic resistance, and epidemiology (Raczniak et al., 2001).

Each of the post-genomic approaches has already contributed to our understanding of specific aspects of the disease process and the development of diagnostic/prognostic clinical applications (Scuturi et al., 2007; The wellcome trust case control consortium, 2007). Cardiovascular disease (Edward et al., 2008 ; Giovane et al., 2008), obesity (Blakemore and Froguel, 2008; Loos et al., 2008; Pietilainen et al., 2007), diabetes (Chen and Hess, 2008; Bougneres and Valleron, 2008; Orešic et al., 2008; Frayling, 2007), autoimmune disease (Su, 2008) and neurodegenerative disorders (Quintana et al., 2008; Papassotiropoulos et al., 2005) are some of the disease areas that have benefited from these types of data. Taking the metabolic syndrome as an example, our knowledge on all aspects of the disease has grown. The metabolic syndrome is the result of a complex bioenergetics problem characterized by disturbances in lipid, carbohydrate and energy metabolisms and blood pressure. In combination, these metabolic factors contribute to an increased susceptibility to cardiovascular disease, morbidity and mortality (Cornier et al., 2008). Genome-wide association (GWA) studies have identified possible genes involved in each aspect of the syndrome: namely type 2 diabetes (Frayling, 2007), obesity (Lindgren and McCarthy, 2008) and hyperlipidaemia (Hegele, 2009). The findings have confirmed the role of certain candidate genes as well as the polygenetic nature of the syndrome. Not surprisingly, replicate GWA studies of type 2 diabetes revealed that the genes associated with disease, among others, are involved in beta-cell function and adipocyte biology (Frayling, 2007; Lindgren and McCarthy, 2008; Perry and Frayling, 2008). In contrast, genes found to be associated with obesity appear to be those that are predominantly involved in central appetite regulation (Li and Loos, 2008) as key contributors to positive energy balance.

Genetic association studies in epidemiology have highlighted a number of issues. Firstly, many common disease states are related to either many genetic polymorphisms of small effect or, in selected cases, to a few of large effect. The involvement of multiple genes with unequal contributions to disease hints of complex gene-gene and gene-environment interactions. The understanding of such interactions becomes a daunting task when other modulating factors remain unknown. Secondly, some common diseases such as type 2 diabetes appear to be relatively less genetically determined compared to diseases such as rheumatoid arthritis and obesity (Maes et al., 1997). In these situations, our understanding of pathophysiology requires additional data outside of genomic information. Thirdly, the initial failures to find robust replicable associations between most of the identified genetic variants and common complex diseases suggest that genomic analysis alone will not account for all of the heritability and phenotypic variation (Maher, 2008). For this reason, there is a growing need to incorporate information derived from environmental studies and post-genomic data into genetic analysis.

7.4 The Post Genomic Era

In the post-genomic era of biology, focus is slowly shifting from the raw sequence data generated from these large multinational projects, to downstream consequence of that sequence, such as gene function, and its relationship to phenotype. As such, the "omics" technologies are enjoying a boom. Transciptomics, the study of gene expression, proteomics, the study of protein expression, and metabolomics, and the snap-shot observation of metabolic status are of great current interest. The new science of proteomics promises exciting insights into the working of the cell. Such technologies offer a powerful insight into molecular mechanisms, and have great utility in functional genomics and diagnostics, to name but two applications. A common technological approach for the studies of proteomics and metabolomics is mass spectrometry, which in combination with chromatography can detect, identify and quantify large numbers of metabolites from a single sample.

One of the greatest challenges in modern medicine is to dissect the cascade of molecular events that lead to the development and progression of tumours. As we emerge into the post-genome era, proteomics finds itself as the driving force field as we translate the nucleic acid information archive into understanding how the cell actually works and how disease processes operate.

The genome-wide study of proteins embodies the exciting promise of proteomics (Mocellin et al., 2004). Even so, the traditionally held view of proteomics as simply cataloging and developing lists of the cellular protein repertoire of a cell are now changing, especially in the sub-discipline of clinical proteomics. The most relevant information archive to clinical applications and drug development involves the elucidation of the information flow of the cell; the "software" of protein pathway networks and circuitry. The deranged circuitry of the cell as the drug target itself as well as the effect of the drug on not just the target, but also the entire network, is what we now are striving towards. Clinical proteomics, as a new and most exciting sub-discipline of proteomics, involves the bench-to-bedside clinical application of proteomic tools. Unlike genomics, proteomics aims ambitiously to study not only protein expression profiles but also protein functions, which should provide researchers with a more comprehensive view of the molecular machinery that governs cellular functions (Chanda and Caldwell, 2003).

Unlike the genome, there are potentially thousands of proteomes: each cell type has its own unique proteome. Moreover, each cell type can alter its proteome depending on the unique tissue microenvironment in which it resides, giving rise to multiple permutations of a single proteome. Since there is no polymerase chain reaction equivalent to proteomics, identifying and discovering the "wiring diagram" of a human diseased cell in a biopsy specimen remains a daunting challenge. New micro-proteomic technologies are being and still need to be developed to drill down into the proteomes of clinically relevant material. Cancer, as a model disease, provides a fertile environment to study the application of proteomics at the bedside. The promise of clinical proteomics and the new technologies that are developed is that we will detect cancer earlier through discovery of biomarkers, we will discover the next generation of targets and imaging biomarkers, and we can then apply this knowledge to patient-tailored therapy (Krieg and Paweletz, 2002). Cancer proteomics is an exciting field that is witnessing many new developments in recent years. It is hoped that these advances will result in decreased cancer death rates, which have not declined dramatically in the last several decades. Some of the problems with current tumor markers include the lack of sensitivity and specificity, factors that prevent their use in population-based screening of disease. Thus, there is an urgent need to identify novel biomarkers that can faithfully detect the disease state. As we are now in the post-genome era, many opportunities have been created. Genomic sequence data are available for human, as well as several other species. We are now poised to mine these data and to determine the functions of the encoded proteins constituting the human genome. Proteomics affords this opportunity by providing enhanced procedures and tools for discovery and also a framework for understanding these components in terms of pathogenesis. New technologies and improvements in existing methodologies will allow for the rapid growth in the identification and characterization of peptides and proteins that are unique to various clinical states. This technology can be successfully applied to clinical specimens for the identification of disease state.

7.5 Advantages of Combining Multiple Types of Data

It is clear that the genetic approach captures only one layer of the complexity inherent within human biology. There is thus a need to integrate multiple "omics" datasets when aiming to unravel the molecular networks underlying common human disease traits (Zhu et al., 2007). Attempts have been made to combine two datasets in relation to the clinical phenotype, and this is reflected in the combination of terms found in the literature, for example metagenomics, pharmacogenomics and epigenetics. Many of the post-genomic approaches linking the genetic association data with other "omics" layers focus on the use of "omics"- derived phenotypic data as quantitative traits. The utility of such approaches has been previously applied, by combining genetics and metabolomics, in plant functional genomics (Raamsdonk et al., 2001). More recently, such approaches have also been applied to human datasets. For example, Papassotiropoulos and colleagues (2005) identified clusters of cholesterol-associated susceptibility genes for Alzheimer's disease by combining genetics with sterol profiling, while Gieger and colleagues (2008) used ratios of metabolites to identify the function of putative genes. In another study, proteomics was linked to quantitative trait loci (QTL) in an attempt to identify changes in function rather than quantity of the protein (Stylianou et al., 2008).

By combining multiple types of techniques, including genetics, transcriptomics, proteomics and metabolomics, we are expecting a shift toward "environmentome" research, where all available information from preconception to disease onset, using both longitudinal and cross-sectional experimental designs, can be obtained (Bougneres and Valleron, 2008). The measurement of traits that are modulated but not encoded by the DNA sequence, commonly referred to as intermediate phenotypes, is of particular interest. These intermediate phenotypes include not only biochemical (metabolites) and genomic (gene expression) traits, but also an individual's microbial (gut microflora) (Turnbaugh et al., 2007; Turnbaugh and Gordon, 2008) and social traits. It is conceivable that by comprehensively examining an individual's "environmentome", we would be able not only to understand both the genetic and environmental determinants of disease, but also to develop "feasible" personalized medicine, that is, tailor specific personalized interventions to the individual's own environmental profile. As a pioneering example of this kind, Orešic Land and colleagues (2008) investigated metabolic profiles of children between birth and type 1 diabetes onset in a large birth cohort, and established that specific metabolic phenotypes, not dependent on human leukocyte antigen (HLA)-associated genetic risk, precede the first autoimmune response. The excitement of this research is the expectation that these early metabolic phenotypes may be validated as specific diagnostic and prognostic markers of disease, with therapeutic implications.

The goal of inferring disease causality and disease mechanisms from integrated data is complicated by the fact that measuring more variables may provide a better characterization of the process but still does not contribute directly to our understanding of cause and effect. In fact, given the progressively increasing number of variables that we can measure, the odds of finding spurious associations that do not reflect true causality are much higher. Confounding and reverse causality are among the main sources of bias for failures to replicate apparently robust associations between risk factors and diseases (Lawlor et al., 2006). Confounding specifically refers to a spurious causal effect inferred from the association between a risk factor and a disease due to the existence of some common causes, that is, confounding factors to both of them. This type of spurious causal effect can be removed if we have enough knowledge about the most likely confounding factor candidates. However, the truth is that for most epidemiological studies confounding factors are unknown and difficult to measure, especially in case-control studies. Reverse causality, the second source of bias, refers to an alternative explanation for the observed association between a risk factor and disease, which states that the "risk factor" is a result of the disease, rather than vice versa. The problem of reverse causality is particularly prevalent in retrospective case-control studies.

One example of a potential confounding association is the established epidemiological evidence of a strong link between obesity and insulin resistance. This association has recently been brought into question from the identification of specific clinical settings where fat mass dissociates from insulin resistance (Garg, 2004; Wildman et al., 2008). This implies that adipose tissue expansion typically associated with obesity per se may not be the cause of metabolic complications. A potential alternative explanation may be related to an individual's ability to optimally store fat. In the presence of caloric excess, a person is likely to remain metabolically healthy despite obesity, provided their adipose tissue can continue to expand and safely store fat. Therefore, while the epidemiological evidence associates the risk of metabolic complication with increased body weight, this relationship may not be direct and may not necessarily reflect a truly biologically relevant process.

7.6 Biomedical Research in the Postgenomic Era

The genomics era has provided great opportunities for deciphering the genes that are mutated in human diseases. Identification of a causative pathogen is essential for the choice of treatment for most infectious diseases. Many FDA approved molecular assays; usually more sensitive and specific compared to traditional tests, have been developed in the last decade. A new trend of high throughput and multiplexing assays are emerging thanks to technological developments for the human genome sequencing project. The enthusiasm for the introduction of genetic screening for single-gene or multigenetic diseases is great in the medical community. However, in general, a reasonable number of individuals need to be screened to prevent one case of disease. Prevention is achieved only by screening for highly penetrant mutations in high-risk families and not for such mutations in the general population or for low-penetrant polymorphisms. The frequency with which the phenotype the gene controls is seen in individuals who carry it describes the penetrance of the gene. The differences observed in the penetrance of a mutation depend on 6 factors (Vineis et al., 2001) role of the protein encoded in metabolism, functional consequence of the mutation, gene-gene interactions, onset of somatic mutations, environmental interactions, and the presence of alternative metabolic pathways that can correct the loss of function caused by the mutation. Proteomics technologies are now offering unique chances to identify new candidate markers. In the diagnostic and the pharmaceutical industry there is a constant need for new diagnostic markers and biomarkers with improved sensitivity and specificity. During the last 5 years, only a few novel diagnostic markers have been introduced into the market. Before a marker can be introduced into the market, three successive developmental phases have to be completed: the discovery phase, in which a variety of proteomics technologies are applied to identify marker candidates; the prototype developmental phase, in which immunological assays are established and validated in defined sample collectives; and finally the product development phase, with assay formats suitable for automated platforms. The hurdles that a potential candidate marker has to pass in each developmental phase before reaching the market are considerable. The costs are increasing from phase to phase, and in industry a number of questions concerning the medical need and the potential return on investment have to be answered before a proteomics discovery project is started.

As the biomedical research community enters the post-genome era, studying gene expression patterns and phenotypes in model organisms will be an important part of analyzing the role of genes in human health and disease. New technologies involving DNA chips will improve the ability to evaluate the differential expression of a large number of genes simultaneously. Also, new approaches for generating mutations in mice will significantly decrease the cost and increase the rate of generating mutant lines that model human disease (Woychik et al., 1998).

Personalized medicine will provide the link between an individual's molecular and clinical profiles, allowing physicians to make the right patient-care decisions and allowing patients the opportunity to make informed and directed lifestyle decisions for their future well-being. Clinical diseases as we know them will be replaced by molecular classification. Therapies directed at the root cause of disease will replace those that simply treat the symptoms of disease. Finally, a pharmacogenomic test that predicts therapy response based on a patient's genomic profile will accompany many drugs. Personalized medicine will involve radical changes in the pharmaceutical industry and medical practice and is likely to affect many aspects of society. Most importantly, the individual whose health is at stake will benefit enormously.

The advent of disease gene patents threatens to constrain the medical community's ability to provide the appropriate standard of care (Balter, 2001). Such a patent has the potential of monopolizing medical testing services and potentially leading to restrictions on research, conflicts of interest, and restricted access to testing services (Merz, 1999).

In the clinical laboratory, DNA or RNA represents the most common starting material for molecular-based assays. Research in the post genomic era will focus on identifying the critical messenger RNAs (mRNAs) (transcriptomics) or proteins (proteomics) and the sequence of interactions required to induce a specific biochemical event. Microarrays or "lab-on-a-chip" may contain a library of genes associated with a specific metabolic pathway that are immobilized in a grid (Lockhart and Winzeler, 2000). The post genomic era will present numerous technological and intellectual challenges. The post genomic era will challenge our traditional problem-solving methods and require more creative, efficient, and accurate techniques for characterizing the clusters of nucleotides within genes and associating their mutation with phenotype and disease treatment.

Mitochondria play a central role in the intracellular signal transduction pathway, leading to apoptosis or programmed cell death (Gottlieb, 2000). This genetically controlled process regulates tissue size in opposition to mitosis. It is characterized by cell shrinkage, membrane blebbing, chromatin condensation, DNA fragmentation into 180– to 200–base pair units and multiples of these units, and phosphatidylserine translocation and activation of the apoptotic signal transduction pathway. This pathway may represent the final common pathway for cell death in neurodegenerative disease (Offen et al., 2000), which is included in diseases characterized by increased apoptotic activity.

The treatment of human disease by delivering genes to human tissues is called gene therapy. Applications in cardiovascular disease include ischemia, restenosis, graft failures, arterial cytoprotection, atherogenesis, and thrombogenesis (O'Brien and Simari, 2000). However, efficacy and toxicity have slowed progress in this field. Complications have included inefficient gene expression and adverse inflammatory reactions. However, encouraging results have been reported in ischemia by the introduction of vascular endothelial growth factor or fibroblast growth factor genes. When these genes are placed in cardiac tissue adjacent to an occluded coronary artery, they stimulate new blood vessel growth (angiogenesis). A different vector was used by Roth et al., 2001 to deliver the factor VIII gene to patients with hemophilia A. They transfected skin fibroblasts from the patient with a plasmid containing factor VIII and injected these cells. Although the results were encouraging, not all patients responded with reduced bleeding and increase in plasma factor VIII activity. Resolution of the problems related to vectors for gene transfer will accelerate investigation in this area. Moldovan and Ferrari describe a silicon-based angiogenesis-assisting device or "angiochip" to release multiple growth factors and other biochemicals to support the growth of endothelial cells grown on the surface. This device would initiate new blood vessel growth without the need for gene therapy.

7.7 Conclusion

The post-genome era has begun, and with it the promise of tailoring the practice of medicine to the individual. This emerging field of personalized medicine encompasses the use of risk algorithms, molecular diagnostics, targeted therapies and pharmacogenomics to improve health care. From SNP maps to individual drug response profiling, the human genome sequence will lead to improved diagnostic testing for disease susceptibility genes and individually tailored treatment regimens for those who have already developed disease symptoms. Molecular diagnostics, the use of DNA-, protein- or mRNA-based biological markers to predict the risk of developing disease or the molecular phenotype of an existing one, will change the way we currently define disease. Genomic analysis of diseases with homogeneous clinical phenotypes will unveil distinct molecular entities that require different treatment strategies for optimal outcomes. The completion of the genome sequence is not the beginning of the end, but is only the end of the beginning.

References

Balter, M., 2001, Transatlantic war over BRCA1 patent. *Science* 292:1818.
Blakemore, A.I., Froguel, P., 2008, Is obesity our genetic legacy? *J Clin Endocrinol Metab* 93(11 Suppl 1):S51–56, doi: 10.1210/jc.2008–1676.
Bougneres, P., Valleron, A.J., 2008, Causes of early-onset type 1 diabetes: toward data-driven environmental approaches. *J Exp Med* 205:2953–2957, doi: 10.1084/jem.20082622.
Chanda, S.K., Caldwell, J.S., 2003, Fulfilling the promise: drug discovery in the post-genomic era. *Drug Discovery Today* 8:168–174.
Chen, X., Hess, S., 2008, Adipose proteome analysis: focus on mediators of insulin resistance. *Expert Rev Proteomics* 5:827–839, doi: 10.1586/14789450.5.6.827.
Collins, F.S., Morgan, M., Patrinos, A., 2003, The Human Genome Project: lessons from large-scale biology. *Science* 300:286–290, doi: 10.1126/science.1084564.
Cornier, M.A., Dabelea, D., Hernandez, T.L., Lindstrom, R.C., Steig, A.J., Stob, N.R., Van Pelt R.E., Wang, H., Eckel, R.H., 2008, The metabolic syndrome. *Endocr Rev* 29:777–822, doi: 10.1210/er.2008–0024.
Edwards, A.V., White, M.Y., Cordwell, S.J., 2008, The role of proteomics in clinical cardiovascular biomarker discovery. *Mol Cell Proteomics* 7:1824–1837. doi: 10.1074/mcp.R800007-MCP200.
Frayling, T.M., 2007, Genome-wide association studies provide new insights into type 2 diabetes aetiology. *Nat Rev Genet* 8:657–662, doi: 10.1038/nrg2178.
Garg, A., 2004, Acquired and inherited lipodystrophies. *N Engl J Med* 350:1220–1234, doi: 10.1056/NEJMra025261.
Gieger, C., Geistlinger, L., Altmaier, E., Hrabé de Angelis, M., Kronenberg, F., Meitinger, T., Mewes, H.W., Wichmann, H.E., Weinberger, K.M., Adamski, J., Illig, T., Suhre, K., 2008, Genetics meets metabolomics: a genome-wide association study of metabolite profiles in human serum. *PLoS Genet* 4:e1000282, doi: 10.1371/journal.pgen.1000282.
Giovane, A., Balestrieri, A., Napoli, C., 2008, New insights into cardiovascular and lipid metabolomics. *J Cell Biochem.* 105:648–654. doi: 10.1002/jcb.21875.
Gottlieb, R.A., 2000, Mitochondria: execution central. *FEBS Lett* 482:6–12.
Hegele, R.A., 2009, Plasma lipoproteins: genetic influences and clinical implications. *Nat Rev Genet* 10:109–121, doi: 10.1038/nrg2481.
Krieg, R.C., Paweletz, C.P., 2002, Clinical proteomics for cancer biomarker discovery and therapeutic targeting. *Technol Cancer Res Treat* 1:263–272.
Lawlor, D.A., Hart, C.L., Hole, D.J., Davey Smith, G., 2006, Reverse causality and confounding and the associations of overweight and obesity with mortality. *Obesity* 14:2294–2304, doi: 10.1038/oby.2006.269.
Li, S., Loos, R.J., 2008, Progress in the genetics of common obesity: size matters. *Curr Opin Lipidol* 19:113–121, doi: 10.1097/MOL.0b013e3282f6a7f3.
Lindgren, C.M., McCarthy, M.I., 2008, Mechanisms of disease: genetic insights into the etiology of type 2 diabetes and obesity. *Nat Clin Pract End Met* 4:156–163, doi: 10.1038/ncpendmet0723.
Lockhart, D.J., Winzeler, E.A., 2000, Genomics, gene expression and DNA arrays. *Nature* 405:827–836.
Loos, R.J., Lindgren, C.M., Li, S., Wheeler, E., Zhao, J.H., Prokopenko, I., Inouye, M., Freathy, R.M., Attwood, A.P., Beckmann, J.S., et al., 2008, Common variants near MC4R are associated with fat mass, weight and risk of obesity. *Nat Genet* 40:768–775, doi: 10.1038/ng.140.
Maes, H.H., Neale, M.C., Eaves, L.J., 1997, Genetic and environmental factors in relative body weight and human adiposity. *Behav Genet* 27:325–351, doi: 10.1023/A:1025635913927.
Maher, B., 2008, Personal genomes: the case of the missing heritability. *Nature* 456:18–21, doi: 10.1038/456018a.
Merz, J.F. 1999, Disease gene patents: overcoming unethical constraints on clinical laboratory medicine. *Clin Chem* 45:324–330.
Mocellin, S., Rossi, C.R., Traldi, P., Nitti, D., Lise, M., 2004, Molecular oncology in the post-genomic era: the challenge of proteomics. *Trends Mol Med* 10:24–32.

Offen, D., Elkon, H., Melamed, E., 2000, Apoptosis as a general cell death pathway in neurodegenerative disease. *J Neurol Transm* 58(suppl):153–166.

Orešic, M., Simell, S., Sysi-Aho, M., Näntö-Salonen, K., Seppänen-Laakso, T., Parikka, V., Katajamaa, M., Hekkala, A., Mattila, I., Keskinen, P., et al., 2008, Dysregulation of lipid and amino acid metabolism precedes islet autoimmunity in children who later progress to type 1 diabetes. *J Exp Med* 205:2975–2984, doi: 10.1084/jem.20081800.

O'Brien, T., Simari, R.D., 2000, Gene therapy for atherosclerotic cardiovascular disease: a time for optimism and caution. *Mayo Clin Proc* 75:831–834.

Papassotiropoulos, A., Wollmer, M.A., Tsolaki, M., Brunner, F., Molyva, D., Lutjohann, D., Nitsch, R.M., Hock, C., 2005, A cluster of cholesterol-related genes confers susceptibility for Alzheimer's disease. *J Clin Psychiatry* 66:940–947.

Peltonen, L., McKusick, V.A., 2001, Dissecting human disease in the postgenomic era. *Science* 291:1224–1229.

Perry, J.R., Frayling, T.M., 2008, New gene variants alter type 2 diabetes risk predominantly through reduced beta-cell function. *Curr Opin Clin Nutr Metab Care* 11:371–377, doi: 10.1097/MCO.0b013e32830349a1.

Pietilainen, K.H., Sysi-Aho, M., Rissanen, A., Seppanen-Laakso, T., Yki-Jarvinen, H., Kaprio, J., Oresic, M., 2007, Acquired obesity is associated with changes in the serum lipidomic profile independent of genetic effects: a monozygotic twin study. *PLoS Genet* 2:218, doi: 10.1371/journal.pone.0000218.

Quintana, F.J., Farez, M.F., Weiner, H.L., 2008, Systems biology approaches for the study of multiple sclerosis. *J Cell Mol Med*12:1087–1093, doi: 10.1111/j.1582-4934.2008.00375.x.

Raamsdonk, L.M., Teusink, B., Broadhurst, D., Zhang, N., Hayes, A., Walsh, M.C., Berden, J.A., Brindle, K.M., Kell, D.B., Rowland, J.J, Westerhoff, H.V, van Dam K., Oliver, S.G., 2001, A functional genomics strategy that uses metabolome data to reveal the phenotype of silent mutations. *Nat Biotechnol* 19:45–50, doi: 10.1038/83496.

Raczniak, G., Ibba, M., Soll, D., 2001, Genomics-based identification of targets in pathogenic bacteria for potential therapeutic and diagnostic use. *Toxicology* 160:181–189.

Roth, D.A., Tawa, N.E., Jr., O'Brien, J.M., Treco, D.A., Selden, R.F., 2001, Factor VIII Transkaryotic Therapy Study Group. *N Engl J Med* 344(23):1735–1742.

Scuteri, A., Sanna, S., Chen, W.M., Uda, M., Albai, G., Strait, J., Najjar, S., Nagaraja, R., Orrú, M., Usala, G., et al., 2007, Genome-wide association scan shows genetic variants in the FTO gene are associated with obesity-related traits. *PLoS Genet* 3:e115, doi: 10.1371/journal.pgen.0030115.

Stylianou, I.M., Affourtit, J.P., Shockley, K.R., Wilpan, R.Y., Abdi, F.A., Bhardwaj, S., Rollins, J., Churchill, G.A., Paigen, B., 2008, Applying gene expression, proteomics and single-nucleotide polymorphism analysis for complex trait gene identification. *Genetics* 178:1795–1805, doi: 10.1534/genetics.107.081216.

Su, L.F., 2008, Updates on high-throughput molecular profiling for the study of rheumatoid arthritis. *Isr Med Assoc J* 10:307–309.

Tang, J., Tan, C.Y., Oresic, M., Vidal-Puig, A., 2009, Integrating post-genomic approaches as a strategy to advance our understanding of health and disease. *Genome Med* 1:35.

The Wellcome Trust Case Control Consortium, 2007, Genome-wide association study of 14,000 cases of seven common diseases and 3,000 shared controls. *Nature* 447:661–678, doi: 10.1038/nature05911.

Turnbaugh, P.J., Gordon, J.I., 2008, An invitation to the marriage of metagenomics and metabolomics. *Cell* 134:708–713, doi: 10.1016/j.cell.2008.08.025.

Turnbaugh, P.J., Ley, R.E., Hamady, M., Fraser-Liggett, C.M., Knight, R., Gordon, J.I., 2007, The human microbiome project. *Nature* 449:804–810, doi: 10.1038/nature06244.

Venter, J.C., Adams, M.D., Myers, E.W., Li, P.W., Mural, R.J., Sutton, G.G., Smith, H.O., Yandell, M., Evans, C.A., Holt, R.A., et al., 2001, The sequence of the human genome. *Science* 291(5507):1304–1351.

Venter, J.C., Adams, M.D., Myers, E.W., Li, P.W., Mural, R.J., Sutton, G.G., Smith, H.O., Yandell, M., Evans, C.A., Holt, R.A., et al., 2008, New insights into cardiovascular and lipid metabolomics. *J Cell Biochem* 105:648–654, doi: 10.1002/jcb.21875.

Vineis, P., Schulte, P., McMichael, A.J., 2001, Misconceptions about the use of genetic tests in populations. *Lancet* 357:709–712.

Wildman, R.P., Muntner, P., Reynolds, K., McGinn, A.P., Rajpathak, S., Wylie-Rosett, J., Sowers, M.R., 2008, The obese without cardiometabolic risk factor clustering and the normal weight with cardiometabolic risk factor clustering: prevalence and correlates of 2 phenotypes among the US population (NHANES 1999–2004). *Arch Intern Med* 168: 1617–1624.

Woychik, R.P., Klebig, M.L., Justice, M.J., Magnuson, T.R., Avner, E.D., 1998, Functional genomics in the post-genome era. *Mutat Res Fund Mol Mech Mut* 400:3–14.

Young, D.B., 2001, Post-genomic perspective. *Nat Med* 7:11–13.

Zhu, J., Wiener, M.C., Zhang, C., Fridman, A., Minch, E., Lum, P.Y., Sachs, J.R., Schadt, E.E., 2007, Increasing the power to detect causal associations by combining genotypic and expression data in segregating populations. *PLoS Comput Biol* 3:e69, doi: 10.1371/journal.pcbi.0030069.

Chapter 8
Ethics, Patents and Regulations

Abstract The ethical, legal, and social implications (ELSI) program was created by the HGP to address issues that arise as a result of the program but that are not directly related to sequencing of the genome. Clinicians, scientists, patient advocacy groups, and government agencies worldwide are debating the optimal legal protections to prevent abuse. Massive amounts of data flowing from the Human Genome Project and other genomics projects have stimulated an avalanche of applications for patents on genes and gene fragments.

Keywords Ethics · Patents · Human genome project · Genetic testing · Diagnostics · Genetic assay · SNP · Single nucleotide polymorphism · Germ line therapy · Somatic gene therapy · Genetic testing · Genomic patents · Expressed sequence tags (ESTs) · Genome technology · Genetically modified organism · HGP

8.1 Prologue

The impact of commercialization and patenting pressure on genomics research is still a topic of considerable debate in academic, policy and popular literature (Murdoch and Caulfield, 2009). Recent advances in molecular genetics challenge the scientific community to understand and implement genetic knowledge. Despite excitement about the potential benefits of new genetic information, concerns have been raised about the inappropriate use of genetic testing, clinicians' incorrect ordering and misinterpretation of test results, and discrimination in employment and insurability based on tests results. Among the public there is fear and mistrust, in part based on horrifying historical events that were gross violations of medical ethical standards. Knowledge of a patient's genetic profile brings great responsibility and many concerns. Who has the right to access this information? How should (or shouldn't) it be used?

8.2 Concept of Ethical, Legal, and Social Issues

The advances in the field of molecular biology and genetics have widened the possibilities for the diagnosis and treatment of hereditary diseases. At the same

M. Debnath et al., *Molecular Diagnostics: Promises and Possibilities*,
DOI 10.1007/978-90-481-3261-4_8,

time research into this field has broken bounds of its legal and ethical regulation (Gutiérrez-Samperio, 2002). Ethical issues are generally defined as those raising questions concerning what is moral or right. Legal issues are those concerning the protections that laws or regulations should provide. Social issues are concerned with how events may affect society as a whole and individuals in society. Clearly, these aspects of the Human genome project and its possible outcomes are not independent of each other. The programs focus on ethical, legal, cultural, social, and psychological consequences that could affect policy. The project's goals included not only identifying all of the genes in the human genome, but also to address the ethical, legal, and social issues which might arise from the availability of genetic information (Durfy, 1993). Five percent of the US annual Human Genome Project (HGP) budget was allocated to address the Ethical Legal Social Imlications (ELSI) arising from the project. Still, many issues, such as the confidentiality of a patient's genetic profile, remain unaddressed by legislation. In addition to providing guidelines for potential new policy, the program examines philosophical, theological, ethical and other social issues in order to ease the entry of genetic information into clinical practice as well as the social consciousness.

8.3 Ethical and Social Issues in Diagnostic Molecular Genetics

Prior to the genome initiative, ethical issues in molecular genetics focused primarily on monitoring and regulating experimentation in recombinant DNA and genetic engineering. Organizers of the Human Genome Project (HGP) understood from the beginning that the scientific activities of mapping and sequencing the human genome would raise ethical, legal, and social issues that would require careful attention by scientists, health care professionals, government officials, and the public (Fig. 8.1). The Ethical, Legal, and Social Issues (ELSI) Program has been established to examine various issues associated with the generation and dissemination of a vast array of genomic information (Meslin et al., 1997; Zneimer, 2002). High-priority issues initially targeted by the Ethical, Legal, and Social Issues Program for development of policies and guidelines included:

1. The integration and impact of new genetic tests and the debate over population screening;
2. Privacy and confidentiality of genetic information;
3. Genetic counseling and reproductive decisions influenced by diagnostic results; and
4. Public education.

In recent years, the Ethical, Legal, and Social Issues Program has also emphasized technical problems such as the potential for genetic discrimination, educating physicians in the advantages and limitations of genetic data, quality control in DNA testing laboratories, and defining guidelines for obtaining informed consent

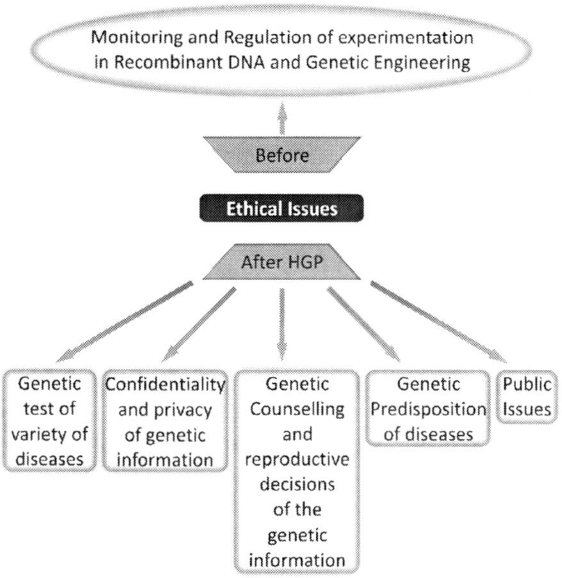

Fig. 8.1 Ethical issues before and after HGP

for genetic research. In addition, these groups are developing clinical guidelines for optimal use. Traditional ethical and legal standards of confidentiality between physicians and their patients are under scrutiny. A new principle, "the duty to warn," is emerging that has applications specific to genetic testing and may conflict with the duty to maintain patient confidentiality. Emerging ethical, legal, and social issues involve the appropriate use and protection of confidential data in tissue and serum banks. Education of the profession and the public at many levels will increase the likelihood that the unraveling of the human genome will maximally benefit society. If fear of genetic testing can be alleviated, selection bias in research could be reduced. Professional and lay organizations concerned with disease should consider a more active role in the public and professional debate, and foster education at all levels.

As new genetic assays are introduced into clinical practice, rigorous adherence to established protocols and quality control assurance are of paramount importance. Attention must be directed toward the debate over implementing population-wide screening programs as routine practice in clinical medicine to detect those at increased genetic risk. The increasing ability to diagnose individuals at risk for genetic diseases for which there are no therapeutic options will require enactment of measures to (1) prevent insurance and/or employment discrimination against asymptomatic carriers and (2) accommodate the psychological needs of those who are likely to develop a late-onset condition. Improvement of noninvasive prenatal and preimplantation diagnostic procedures is creating an immediate need to explore the ethical dilemmas and difficult reproductive choices faced by prospective parents known to carry disease-associated genes.

Increasing public awareness of their availability, benefits, and limitations of molecular diagnostic tests is anticipated to simultaneously improve health care delivery while minimizing the potential for psychological and social stigmatization. The impending explosion in the number of well-characterized human disease genes and new abilities to diagnose genetic disorders will likely necessitate development of novel avenues for education and genetic testing (McCunney, 2002). Careful integration of genetic information allow a practical system for characterizing and resolving ethical and social issues and provide future directions for the fields of molecular genetics and clinical medicine.

8.4 Confronting the Ethical, Legal and Social Issues

James Watson, who won the Nobel Prize in Physiology in Medicine in 1962 for co-discovering the structure of DNA, made a seminal contribution to the HGP when he recognized that knowledge derived from genome studies has broader medical and societal implications. This led directly to the establishment of a program devoted to the ethical, legal, and social implications (ELSI) of genome research. The ELSI program (Masui and Takada, 2003) focused on the possible consequences of genomic research in four main areas:

- Privacy and fairness in the use of genetic information, including the potential for genetic discrimination in employment and insurance.
- The integration of new genetic technologies, such as genetic testing, into the practice of clinical medicine.
- Ethical issues surrounding the design and conduct of genetic research with people, including the process of informed consent.
- The education of healthcare professionals, policy makers, students, and the public about genetics and the complex issues that result from genomic research.

One goal of the ELSI program is to address the implications of vastly increased genetic information and protocols on individuals and society. Another ELSI goal is to identify and develop appropriate policy options to confront and contain future ELSI problems. Because we know that "genetic information" has been misused previously in the United States and other countries, we must ensure that such mistakes are never repeated. Both the DOE and the NIH are optimistic that the ELSI program can contribute to the integration of HGP results in ways that are less disruptive, painful, or destructive than those in the past (Jin, 2000).

The list of ELSI issues is long and virtually all of them have legal ramifications. They include the fair use of genetic information; the impact on genetic counseling and medical practice; the effects on personal reproductive decisions; past uses and misuses of genetic information; privacy implications of personal genetic information in various settings, e.g., the work place, schools, or in the context of adoptions; issues of the commercialization and intellectual property protection of genome results, including DNA sequences; conceptual and philosophical implications; implications of personal genetic variation; and genetic literacy and the understanding of genetic information, particularly information related to complex conditions that involve multiple genes and genetic-environmental interactions (Tsai and Chen, 2000). This last category, involving health issues like mental illness, heart disease, liver disease, diabetes, or cancer (Ellerin et al., 2005), represents the most complex of ELSI issues because the underlying science is poorly understood (Van Leeuwen and Bernat, 2006). For example, informing a woman that she carries an allele at BRCA1 that is associated with a high lifetime risk of developing breast cancer is a serious issue, particularly if treatment options are difficult, painful, debilitating, or oftentimes less than successful. However, other individuals or entities may want to know about such conditions, including insurers and employers (who often are responsible for insuring their employees). The extent of this protection is undoubtedly an issue that courts will have to adjudicate.

We see many ELSI challenges in the future. Informed consent for participants in genetic research will remain an important issue. Genetics involves shared familial information, and the diagnosis of one person has direct implications for his or her family members. It is extremely important that patients and research participants understand what information and future predictive insights about them may emerge from genetic studies, particularly when they involve genetic testing or screening for multigenic and predisposition diseases. For example, over 600 mutations in the gene for the cystic fibrosis transmembrane regulator can lead to cystic fibrosis. Many experts think that only seven of these mutations are responsible for 85% of the cases of cystic fibrosis seen clinically, and it is these seven for which most people are tested. However, a negative test for cystic fibrosis disease-associated alleles does not necessarily mean that a person does not have a risk for cystic fibrosis. The gene for breast cancer susceptibility (BRCA1) is another case to point. Over 150 alleles in the gene have been discovered. The BRCA1 region is very large, and the number of alleles that actually predispose to breast cancer is not yet known. We also know very little about other influences that are necessary (along with one of these mutations) for breast cancer. What do you tell someone who tests positive for a disease-associated allele when you can only be vague about its clinical implications? What responsibilities do physicians and counselors have in the communication of risk information to patients when the risks themselves are poorly understood? What liability issues accompany genetic information? Can genetic information be "owned" and, if so, by whom and under what circumstances? These and other issues that arise from genetic information will challenge the courts and will be exacerbated as we get better at "reading" and interpreting the content of our genomes.

8.5 Genes and Disease

The Human Genome Project is an international effort to complete the sequencing of the 100,000 genes that comprise the human genome. Upon conclusion of the Project, the genes responsible for single gene deficiency diseases, e.g., Huntington's disease and cystic fibrosis, as well as for multifactorial diseases, e.g., atherosclerosis and cancer have been identified. The Project seems almost like science fiction. It is easy to think of it as not having immediate policy implications.

Gene mutation and variation may result in hereditary disease, cancer, hypertension, and even susceptibility to infectious diseases. A complete compilation of all human genes (the human genome) now allow a better understanding of the role of specific genes in diseases and, consequently, better design of effective treatments. Approximately 5,000 human diseases are currently thought to have some genetic component. Some of the fundamental causes are easy to identify. In the case of Down's syndrome for instance, there is an extra copy of chromosome 21, which is visible under the light microscope. Other genetic diseases that are strongly under the control of a single gene have also been relatively easy to identify. In the case of sickle cell anemia (SCA), a change in a single nucleotide of one gene alters a codon so that a different amino acid is incorporated into hemoglobin, the protein responsible for carrying oxygen in the blood. Such a change, which is called a single nucleotide polymorphism (SNP), may cause a change in an amino acid that can have a range of effects depending on where the amino acid is located in the protein and the change in amino acid. An SNP can also occur without changing the amino acid because many amino acids have more than one codon. In Huntington's disease, another disease that is strongly determined by genetics, a single gene has a series of nucleotides repeated many times.

In the case of SCA the gene is known as "recessive" because it requires two copies of the gene, one from each parent, to produce the disease in the offspring. If one parent donates the gene for the disease and the other does not, the child will not get the disease. In the case of Huntington's, the gene is "dominant"; that is, if one parent has the gene, then half of his or/her sperm eggs will carry the disease and a child who receives the gene will get the disease. Thus, a child whose parent has Huntington's has a 50% chance of getting the disease.

As a result of strong single gene control over some diseases, close to 1,000 genes for a variety of diseases have been identified and localized to a chromosome through inheritance patterns. The genes involved in other diseases with some genetic component, such as cancer, heart disease and mental illness, are more difficult to identify because they involve multiple genes that are also heavily influenced by environmental factors. Environment is only one of many factors that control the degree to which harmful changes in a gene, or combination of genes, may cause health problems. Potentially harmful changes in genes may not result in problems if:

- associated environmental factors do not turn these genes on;
- there is only a partial set of the combination of genes underlying a disease;
- the gene or genes are not expressed strongly;
- the change leads to only a mild form of the disease;
- the gene is recessive and only one copy is present;
- Genetic damage does not result from environmental substances or ageing.

The extent to which health problems result from potentially harmful changes to genes is therefore very difficult to predict. At one extreme, we know that diseases such as Huntington's and SCA are strongly determined by changes in a single gene. As more genes become involved and environmental factors come into play, the genetic disposition to a disease becomes less and less obvious, however, diseases that are heavily influenced by environmental factors and interaction with other genes are likely to be far more common than diseases with very strong genetic components.

Many of the ELSI implications are not new. The gene for Huntington's disease was discovered in 1993, after a 10-year search following the localization of the gene to chromosome 4 in 1983. A test for the disease was developed soon after. Many of the questions currently being addressed by the ELS issues program of the HGP have, therefore, been familiar for many years to families afflicted with Huntington's. As a result of the HGP, however, society as a whole will have to deal much more frequently with issues arising from knowledge of the human genome. Moreover, the implications may be less clear in the case of genes identified for diseases that have strong environmental aspects and involve interaction with many other genes. Even strongly deterministic genes, such as those for Huntington's, have a range of effects; for example, some people with the Huntington's gene live much longer than others.

Clinical issues include the education of doctors and other health service providers, patients, and the general public in genetic capabilities, scientific limitations, and social risks; and implementation of standards and quality-control measures in testing procedures. Uncertainties are associated with gene tests for susceptibilities and complex conditions (e.g., heart disease) linked to multiple genes and gene-environment interactions.

8.6 The Existence of Genetic Information

The existence of genetic information with respect to individuals and the human population as a whole will have a profound impact on our day-to-day lives and may well change how we regard ourselves and one another. The knowledge of predisposition to a certain disease and the ability to design "tailor made" therapies may greatly help in the treatment of disease. Already a company in Great Britain has applied for a patent on a device that can apparently detect different forms of over 2,500 genes said to be associated with traits including behavior and intelligence. It has been argued, however, that it is not proper, particularly at this juncture in history, to search for such knowledge. For example, some have pointed out that science has often been co-opted as a tool to accentuate racial differences and to defend racist practices. Given that humans are far from resolving issues of race, it is thought that information from the HGP, and such follow-up projects as the Human Genome Diversity Project, may have the potential to inflame racism in an already overly racist world.

Equally, some feel that if the goal of the HGP is to prevent disability and disease, increase life spans, decrease infant mortality, and increase intelligence, the money would be far better spent elsewhere. Given that we already know that environmental and social factors can influence such diseases as diabetes in aboriginal populations and drug addiction among the socially marginalized, some consider it unconscionable to dispense limited resources looking for genetic causes for these diseases. The legal aspects of knowledge of the human genome are enormous. Already DNA evidence is being used as a powerful legal tool, particularly in exonerating wrongly accused individuals. Does this mean that the criminal system should be able to keep a bank of DNA information on anyone accused and/or convicted of a crime? Could the database be used for other purposes than simply identifying and eliminating suspects? A DNA database could contain much more information on individuals, both guilty and innocent, than does the current system of taking fingerprints.

On a more hypothetical note, should genes leading to a propensity for criminal activity be found, could they be used as prosecution or defense evidence in a trial? For instance, is a suspect who knows that he or she has a genetic disposition toward criminal behavior and does nothing to avoid provoking such behavior, guilty of a more serious crime than a suspect who is ignorant of having such a propensity? On the other hand, could genetic disposition be used as a defense on the grounds that the crime was really the fault of the gene, not the person? When a patient tests positive for a gene linked to risk of disease, does the physician (or the patient) have a legal responsibility to inform the patients' relatives of their own risks? Suppose a patient finds out that she has a genetic propensity for breast cancer, but neither she nor her doctor informs her relatives; would a relative who later developed that form of cancer be able to sue, on the grounds that the genetic information had not been disclosed?

Ensuring that the judge and jury in a trial are sufficiently educated to deal with these issues is yet another problem with which the legal system will have to deal. On a larger social scale, knowledge of the human genome could be used to emphasize the similarities among all humans. The genetic differences between people within an identified group have already been shown to be greater than the differences between groups. In other words, people within an "ethnic" population are more different from each other than the group as a whole is different from other "ethnic" groups. This fact is unlikely, however, to deter those who wish to emphasize any ethnic differences that may be found. On a more individual level, the results of the HGP might encourage people to view themselves as being wholly under the control of their genes. What has traditionally been viewed as the human spirit might in future be seen as limited by pre-programming at birth? Thus, though we cannot predict exactly how knowledge of the human genome will affect society, it could clearly have important consequences.

Individual decisions, such as choices with respect to mates and reproduction, could also be influenced by knowledge of genetic makeup. Awareness of personal genetic differences from a perceived norm might lead to confusion and uncertainty about the potential for disease, particularly in the absence of adequate professional consultation. Genetic analysis might reveal a myriad of genetic flaws that may or may not lead to disease, depending on what they are and how they interact with the environment. How will individuals select from a debilitating array of lifestyle choices, none of which has a certain outcome? Again, analysis of one's own genetic makeup could reveal the genetic makeup of parents and siblings, including, for example, unsuspected information about paternity. How willing would people be to share this knowledge and, if they decided to withhold it, how would they be affected by living with the secret?

8.6.1 Genetic Treatment of Disease

From the outset, one of the defining goals of the HGP has been its potential for molecular medicine. The concept is that, once the functions of genes are known and we understand the effects of malfunctioning genes, we will be able to correct the problem either through the use of designer drugs or by replacing the faulty gene. It is the latter option that has created the most controversy.

There are two routes to replacing a faulty gene. The first route, germ line therapy, has the goal of replacing a harmful gene in a fertilized human egg with a properly functioning gene that would be passed on to future generations. The other route, somatic gene therapy, aims to replace the gene in target organs or tissues of an adult, so as to fix the symptoms in that individual but not in the next generation. Germ line therapy has the more profound ethical, legal and social implications.

As yet germ-line therapy in humans is not possible and some have argued that it will continue to be so for the foreseeable future. While this kind of therapy may be a long way off, it would bring, on the one hand, the hope of eradicating some genetic diseases but, on the other hand, the spectra of eugenics.

The eradication of disease through germ-line therapy might not seem, by itself, to raise many ethical questions. After all, humans have eradicated the smallpox virus from the world, why not eradicate diseases with genetic components? Do doctors not have the moral obligation to provide the very best treatment to their patients and would not the eradication of the disease be more cost effective in the long run than continually treating adults with somatic gene therapy? The main ethical problem arises in defining a "treatable" disease.

Some might say that eradication of a genetic disease for which there is no treatment and which is always fatal, should be pursued with all means possible. Others say that this would be the start of a slippery slope moving on toward the treatment of less obvious diseases and then to genetic enhancement. Some argue that if the technology is advanced in order to eradicate some diseases, it will inevitably be used by parents wishing to "enhance" their children, giving them the genes for raven black hair and blue eyes or athletic prowess. It was serious ethical concerns about genetic enhancement. Another ethical consideration with respect to germ-line therapy is defining what is normal, what a disability is, and what a disease is. Which of the genetic variations within a population ought to be eradicated, if any? In trying to eradicate a certain variation, are we demeaning those in the population who currently carry the gene?

Somatic gene therapy has its own, less controversial, set of ELSI implications. These may be less ominous than eugenics but are of perhaps more immediate concern, given the more advanced state of the technology. Effectively, gene therapy involves the introduction of a properly functioning gene into target tissues in the hopes that it will be translated into a properly functioning protein, which will mask the malfunctioning protein. Often the new gene is placed into a modified virus, which is then introduced into a patient in the hope that the gene will be introduced into a tissue and properly expressed.

Such types of therapy, after much research on laboratory animals, have now reached the clinical trial stage. Unfortunately, what works for a mouse does not always work for a human being. In one highly publicized case, a patient, Jesse Gelsinger, was given an injection of a virus in the hope of introducing a protein into the liver. Mouse studies showed good absorption of the gene into the liver; however, the mouse has a much higher concentration of viral receptors on its liver cells than do humans. The virus did not absorb well into the human patient and, for still unknown reasons, created a massive immune response, causing the patient to die. The original plan for the trials had been to use the virus only on children in a coma caused by the lack of the particular liver enzyme; however, ethical and safety reviews caused the researchers to change the trial direction and use adults only. Many questions are now being asked regarding the ethics and scientific judgment of those performing such clinical trials. How well are "volunteer" patients informed of the possible risks and benefits? How objective are investigators who have equity in the companies that are funding the trials? One of the risks at this stage of gene therapy is the excessive public anticipation, created in part by some researchers, with respect to future benefits. This anticipation may turn to public distrust of science, if the benefits fail to be realized and problems such as that in the Gelsinger case continue to occur. Some clinical trials have shown positive results, and so there is still hope that somatic gene therapy will become a powerful medical tool.

We believe medicine can eventually find a cure for everything that ails us. This is why the HGP genetics treatments will, of course, continue their progression. To prevent unethical prejudice, discrimination, or abuse that may occur because of genetics treatments, we must explore all ethical issues and create and implement policies on genetic research and the extent in which we allow genetics treatments to change the population.

8.6.2 Genetic Testing

The ELSI goals included in the HGP are composed of issues that will arise upon completion of the human DNA sequenced. The issues include integrating the new genetic information into health care and non-clinical settings, the philosophical, theological, and ethical aspects that will arise with increasing advancements, and finally how race, ethnicity and socio-economic factors will influence the development and use of genetic services.

With continued advancement of the HGP, there are many benefits and arguments in favor of the HGP. The knowledge about the effects of DNA variations between individual has lead to revolutionary new ways to diagnose, treat, and possibly prevent disorders. Many new genetic disorders have been discovered and with these, testing is becoming increasingly available. A good example is prenatal genetic testing. An expectant mother may have genetic testing done to determine if her child has a mutation that causes a genetic disorder. Newborn screening practices have changed since breakthroughs have occurred in genetics and mapping of the human genome. Although newborn screening has been in existence since the 1960s, today's newborn screening practices are subsumed primarily under the umbrella of genetic testing (Kenner et al., 2008).

With the knowledge from the HGP, persons may have their genetic profile available to them. Their profile could be helpful in many ways. This information can help the person to build a suitable behaviour to be followed throughout their life to remain healthy. Their genetic profile may also tip the person off to have frequent check-ups if they are at risk for a certain disease. For example, if a person knows from their genetic profile that they are at risk for heart disease, they may have a modified diet and exercise program and frequent doctor's visits to help prevent heart disease.

Genetic treatment has come about from advances in genetics projects. We can improve crops by making them more resistant to insects and disease. We can make oversized and faster growing animals by inserting growth hormones in mice, pigs, sheep, and fish, thereby increasing food production. We can also increase food production by developing animals that have larger litters and lay more eggs, engineering fruits and vegetables to stay ripe longer, and developing crops and livestock that con survive harsh conditions. In addition, we can put human genes into pigs and baboons and use their organs for transplants into humans with less rejection from the human body.

Genetic treatment can be used in a variety of ways to improve the human condition. In the future it is possible we may be able to correct medical disorders by going straight to their source, the genes. If a disorder is caused by a faulty or mutated gene, doctors may be able to insert new DNA into the cells. This new DNA would replace, or knockout the action of the mutation and restore the function that was damaged by the mutation. Cystic fibrosis (CF) is one of the diseases researchers are trying to cure through this type of genetic treatment. CF is a disease in which the lungs become clogged and make respiration difficult. The disease is a result of a mutation that restricts production of an enzyme that helps make the lining of the lungs.

In this treatment, doctors would make copies of the corrected gene and insert it into a virus. The virus is used as a sort of transfer device. The virus would be sprayed into the nostrils of CF patients. Here the virus would reproduce by injecting the corrected gene into other cells of the patient's body. Doctors hope the virus would reproduce enough to help the body make the needed enzyme. Although there are many advantages that supporters of the HGP can give for reasons to continue research, there are also those that argue there are many risks and disadvantages to the HGP and the use of genetic treatment. Genetic treatment may easily be twisted to support prejudice and discrimination against others. Who is to decide if a trait is good or bad? For example, a doctor has two children patients with the same problem, but she is unsure of whether to treat them both. The two children, Sam and John, are both very short for their age. Sam does not produce enough growth hormone and will never grow above five feet. Both of his parents are close to six feet tall. John, on the other hand, inherited his height from his parents who are both barely over five feet. A growth hormone has been produced and is available to both patients. Both sets of parents feel their child should receive the treatment to help them grow taller. They want the treatment for their child because they believe there are advantages to being tall. The doctor has to decide whether to give the treatment to only Sam, since his height is a result of a mutation, or to give it to both children. How does the doctor decide what is fair? Neither of the children is sick. Some argue that instead of changing those traits society has deemed as undesirable, we should change the idea that different is undesirable.

Genetic treatment may be used in many ways to treat, cure, and maybe even prevent many diseases, but there is a concern its use may not be limited to health. With the growing knowledge involved in genetics treatment, there is an increased opportunity for people to be able to actually change how they or their children look. People may use genetic treatment to make them look younger, thinner, or even smarter. As a result, genetic treatment may also be equated with eugenics in the future. The term has a negative connotation because it implies some people or traits are better than others are.

8.6.3 Discrimination

One of the problems some fear might result from knowledge of the human genome is the emergence of a whole population of socially marginalized individuals, unable to obtain a job, a family, insurance, or health care and stigmatized by the rest of society. Insurance companies already insist that those identified at risk of Huntington's disease must take a genetic test. If the results are positive, insurance is frequently refused. Insurance companies are on record as saying that if genetic information was available, they would use it in their risk assessment. In Canada, the refusal to insure a Huntington's patient does not have dire consequences; in general, public insurance covers many aspects of care, though the level of care varies across the country and the coverage for pharmaceuticals is less clear. In countries without a public health insurance system, however, the plight of such a non-insured person can be a nightmare. Care may be available but finding it is very difficult. As more genetic tests become available, insurance is likely to be more and more expensive for those carrying what the insurance companies deem to be risky genes. The public insurance schemes may also start to feel the pressure for such genetic testing, and be forced to make policy decisions based on the funding available and the knowledge of genetic predisposition to disease within populations.

Gene therapy is at the experimental stage at this point but will certainly be very expensive when it first comes into regular use. Who will pay for it? If not public insurance, will the therapy be available only to rich people, thus creating an ever widening gap between groups in society, based on both money and genetic inheritance?

Employers may also want access to genetic information. Some genes might reveal a susceptibility to environmental damage that was incompatible with a certain workplace environment. Employers might choose to screen out workers carrying that gene rather than trying to improve the environment. Individuals with genes associated with certain behavioral traits might also be excluded from the workplace. The human genome project has led to impressive scientific advances in understanding the genetic basis of disease. To date, genetic risks associated with occupational illnesses are not well understood. Recent research, however, has uncovered an allele that appears directly related to the risk of contracting chronic beryllium disease; other chromosomal abnormalities have been identified in association with cancer. Progress from the human genome project has potential implications for predicting, screening, and diagnosing occupational diseases. Ethical issues associated with the use of genetic testing in the workplace will present employers, insurers, and physicians with challenging decisions related to promoting health in the workplace while avoiding potential misuse of sensitive genetic information.

8.7 Impact of the Human Genome Project at the Interface Between Patent and FDA Law

The debate over biotechnology policy is at heart a debate over information policy. At one level, the debate covers how to provide intellectual property protection to the tools of biotechnology and the valuable information they produce, such as gene sequences and chromosome maps. At a different level, the debate shifts to questions of how best to distribute information to empower others and to prevent information misuses and loss of privacy. Aided by the new tools of the computer age, biotechnology is developing faster than any previous technology. In the process of development, biotechnology is creating a wider gap between practice and policy. Our growing ability to transform genetic information into new products and organisms is intended to enhance agriculture, fight pollution, and alter hereditary diseases which make biotechnology a powerful and threatening tool.

Development of genome technology will outpace needed public policy which will be required to cope with consequences. No matter how safe and well controlled technology may be, public opinion will be driven by lack of knowledge, lack of trust of science and government and a tendency to believe that things are worse than what the public is told. Therefore, the question is: How does society ensure that its governmental and social institutions manage the future course of evolution with enough wisdom to avoid catastrophic mistakes?

Two kinds of currently available genomic patents may significantly interfere with medical research: (1) patents such as those on specific single nucleotide polymorphisms (SNPs), which may include claims that control the inference of phenotypic characteristics from specific genotypes, and (2) patents on computer-based genomic information, databases, or manipulation procedures. These will create more serious encumbrances than will patents on expressed sequence tags (ESTs). Two approaches should be considered vis-à-vis these genomic patents:

(1) Reconsideration and redefinition of the recent extensions of patentable subject matter into more and more intangible areas. This could be pursued by legislation or by test litigation to seek Supreme Court reversal of certain of the decisions of the Court of Appeals for the Federal Circuit (CAFC).
(2) A narrow legislative exemption protecting the ability to use SNPs and phenotypic-genotypic relationships in medical research, including contexts in which medical research and clinical practice are substantially intertwined (Barton, 2002). Patents and licenses on gene-based diagnostic tests are properly enforceable and do not permit academic or other clinical practitioners to practice such tests without license or other authorization from the patent holder (Goldstein and Golod, 2002).

The real risk of genome technology is not whether appropriate clinical trials can be designed, nor how or where inserted DNA integrates into the host cell or any other such events. The real risk may be illustrated by recalling the incident which led to the Asylomar Conference and creation of the NIH's Guidelines for Research Involving Recombinant DNA Molecules (Guidelines) viz., The potential benefits and harms of DNA patenting for biomedical research and medical practice. It argues that, all things considered, the benefits of patenting DNA outweigh the harms, although societies should adopt policies designed to prevent or mitigate the harms associated with patenting. Some of these policies include:

(1) Reinforcing the research exemption for academic researchers,
(2) Raising the "bar" for the criteria of patentability,
(3) Restricting the scope of patents,
(4) Disclosing conflicts of interest related to DNA patents,
(5) Sharing the economic benefits of patenting with patients, and
(6) Providing insurance coverage for some types of genetic tests (Resnick, 2003).

The first patent for a genetically modified organism (GMO) issued in 1988 on a transgenic mouse modified to be useful in studying cancer. Patent and FDA law may be said to "interface" in a couple of different ways. Legislative enactments such as the Orphan Drug Act 9 and the Patent Term Restoration Act10 "interface" in the sense that each modifies patent law and FDA law. However, a different kind of interface may be found in the tension between the differing policies which underlie patent and FDA law. The patent system is intended to foster technological innovation and economic progress. FDA law serves different policy objectives because it is intended to protect the public's safety and welfare. To speak of the interface between patent law and FDA law is to speak of balancing competing policy considerations.

There is little reason to believe that differences in opinion about the appropriateness of human gene therapy will resolve spontaneously, or even after extensive public discussion. Where there is no agreement on what decision to make, the only alternative is a process for making the decision. Government agencies must demonstrate that the process is rational and fair. Whatever process is adopted must involve professionals from many disciplines, must be open to the public and must be insulated from political pressures. The best hope of setting policy to deal with the implications of genome technology is to use an effective process for gathering necessary information and subjecting that information to appropriate and thorough consideration to formulate recommendations for appropriate bodies.

The biotechnology industry can and should do better. It ought to be at the forefront of fostering a rational discussion of genome technology and its implications for social policy. If such a discussion does not begin soon, events will overtake the opportunity. Announcements of experiments made too soon, whether successful or unfortunate, will catch the public's attention with the potentially destructive consequences of fear, mistrust and backlash.

8.8 Regulatory Process

Regulation of gene transfer therapies by the Food and Development Authority (FDA) has been an exercise of fitting the existing system to the new technology. The FDA's regulatory approach may be modified in light of additional knowledge about risks and benefits. The products for somatic-cell and gene therapy are more prudent than waiting until the field has matured.

The Recombinant DNA Advisory Committee (RAC) reviews all proposals for National Institute of Health (NIH) funded research projects pursuant to the Guidelines. The Guidelines were intended to monitor biotechnology research until more was known about the safety of the organisms produced through genetic engineering. These Guidelines have been revised over the years in a manner characterized as demonstrating the gradual rise of technology to prominence over policy development to the point that the RAC has virtually relaxed itself out of a job.

The RAC responded to the recommendations of the President's Commission in its 1982 report, Splicing Life, by establishing the Human Gene Therapy Subcommittee to review proposals involving the use of rDNA techniques on human beings and to continue to explore issues that would be posed by the extension of these techniques into genetic enhancement and germline gene therapy.

Thus for the short term, technologies such as gene therapy will be dealt with as drugs or biologics in the same manner as other therapies, although approval to test gene therapies in patients has been more rigorous than any other medical area. A synthetic polynucleotide sequence intended to alter a genetic sequence in human somatic cells after administration to the patient is classified as a drug, so that a new drug application (NDA) is required. A retroviral vector containing a gene to be administered intravenously into the patient will be classified as a biologic, requiring a product license application (PLA) and ELA (establishment license application). A retroviral vector containing a gene and intended to modify cells ex vivo is considered a biologic intended for further manufacture; hence, both a PLA and ELA will be required.

Issues raised by biotechnology in the 1980s were trivial compared to those to be raised by genome technology. To resolve those issues effectively and fairly, we must develop a coherent moral and ethical basis for decisions in several areas, including patent and FDA law (Kessler et al., 1993).

8.9 Conclusion

Searching for medically useful, and therefore potentially profitable, genes also raises many ethical questions. Heritable disease patterns sometimes emerge in populations that have not mixed extensively with other populations. How to regulate the gene hunters without scaring off investment is a familiar problem to governments that already have experience with charging royalties and regulating natural resource operations. Gene "mining" companies, however, present a much more complex and emotional set of ethical issues than does the natural resources sector.

Many of the ethical, legal and social issues that are being discussed with respect to the Human Genome Project are not new. Genetic tests for a variety of diseases are currently available and some people are already struggling with the ethical and practical implications. What will change over the next few years, as a result of the Human Genome Project, is the scale of the issues and how society will have to cope with the greyer areas of genetic disease and disability. Dealing with a single gene that causes death or chronic disability is one issue; dealing with whole sets of genes whose impacts vary depending on environmental interactions is another. The rate of scientific advancement has tended to outstrip the legislative capacity of governing bodies and there has been some media "over hype" with respect to genetic research and its potential for treatment of disease. It will be years before many of the genetic tests are available and before genetic diseases can be treated. Society as a whole must use this time to discuss and decide on how genetic information ought to be used, before the choices are made for them. It is a discussion that those with genetic dispositions to diseases such as Huntington's have long wanted to make more public.

Understanding the ethical and moral implications of genetic information and technology is crucial towards ensuring the proper use of genetic data. The ELSI programs have had a positive influence in understanding problematic areas surrounding the HGP by acting as a center for discussion for many bioethicists and scientists. However, ELSI is often too passive and does not provide enough practical guidance to the public on the complex implications of the HGP. As it stands now, there is room for major improvement such as increased cooperation between ELSI and non-government organizations. NGO's would serve as bridges between ELSI and the public, and collaboration would enable in-depth probing and more comprehensive analysis of issues of public concern.

New knowledge in the area generated by HGP can result in new ways to diagnose, treat, and prevent many genetic disorders. Apart from medical applications, knowing about our DNA can affect the world's condition. It can lead to problem solving in energy sources and environmental cleanup plus better health care. Learning about DNA and gaining an understanding of human biology will give us a better world and a better life, as long as we keep it within the boundaries of ethical, legal, and social concern.

References

Barton, J.H., 2002, Patents, genomics, research and diagnostics. *Acad Med* 77:1348–1367.

Durfy, S.J., 1993, Ethics and the Human Genome Project. *Arch Pathol Lab Med* 117(5):466–469.

Ellerin, B.E., Schneider, R.J., Stern, A., Toniolo, P.G., Formenti, S.C., 2005, Ethical, legal, and social issues related to genomics and cancer research: the impending crisis. *J Am Coll Radiol* 2(11):919–926.

Goldstein, J.A., Golod, E., 2002, Human gene patents. *Acad Med* 77(12Pt 2):1315–1328.

Gutiérrez-Samperio, C., 2002, Bioethics in genetic engineering. *Gac Med Mex* 138(1):109–119.

Jin, J., 2000, An evaluation of the ethical, legal and social implications program of the U.S. Human Genome Project. *Princet J Bioeth Spring* 3(1):35–50.

Kenner, C., Lewis, J.A., Pressler, J.L., Little, C.M., 2008, Neonatal genetic testing is more than screening. *Crit Care Nurs Clin North Am* 20(2):233–237.

Kessler, D.A., Siegel, J.P., Noguchi, P.D., Zoon, K.C., Feiden, K.L., Woodcock, J., 1993, Regulation of somatic-cell therapy and gene therapy by the food and drug administration. *New Eng J Med* 329(16):1169–1173.

Masui, T., Takada, Y., 2003, Ethical, legal, and social issues of genome research – new phase of genome research desperately requires social understanding and safeguards on the use of medical records and other personal information. *Yakugaku Zasshi* 123(3):107–119.

McCunney, R.J., 2002, Genetic testing: ethical implications in the workplace. *Occup Med* 17(4):665–672.

Meslin, E.M., Thomson, E.J., Boyer, J.T., 1997, The Ethical, Legal, and Social Implications Research Program at the National Human Genome Research Institute. *Kennedy Inst Ethics J* 7(3):291–298.

Murdoch, C., Caulfield, T., 2009, Commercialization, patenting and genomics: researcher perspectives. *Genome Med* 1(2):22.

Resnik, D.B., 2003, Are DNA patents bad for medicine? *Health policy* 65(2):181–197.

Tsai, S.F., Chen, P.J., 2000, Impact of human genome research on medicine – the initial Taiwan experience. *J Formos Med Assoc* 99(2):107–115.

Van Leeuwen, D.J., Bernat, J.L., 2006, Ethical, social and legal implications of genetic testing in liver disease. *Hepatology* 43(6):1195–1201

Zneimer, S.M., 2002, The human genome project: exploring its progress and successes and the ethical, legal, and social implications. *Clin Leadersh Manag Rev* 16(3):151–157.

Chapter 9
Polymerase Chain Reaction

Abstract Kary Mullis devised a method of replicating genes called "PCR" (polymerase chain reaction). A DNA sequence less than one part in a million of the total sample can be cloned. In fact, a single gene can be amplified into millions of duplicate copies. In order to determine the exact DNA sequence of a gene or section of DNA, it is necessary to have an adequate sample of the particular gene to work with. PCR allows a researcher to replicate a gene into a workable amount. PCR can achieve more sensitive detection and higher levels of amplification of specific sequences in less time than previously used methods. These features make the technique extremely useful, not only in basic research, but also in commercial uses, including genetic identity testing, forensics, industrial quality control and in vitro diagnostics. Basic PCR has become commonplace in many molecular biology labs where it is used to amplify DNA fragments and detect DNA or RNA sequences within a cell or environment. However, PCR has evolved far beyond simple amplification and detection, and many extensions of the original PCR method have been described.

Keywords Amplification · DNA sequencing · Site specific mutation · Polymerisation · PCR · Taq polymerase · *Thermus auquaticus* · Polymerase chain reaction · Primer design · GC-content · Melting temperature · T_m · Template · Real-time PCR · BRCA1 · Cystic fibrosis · RT-PCR

9.1 Prologue

On 10th of December 1993, Kary Mullis received the Noble prize in Chemistry from King Carl XVI Gustaf of Sweden for his invention of polymerase chain reaction. Mullis had published and patented his work in 1985. The applications of this method have revolutionised both basic and applied sciences. The Polymerase Chain Reaction (PCR) provides an extremely sensitive means of amplifying small quantities of DNA. The development of this technique resulted in an explosion of new techniques in molecular biology as more and more applications of the method were published. The technique was made possible by the discovery of Taq polymerase obtained from *Thermus auquaticus*. This DNA polymerase is stable at the high

M. Debnath et al., *Molecular Diagnostics: Promises and Possibilities*,
DOI 10.1007/978-90-481-3261-4_9,

temperatures needed to perform the amplification, whereas other DNA polymerases become denatured. Mullis defined PCR not as a specific technique but rather as a concept.

Since this technique involves amplification of DNA, the most obvious application of the method is in the detection of minuscule amounts of specific DNA. This is important in the detection of low level bacterial infections or rapid changes in transcription at the single cell level, as well as the detection of a specific individual's DNA. It can also be used in DNA sequencing, screening for genetic disorders, site specific mutation of DNA, or cloning or subcloning of cDNAs.

9.2 The Concept of Polymerase Chain Reaction

PCR, like DNA sequencing, is based on the DNA polymerization reaction. PCR is used in genome sequencing, including the Human genome project. Using the randomn primers, the entire sequence of a genome can be amplified in pieces. A primer and dNTPs are added along with a DNA template and the DNA polymerase (in this case, *Taq*). The main concern with PCR is that, in addition to using a primer that sits on the 5′ end of the gene and makes a new strand in that direction, a primer is made to the opposite strand to go in the other direction. The original template is melted (at 94°C), the primers anneal (@ 45–55°C) and the polymerase makes two new strands (@ 72°C), doubling the amount of DNA present. This provides 2 new templates for the next cycle. The DNA is again melted, primers anneal, and the Taq makes 4 new strands. A wrong temperature during the annealing step can result in primers not binding to the template DNA at all, or binding at random. The elongation temperature depends on the DNA Polymerase. The time for this step depends both on the DNA polymerase itself and on the length of the DNA fragment to be amplified.

Prior to the use of thermostable DNA polymerases in PCR, researchers had to laboriously replenish the reaction with fresh enzyme (such as Klenow or T4 DNA polymerase) after each denaturation cycle. Thermostable DNA polymerases revolutionized and popularized PCR because of their ability to withstand the high denaturation temperatures. The use of thermostable DNA polymerases also allowed higher annealing temperatures, which improved the stringency of primer annealing.

Thermostable DNA polymerases can also be used for either one-enzyme or two-enzyme RT-PCR (Myers and Gelfand, 1991; Chiocchia and Smith, 1997). For example, *Tth* DNA polymerase can act as a reverse transcriptase in the presence of Mn^{2+} for one-enzyme RT-PCR (Myers and Gelfand, 1991). All of the DNA polymerases mentioned can be used for amplification of the first-strand cDNA produced by a reverse transcriptase, such as AMV RT, in two-enzyme RT-PCR. The thermostable DNA polymerases can be divided into two groups, those with a 3′→5′ exonuclease (proofreading) activity, such as *Pfu* DNA polymerase, and those without the proofreading function, such as *Taq* DNA polymerase. These two groups have some important differences. Proofreading DNA polymerases are more accurate than nonproofreading polymerases due to the 3′→5′ exonuclease activity, which

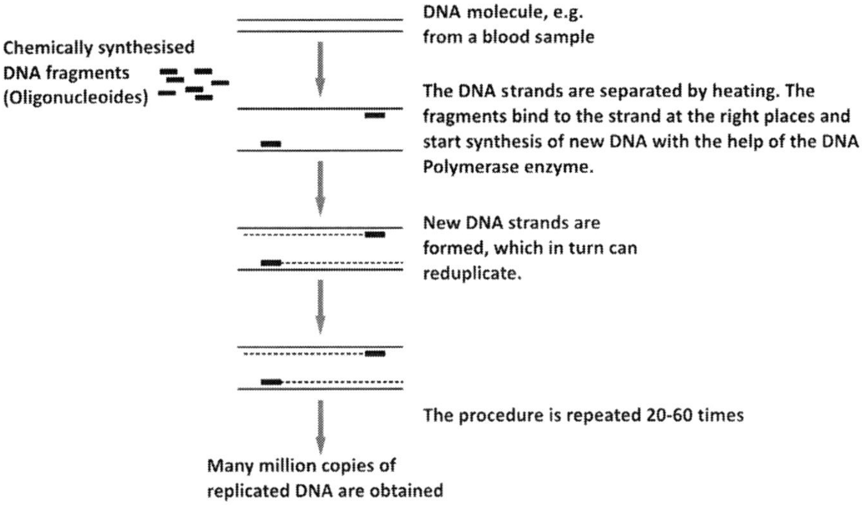

Fig. 9.1 Concept of PCR

can remove a misincorporated nucleotide from a growing chain of DNA. When the amplified product is to be cloned, expressed or used in mutation analysis, *Pfu* DNA polymerase is a much better choice due to its high fidelity. However, for routine PCR, where simple detection of an amplification product is the goal, *Taq* DNA polymerase is the most commonly used enzyme because yields tend to be higher with a nonproofreading DNA polymerase (Fig. 9.1).

Taq DNA polymerase (*Taq*) can amplify long DNA fragments, if the conditions were set properly. Taq could successfully perform the "long PCR" up to 24 kb. The conditions include: (1) longer primers, (2) a 2-step cycling, and (3) a "long buffer." The most important requirements are the survival rate of *Taq* at high temperatures and that of the primers against the 5′–3′ exonuclease activity of *Taq* (Lee et al., 2009).

Amplification with nonproofreading DNA polymerases results in the template-independent addition of a single nucleotide to the 3′-end of the PCR product, whereas the use of proofreading DNA polymerases results in blunt-ended PCR products (Hu, 1993). The single-nucleotide overhang can simplify the cloning of PCR products. Proofreading DNA polymerases are also used in blends with nonproofreading DNA polymerases, or amino-terminally truncated versions of *Taq* DNA polymerase, to amplify longer stretches of DNA with greater accuracy than the nonproofreading DNA polymerase alone (Cline et al., 1996).

A disadvantage of *Taq* is that it sometimes makes mistakes when copying DNA, leading to mutations (errors) in the DNA sequence, since it lacks 3′→5′ proofreading exonuclease activity. Polymerases such as *Pwo* or *Pfu*, obtained from *Archaea*, have *proofreading mechanisms* (mechanisms that check for errors) and can significantly reduce the number of mutations that occur in the copied DNA sequence. However these enzymes polymerize DNA at a much slower rate than Taq.

Combinations of both *Taq* and *Pfu* are available nowadays that provide both high processivity (fast polymerization) and high fidelity (accurate duplication of DNA).

Polymerase chain reaction (PCR) is widely used in different areas of science and biotechnology. Grouping of PCR primer sets for amplification in a single tube (multiplexing) provides substantial savings in terms of time, chemicals and, most importantly sample materials. Thus, multiplexing is a powerful way for optimizing the cost of genetic analysis and some procedures of PCR are patented (Piepenbrock et al., 2004), which further demonstrates its importance. While a lot of effort has been put into the experimental optimization of multiplex PCR conditions (Henegariu et al., 1997; Zangenberg et al., 1999), little attention has been paid to the prediction of primer compatibility in multiplex reactions. Smaller multiplex groups are often composed manually and verified by trial and error, but this kind of approach is suboptimal for large datasets. Although there are software programs available for testing the compatibility of PCR primer pairs for multiplexing (Rychlik, 1995; Vallone and Butler, 2004), these do not perform automatic grouping and are unable to handle large datasets automatically.

9.3 PCR Optimisation

The Polymerase Chain Reaction (PCR) provides an extremely sensitive means of amplifying small quantities of DNA. Each step of the cycle should be optimized for each template and primer pair combination. Successful amplification depends on many factors viz. DNA template quantity and quality, magnesium ions as cofactors of Taq polymerase etc.

9.3.1 Magnesium Concentration

Magnesium is a required cofactor for thermostable DNA polymerases, and magnesium concentration is a crucial factor that can affect the success of the amplification. Template DNA concentration, chelating agents present in the sample (e.g., EDTA or citrate), dNTP concentration and the presence of proteins can all affect the amount of free magnesium in the reaction. Excess free magnesium reduces enzyme fidelity (Eckert and Kunkel, 1990) and may increase the level of nonspecific amplification (Williams, 1989; Ellsworth et al., 1993).

9.3.2 Buffer Concentration

Most reaction buffers consist of a buffering agent, most often a Tris-based buffer, and salt, commonly KCl. The buffer regulates the pH of the reaction, which affects the DNA polymerase activity and fidelity. Modest concentrations of KCl will increase DNA polymerase activity by 50–60% over activities in the absence of KCl; 50 mM KCl is considered optimal (Gelfand and White, 1990).

9.3.3 Enzyme Concentration

1–1.25 units of *Taq* DNA polymerase in a 50 μl amplification reaction is usually recommended. In most cases, this is an excess of enzyme, and adding more enzyme will not significantly increase product yield. In fact, increased amounts of enzyme increase the likelihood of generating artifacts associated with the intrinsic 5′→3′ exonuclease activity of *Taq* DNA polymerase, resulting in smeared bands in an agarose gel (Longley et al., 1990; Bell and DeMarini, 1991).

9.3.4 PCR Primer Design

PCR primers define the target region to be amplified and generally range in length from 15–30 bases. Ideally primers will have a GC-content of 40–60%. Avoid three G or C residues in a row near the 3′-end of the primer to minimize nonspecific primer annealing. Also, avoid primers with intra- or intermolecular complementary sequences to minimize the production of primer-dimer. Intramolecular regions of secondary structure can interfere with primer annealing to the template and should be avoided.

Ideally, the melting temperature (T_m), the temperature at which 50% of the primer molecules are annealed to the complementary sequence, of the two primers will be within 5°C, so the primers anneal efficiently at the same temperature. Primers can be designed to include sequences that can be useful for downstream applications. For example, restriction enzyme sites can be placed at the 5′-ends of the PCR primers to facilitate subsequent cloning of the PCR product, or a T7 RNA polymerase promoter can be added to allow in vitro transcription without the need to subclone the PCR product into a vector.

9.3.5 Template Quality

Successful amplification depends on DNA template quantity and quality. Reagents commonly used in the purification of nucleic acids (salts, guanidine, proteases, organic solvents and SDS) are potent inactivators of DNA polymerases. For example, 0.01% SDS will inhibit *Taq* DNA polymerase by 90%, while 0.1% SDS will inhibit *Taq* DNA polymerase by 99.9% (Konat et al., 1994). A few other examples of PCR inhibitors are phenol (Katcher and Schwartz, 1994), heparin (Beutler et al., 1990; Holodniy et al.,1991), xylene cyanol, bromophenol blue (Hoppe et al., 1992), plant polysaccharides (Demeke and Adams, 1992), and the polyamines spermine and spermidine (Ahokas and Erkkilä, 1993). In some cases, the inhibitor is not introduced into the reaction with the nucleic acid template.

If an amplification reaction fails and you suspect the DNA template is contaminated with an inhibitor, add a control DNA and primer pair that has amplified well in the past to the amplification reaction with the suspect DNA preparation. Failure to amplify the control DNA usually indicates the presence of an inhibitor. Additional steps to clean up the DNA preparation, such as phenol/chloroform extraction or ethanol precipitation, may be necessary.

9.3.6 Template Quantity

The amount of template required for successful amplification depends upon the complexity of the DNA sample. Common mistakes include using too much plasmid DNA, too much PCR product or too little genomic DNA as the template. Reactions with too little DNA template will have low yields, while reactions with too much DNA template can be plagued by nonspecific amplification.

9.3.7 Cycling Parameter

The two most commonly altered cycling parameters are annealing temperature and extension time. The lengths and temperatures for the other steps of a PCR cycle do not usually vary significantly. However in some cases, the denaturation cycle can be shortened or a lower denaturation temperature used to reduce the number of depurination events, which can lead to mutations in the PCR products.

Primer sequence is a major factor that determines the optimal annealing temperature, which is often within 5°C of the melting temperature (T_m) of the primers. Using an annealing temperature slightly higher than the primer T_m will increase annealing stringency and can minimize nonspecific primer annealing, decreasing the amount of undesired products synthesized. However, using an annealing temperature lower than the primer T_m can result in higher yields, as the primers anneal more efficiently at the lower temperature. Numerous formulas exist to determine the theoretical T_m of nucleic acids (Baldino, Jr. et al., 1989; Rychlik et al., 1990). The formula below can be used to estimate the melting temperature for oligonucleotides:

$$T_m = 81.5 + 16.6 \times (\log_{10}[Na^+]) + 0.41 \times (\%G{+}C) - 675/n$$

where $[Na^+]$ is the molar salt concentration, $[K^+] = [Na^+]$ and n = number of bases in the oligonucleotide.

The length of the extension cycle, which may also need to be optimized, depends on the size of the PCR product and the DNA polymerase being used. In general, allow approximately 1 min for every 1 kb of amplicon (minimum extension time = 1 min) for nonproofreading DNA polymerases and 2 min for every 1 kb of amplicon for proofreading DNA polymerases. Avoid excessively long extension times, as they can increase the likelihood of generating artifacts associated with the intrinsic $5' \rightarrow 3'$ exonuclease activity of *Taq* DNA polymerase (Longley et al., 1990; Bell and DeMarini, 1991).

PCR typically involves 25–35 cycles of amplification. The risk of undesirable PCR products appearing in the reaction increases as the number of cycles increases, so we recommend performing only enough cycles to synthesize the desired amount of product. If nonspecific amplification products accumulate before sufficient amounts of PCR product can be synthesized, consider diluting the products of the first reaction and performing a second amplification with the same primers or primers that anneal to sequences within the desired PCR product (nested primers).

9.3.8 PCR Enhancers and Additives

The addition of PCR-enhancing agents can increase yield of the desired PCR product or decrease the production of undesired products. There are many PCR enhancers, which can act through a number of different mechanisms. These reagents will not enhance all PCR reactions; the beneficial effects are often template- and primer-specific and will need to be determined empirically. Some of the more common enhancing agents are discussed below.

The addition of betaine, DMSO and formamide can be helpful when amplifying GC-rich templates and templates that form strong secondary structures, which can cause DNA polymerases to stall. GC-rich templates can be problematic due to inefficient separation of the two strands of DNA or the tendency for the complementary, GC-rich primers to form intermolecular secondary structures, which will compete with primer annealing to the template. Betaine reduces the amount of energy required to separate the strands of DNA templates (Rees et al., 1993). DMSO and formamide are thought to aid in amplification in a similar manner by interfering with the formation of hydrogen bonds between the two strands of DNA (Geiduschek and Herskovitzs, 1961).

Some reactions that amplify poorly in the absence of enhancers will give a higher yield of PCR product when betaine (1 M), DMSO (1–10%) or formamide (1–10%) are added. Concentrations of DMSO greater than 10% and formamide greater than 5% can inhibit *Taq* DNA polymerase and presumably other DNA polymerases as well (Varadaraj and Skinner, 1994). Specific examples of the effects of DMSO and betaine have been published (Frackman et al., 1998).

In some cases, general stabilizing agents such as BSA (0.1 mg/ml), gelatin (0.1–1.0%) and nonionic detergents (0–0.5%) can overcome failures to amplify a region of DNA. These additives can increase DNA polymerase stability and reduce the loss of reagents through adsorption to the tube walls. BSA has also been shown to overcome the inhibitory effects of melanin on RT-PCR (Giambernardi et al., 1998). Nonionic detergents, such as Tween®-20, NP-40, and Triton® X-100, have the added benefit of overcoming the inhibitory effects of trace amounts of strong ionic detergents, such as 0.01% SDS (Gelfand and White, 1990). Ammonium ions can make an amplification reaction more tolerant of nonoptimal conditions. For this reason, some PCR reagents include 10–20 mM $(NH_4)_2SO_4$. Other PCR enhancers include glycerol (5–20%), polyethylene glycol (5–15%) and tetramethyl ammonium chloride (60 mM).

9.3.9 Nucleic Acid Cross-Contamination

It is important to minimize cross-contamination between samples and prevent carryover of RNA and DNA from one experiment to the next. Use separate work areas and pipettors for pre- and post-amplification steps. Use positive displacement pipets or aerosol-resistant tips to reduce cross-contamination during pipetting. Wear gloves and change them often.

There are a number of techniques that can be used to prevent the amplification of DNA contaminants. PCR reagents can be treated with isopsoralen, a photo-activated, cross-linking reagent that intercalates into double-stranded DNA molecules and forms covalent, interstrand crosslinks, to prevent DNA denaturation and replication. These interstrand crosslinks effectively render contaminating DNA unamplifiable.

Treatment of the PCR reagents with uracil-N-glycosylase (UNG), a DNA repair enzyme that hydrolyzes the base-ribose bond at uracil residues, eliminates one of the most common sources of DNA contamination: previously amplified PCR products. UNG treatment prevents replication of uracil-containing DNA by causing the DNA polymerase to stall at the resulting abasic sites. For UNG to be an effective safeguard against contamination, the products of previous amplifications must have been synthesized in the presence of dUTP. This is easily accomplished by substituting dUTP for some or all of the dTTP in the reaction. Nonproofreading polymerases will readily incorporate dUTP into a PCR product, but proofreading polymerases incorporate dUTP much less efficiently (Slupphaug et al., 1993; Greagg et al., 1999; Lasken et al., 1996). Since the incorporation of dUTP has no noticeable effect on the intensity of ethidium bromide staining or on the electrophoretic mobility of the PCR product, the reactions can be analyzed by standard agarose gel electrophoresis. While both methods are effective (Rys and Persing, 1993), UNG treatment has the advantage that both single-stranded and double-stranded DNA templates will be rendered unamplifiable (Longo et al., 1990).

9.4 Advances in the PCR Technique

The advent of the polymerase chain reaction (Hooberman, 1992; Chan and Greiner, 1994) has provided an alternative to Southern blot analysis and has the following advantages: it is technically simpler and has a much faster turnaround time; the amount of clinical materials required is much smaller; and in most situations, the test can be performed on archival, formalin-fixed, paraffin-embedded materials. Modifications of the technique to include reverse transcription make it possible to use RNA as the starting material (reverse transcription polymerase chain reaction; Elmberger et al., 1995). With PCR, it is routinely possible to amplify enough DNA from a single hair follicle for DNA typing. Some workers have successfully amplified DNA from a single sperm cell. The PCR technique has even made it possible to analyze DNA from microscope slides of tissue preserved years before. However, the great sensitivity of PCR makes contamination by extraneous DNA a constant problem. The polymerase chain reaction is a technique for quickly "cloning" a particular piece of DNA in the test tube (rather than in living cells like *E. coli*).

PCR is used also used to detect mutations after sequencing. Recently DNA genotyping with mutation specific TaqMan® probes has been reported in detection of single nucleotide polymorphism (Li, 2009) using coamplification at low denaturation temperature – PCR (COLD-PCR) .This helped to identify low level mutation in clinical samples and to identify tumor protein p53 (TP 53) and epidermal growth factor receptor (EGFR) mutations in tumors.

Recent technical advances have allowed the accurate quantitation of the template in real time (Erlich et al., 1991; Gerrard et al., 1998; Orlando et al., 1998; Luthra et al., 1998). The latter has major implications in the monitoring of residual disease.

9.4.1 RT-PCR

The discovery of reverse transcriptases, or RNA-dependent DNA polymerases, and their role in retrovirus infection (Baltimore, 1970; Temin and Mizutani, 1970) altered molecular biology's central dogma of DNA→RNA→protein. Reverse transcriptases use an RNA template to synthesize DNA and require a primer for synthesis, like other DNA polymerases. RT-PCR (reverse transcription-polymerase chain reaction) is the most sensitive technique for mRNA detection and quantitation currently available. Compared to the two other commonly used techniques for quantifying mRNA levels, Northern blot analysis and RNase protection assay, RT-PCR can be used to quantify mRNA levels from much smaller samples. In fact, this technique is sensitive enough to enable quantitation of RNA from a single cell.

9.4.2 "Hot Start" PCR

Hot Start PCR allows the inhibition of polymerase activity during PCR reaction preparation (Birch, 1996). By limiting polymerase activity prior to PCR cycling, Hot Start PCR reduces non-specific amplification and increases PCR product target yield. To perform hot-start PCR, the reactions are assembled on ice or at room temperature, but one critical component is omitted until the reaction has been heated to 60–65°C, at which point the missing reagent is added. This omission prevents the polymerase from extending primers until the critical component is added at the higher temperature where primer annealing is more stringent. Hot Start PCR is commonly performed by using included chemical modifications (Kainz et al., 2000), wax-barrier methods (Horton et al., 1994, Bassam and Caetano-Anolles, 1993) and inhibition by a Taq-directed antibody (Takahashi et al.,1996).

However, this method is tedious and increases the risk of contamination. A second, less labor-intensive approach involves the reversible inactivation or physical separation of one or more critical components in the reaction. For example, the magnesium or DNA polymerase can be sequestered in a wax bead, which melts as the reaction is heated to 94°C during the denaturation step, releasing the component only at higher temperatures (Krishnan et al., 1991). Alternatively, the DNA polymerase can be kept in an inactive state by binding to an oligonucleotide, also known as an aptamer, (Lin and Jayasena, 1997) or an antibody (Scalice et al., 1994).

This bond is disrupted at the higher temperatures, releasing the functional DNA polymerase.

9.4.3 Long Range PCR

Long range PCR allows the amplification of PCR products, which are much larger than those achieved with conventional Taq polymerases. Up to 27 kb fragments are possible from good quality genomic DNA, although 10–20 kb fragments are routinely achievable, given the appropriate conditions. The method relies on a mixture of thermostable DNA polymerases, usually Taq DNA polymerase for high processivity (i.e. 5′-3′ polymerase activity) and another DNA polymerase with 3'-5' proofreading abilities (usually *Pwo*). This combination of features allows longer primer extension than can be achieved with Taq alone. Amplification of long DNA fragments is desirable for numerous applications such as physical mapping applications (Rose, 1991) and direct cloning from genomes.

9.4.4 Inverse PCR

Inverse PCR is one of numerous tools utilized for furthering our understanding of the complexities of DNA, and facilitates cloning and sequencing of otherwise unknown elements of a genome. Inverse PCR (IPCR), described by Ochman et al. in 1988, is a method for the rapid in vitro amplification of DNA sequences that flank a region of known sequence. The method uses the polymerase chain reaction (PCR), but it has the primers oriented in the reverse direction of the usual orientation. The template for the reverse primers is a restriction fragment that has been ligated upon itself to form a circle. Inverse PCR has many applications in molecular genetics, for example, the amplification and identification of sequences flanking transposable elements. Mostly it is done when bordered sequences of genes are not known. The borders of the vectors are used as primers and thus polymerisation proceeds in inverse direction i.e. away from the vector sequence flanked by the primers towards the DNA sequence of inserted segment.

9.4.5 Anchored PCR

This is applicable when only one of the two ends of the DNA sequence is to be amplified. In such cases only one primer instead of two primers are used .Variety of polymerase chain reaction in which only enough information is known to make a single primer. A known sequence is thus added to the end of the DNA, perhaps by enzymic addition of a polynucleotide stretch or by ligation of a known piece of DNA. The PCR can then be performed with the gene-specific primer and the anchor primer.

9.4.6 Nested Primer PCR

PCR amplification is performed with one set of primers and then some product is taken with or without removal of reagents for re-amplification with an internally-situated, "nested"set of primers. This process adds another level of specificity, meaning that all products non-specifically amplified in the first round will not be amplified in the second.

Nested PCR uses two sets of amplification primers (Gong et al., 2002). The target DNA sequence of one set of primers (termed "inner" primers) is located within the target sequence of the second set of primers (termed "outer" primers). In practice, a standard PCR reaction is first run with the patient sample using the "outer primers". Then a second PCR reaction is run with the "inner primers" using the product of the first reaction as the amplification target. This procedure increases the sensitivity of the assay by reamplifying the product of the first reaction in a second reaction. The specificity of the assay is increased because the inner primers amplify only if the first PCR reaction yielded a specific product.

9.4.7 Colony PCR

Colony PCR is the screening of bacterial (*E. coli*) or yeast clones for correct ligation or plasmid products (Plourde-Owobi et al., 2005). Selected colonies of bacteria or yeast are picked with a sterile toothpick or pipette tip from a growth (agarose) plate. This is then inserted into the PCR master mix or pre-inserted into autoclaved water. PCR is then conducted to determine if the colony contains the DNA fragment or plasmid of interest.

9.4.8 Quantitative PCR

PCR and RT-PCR are generally used in a qualitative format for evaluating biological samples. However, a wide variety of applications, such as the determining viral load, measuring responses to therapeutic agents and characterizing gene expression, would be improved by quantitative determination of target abundance. Theoretically, this should be easy to achieve, given the exponential nature of PCR, because a linear relationship exists between the number of amplification cycles and the logarithm of the number of molecules. In practice however, the actual efficiency of amplification is decreased because of contaminants (inhibitors), competitive reactions, substrate exhaustion, inactivation of the polymerase and target reannealing. As the number of cycles increases, the amplification efficiency decreases, eventually resulting in a plateau effect.

Normally, quantitative PCR requires the measurement to be taken before the plateau phase, so the relationship between the number of cycles and molecules is relatively linear. This point must be determined empirically for different reactions because of the numerous factors that can affect the amplification efficiency. Because the measurement is taken prior to the reaction plateau, quantitative PCR uses fewer amplification cycles than basic PCR. This can cause problems in the detection of the final product because there is less product to detect.

To monitor the efficiency of amplification, many applications are designed to include an internal standard in the PCR. One such approach includes a second primer pair that is specific for a "housekeeping" gene (i.e., a gene that has constant expression levels among the samples compared) in the reaction (Gaudette and Crain, 1991). Amplification of housekeeping genes verifies that the target nucleic acid and reaction components were of acceptable quality but does not account for differences in amplification efficiencies due to differences in product size or primer annealing efficiency between the internal standard and target being quantified.

The concept of competitive PCR, a variation of quantitative PCR is a response to this limitation. In competitive PCR, a known amount of a control template is added to the reaction. This template is amplified using the same primer pair as the experimental target molecule but yields a distinguishable product (e.g., different size, restriction digest pattern, etc.). The amounts of control and test product are compared after amplification. While these approaches control for the quality of the target nucleic acid, buffer components and primer annealing efficiencies, they have their own limitations (Siebert and Larrick, 1993; McCulloch et al., 1995), including the fact that many depend on final analysis by electrophoresis.

Numerous fluorescent solution and solid-phase assays have been described to measure the amount of amplification product generated in each reaction, but they can fail to discriminate amplified DNA of interest from nonspecific amplification products. Some of these analyses rely on blotting techniques, which introduces another variable due to nucleic acid transfer efficiencies, while other assays have been developed to eliminate the need for gel electrophoresis yet provide the requisite specificity. Real-time PCR, which provides the ability to view the results of each amplification cycle, is a popular way of overcoming the need for analysis by electrophoresis.

9.4.9 Real-Time PCR

The use of fluorescently labeled oligonucleotide probes or DNA-binding fluorescent dyes, such as SYBR® green, to detect and quantitate a PCR product allows quantitative PCR to be performed in real time (Antwerpen, 2008). DNA-binding dyes are easy to use but do not differentiate between specific and nonspecific PCR products. Fluorescently labeled nucleic acid probes have the advantage that they react with only specific PCR products. These probes can also be used to detect single nucleotide polymorphisms (Bernard et al., 1998).

Real-time PCR using labeled oligonucleotide probes employs two different fluorescent reporters and relies on the transfer of energy from one reporter (the energy donor) to the second reporter (the energy acceptor) when the reporters are in close proximity. The second reporter can be a quencher or a fluor. If the second reporter is a quencher, the energy from the first reporter is absorbed but re-emitted as heat rather than light, leading to a decrease in the fluorescent signal. Alternatively, if the second reporter is a fluor, the energy can be absorbed and re-emitted at another wavelength through fluorescent resonance energy transfer (FRET), and the progress of the reaction can be monitored by the decrease in fluorescence of the energy donor or the increase in fluorescence of the energy acceptor. During the exponential phase of PCR, the change in fluorescence is proportional to the accumulation of PCR product. To simplify quantitation, specially designed instruments perform both the thermal cycling steps to amplify the target and the fluorescence detection to monitor the change in fluorescence in real time during each PCR cycle. Real time reverse PCR is a modification of the RT-PCR and is different from other quantitative PCR as it quantitates the initial amount of the template instead of detecting the amount of final amplified product.

9.4.10 Rapid Amplification of cDNA Ends (RACE)

Rapid amplification of cDNA ends (RACE) is a variation of RT-PCR that amplifies unknown cDNA sequences corresponding to the 3′- or 5′-end of the RNA (Hubank and Schatz, 1994). Numerous variations of the original protocols have been published (Troutt et al., 1992; Edwards et al., 1991; Edwards, 1993; Liu and Gorovsky, 1993; Fromont-Racine et al., 1993; reviewed in Schaefer, 1995) but will not be discussed in detail here.

Two general RACE strategies exist: one amplifies 5′ cDNA ends (5′ RACE) and the other captures 3′ cDNA end sequences (3′ RACE). In either strategy, the first step in the RACE reaction involves the conversion of RNA into single-stranded cDNA using a reverse transcriptase. For the subsequent amplification reaction, two PCR primers are designed to flank the unknown sequence. One PCR primer is complementary to known sequences within the gene, and a second primer is complementary to an "anchor" site (anchor primer). The anchor site may be present naturally, such as the poly(A) tail of most mRNAs, or can be added in vitro after completion of the reverse transcription step. The anchor primer can also carry adaptor sequences, such as restriction enzyme recognition sites, to facilitate subsequent cloning of the amplified product. Amplification using these two PCR primers results in a product that spans the unknown 5′ or 3′ cDNA sequence, and sequencing this product will reveal the unknown sequence. The information obtained from partial cDNA sequences can then be used to assemble the sequence of the full-length cDNA (Frohman et al., 1988; Loh et al., 1989; Ohara et al., 1989).

9.4.11 AFLP

Amplified Fragment Length Polymorphism (AFLP) is a highly sensitive PCR-based method for detecting polymorphisms in DNA. AFLP can be also used for genotyping individuals for a large number of loci using a minimal number of PCR reactions (Vos et al., 1995). AFLP is composed of 3 steps:

1. Cellular DNA is digested with one or more restriction enzymes. Typically this involves a combination of two restriction enzymes: a 4 base cutter (MseI) and a 6 base cutter (EcoRI). Ligation of linkers (restriction half-site specific adaptors) to all restriction fragments.
2. Pre-selective PCR is performed using primers which match the linkers and restriction site specific sequences.
3. Electrophoretic separation and amplicons on a gel matrix, followed by visualisation of the band pattern. The aim of this tool is to perform a theoretical AFLP-PCR experiment by using the same principles, and to suggest the adaptors and primers needed in the experiment.

9.4.12 In Situ PCR

In situ PCR, first described in 1990, combines the sensitivity of PCR or RT-PCR amplification with the cellular or histological localization associated with in situ hybridization techniques (Haase et al., 1990). These features make in situ PCR a powerful tool for detecting proviral DNA, oncogenesis and localization of rare messages.

The technique is amenable to analysis of fixed cells or tissue cross-sections. Detection of amplified products can be accomplished indirectly by subsequent hybridization using either radiolabeled, fluorescently labeled or biotin-labeled nucleic acid probes. PCR products can also be detected directly by the incorporation of a labeled nucleotide, although this method is subject to higher background levels.

The use of in situ PCR requires altering some of the reaction parameters typical of basic PCR (Nuovo et al., 1993; Thaker, 1999). For example, increased Mg^{2+} concentrations (approximately 4.5 mM versus the normal 1.5–2.5 mM) are used for in situ PCR. An increased amount of DNA polymerase is also required unless BSA is added to the reaction, presumably because the polymerase binds to the glass plate and coverslip.

9.5 Cloning PCR Products

9.5.1 T-A Cloning Strategy

Taq and other polymerases seem to have a terminal transferase activity which results in the non-templated addition of a single nucleotide to the 3′-ends of PCR products. In the presence of all 4 dNTPs, dA is preferentially added; however, use of a single dNTP in a reaction mix results in (relatively inefficient) addition of that nucleotide. This complicates cloning, as the supposedly blunt-ended PCR product often is not, and blunt-ended cloning protocols often do not work or are very inefficient. This can be remedied by incubation of PCR products with T4 DNA pol or Klenow pol, which "polishes" the ends due to a $3' \rightarrow 5'$ exonuclease activity. However, this terminal transferase activity is also the basis of a clever cloning strategy: this uses Taq pol to add a single dT to the 3′-ends of a blunt-cut cloning vector such as pUC or pBluescriptTM, and simple ligation of the PCR product into the now "sticky-ended" plasmid.

9.5.2 Incorporation of Restriction Sites in Primers

Although this may be rendered simple by incorporating the same or different restriction sites at the 5′-ends of PCR primers which allows generation of sticky ends and straightforward cloning into appropriate vectors – these should have at least two additional bases 5′ to the recognition sequence to ensure that the enzymes will in fact recognise the sequence – and it is often found that even when this is done, the efficiency of cutting of fresh product is next to zero. This can sometimes be remedied by incubating fresh product with Proteinase K (to digest off tightly-attached Taq pol), but often is not. A solution to the problem is to use the "Klenow-Kinase-Ligase" (KKL) method: this involves "polishing" products with Klenow, kinasing them to get 5′-phosphorylation, ligating the fragments together to get concatemers, then restricting these with the appropriate restriction enzymes to generate the sticky-ended fragments suitable for cloning.

9.6 PCR as a Diagnostic Tool

The Human Genome Project is a worldwide endeavor to map the DNA base sequence of every gene in the human genome. As of February 2001, the total number of functional genes is considerably less than expected, about 30,000 genes per cell compared with previous estimates of 100,000 genes. It has been estimated that a human somatic cell contains about 5 billion base pairs. If the average gene contains 1,500 bases, then 30,000 functional genes is only about 1% of the total DNA per cell. Although there is an estimated six feet of DNA per human cell, only a small fraction of this amount consititutes the actual protein-coding genes.

PCR can be used in the analysis of disease genes by being able to amplify detectable amounts of specific fragments of DNA. The amplified fragments from disease genes may be larger, due to insertions, or smaller, due to deletions. The dramatic amplification of DNA by PCR allows the analysis of disease genes in extremely small samples of DNA. For example, only a small number of fetal cells need be extracted from amniotic fluid in order to analyze for the presence of specific disease genes. Additionally, single point mutations can be detected by modified PCR techniques such as the ligase chain reaction (LCR) and PCR-single-strand conformational polymorphisms (PCR-SSCP) analysis. The PCR technique also can be used to identify the level of expression of genes in extremely small samples of material, e.g. tissues or cells from the body.

PCR is an extremely valuable technique in forensic criminology involving rape, murder and disputed parentage. DNA can be identified from small samples of blood, saliva, skin, hair follicles and semen. When amplifying genes using PCR, it is imperative that the sample not be contaminated with any foreign DNA, otherwise the foreign genes may be inadvertently amplified.

PCR is also used for genetic testing to determine whether patients can carry a genetic mutation that could be passed on to their children (e.g. the mutation that causes cystic fibrosis) or to determine disease risk in patients themselves (e.g. a mutation in the gene BRCA1 predisposes a women to breast or ovarian cancer).

Much of the DNA of humans is referred to as "variable number tandem repeats" (VNTRs) rather than specific protein-coding genes. The greatest variation in the DNA of two individuals is not in the protein-coding genes, but in the nonprotein-coding sections of their DNA. Natural selection has resulted in some time-tested DNA sequences called genes which are identical in normal individuals. The exact number and order of amino acids in protein molecules are determined by the DNA base sequences of genes, and genetic mutations are essentially "misspelled" genes. Genetic mutations, including variations in the base sequences of vital genes, may be fatal if they fail to code for a vital enzyme. For example, the dominant gene for hemoglobin is a time-tested sequence of DNA bases that is essential for the production of this life-giving pigment. Hemoglobin is a quaternary protein composed of four polypeptides and 484 amino acids. The substitution of valine for glutamic acid (glutamate) in the beta polypeptide changes the oxygen-carrying potential of this vital blood cell pigment, and is the biochemical explanation for the genetic disease called sickle-cell anemia.

PCR has a major place in human diagnostics, and is gradually emerging as a tool in plant diagnostics, and is now a routine tool in epidemiological and ecological studies, and for genetic mapping – RAPD and AFLP. The molecular diagnostic laboratory is responsible for the development and performance of molecular diagnostic tests for nucleic acid targets found in a variety of settings in medicine. PCR can become an important tool for medical diagnosis (Table 9.1). PCR can detect and identify bacteria and viruses that cause infections such a tuberculosis, chlamydia, viral meningitis, viral hepatitis, HIV, cytomegalovirus and many others. Using various types of PCR, the three broad areas of testing are genetics, hematopathology, and infectious disease.

PCR is a powerful and reliable technique for rapid diagnosis of *M. tuberculosis* (Baek et al., 2000). The usefulness of PCR in the diagnosis of TB by using a variety of unselected clinical specimens is not clear as studies have differed in techniques including lysing method and target nucleic acid to detect products as well as the number and type of samples used, making the reported sensitivities and specificities difficult to compare (Pahwa and Hedau, 2005). A recent study comparing four conventional techniques, FNA cytology, ZN staining, culture and lymph node biopsies, and TB PCR indicated 94.8% diagnosis but PCR was found to be highly sensitive (94.4%) though less specific (38.2%) (Goel et al., 2001).

Two types of JC virus (JCV) are found in infected brain and kidney tissues. A highly reliable allele-specific PCR is used to detect point mutations in cellular genes. (Ault et al., 1994).

Table 9.1 Molecular diagnostic tests using PCR currently performed in a standard laboratory

Genetics		
Factor V Leiden (R506Q) mutation	Allele-specific PCR	APC resistance
MTHFR A233V mutations	PCR-RFLP	Hyperhomocysteinemia
Prothrombin G20210A mutation	PCR-RFLP	Hypercoagulability
Human platelet antigen (HPA0) ($P1_{A1}$ and $P1_{A2}$)	Allele-specific PCR	Coronary artery thrombosis
Hemochromatosis (HFE) C282Y and H63D mutations	PCR-RFLP	Iron overload
Cystic fibrosis (DF508) mutation	Allele-specific PCR	Newborn screening
Hematopathology		
IgH gene rearrangement	PCR	B cell clonality
T-cell receptor gene rearrangement, gamma/beta chains	PCR	B cell clonality
t(9;22) (BCR/ABL)	RT-PCR	Diagnosis of CML
Infectious disease		
Epstein-Barr virus	PCR	Quantitative
Epstein-Barr virus	In situ hybridization	Presense in tissue
Bacterial identification	PCR/DNA sequencing	
Virology		
Adenovirus	PCR (TaqMan)	Qualitative and quantitative
CMV DNA	PCR (TaqMan)	Qualitative and quantitative
EBV DNA	PCR (TaqMan)	Qualitative
Enterovirus RNA	RT-PCR (TaqMan)	Qualitative
HBV DNA	PCR; Roche	Quantitative
HIV-1 RNA	RT-PCR; Roche	Quantitative
HIV-1 DNA	PCR; Roche	Qualitative
HIV-1 resistance genotyping	TruGene sequencing; Bayer	Detects drug resistance mutations
HCV RNA	RT-PCR; Roche	Quantitative
HCV genotype	Invader assay; Third Wave	Qualitative
HSV DNA	PCR (TaqMan)	Qualitative
Human metapneumovirus	RT-PCR (TaqMan)	Qualitative
Influenza A and B	RT-PCR (TaqMan)	Qualitative
Influenza A H5N1	RT-PCR (TaqMan)	Qualitative
Norovirus genogroups I and II	RT-PCR (TaqMan)	Qualitative
Parvovirus B19	PCR (TaqMan)	Qualitative
Polyomavirus BK	PCR (TaqMan)	Qualitative and Quantitative
Polyomavirus JC virus	PCR (TaqMan)	Qualitative
VZV DNA	PCR (TaqMan)	Qualitative

Diagnosis of bacterial disease has improved in recent years, due to the routine use of PCR in the clinical laboratory (Clarke, 2002). The multiplex PCR is a sensitive assay with the ability to detect a mean of one to two genome copy units per 100 μl clinical sample and allows non-culture-based diagnosis of the disease-causing organism within 3 h. A fluorescence-based multiplex PCR was automated for the simultaneous detection of *Neisseria meningitidis*, *Streptococcus pneumoniae* and *Haemophilus influenzae* in clinical samples from patients with suspected meningitis (Smith et al., 2004).

Specific identification of *Bacillus anthracis* and differentiation from closely related *Bacillus cereus* and *Bacillus thuringiensis* strains is still a major diagnostic problem. Commercially available diagnostic kits targeting plasmid-markers cannot differentiate between *B. anthracis*, non-anthracis *Bacillus* species harbouring anthrax-specific virulence plasmids, and plasmidless *B. anthracis* strains. A specific TaqMan PCR assay was successfully designed targeting sequences of gene locus BA 5345 of the *B. anthracis* strain based on a chromosomal marker. In another recent study by Lee and his collegues (2009) the multiplex real-time PCR assay was developed that could sensitively detect *Salmonella* spp. and specifically differentiate *S. typhimurium* from *S. enteritidis* in meats.

PCR-denaturing gradient gel electrophoresis (PCR-DGGE) method is used for the detection and identification of *Campylobacter*, *Helicobacter* and *Arcobacter* species (Epsilobacteria) in clinical samples and evaluates its efficacy on saliva samples from humans and domestic pets. The PCR-DGGE method allows determination of the true prevalence and diversity of Epsilobacteria in clinical and other samples. Contact with the oral cavity of domestic pets may represent a route of transmission for epsilobacterial enteric diseases.

Using "hot start" polymerase chain reaction (PCR) the distribution pattern of *human papillomavirus* (HPV) in vulvar lesions was analysed. After amplification, in situ analysis showed that many of the cells that lacked halos contain HPV DNA and that the hybridization signal often localized to areas where there was a thickened granular layer. HPV DNA was not noted in the basal cells. The one copy of HPV 16 was detectable after PCR with a single primer pair by in situ analysis only if the hot start modification was employed (Nuovo et al., 1992). Gong et al. (2002) employed nested PCR for diagnosis of tuberculous lymphadenitis and PCR-SSCP for identification of rifampicin resistance in fine-needle aspirates.

Real-Time PCR can be applied to traditional PCR applications as well as new applications that would have been less effective with traditional PCR. A fluorogenic-probe hydrolysis (TaqMan)-reverse transcriptase (RT) PCR has been used for evaluation of diagnostic sensitivity of classical swine fever virus using clinical samples (Risatti et al., 2005). With the ability to collect data in the exponential growth phase, the power of PCR has been expanded into applications such as viral quantitation, pathogen detection including CMV detection, rapid diagnosis of meningococcal infection, penicillin susceptibility of *Streptococcus pneumoniae*, *Mycobacterium tuberculosis* and its resistant strains, waterborne microbial pathogens in the environment, genotyping (allelic discrimination) by melting-curve analysis or specific probes/beacons, Trisomies, single-gene copy numbers, microdeletion genotypes, haplotyping, quantitative microsatellite analysis, prenatal diagnosis/sex determination using single cell isolated from maternal blood or fetal DNA in maternal circulation and prenatal diagnosis of haemoglobinopathies intraoperative cancer diagnostics (Wittwer, 2001). Linear-after-the-exponential (LATE)-PCR is a new method for real-time quantitative analysis of target numbers in small samples. It is adaptable to high throughput applications in clinical diagnostics, biodefense, forensics, and DNA sequencing.

Real-time PCR technology may also detect reciprocal duplications of common micro deletions. Duplications might be under-ascertained either due to incapacity of conventional PCR to detect duplications (as opposed to deletions), or because they could be associated with a mild, or absent, phenotype. The AZFa and AZFc duplications have already been identified as reciprocal deletion/duplication syndrome, and have been shown to be compatible with male fertility. For the rapid, simple and reproducible detection of antibiotic resistance genes in strains of methicillin-resistant *Staphylococcus aureus* (MRSA) and *Enterococcus*, colony PCR using intact cells as the template DNA source in colony PCR was found efficient (Tsuchizaki, 2000).

Pulsed-field gel electrophoresis and colony PCR were applied for the first time by Plourde-Owobi et al. (2005) for the molecular characterization of *Clostridium tetani*. Among five strains tested, one turned out to contain a mixture of two genetically different clones and two (D11 and G761) to contain bacteria differing by the presence or absence of the 74-kb plasmid harboring the tetX gene.

Genetic analysis of organisms (animals, plants and bacteria) at the molecular level is an important and widely practiced area of genetic science. A number of techniques developed over more than a decade offers the opportunity to identify each individual or type of individual in a species uniquely and unambiguously. PCR has impacted this area of analysis because of its ease of use and simplicity over traditional VNTR- and RFLP-based methods (Jeffreys et al., 1985; Sambrook and Russell, 2001).

One important PCR-based genetic analysis is random amplified polymorphic DNA analysis (RAPD; reviewed in McClelland and Welsh, 1994; Power, 1996; Black, 1993). RAPD uses small, nonspecific primers for the amplification of regions of genomic DNA. Successful primer pairs produce different banding profiles of PCR products between individuals, strains, cultivars or species when analyzed by gel electrophoresis. Slight modifications to the basic PCR method are made for RAPD. First, the primers are approximately 10 bases in length compared to 17–23 base primer length of normal PCR. Because the primers are shorter, the temperature of the annealing reaction is reduced to less than 40°C.

As with most PCR techniques, RAPD requires very little material for analysis and is relatively insensitive to the integrity of the material. No blotting techniques are required, thus eliminating the use of ^{32}P, bypassing probe generation and decreasing the amount of time required to obtain results.

9.7 Conclusion

The use of PCR in molecular diagnostics has increased to the point where it is now accepted as the gold standard for detecting nucleic acids from a number of origins and it has become a essential tool in research laboratories. Although conventional diagnostic techniques remain the method of choice in regions with low resource, PCR may be employed in cases with strong clinical suspicion and equivocal results, especially at an early stage of the disease for better diagnosis, management and treatment.

References

Ahokas, H., Erkkilä, M.J., 1993, Interference of PCR amplification by the polyamines, spermine and spermidine. *PCR Methods Appl* 3:65–68.

Antwerpen, M.H., 2008, Real-time PCR system targeting a chromosomal marker specific for Bacillus anthracis. *Mol Cell Probes* 22(5–6):313–315.

Ault, G.S., Ryschkewitsch, C.F., Stoner, G.L., 1994, Type-specific amplification of viral DNA using touchdown and hot start PCR. *J Virol Methods* 46(2):145–156.

Baek, C.H., Kim, S.I., Ko, Y.H., Chu, K.C., 2000, Polymerase chain reaction detection of Mycobacterium tuberculosis from fine-needle aspirate for the diagnosis of cervical tuberculous lymphadenitis. *Laryngoscope* 110(1):30–34.

Baldino, F., Jr, Chesselet, M.F., Lewis, M.E., 1989, High-resolution in situ hybridization histochemistry. *Meth Enzymol* 168:761–777.

Baltimore, D., 1970, RNA-dependent DNA polymerase in virions of RNA tumour viruses. *Nature* 226:1209–1211.

Bassam, B.J., Caetano-Anollés, G., 1993, Automated "hot start" PCR using mineral oil and paraffin wax. *Biotechniques* 14(1):30–34.

Bell, D.A., DeMarini, D.M., 1991, Excessive cycling converts PCR products to random-length higher molecular weight fragments. *Nucl Acids Res* 19:5079.

Bernard, P.S., Lay, M.J., Wittwer, C.T., 1998, Integrated amplification and detection of the C677T point mutation in the methylenetetrahydrofolate reductase gene by fluorescence resonance energy transfer and probe melting curves. *Anal. Biochem.* 255:101–107.

Beutler, E., Gelbart, T., Kuhl, W., 1990, Interference of heparin with the polymerase chain reaction. *Biotechniques* 9(2):166.

Birch, D.E. 1996, Simplified hot start PCR. *Nature* 30;381(6581):445–446.

Black, W.C., 1993, PCR with arbitrary primers: Approach with care. *Insect Mol Biol* 2:1–6.

Chan, W.C., Greiner, T.C., 1994, Diagnosis of lymphomas by the polymerase chain reaction. *Am J Clin Pathol* 102:273–274.

Chiocchia, G., Smith, K.A., 1997, Highly sensitive method to detect mRNAs in individual cells by direct RT-PCR using *Tth* DNA polymerase. *Biotechniques* 22:312–318.

Clarke, S.C., 2002, Nucleotide sequence-based typing of bacteria and the impact of automation. *Bioessays* 24:858–862.

Cline, J., Braman, J.C., Hogrefe, H.H., 1996, PCR fidelity of Pfu DNA polymerase and other thermostable DNA polymerases. *Nucl Acids Res* 24:3546–3551.

Demeke, T., Adams, R., 1992, The effects of plant polysaccharides and buffer additives on PCR. *Biotechniques* 12: 332–334.

Eckert, K.A., Kunkel, T.A., 1990, High fidelity DNA synthesis by the Thermus aquaticus DNA polymerase. *Nucl Acids Res* 18:3739–3744.

Edwards, J.B., 1993, Methods in Molecular Biology (Vol. 15), White, B.A., ed., Humana Press, Totowa, NJ.

Edwards, J.B., Delort, J., Mallet, J., 1991, Oligodeoxyribonucleotide ligation to single-stranded cDNAs: a new tool for cloning 5? ends of mRNAs and for constructing cDNA libraries by in vitro amplification. *Nucl Acids Res* 19:5227–5232.

Ellsworth, D.L., Rittenhouse, K.D., Honeycutt, R.L., 1993, Artifactual variation in randomly amplified polymorphic DNA banding patterns. *Biotechniques* 14(2): 214–217.

Elmberger, P.G., Lozano, M.D., Weisenburger, D.D., Sanger, W., Chan, W.C., 1995, Transcripts of the npm-alk fusion gene in anaplastic large cell lymphoma, Hodgkin disease, and reactive lymphoid lesions. *Blood* 86(9):3517–3521.

Erlich, H.A., Gelfand, D., Sninsky, J.J., 1991, Recent advances in the polymerase chain reaction. *Science* 252(5013):1643–1651.

Frackman, S., Kobs, G., Simpson, D., Storts, D., Promega Corporation, 1998, Betaine and DMSO: enhancing agents for PCR. *Promega Notes* 65:27–30.

Frohman, M.A., Dush, M.K., Martin, G.R., 1988, Rapid production of full-length cDNAs from rare transcripts: amplification using a single gene-specific oligonucleotide primer. *Proc Natl Acad Sci USA* 85:8998–9002.

Fromont-Racine, M., Bertrand, E., Pictet, R., Grange, T., 1993, A highly sensitive method for mapping the 5? termini of mRNAs. *Nucl Acids Res* 21(7):1683–1684.

Gaudette, M.F., Crain, W.R., 1991, A simple method for quantifying specific mRNAs in small numbers of early mouse embryos. *Nucl Acids Res* 19:1879–1884.

Geiduschek, E.P., Herskovits, T.T., 1961, Nonaqueous solutions of DNA. Reversible and irreversible denaturation in methanol. *Arch Biochem Biophys* 95:114–129.

Gelfand, D.H., White, T.J., 1990, PCR Protocols: A Guide to Methods and Applications, Innis, M.A., Gelfand, D.H., Sninsky, J.J., White, T.J., eds., Academic Press, San Diego, CA, pp. 129–141.

Gerard, C.J., Olsson, K., Ramanathan, R., Reading, C., Hanania, E.G., 1998, Improved quantitation of minimal residual disease in multiple myeloma using real-time polymerase chain reaction and plasmid-DNA complementarity determining region III standards. *Cancer Res* 58(17): 3957–3964.

Giambernardi, T.A., Ulrich, R., Klebe, R.J., 1998, Bovine serum albumin reverses inhibition of RT-PCR by melanin. *Biotechniques* 25, 564–566.

Goel, M.M., Ranjan, V., Dhole, T.N., Srivastava, A.N., Mehrotra, A., Kushwaha, M.R., Jain, A., 2001, Polymerase chain reaction vs. conventional diagnosis in fine needle aspirates of tuberculous lymph nodes. *Acta Cytol* 45(3):333–340.

Gong, G., Lee, H., Kang, G.H., Shim, Y.H., Huh, J., Khang, S.K., 2002, Nested PCR for diagnosis of tuberculous lymphadenitis and PCR-SSCP for identification of rifampicin resistance in fine-needle aspirates. *Diagn Cytopathol* 26(4):228–231.

Greagg, M.A., Fogg, M.J., Panayotou, G., Evans, S.J., Connolly, B.A., Pearl, L.H., 1999, A read-ahead function in archaeal DNA polymerases detects promutagenic template-strand uracil. *Proc Natl Acad Sci USA* 96(16):9045–9050.

Haase, A.T., Retzel, E.F., Staskus, K.A., 1990, Amplification and detection of lentiviral DNA inside cells. *Proc Natl Acad Sci USA* 87(13):4971–4975.

Henegariu, O., Heerema, N.A., Dlouhy, S.R., Vance, G.H., Vogt, P.H., 1997, Multiplex PCR: critical parameters and step-by-step protocol. *Biotechniques* 23(3):504–511.

Holodniy, M., Kim, S., Katzenstein, D., Konrad, M., Groves, E., Merigan, T.C., 1991, Inhibition of human immunodeficiency virus gene amplification by heparin. *J Clin Microbiol* 29(4): 676–679.

Hooberman, A.L., 1992, The use of the polymerase chain reaction in clinical oncology. *Oncology* (Huntingt) 6:25–34.

Hoppe, B.L., Conti-Tronconi, B.M., Horton, R.M.. 1992, Gel-loading dyes compatible with PCR. *Biotechniques* 12:679–680.

Horton, R.M., Hoppe, B.L., Conti-Tronconi, B.M., 1994, Ampli Grease: "hot start" PCR using petroleum jelly. *Biotechniques* 16(1):42–43.

Hu, G., 1993, DNA polymerase-catalyzed addition of nontemplated extra nucleotides to the 3' end of a DNA fragment. *DNA Cell Biol* 12:763–770.

Hubank, M., Schatz, D.G., 1994, Identifying differences in mRNA expression by representational difference analysis of cDNA. *Nucl Acids Res* 22:5640–5648.

Jeffreys, A.J., Wislon, V., Thein, S.L., 1985, Individual-specific 'fingerprints' of human DNA. *Nature* 316:76–79.

Katcher, H.L., Schwartz, I., 1994, A distinctive property of *Tth* DNA polymerase: Enzymatic amplification in the presence of phenol. *Biotechniques* 16: 84–92.
Kainz, P., Schmiedlechner, A., Strack, H.B., 2000, Specificity-enhanced hot-start PCR: addition of double-stranded DNA fragments adapted to the annealing temperature. *Biotechniques* 28(2):278–282.
Konat, G.W., Laszkiewicz, I., Grubinska, G., Wiggins, R.C., 1994, PCR Technology: Current Innovations, Griffin, H.G., Griffin, A.M., eds., CRC Press, Inc., Boca Raton, Florida.
Krishnan, B.R., Blakesley, R.W., Berg, D.E., 1991, Linear amplification DNA sequencing directly from single phage plaques and bacterial colonies. *Nucl Acids Res* 19(5):1153.
Lasken, R.S., Schuster, D.M., Rashtchian, A., 1996, Archaebacterial DNA polymerases tightly bind uracil-containing DNA. *J Biol Chem* 271(30):17692–17696.
Lee, S.H., Jung, B.Y., Rayamahji, N., Lee, H.S., Jeon, W.J., Choi, K.S., Kweon, C.H., Yoo, H.S., 2009, A multiplex real-time PCR for differential detection and quantification of Salmonella spp., Salmonella enterica serovar Typhimurium and Enteritidis in meats. *J Vet Sci* 10(1):43–51.
Lee, H., Kim, K.N., Kee Chae, Y., 2007, Reevaluating the capability of Taq DNA polymerase: long PCR amplification. *Protein Pept Lett* 14(4):321–323.
Li, J., 2009, Coamplification at lower denaturation temperature-PCR increases mutation-detection selectivity of TaqMan-based real-time PCR. *Clin Chem* 55:748–756.
Lin, Y., Jayasena, S.D., 1997, Inhibition of multiple thermostable DNA polymerases by a heterodimeric aptamer. *J Mol Biol* 271:100–111.
Liu, X., Gorovsky M.A., 1993, Mapping the 5′ and 3′ ends of Tetrahymena thermophila mRNAs using RNA ligase mediated amplification of cDNA ends (RLM-RACE). *Nucl Acids Res* 21:4954–4960.
Loh, E.Y., Elliott, J.F., Cwirla, S., Lanier, L.L., Davis, M.M., 1989, Polymerase chain reaction with single-sided specificity: analysis of T cell receptor delta chain. *Science* 243(4888): 217–220.
Longley, M.J., Bennett, S.E., Mosbaugh, D.W., 1990, Characterization of the 5′ to 3′ exonuclease associated with Thermus aquaticus DNA polymerase. *Nucl Acids Res* 18(24): 7317–7322.
Longo, M.C., Berninger, M.S., Hartley, J.L., 1990, Use of uracil DNA glycosylase to control carry-over contamination in polymerase chain reactions. *Gene* 93(1):125–128.
Luthra, R., McBride, J.A., Cabanillas, F., Sarris, A., 1998, Novel 5′ exonuclease-based real-time PCR assay for the detection of t(14;18) (q32;q21) in patients with follicular lymphoma. *Am J Pathol* 153(1):63–68.
McClelland, M., Welsh, J., 1994, DNA fingerprinting by arbitrarily primed PCR. *PCR Methods Appl* 4:S59–S65.
McCulloch, R.K., Choong, C.S., Hurley, D.M., 1995, An evaluation of competitor type and size for use in the determination of mRNA by competitive PCR. *PCR Methods Appl* 4(4):219–226.
Myers, T.W., Gelfand, D.H., 1991, Reverse transcription and DNA amplification by a Thermus thermophilus DNA polymerase. *Biochemistry* 30:7661–7666.
Nuovo, G.J., Gallery, F., Hom, R., MacConnell, P., Bloch, W., 1993, Importance of different variables for enhancing in situ detection of PCR-amplified DNA. *PCR Methods Appl* 2(4):305–312.
Nuovo, G.J., Gallery, F., MacConnell, P., 1992, Detection of amplified HPV 6 and 11 DNA in vulvar lesions by hot start PCR in situ hybridization. *Mod Pathol* 5(4):444–448.
Ohara, O., Dorit, R.L., Gilbert, W., 1989, One-sided polymerase chain reaction: the amplification of cDNA. *Proc Natl Acad Sci USA* 86(15):5673–5677.
Orlando, C., Pinzani, P., Pazzagli, M., 1998, Developments in quantitative PCR. *Clin Chem Lab Med* 36:255–269.
Pahwa, R., Hedau, S., 2005, Assessment of possible tuberculous lymphadenopathy by PCR compared to non-molecular methods. *J Med Microbiol* 54:873–878.
Piepenbrock, C., Adorjan, P., Schmitt, A., Rujan, T., 2004, Method for amplification of nucleic acids of low complexity. *European patent WO2004015139.*

Plourde-Owobi, L., Seguin, D., Baudin, M.A., Moste, C., Rokbi, B., 2005, Molecular characterization of Clostridium tetani strains by pulsed-field gel electrophoresis and colony PCR. *Appl Environ Microbiol* 71(9):5604–5606.

Power, E.G.M., 1996, RAPD typing in microbiologyema technical review. *J Hosp Infect* 34: 247–265.

Rees, W.A., Yager, T.D., Korte, J., von Hippel, P.H., 1993, Betaine can eliminate the base pair composition dependence of DNA melting. *Biochemistry* 32(1):137–144.

Risatti, G., Holinka, L., Lu, Z., Kutish, G., Callahan, J.D., Nelson, W.M., Brea Tió, E., Borca, M.V., 2005, Diagnostic evaluation of a real-time reverse transcriptase PCR assay for detection of classical swine fever virus. *J Clin Microbiol* 43(1): 468–471.

Rose, E.A.,1991, Applications of the polymerase chain reaction to genome analysis. *FASEB J* 5:46–54.

Rychlik, W., 1995, Selection of primers for polymerase chain reaction. *Mol Biotechnol* 3:129–134.

Rychlik, W., Spencer, W.J., Rhoads, R.E., 1990, Optimization of the annealing temperature for DNA amplification in vitro. *Nucl Acids Res* 18(21):6409–6412.

Rys, P.N., Persing, D.H., 1993, Preventing false positives: quantitative evaluation of three protocols for inactivation of polymerase chain reaction amplification products. *J Clin Microbiol* 31: 2356–2360.

Sambrook, J., Russell, D.W., 2001, Molecular Cloning: A Laboratory Manual, 3rd edition. Cold Spring Harbor Laboratory Press, Cold Spring Harbor, NY.

Scalice, E.R., Sharkey, D.J., Daiss, J.L., 1994, Monoclonal antibodies prepared against the DNA polymerase from Thermus aquaticus are potent inhibitors of enzyme activity. *J Immunol Methods* 172(2):147–163.

Schaefer, B.C., 1995, Revolutions in rapid amplification of cDNA ends: new strategies for polymerase chain reaction cloning of full-length cDNA ends. *Anal Biochem* 227:255–273.

Siebert, P.D., Larrick, J.W., 1993, PCR MIMICS: competitive DNA fragments for use as internal standards in quantitative PCR. *Biotechniques* 14:244–249.

Slupphaug, G., Alseth, I., Eftedal, I., Volden, G., Krokan, H.E., 1993, Low incorporation of dUMP by some thermostable DNA polymerases may limit their use in PCR amplifications. *Anal Biochem* 211(1):164–169.

Smith, K., Diggle, M.A., Clarke, S.C., 2004, Automation of a fluorescence-based multiplex PCR for the laboratory confirmation of common bacterial pathogens. *J Med Microbiol* 53(Pt 2): 115–117.

Takahashi, M., Inoue, H., Kawakami, B., Kawamura, Y., 1996, Hot-start PCR by using antibody. *Tanpakushitsu Kakusan Koso* 41(5):472–476.

Temin, H., Mizutani, S., 1970, RNA-dependent DNA polymerase in virions of Rous sarcoma virus. *Nature* 226:1211–1213.

Thaker, V., 1999, In situ RT-PCR and hybridization techniques. *Methods Mol Biol* 115:379–402.

Troutt, A.B., McHeyzer-Williams, M.G., Pulendran, B., Nossal, G.J., 1992, Ligation-anchored PCR: a simple amplification technique with single-sided specificity. *Proc Natl Acad Sci USA* 89(20):9823–9825.

Tsuchizaki, N., Ishikawa, J., Hotta, K., 2000, Colony PCR for rapid detection of antibiotic resistance genes in MRSA and enterococci. *Jpn J Antibiot* 53(6):422–429

Varadaraj, K., Skinner, D.M., 1994, Denaturants or cosolvents improve the specificity of PCR amplification of a G + C-rich DNA using genetically engineered DNA polymerases. *Gene* 140: 1–5.

Vallone, P.M., Butler, J.M., 2004, Auto dimer: a screening tool for primer-dimer and hairpin structures. *Biotechniques* 37:226–231.

Vos, P., Hogers, R., Bleeker, M., Reijans, M., van de Lee T., Hornes, M., Frijters, A., Pot, J., Peleman, J., Kuiper, M., 1995, AFLP: a new technique for DNA fingerprinting. *Nucl Acids Res* 23:(21)4407–4414.

Williams, J.F., 1989, Optimization strategies for the polymerase chain reaction. *Biotechniques* 7:762–769.

Wittwer, C.T., Herrmann, M.G., Gundry, C.N., Elenitoba-Johnson, K.S., 2001, Real-time multiplex PCR assays. *Methods* 25: 430–442.

Zangenberg, G., Saiki, R.K., Reynolds, R., 1999, Multiplex PCR Optimization guidelines. In Innis, M., Gelfand, G.H., Sninsky, J.J., eds., PCR Application: Protocols for Functional Genomics, Academic Press, London, pp. 73–94.

Chapter 10
In Situ Hybridization

Abstract Nucleic acid hybridization techniques have contributed significantly to the understanding of gene organization, regulation and expression. In situ hybridization is a method for detecting specific nucleotide sequences by using a labeled complementary nucleic acid probe. The power of in situ hybridization can be greatly extended by the simultaneous use of multiple fluorescent colors. Multicolor fluorescence in situ hybridization (FISH), in its simplest form, can be used to identify as many labeled features as there are different fluorophores used in the hybridization. Key methodological advances have allowed facile preparation of low-noise hybridization probes, and technological breakthroughs now permit multi-target visualization and quantitative analysis – both factors that have made FISH accessible to all and applicable to any investigation of nucleic acids. In the future, this technique is likely to have significant further impact on live-cell imaging and on medical diagnostics.

Keywords In situ hybridization · Clinical diagnostics · Chromosome · Fluorescence · FISH · Nucleic acid probe · Nucleic acid hybridization · Chromosome painting · Cytogenetics · End labeling · Probe

10.1 Prologue

Nearly a quarter-century has passed since the first research articles introducing in situ hybridization as a method of detecting and studying DNA sequences in chromosomes and cells appeared in the literature (Gall and Pardue, 1969). In situ hybridization presents a unique set of problems as the sequence to be detected will be at a lower concentration, be masked because of associated protein, or protected within a cell or cellular structure. Therefore, in order to probe the tissue or cells of interest one has to increase the permeability of the cell and the visibility of the nucleotide sequence to the probe without destroying the structural integrity of the cell or tissue. In situ hybridization (ISH) with radiolabelled probes is a long established method in cytogenetics. This is distinct from immunohistochemistry, which localizes proteins in tissue sections. The technique of in situ hybridization is vital

M. Debnath et al., *Molecular Diagnostics: Promises and Possibilities*,
DOI 10.1007/978-90-481-3261-4_10,

to molecular biologists and their understanding of the pathophysiology of cellular functions.

10.2 The Concept of In Situ Hybridization

"In situ" means "in the original place" in Latin, so ISH involves a labeled nucleic acid probe hybridizing with a DNA or RNA sequence in situ (in the cells). Normal hybridization requires the isolation of DNA or RNA, separating it on a gel, blotting it onto nitrocellulose and probing it with a complementary sequence. The basic principles for in situ hybridization are the same, except one is utilizing the probe to detect specific nucleotide sequences within cells and tissues (Hougaard et al., 1997). The sensitivity of the technique is such that threshold levels of detection are in the region of 10–20 copies of DNA/mRNA per cell.

10.2.1 The Process

There are almost as many methods for carrying out in situ hybridization as there are tissues that have been probed. So more important than having a method is to have an understanding of the different stages in the process and their purpose (Litchter et al., 1988).

10.2.1.1 Preparation of Material

The most common tissue sections used with in situ hybridization are frozen sections, paraffin embedded sections and cells in suspension. Cells can also be cytospun onto glass slides and fixed with methanol.

10.2.1.2 Choice of Probe

Probes are complimentary sequences of nucleotide bases to the specific mRNA sequence of interest. These probes can be as small as 20–40 base pairs or be up to 1,000 bp. The strength of the bonds between the probe and the target plays an important role. The strength decreases in the order RNA-RNA to DNA-RNA. This stability is in turn influenced by the various hybridization conditions such as concentration of formamide, salt concentration, hybridization temperature, and pH.

10.2.1.3 Probe Types

There are essentially four types of probes that can be used for performing in situ hybridization. Correctly designed and purified oligonucleotide probes will represent the easiest, fastest, least labor intensive and most inexpensive method. *Oligonucleotide probes* are produced synthetically by an automated chemical synthesis. The method utilizes readily available deoxynucleotides which are economical, but of course requires that you know the specific nucleotide sequence you wish to prepare (Hougaard et al., 1997). These probes have the advantages of being resistant to RNases and are small, generally around 40–50 base-pairs. This is ideal for in situ hybridization because of its stability, availability, faster and less expensive to use, easier to work with, more specific, better tissue penetration, better reproducibility and a wide range of labeling methods that do not interfere with target detection.

The other probes are *single stranded DNA probes, double stranded DNA probes* and *RNA probes* (cRNA probes or riboprobes). They require time to prepare, expensive reagents are used during their preparation and a good repertoire of molecular skills is required for their use.

10.2.1.4 Labeling the Oligonucleotide

The probe is hybridized (bound) within the tissue section or within cells and thus determines where the target gene is being expressed (Fig. 10.1). This can be detected by attaching to the probe an easily detectible substance or "labeling" before hybridization. Classically oligonucleotide probes have either 5′ or 3′ end-labeled or 3′ tailed with modified nucleotides that have a "radiolabel" attached that can be

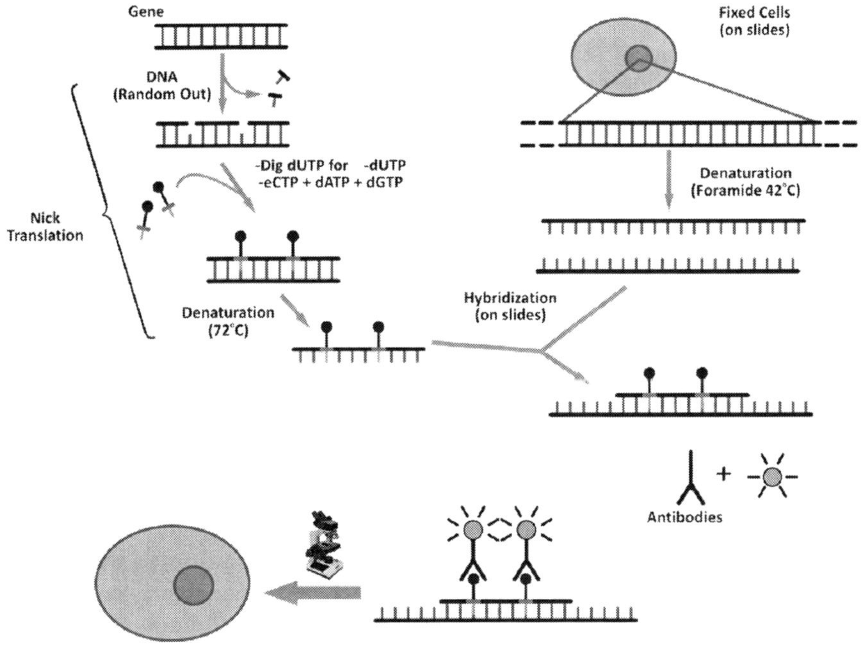

Fig. 10.1 The molecular mechanism of hybridization and detection by in situ hybridization technique

detected after the probe has hybridized to its target. With end-labeling a single modified ddNTP (that incorporates the label) is added to either the 5′ or the 3′ end of the molecule enzymatically or during probe synthesis. 3′ tailing involves addition of a tail (on average 5–50 nucleotides long of modified dNTPs depending on the method used) using the enzyme terminal transferase (TdT). The degree of specificity to which the probe hybridizes to the target sequence can be controlled by the design of the probe and the conditions of the buffer solution, including temperature, pH, and salt concentration. Hybridization mixtures usually have a small volume (about 10–20 μl total) with 50% formamide and hybridization typically occurs between 37–60°C (Palotie et al., 1996).

10.2.1.5 Detection

As mentioned, radiolabeled probes are detectable using either photographic film or photographic emulsion using the technique of autoradiography.

While in situ hybridization is undoubtedly a very powerful technique, for the average laboratory it is expensive to undertake, is time consuming, requires detailed molecular biological knowledge of sub cloning, in vitro transcription and bacterial expression (Hopman et al., 1986).

10.3 Disadavantages of Radioactive Probes

In the early versions of in situ hybridization the probe was radioactively labeled probe but this procedure was unsatisfactory. With radioactive label it is difficult to achieve sensitivity and resolution, the two critical requirements for successful hybridization. Sensitivity requires that the radioactive probe has high emission energy, but if the radiolabel has high emission energy then it scatters its signal and gives poor resolution. High resolution is possible if a radiolabel with low emission energy, such as 3H, is used, but these have such low sensitivity that lengthy exposures are needed leading to a high background and difficulties in discerning the genuine signal (Landegent et al., 1987).

10.4 Solving the Problem: Advent of FISH Technique

Over the past 15 years, however, a revolution in light microscopy has occurred through the development of fluorescence techniques that allow unprecedented ease, precision, and accuracy in locating, identifying, and recording data on the genetic makeup of biomedical samples. The problem with in situ hybridization with the use of radioactive probes was solved in the late 1980s by the development of nonradioactive, fluorescent DNA labels. These labels combine high sensitivity with high resolution and are ideal for in situ hybridization. Fluorescence in situ hybridization, the assay of choice for localization of specific nucleic acids sequences in native context, is a 20-year-old technology that has developed continuously. Over its maturation, various methodologies and modifications have been introduced to optimize the detection of DNA and RNA. The power of in situ hybridization can be greatly extended by the simultaneous use of multiple fluorescent colors. Multicolor fluorescence in situ hybridization (FISH), in its simplest form, can be used to identify as many labeled features as there are different fluorophores used in the hybridization enabling the relative positions of the probe sequences to be mapped. By using not only single colors, but also combinations of colors, many more labeled features can be simultaneously detected in individual cells using digital imaging microscopy.

The pervasiveness of this technique is largely because of its wide variety of applications and the relative ease of implementation and performance of in situ studies. Although the basic principles of FISH have remained unchanged, high-sensitivity detection, simultaneous assay of multiple species, and automated data collection and analysis have advanced the field significantly. The introduction of FISH surpassed previously available technology to become a foremost biological assay.

10.4.1 Fluorescence In Situ Hybridization

Fluorescence in situ hybridization is a molecular cytogenetic technique in which fluorescently labeled DNA probes are hybridized to metaphase spreads or interphase nuclei to detect or confirm gene or chromosome abnormalities that are generally beyond the resolution of routine cytogenetics. FISH have much higher rates of sensitivity and specificity. FISH also provides results more quickly because no cell culture is required. The first application of fluorescence in situ detection came in 1980, when RNA that was directly labeled on the 3′ end with fluorophore was used as a probe for specific DNA sequences (Bauman et al., 1980). Enzymatic incorporation of fluorophore-modified bases throughout the length of the probe has been widely used for the preparation of fluorescent probes; one color is synthesized at a time (Wiegant et al., 1991).

The use of amino-allyl modified bases, which could later be conjugated to any sort of hapten or fluorophore, was critical for the development of in situ technologies because it allowed production of an array of low-noise probes by simple chemistry (Wiegant et al., 1991). Methods of indirect detection allowed signal output to be increased artificially by the use of secondary reporters that bind to the hybridization probes. In the early 1980s, assays featuring nick-translated, biotinylated probes, and secondary detection by fluorescent streptavidin conjugates were used for detection of DNA (Manuelidis et al., 1982) and mRNA (Singer and Ward, 1982) targets. Approximately a decade later, improved labeling of synthetic, single-stranded DNA probes allowed the chemical preparation of hybridization probes carrying enough fluorescent molecules to allow direct detection. Many variations on these themes of indirect and direct labeling have since been introduced, giving a wide spectrum of detection schemes from which to choose (Pachmann, 1987; Rudkin and Stollar, 1977); the specifications, sensitivity and resolution of these techniques are well described elsewhere (Raap, 1998).

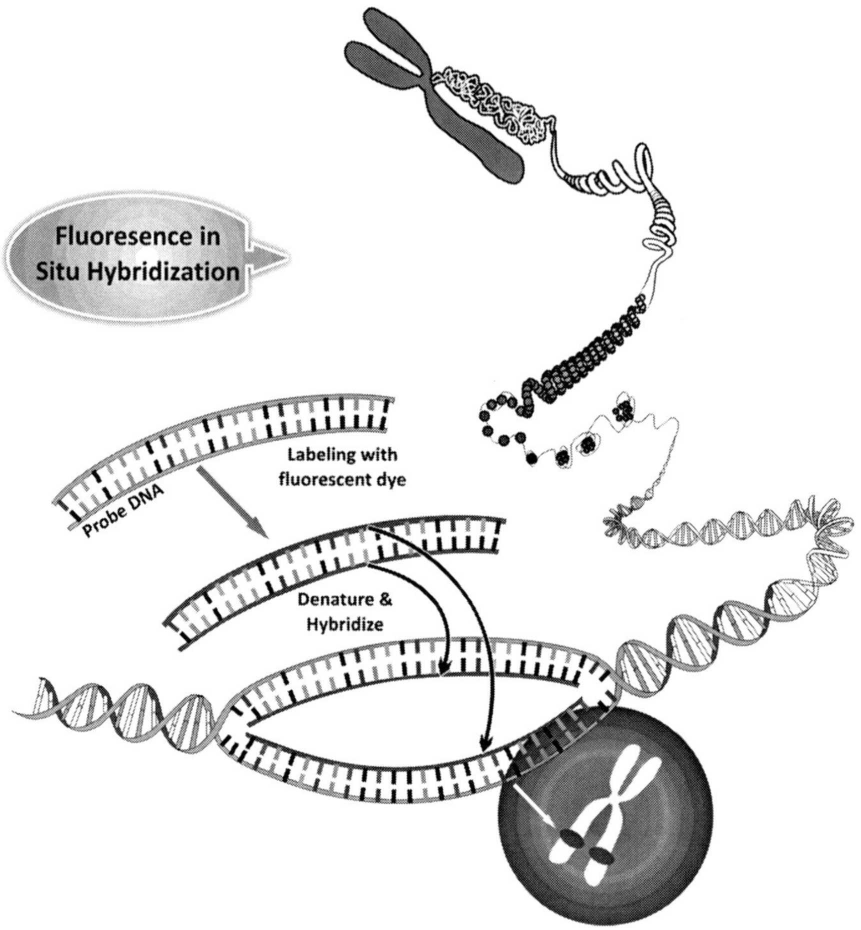

Fig. 10.2 The schematic representation of fluorescence in situ hybridization technique

The technique of in situ hybridization is used to locate the chromosomal location of a specific DNA (or RNA) probe (Femino et al., 1998). The theory is the same as for Southern hybridization, except that the DNA to which the probe will hybridize is the actual chromosome. The probe is labeled with a flourescent probe. The sample DNA (metaphase chromosomes or interphase nuclei) is first *denatured* a process that separates the complimentary strands within the DNA double helix structure (Fig. 10.2). The fluorescently labeled probe of interest is then added to the denatured sample mixture and hybridizes with the sample DNA at the target site as it *reanneals* (or reforms itself) back into a double helix. The probe signal can then be seen through a fluorescent microscope and the sample DNA scored for the presence or absence of the signal. If the species has been characterized cytogenetically, the marker can be assigned to the appropriate chromosome. Because this technique uses a fluorescent probe it is called fluorescence in situ hybridization or FISH. It is also called chromosome painting.

10.4.2 FISH in Action

Chromosome analyses by FISH have led to marked progress in cytogenetics research (Pinkel et al., 1986). A prime example of the power of hybridization approaches in genome investigation is comparative genomic hybridization, during which deletions and duplications of chromosomal regions are detected by differential fluorescence signals (Forozan et al., 1997). However, because the assay does not benefit from preservation of tissue structure or cellular architecture, its future applications are more likely to be in silico than in situ (Lichter et al., 1988). Initially, RNA assays could reliably detect only rather abundant messages, using clone-derived probes (Lawrence et al. 1989).

Enhancements in detection and computer processing algorithms have subsequently allowed detection of smaller and smaller targets (Tanke et al., 1995; Piper et al., 1990; 1994). Advances in microscope and detector hardware have allowed the low light level produced by FISH to be recorded and analyzed with increasing sensitivity (Tanke et al., 1999). Mathematical image-processing algorithms have built on this process to yield super-resolution technology to probe at submicroscopic resolution, using digital image stacks (Carrington et al., 1995).

New targets led to new applications of the FISH procedure. The new avenues of research opened by these applications required that more and more species be simultaneously detected (Fauth and Speicher, 2001). A major milestone in the detection of chromosome targets was the discrimination of all human chromosomes simultaneously, using computed interpretation of a 5-color scheme (Galbraith et al., 1991) Speicher et al., 1996). Although mRNAs can also be visualized in a multiplex fashion (Levsky et al., 2003), FISH analysis of the entire transcriptome in situ is a daunting task. One can only speculate that future technologies will feature increasingly higher-order multiplexing, until the number of interesting nucleic acid targets is reached. The technical means for color coding such a large number of entities is already in place (Nederlof et al., 1989, 1990) although reduction to practice will be difficult and a means of deciphering spatially overlapping signals will need to be developed.

10.4.2.1 FISH with Metaphase Chromosome

FISH was originally used with metaphase chromosomes. These chromosomes, prepared from nuclei that are undergoing division are highly condensed with each chromosome in a set taking up a recognizable appearance characterized by the position of centromere and the banding pattern that emerges after the chromosome preparation is stained. With metaphase chromosomes, a fluorescent signal obtained by FISH is mapped by measuring its position relative to the end of the short arm of the chromosome.

A disadvantage is that the highly condensed nature of the metaphase chromosomes means that only low resolution mapping is possible, two markers needing to be at least 1 Mb apart to be resolved as separate hybridization signals. This degree of resolution is insufficient for the construction of useful chromosome maps, and the main application of metaphase FISH has been in determining which chromosome a new marker is located on, and providing a rough idea of its map position, as a preliminary to a finer scale mapping by other methods. For several years these "other methods" did not involve any form of FISH, but since 1995 a range of higher resolution FISH is not restricted to intact chromosomes. It also works well on interphase nuclei, where it is used to visualize the order of closely spaced (from 100 to 500 kb apart) probes on a chromosome and to measure their distances apart.

10.4.2.2 FISH with Extended Chromosomes and DNA Fibers

The resolution of FISH is determined not by the hybridization technique itself but by the nature of the chromosomal preparation being studied. If metaphase chromosomes are too condensed for fine scale mapping then chromosomes that are more extended be used. Mechanically stretched chromosomes can be obtained by modifying the preparative method used to isolate chromosomes from metaphase nuclei. Centrifugations generate shear forces which can result in the chromosomes becoming stretched upto 20 times their normal length. Individual chromosomes are still recognizable and FISH signals can be mapped in the same way as with normal metaphase chromosomes. The resolution is significantly improved and markers that are 200–300 Kb apart can be distinguished.

Nonmetaphase chromosomes can be used (Lawrence et al., 1989; Siebert et al., 1998) because it is only during metaphase that chromosomes are highly condensed, at other stages of the cell cycle the chromosomes are naturally unpacked. Attempts have been made to use prophase nuclei as in these the chromosomes are still sufficiently condensed for individual ones to be identified. In practice these preparations provide no advantage over mechanically stretched chromosomes. Interphase chromosomes are more useful, as this stage of cell cycle (between nuclear divisions) is when the chromosomes are most unpacked. Resolution down to 25 Kb is possible, but chromosome morphology is lost so there are no external reference points against which to map the position of the probe. This technique is therefore used after preliminary map information has been obtained usually as a means of determining the order of a series of markers in a small region of a chromosome. Most excitingly, FISH can be used on stretched extended DNA fibers where it has a resolution of less than 5 kb.

10.5 Applications of FISH as a Diagnostic Tool for Research

The development of in situ technologies has provided us with a wealth of information regarding the locations and expression patterns of genes in single cells. In situ hybridization (ISH) is an invaluable tool for research and diagnostics, dramatically advancing the study of cell- and tissue-specific expression of many genes. Some of ISH applications include determination of chromosome structure, function, and evolution, chromosomal gene mapping, expression of genes, localization of viral DNA sequences, diagnosis of viral diseases (Ahtiluoto et al., 2000), and localization of oncogenes (Suciu et al., 2008) sex determination and determination of chromosomal abnormalities (Siffroi et al., 2000; Brown et al., 2000). The uses and different approaches for ISH continue to increase, thus impacting many different research fields. New labeling techniques for probes, new detection systems, and advanced computer software increase the availability and efficiency of ISH (Leitch et al., 1994).

More complete gene expression profiles of single cells will provide a new level of insight into the correlation of gene expression patterns with particular cellular phenotypes. This will be particularly important in studies of development and disease progression, where complicated; finely demarcated gene expression programs are in play.

10.5.1 To Analyze the Onset of Specific Gene Expression

This technique has been utilized by neuroscientists to analyze the onset of specific gene expression during nervous system morphogenesis and to assess the cellular distribution of mRNAs encoding many important neuronal proteins. In the past decade, a number of genes related to nervous system induction, specification, and regionalization have been identified.

In situ hybridization is an important tool for analyzing gene expression and developing hypotheses about gene functions. The discovery of hundreds of microRNA (miRNA) genes in animals has provided new challenges for analyzing gene expression and functions. The small size of the mature miRNAs (~20–24 nucleotides in length) presents difficulties for conventional in situ hybridization methods (Thompson et al., 2007).

10.5.2 Analysis of the Chromosome Structure

The FISH technique is widely used in analysis of the chromosome structure, but is also gaining ground in areas as the experimental and routine karyotype analysis (Macville et al., 1997) of tumors. The method can be fine tuned to screen single tumor cells for genetic aberrations, as well as identify candidate tumor suppressor genes. Fluorescence in situ hybridization was used to map leukemia-related deletions of the long arm of chromosome 6 (6q) (Sinclair et al., 2004). Fluorescence in situ hybridization of telomere repeats has been used to calculate telomere length, a method called quantitative (Q)-FISH (Hultdin et al., 1998) using a fluorescein-labeled peptide nucleic acid (PNA) (CCCTAA) probe and DNA staining with propidium iodide.

10.5.3 Localisation of RNA Transcripts

Another application of in situ hybridisation is the localization of RNA transcripts in tissue, in analogy to the histochemical analysis of proteins. Probes that is labeled either radioactively or optically are hybridized with the RNA of a histological tissue section or preparation.

10.5.4 FISH as a Molecular Cytogenetic Technique to Understand Diseases

Cytogenetics analysis is at present the basic element of the diagnostic process of genetic disorders which are caused by chromosomal abnormalities. Since the chromosome banding technique has been introduced in the 1970s, it has been available as a diagnostic tool of a number of clinical syndromes. However, since banding resolution is not always sufficient for the identification of chromosomal abnormalities, additional techniques for solving diagnostic dilemma of classical cytogenetics are needed. The basic method of molecular cytogenetics is fluorescence in situ hybridization. It enables a specific detection of unique sequences, chromosomal regions or entire chromosomes in metaphase, interphase cells or in tissue sections. Molecular cytogenetic approaches based on fluorescence in situ hybridization with chromosome-specific probes have been increased in recent years. They become a powerful tool for chromosomal diagnosis in prenatal, constitutional and cancers genetic disorders. In fact, various procedures are now available for rapid identification of numerical and structural chromosome aberrations that escape to conventional chromosome banding analysis.

Fluorescence in situ hybridization is a molecular cytogenetic technique in which fluorescently labeled DNA probes are hybridized to metaphase spreads or interphase nuclei. Applications include identification of structurally abnormal chromosomes, including several of the cancer translocations, such as BCR/ABL and TEL/AML1 translocations; identification of marker chromosomes; detection of very small deletions (microdeletions); and rapid detection of Down syndrome and other numerical chromosome abnormalities; and the rapid detection of sex chromosomes and the SRY gene. FISH should be used in conjunction with G-banded chromosome analysis. FISH is also utilized to confirm microdeletions identified during high resolution chromosome analysis and to aid in identification of abnormal chromosomes. Interphase FISH is especially useful in bone marrow/cancer analyses when there is poor or no growth of the specimen (Tubbs et al., 2007).

Chromosomal abnormalities are responsible for a considerable number of birth defects, and more than 50% of spontaneous abortions. These numbers have been significantly higher since the advent of FISH technology that allows the detection of submicroscopic chromosome alterations.

10.5.4.1 Microdeletion Syndromes

A number of genetic syndromes are caused by the deletion of a small region of a particular chromosome. Often these deletions are too small to be picked up by standard or high resolution chromosome analysis, in which case, microdeletion syndrome probes must be used to elucidate the chromosome abnormality. These probes are pieces of DNA specific for the region deleted in the specified syndrome, and usually include a control probe which identifies the chromosome of interest.

Wolf-Hirschhorn (4p-); Cri-du-chat (5p-);Williams syndrome (7q11.23); Prader-Willi syndrome (15q11.2-q13); Angelman syndrome (15q11.2-q13); Miller-Dieker syndrome (17p13.3); Smith-Magenis syndrome (17p11.2); DiGeorge and Velo-cardio-facial syndromes (22q11.2); Kallman syndrome (Xp22.3); Steroid Sulfatase Deficiency (STS) (Xp22.3); X-Linked Ichthiosis (Xp22.3); Retinoblastoma (13q14); Trisomy Detection and Sex Determination; Probes for chromosomes 13, 18, 21, X, Y and SRY. These probes are used to screen interphase (uncultured) cells for trisomy 13, trisomy 18, trisomy 21, chromosome number for sex chromosomes (X and Y) and for the presence of the male determining gene SRY. FISH can be performed as an initial screening test in certain high risk situations where trisomy is suspected, such as in amniocytes from a pregnancy with an abnormal ultrasound or uncultured lymphocytes from an infant with ambiguous genitalia.

10.5.4.2 FISH For Oncology Screening

FISH analysis is available to rule out certain common oncology related translocations, deletions and amplifications. This analysis is particularly useful when a specific hematologic disease is highly suspected (i.e. Philadelphia chromosome in chronic myelogenous leukemia) and/or cells fail to grow in culture. FISH can be used to look for minimal residual disease in patients undergoing treatment or in patients thought to be going into or coming out of remission. It can be used to detect unusual variants of the Philadelphia chromosome translocation, and to follow bone marrow transplants in certain patients. Since most of these oncology probes are used on interphase cells, standard cytogenetics is still necessary to look for other chromosomal aberrations and to detect clonal evolution of disease, an important prognostic indicator.

Single Gene Probes (deletion or amplification)
P58 CLK-1 Locus (1p36)
D7S486 (7q31)
Retinoblastoma (13q14)
p53 (17p13.1)
Her-2/neu (17q11.2-q12)
Enumeration probes for all chromosomes
Dual Color Translocation Probes
Bcr/abl translocation t (9; 22) (q34; q11.2) (both major and minor breakpoints)
M-bcr/abl translocation t (9; 22) (q34; q11.2) (major breakpoint)
IGH/CCND1 translocation t (11; 14) (q13; q32)
PML/RARA translocation t (15; 17) (q22; q21.1)
TEL/AML1 translocation t (12; 21) (p13; q22)
Amniotic Fluid

Aneuploid detection by FISH for chromosomes 13, 18, 21, X and Y may be indicated in situations where there is a risk of a numerical chromosome abnormality. This risk is usually based on abnormal ultrasound findings. FISH may also be indicated in cases of late gestational age when a rapid result is required. FISH should be followed by a complete karyotype analysis and no clinical action should be taken based solely upon FISH results. FISH is best performed prior to 22 weeks gestation. After 22 weeks, considerable cellular debris is present and may lead to inconclusive results. FISH assay also provides a feasible and sensitive tool for the routine detection of the translocation t (8; 14) in interphase cells which might also offer new insights into the biology of high-grade B-cell lymphomas (Siebert et al., 1998). This translocation t (8; 14) (q24; q32) is the characteristic chromosomal aberration of Burkitt's-type lymphomas and leukemias (BLs). As to the diagnostic and clinical relevance of the classical Burkitt translocation, the diagnosis of t (8; 14) is an important aim in the management of patients with BL.

10.5.4.3 FISH Diagnosis of Prenatal, Post Natal and New Born Diseases

The clinical application of FISH technology in postnatal, prenatal, and preimplantation diagnoses has been playing an important role in the diagnosis and prevention of birth defects. As new technologies evolve, more and more new FISH techniques-such as subtelomeric FISH, multicolor FISH (M-FISH), comparative genomic hybridization (CGH), and microarray are used in clinical diagnoses, the role of FISH technology in both research and clinical aspects of birth defects will surely continue to expand.

Rapid aneuploidy detection methods allow prenatal diagnosis results to be released within 48 h, but not on the same day as the invasive test. A rapid fluorescence in situ hybridization (FISH) method (*FastFISH*) that releases accurate results on the same day as amniocentesis was developed recently (Choolani et al., 2007). *Fast FISH* was optimized to be completed within 2 h of sample collection using CEP and LSI probes for chromosomes 13, 18, 21, X, Y and DiGeorge syndrome (DGS).

10.5.4.4 FISH in Diagnosing Neoplasia

Fluorescence in situ hybridization has become an important tool for diagnosing neoplasia in children. With probes designed to identify specific chromosomes and chromosomal regions, FISH is commonly used to detect the specific chromosomal abnormalities associated with hematologic diseases and solid tumors. Variations of FISH currently being investigated, such as comparative genomic hybridization, multicolor FISH, and microchip arrays, will probably result in additional uses of FISH in both research and clinical cytogenetic laboratories. Although FISH has disadvantages when compared with conventional cytogenetics and molecular methods, FISH will continue to be important in analyzing chromosomal abnormalities of tumors in children. Molecular probes for FISH are used for diagnosis of pediatric neoplastic diseases.

10.5.5 FISH for Detection of Pathogens

Human herpes virus 6 (HHV-6) DNA was detected in pontine nuclei of the brain cell of a patient by in situ hybridization (Ahtiluoto et al., 2000). In another study, Forrest et al. (2006) reported identification of *Candida albicans* in blood isolates by peptide nucleic acid fluorescence in situ hybridization (PNA FISH). This technique was 100% sensitive and specific in the rapid identification of 31 out of 72 candidemias as *C. albicans* and resulted in a significant reduction of caspofungin usage, with an overall cost savings of $1,729 per patient. In food, detection and identification of Salmonella species with the use of fluorescent in situ hybridization with 23S rRNA-targeted oligonucleotide probes was also reported by Qiang et al. (2003) and Fang et al. (2003).

10.6 Recent Advances of In Situ Hybridisation Technology

Lately, fluorescence in situ hybridization technology has introduced the possibility of chromosome classification based on 24-color chromosome painting (M-FISH). A clarified cerotype for K562 obtained by a combination of the following molecular cytogenesis techniques: comparative genomic hybridization (CGH), FISH mapping using locus-specific probes, and M-FISH was reported recently. Multicolor FISH has identified the marker structures in this cell line. The characteristic marker chromosome in K562 has been confirmed by this study to be a der(18)t(1; 18). Multicolor FISH further confirmed the identity of marker structures partially identified by G-banding as der(6)t(6;6), der(17)t(9;17), der(21)t(1;21), der(5)t(5;6). In addition M-FISH has revealed a deleted 20q and a complex small metacentric marker comprised of material from chromosomes 1, 6, and 20. A cryptic rearrangement was revealed between chromosomes 12 and 21 that produced a structure that looks like a normal chromosome 12 homologue by G-banding analysis. Finally, M-FISH detected regions from chromosome 13 intercalated into two acrocentric markers. Using fluorescent in situ hybridization, three human BAC clones), localized in the terminal region of human chromosome 17p (HSA17p13; 1.44–3.68 Mp), were mapped to chromosome 8p of American mink (MVI8p) (Zhdanova, 2007).

Reciprocal chromosome painting and hybridizations with probes such as yeast artificial chromosomes, cosmids, and fibre fluorescence in situ hybridisation allow subchromosomal assignments of chromosome regions and can identify breakpoints of rearranged chromosomes. Recent advances in fluorescence in situ hybridisation using probes specific for entire chromosomes or chromosome segments can quickly and economically provide a cytogenesis map of homologies (Lawrence et al., 1988). These probes cover large segments on chromosomes of various species and confirm previous hypotheses of extensive linkage conservation in mammals.

Cytopathological diagnosis of Merkel cell carcinoma using interphase fluorescent in situ hybridization analysis was performed for research purposes using centromeric probes of chromosomes 6 and 8 (Suciu et al., 2008). Trisomy of chromosome 6 was found in 85% of tumour cells in the first case of MCC and case 2 exhibited trisomy 8 in 77% of tumour cells. In the absence of specific molecular markers, detection of trisomy 6 and/or trisomy 8 could help in identifying MCC. FISH analysis is easily and quickly performed on interphase nuclei may be extended to the study of other relevant genetic abnormalities. Recently using FISH analysis gene copy number alterations of circulating tumor cells in metastasis prostrate cancer patients have been reported (Leversha et al., 2009). FISH assay may be a promising tool for diagnosis, surveillance and monitoring of carcinoma in the upper urinary tract (Luo et al., 2009).

10.7 Conclusion

The recently developed tissue microarray technology is an ideal platform for the introduction of high-throughput molecular profiling of tumor specimens at the single cell level. To construct a tissue microarray, small core biopsies are taken from representative areas of paraffin-embedded tumor tissues and assembled in a single block. Microtome sections are taken from the tissue microarray and placed on glass slides for rapid and efficient molecular analyses. In addition to pathological specimens such as tumor tissues, microarrays generally contain corresponding normal tissue and internal controls. The entire group of samples is analyzed simultaneously in one experiment, providing enormous amounts of correlative information about specific biomarkers, in the context of rigorous procedural controls. The next challenge will be to apply multi-gene FISH technology to these samples to correlate putative genes of prognostic value with specific morphological features initially, and then extend studies to samples where the morphology is not sufficiently informative. Certain genes can then be associated with the pre-cancerous state, for instance. Through such developments, one can foresee how molecular pathology could eventually surpass the limitations of morphological pathology. FISH has already colored the way that we visualize and conceptualize genes, chromosomes, transcription and nucleic acid movements. What remains to be seen is how exhaustive molecular analysis of single cells and tissue samples will impact how we identify, diagnose and alter the course of genetic pathology. Over the long term, it is expected that databases correlating gene expression patterns on the single cell level will accumulate as investigators and industries employ the technology of FISH with their favored biomarkers. Ultimately, FISH will be the preferred approach to anticipate the complicated components of gene expression leading to disease. These features should ensure greater use of in situ hybridization techniques in future research studies (Smith, 1997).

When the tools become available for us to visualize multiple gene expression patterns in living cells, we will finally be able to fulfill the promise of FISH technology by building and testing models of molecular transcriptional dynamics within the true native context of the cell.

References

Ahtiluoto, S., Mannonen, L., Paetau, A., Vaheri, A., Koskiniemi, M., Rautiainen, P., Muttilainen, M., 2000, In situ hybridization detection of human herpesvirus 6 in brain tissue from fatal encephalitis. *Pediatrics* 105:431–433.

Bauman, J.G., Wiegant, J., Borst, P., van Duijn, P., 1980, A new method for fluorescence microscopical localization of specific DNA sequences by in situ hybridization of fluorochrome-labelled RNA. *Exp Cell Res* 128:485–490.

Brown, J., Horsley, S.W., Jung, C., Saracoglu, K., Janssen, B., Brough, M., Daschner, M., Beedgen, B., Kerkhoffs, G., Eils, R., Harris, P.C., Jauch, A., Kearney, L., 2000, Identification of a subtle t(16;19)(p13.3;p13.3) in an infant with multiple congenital abnormalities using a 12-colour multiplex FISH telomere assay, M-TEL. *Eur J Hum Genet* 8:903–910.

Carrington, W.A., Fogarty, K.E., Lifschitz, L., Fay, F.S., 1995, Superresolution three-dimensional images of fluorescence in cells with minimal light exposure. *Science* 268:1483–1487.

Choolani, M., Ho, S.S., Razvi, K., Ponnusamy, S., Baig, S., Fisk, N.M., Biswas, A., Rapid Molecular Testing in Prenatal Diagnosis Group, 2007, FastFISH: technique for ultrarapid fluorescence in situ hybridization on uncultured amniocytes yielding results within 2 h of amniocentesis. *Mol Hum Reprod* 13:355–359.

Fang, Q., Brockmann, S., Botzenhart, K., Wiedenmann, A., 2003, Improved detection of *Salmonella* spp. in foods by fluorescent in situ hybridization with 23S rRNA probes: a comparison with conventional culture methods. *J Food Protect* 66:723–731.

Fauth, C., Speicher, M.R., 2001, Classifying by colors: FISH-based genome analysis. Cytogenet. *Cell Genet* 93:1–10.

Femino, A.M., Fay, F.S., Fogarty, K., Singer, R.H., 1998, Visualization of single RNA transcripts in situ. *Science* 280:585–590.

Forozan, F., Karhu, R., Kononen, J., Kallioniemi, A., Kallioniemi, O.P., 1997, Genome screening by comparative genomic hybridization. *Trends Genet* 13:405–409.

Forrest, G.N., Mankes, K., Jabra-Rizk, M.A., Weekes, E., Johnson, J.K., Lincalis, D.P., Venezia, R.A., 2006, Peptide nucleic acid fluorescence in situ hybridization-based identification of candida albicans and its impact on mortality and antifungal therapy costs. *J Clinl Microbiol* 44:3381–3383.

Galbraith, W., Wagner, M.C., Chao, J., Abaza, M., Ernst, L.A., Nederlof, M.A., Hartsock, R.J., Taylor, D.L., Waggoner, A.S., 1991, Imaging cytometry by multiparameter fluorescence. *Cytometry* 12:579–596.

Gall, J.G., Pardue, M.L., 1969, Formation and detection of RNA-DNA hybrid molecules in cytological preparations. *Proc Natl Acad Sci* 63:378–383.

Hopman, A.H., Wiegant, J., Raap, A.K., Landegent, J.E., van der Ploeg, M., van Duijn, P., 1986, Bi-color detection of two target DNAs by non-radioactive in situ hybridization. *Histochemistry* 85:1–4.

Hougaard, D.M., Hansen, H., Larsson, L.I., 1997, Non-radioactive in situ hybridization for mRNA with emphasis on the use of oligodeoxynucleotide probes. *Histochem Cell Biol* 108: 335–344.

Hultdin, M., Grönlund, E., Norrback, K., Eriksson-Lindström, E., Just, T., Roos, G., 1998, Telomere analysis by fluorescence in situ hybridization and flow cytometry. *Nucl Acid Res*, 26:3651–3656.

Landegent, J.E., Jansen in de Wal, N., Dirks, R.W., Baao, F., van der Ploeg, M., 1987, Use of whole cosmid cloned genomic sequences for chromosomal localization by non-radioactive in situ hybridization. *Hum Genet* 77:366–370.

Lawrence, J.B., Singer, R.H., Marselle, L.M., 1989, Highly localized tracks of specific transcripts within interphase nuclei visualized by in situ hybridization. *Cell* 57:493–502.

Lawrence, J.B., Singer, R.H., Villnave, C.A., Stein, J.L., Stein, G.S., 1988, Intracellular distribution of histone mRNAs in human fibroblasts studied by in situ hybridization. *Proc Natl Acad Sci USA* 85:463–467.

Leitch, A.R., Schwarzacher, T., Jackson, D., Leitch, I.J., 1994, *In situ hybridization : a practical guide. RMS microscopy handbooks, no. 27.* Oxford: Bios Scientic® Publishers Ltd.
Leversha, M.A., Han, J., Asga, Z., Danila, D.C., Lin, O., Gonzalez-Espinoza, R., Anand, A., Lilja, H., Heller, G., Fleisher, M., Scher, H.I., 2009, Fluorescence in situ hybridisation analysis of circulating tumor cells in metastatic prostrate cancer. *Clin Cancer Res* 15:2091–2097.
Levsky, J.M., Braut, S.A., Singer, R.H., 2003, Single cell gene expression profiling by multiplexed expression fluorescence in situ hybridization: application to the analysis of cultured cells. In J.E. Celis, ed., Cell Biology: A Laboratory Handbook, Academic Press, Amsterdam.
Lichter, P., Cremer, T., Borden, J., Manuelidis, L., Ward, D.C., 1988, Delineation of individual human chromosomes in metaphase and interphase cells by in situ suppression hybridization using recombinant DNA libraries. *Hum Genet* 80:224–234.
Luo, B., Li, W., Deng, C.H., Zheng, F.F., Sun, X.Z., Wang, D.H., Dai, Y.P., 2009, Utility of fluorescence in situ hybridisation in the diagnosis of upper urinary tract urothelial carcinomas. *Cancer Genet Cytogenet* 189:93–97.
Macville, M., Veldman, T., Padilla-Nash, H., Wangsa, D., O'Brien, P., Schröck, E., Ried, T., 1997, Spectral karyotyping, a 24-colour FISH technique for the identification of chromosomal rearrangements. *Histochem Cell Biol* 108:299–305.
Manuelidis, L., Langer-Safer, P.R., Ward, D.C., 1982, High-resolution mapping of satellite DNA using biotin-labeled DNA probes. *J Cell Biol* 95:619–625.
Nederlof, P.M., Robinson, D., Abuknesha, R., Wiegant, J., Hopman, A.H., Tanke, H.J., Raap, A.K., 1989, Three-color fluorescence in situ hybridization for the simultaneous detection of multiple nucleic acid sequences. *Cytometry* 10:20–27.
Nederlof, P.M., van der Flier, S., Wiegant, J., Raap, A.K., Tanke, H.J., Ploem, J.S., van der Ploeg, M., 1990, Multiple fluorescence in situ hybridization. *Cytometry* 11:126–131.
Pachmann, K., 1987, In situ hybridization with fluorochrome-labeled cloned DNA for quantitative determination of the homologous mRNA in individual cells. *J Mol Cell Immunol* 3:13–19.
Palotie, A., Heiskanen, M., Laan, M., Horelli-Kuitunen, N., 1996, High-resolution fluorescence in situ hybridization: a new approach in genome mapping. *Ann Med* 28:101–106.
Pinkel, D., Straume, T., Gray, J.W., 1986, Cytogenetic analysis using quantitative, high-sensitivity, fluorescence hybridization. *Proc Natl Acad Sci USA* 83:2934–2938.
Piper, J., Fantes, J., Gosden, J., Ji, L.A., 1990, Automatic detection of fragile X chromosomes using an X centromere probe. *Cytometry* 11:73–79.
Piper, J., Poggensee, M., Hill, W., Jensen, R., Ji, L., Poole, I., Stark, M., Sudar, D., 1994, Automatic fluorescence metaphase finder speeds translocation scoring in FISH painted chromosomes. *Cytometry* 16:7–16.
Qiang, F., Brockmann, S., Botzenhart, K., Wiedenmann, A., 2003, Improved detection of Salmonella spp. in foods by fluorescent in situ hybridization with 23S rRNA probes: A comparison with conventional culture methods. *Journal of food protection* 66(5):723–731
Raap, A.K., 1998, Advances in fluorescence in situ hybridization. *Mutat Res* 400:28–298.
Rudkin, G.T., Stollar, B.D., 1977, High resolution detection of DNA RNA hybrids in situ by indirect immunofluorescence. *Nature* 265:472–473.
Siebert, R., Matthiesen, P., Harder, S., Zhang, Y., Borowski, A., Zühlke-Jenisch, R., Metzke, S., Joos, S., Weber-Matthiesen, K., Grote, W., Schlegelberger, B., 1998, Application of interphase fluorescence in situ hybridization for the detection of the Burkitt translocation t(8;14)(q24;q32) in B-cell lymphomas. *Blood* 91:984–990.
Siffroi, J.P., Dupuy, O., Joye, N., Le Bourhis, C., Benzacken, B., Portnoi, M., Berkane, N., Franco, J.C., Studer, C., Carbonne, B., Gonzales, M., Bucourt, M., Uzan, S., Uzan, M., Milliez, J., Wolf, J.P., Taillemite, J., Dadoune, J.P., 2000, Usefulness of fluorescence in situ hybridization for the diagnosis of Turner mosaic fetuses with small ring X chromosomes fetal. *Diagn Ther* 15:229–233.
Sinclair, P.B., Sorour, A., Martineau, M., Harrison, C.J., Mitchell, W.A., O'Neill, E., Foroni, L., 2004, A fluorescence in situ hybridization map of 6q deletions in acute lymphocytic leukemia, identification and analysis of a candidate tumor suppressor gene. *Cancer Res* 64:4089–4098.

Singer, R.H., Ward, D.C., 1982, Actin gene expression visualized in chicken muscle tissue culture by using in situ hybridization with a biotinated nucleotide analog. *Proc Natl Acad Sci USA* 79:7331–7335.

Smith, M.D., Triantafillou, S., Parker, A., Coleman, M., 1997, A nonradioactive method of in situ hybridization that uses riboprobes and paraffin-embedded tissue and its combination with immunohistochemistry. *Diagnostic Mol Pathol* 6(1).

Speicher, M.R., Gwyn Ballard, S., Ward, D.C., 1996, Karyotyping human chromosomes by combinatorial multi-fluor FISH. *Nat Genet* 12:368–375.

Suciu, V., Botan, E., Valent, A., Chami, L., Spatz, A., Vielh, P., 2008, The potential contribution of fluorescent in situ hybridization analysis to the cytopathological diagnosis of Merkel cell carcinoma. *Cytopathology* 19:48–51.

Tanke, H.J., Florijn, R.J., Wiegant, J., Raap, A.K., Vrolijk, J., 1995, CCD microscopy and image analysis of cells and chromosomes stained by fluorescence in situ hybridization. *Histochem J* 27:4–14.

Tanke, H.J., Wiegant, J., van Gijlswijk, R.P., Bezrookove, V., Pattenier, H., Heetebrij, R.J., Talman, E.G., Raap, A.K., Vrolijk, J., 1999, New strategy for multi-colour fluorescence in situ hybridisation: COBRA: COmbined Binary Ratio labelling. *Eur J Hum Genet* 7:2–11.

Thompson, R.C., Deo, M., Turner, D.L., 2007, Analysis of microRNA expression by in situ hybridization with RNA oligonucleotide probes. *Methods* 43:153–161.

Tubbs, R.R., Hicks, D.G., Cook, J., Downs-Kelly, E., Pettay, J., Hartke, M.B., Hood, L., Neelon, R., Myles, J., Budd, G.T., Moore, H.C., Andresen, S., Crowe, J.P., 2007, Fluorescence in situ hybridization (FISH) as primary methodology for the assessment of HER2 status in adenocarcinoma of the breast: a single institution experience. *Diagn Mol Pathol* 16:207–210.

Wiegant, J., Ried, T., Nederlof, P.M., van der Ploeg, M., Tanke, H.J., Raap, A.K., 1991, In situ hybridization with fluoresceinated DNA. *Nucl Acids Res* 19:3237–3241.

Zhdanova, N.S., 2007, Comparative mapping of mink (Mustela vison) chromosome 8p: localization of three human BAC clones. *Genetika* 43:1074–1078.

Chapter 11
Immunoassay

Abstract Antibody-based detection systems for specific Ags are versatile and powerful tools for various molecular and cellular analyses, as well as clinical diagnostics. The power of such systems originates from the considerable specificity of Abs for particular antigenic epitopes. There are, however, numerous examples where important biological markers for cancer, infectious disease, or biochemical processes are present at too low a concentration in body fluids or tissues to be detected by using conventional immunoassays.

Keywords Antigen · Antibody · Immunoassay · ELISA · Radioimmunoassay · Immunogenicity · Immunoassays · Antigen-antibody interactions · Enzyme-linked immunosorbant assay · ImmunoRCA · nanoDLSA · Antibody-based microarray techniques · Protein microchips

11.1 Prologue

Immunogenicity has always been an important consideration in the evaluation of pharmaceutical protein biologics (Geng et al., 2005). Immunoassays are bioanalytical methods in which quantitation of the analyte depends on the reaction of an antigen (analyte) and an antibody. Although applicable to the analysis of both low molecular weight xenobiotic and macromolecular drugs, these procedures currently find most consistent application in the pharmaceutical industry to the quantitation of protein molecules. Immunoassays are also frequently applied in such important areas as the quantitation of biomarker molecules which indicate disease progression or regression, and antibodies elicited in response to treatment with macromolecular therapeutic drug candidates. Currently available guidance documents dealing with the validation of bioanalytical methods address immunoassays in only a limited way (Findlay et al., 2000).

M. Debnath et al., *Molecular Diagnostics: Promises and Possibilities*,
DOI 10.1007/978-90-481-3261-4_11,

11.2 Concept

All assay procedures which do not rely on the direct detection of a specific property of the unknown substance must depend on the reaction of an unknown with a reagent to give a product which can then be assayed. The product can be estimated directly or indirectly (for example, as a result of a further reaction). If the unknown is regenerated after the primary reaction it may then react with more reagent and a cycling assay is established. The advantage of such a system is that one molecule of the unknown gives rise to several molecules of product with a consequent increase in sensitivity. The reagent can be as diverse as a dye, neutron irradiation, an enzyme substrate or an antibody. Adequate assay systems must be shown to have suitable specificity, sensitivity, precision, range and convenience. In order to obtain maximum sensitivity and precision: (1) all the unknown should be reacted at least once and preferably several times; (2) the amount of product should be assayed by a procedure which gives a low background and shows changes in direct proportion to the change in product concentration; (3) the property measured should be capable of detection at very low concentrations of the product. The use of one or more cycling reactions is one well recognized way of achieving suitable amplification.

An immunoassay is a biochemical test that measures the concentration of a substance in a biological liquid, typically serum or urine, using the reaction of an antibody or antibodies to its antigen. The assay takes advantage of the specific binding of an antibody to its antigen. Monoclonal antibodies are often used as they only usually bind to one site of a particular molecule, and therefore provide a more specific and accurate test, which is less easily confused by the presence of other molecules (Sipponen et al., 1976; Uotila, 1981). The antibodies picked must have a high affinity for the antigen (if there is antigen available, a very high proportion of it must bind to the antibody).

Both the presence of antigen or antibodies can be measured. For instance, when detecting infection the presence of antibody against the pathogen is measured. For measuring hormones such as insulin, the insulin acts as the antigen. For numerical results, the response of the fluid being measured must be compared to standards of a known concentration. This is usually done though the plotting of a standard curve on a graph, the position of the curve at response of the unknown is then examined, and so the quantity of the unknown found. Detecting the quantity of antibody or antigen can be achieved by a variety of methods. One of the most common is to label either the antigen or antibody. The label may consist of an enzyme such as enzyme immunoassay (EIA), radioisotopes such as I-125 (Radioimmunoassay – RIA), magnetic labels (magnetic immunoassay – MIA) or fluorescence. Other techniques include agglutination, nephelometry, turbidimetry and Western Blot.

11.3 Types of Immunoassay

Several methods for the detection of an antibody response to a protein are described along with their limitations (Van Cleave, 2003). Immunoassays can be divided into those that involve labelled reagents and those which involve non-labelled reagents. Those which involve labelled reagents are divided into homogenous and heterogeneous (which involved a separation step) immunoassays. Heterogeneous immunoassays can be competitive or non-competitive.

In a *competitive* immunoassay, the antigen in the unknown sample competes with labeled antigen to bind with antibodies. Whereas in *noncompetitive* immunoassays, the "sandwich assay," antigen in the unknown is bound to the antibody site, then labeled antibody is bound to the antigen. The amount of labeled antibody on the site is then measured. Unlike the competitive method, the results of the noncompetitive method (Howell et al., 1981) will be directly proportional to the concentration of the antigen.

Antigen-antibody interactions involving the formation of tertiary antibody-antigen-antibody complex has also been reported (Rubina et al., 2005). Sandwich assay on microchips with immobilized antibodies provided the highest sensitivity of detection. Antibodies labeled with fluorescent dyes, horseradish peroxidase conjugates, or biotinylated antibodies with subsequent treatment with labeled avidin were used as developing antibodies. The results of immunoassays can be recorded using fluorescence, chemiluminescence, or matrix-assisted laser desorption ionization mass spectrometry directly from microchip gel elements. Gel microchips with immobilized capture antibodies were used to analyze the sample simultaneously for the presence of all six biotoxins with the same sensitivity as that for any single toxin (Arenkov et al., 2000).

The ELISA (enzyme-linked immunosorbant assay) is a powerful immunological method for detecting specific proteins in complex protein mixtures (Engvall et al., 1971; Von der Waat, 1978; Wolters et al., 1976). The ELISA has become an important tool for the cell and molecular biologist. It is increasingly being applied in clinical medicine for detecting proteins associated with disease including antibodies produced in response to infection by the HIV virus (Engvall and Perlmann, 1971). The ELISA is a fundamental tool of clinical immunology (Bosch et al., 1975; Engvall and Wewer, 2003; Mida et al., 2003; Schuurs and Van Weemen, 1980 Van weemen and schuurs 1974), and is used as an initial screen for HIV detection. Based on the principle of antibody-antibody interaction, this test allows for easy visualization of results and can be completed without the additional concern of radioactive materials use.

ELISA measurement tool has also been used in parasitology e.g., malaria and trichinosis, microbiology, and oncology (Seppala et al., 1978); biochemistry of tissues, e.g., fibronectin, laminin, integrins, and muscular dystrophies. Currently, investigation using ELISA has been performed on the differentiation factors for muscle regeneration and myogenic cells from nonmuscle tissues for muscle cell replacement.

During the late 1960s and early 1970s, many RIA test systems were essentially "home-brew" methods developed by individual researchers who could not keep pace (particularly financially) with the possibilities and facilities of commercial manufacturers such as Boehringer-Mannheim (Germany), Abbott (United States), and Organon Teknika (The Netherlands), to name only a few. Commercialization of EIA/ELISA test kits had started. Solid-phase techniques were used in the development of microtiter plates (96 wells) in which either an antigen or an antibody is noncovalently bound to a solid-phase support (Catt and Tregear, 1967). Technical advances led to automated pipetting devices (Micromedics; Hamilton), multi-channel pipettes (Lab Systems), and microtiter plate readers and washers and in the 1980s fully automated test instruments were manufactured by Boehringer-Mannheim and Abbott, among others. Such automated systems have come to stay in medical laboratories.

In the early 1970s, blood-bank screening for virologic diseases such as hepatitis B antigen was done either by (semi)automated RIA or nonradioactive but rather cumbersome hemagglutination tests. In 1976, Organon Teknika developed and marketed a highly successful EIA system for the hepatitis B surface antigen (HbsAg), featuring a 96-well microtiter plate format. This test became the first commercially available EIA. Other microbiological and virologic diagnostic tests soon followed, e.g., for hepatitis B "e" (HBe) antigens, rubella antibodies, toxoplasma antibodies, and in the 1980s, an EIA system for detection of human immunodeficiency virus antibodies.

11.4 Recent Advances in the Field of Immunodiagnostics

Recent advances in the field of low-level Ag detection include the development of stronger fluorochromes and chemiluminescent substrates for use in ELISAs, immunofluorescence-based staining and immunoblotting, and the application of signal amplification methods such as tyramide deposition. Although these techniques can be quite powerful, greater sensitivity and specificity are often required, particularly when working with limited amounts of sample material or when Ag density is extremely low. With these needs in mind, rolling circle amplification (RCA) reporter system for the detection of protein Ags have been developed. RCA driven by DNA polymerase can replicate circularized oligonucleotide probes with either linear or geometric kinetics under isothermal conditions. Using a single primer, RCA generates hundreds of tandemly linked copies of the circular template within a few minutes. In ImmunoRCA, the 5′ end of this primer is attached to an Ab. In the presence of circular DNA, DNA polymerase, and nucleotides, the rolling circle reaction results in a DNA molecule consisting of multiple copies of the circle DNA sequence that remains attached to the Ab. The amplified DNA can be detected in a variety of ways, including direct incorporation of hapten-labeled or fluorescently labeled nucleotides, or by hybridization of fluor-labeled or enzymatically labeled complementary oligonucleotide probes (Schweitzer et al., 2000). ImmunoRCA, therefore, represents a novel approach for signal amplification of Ab–Ag recognition events. ImmunoRCA-profiling based on the simultaneous quantitation of multiple Ags should expand the power of immunoassays by exploiting the increased information content of ratio-based expression analysis.

In immuno-PCR, a unique DNA sequence tag is associated with a specific Ab using streptavidin-biotin interactions, alternative bridging moieties, or covalent linkage. Abs bound to Ag are then detected by PCR amplification of the associated

DNA tag. Multiple Abs and multiple DNA tags have been used to analyze several Ags simultaneously. Although immuno-PCR was shown to be significantly more sensitive than ELISA, gel electrophoresis was required after DNA amplification in solution to separate and/or quantitate the different amplified DNA tags. The requirements for thermal cycling and product separation by gel electrophoresis have restricted the widespread adoption of immuno-PCR as an alternative to ELISA and have precluded its utility in immunohistochemical or array formats.

A research group has developed a highly sensitive homogeneous one-step immunoassay, nanoDLSA, for cancer marker detection. To prepare the immunoassay, a pair of monocolonal antibodies that can specifically bind with a cancer marker antigen, are conjugated with two gold nanoparticle probes. To conduct the assay, one simply mixes the two nanoprobes with a sample solution. The antigen-antibody binding will introduce a nanoparticle aggregation. By measuring the degree of nanoparticle aggregation in the assay solution using dynamic light scattering (DLS), the concentration of cancer marker antigen in the sample can be quantitatively determined. In a recent study, the research group has analyzed the free-PSA level of a few human serum samples collected from prostate cancer patients using both nanoDLSA and ELISA. The preliminary results revealed an excellent correlation between nanoDLSA and ELISA (Liu et al., 2008).

The nanoDLSA immunoassay is fast, highly sensitive, accurate, and extremely easy to conduct. It requires a much smaller amount (at least 100 times less) of blood samples and antibody probes to conduct the assay compared to ELISA. The cost reduction of nanoDLSA compared to other immunoassays is tremendous. Because of the minute amount of sample that is required by nanoDLSA, it is possible to conduct the detection and measurement of one or multiple cancer markers from a single drop of human blood sample using this new immunoassay technology.

11.5 Clinical Applications of Immunoassay

Most biopharmaceutical therapeutics elicits some level of antibody response against the product. This antibody response can, in some cases, lead to potentially serious side effects and/or loss of efficacy. Therefore, the immunogenicity of therapeutic proteins is a concern for clinicians, manufacturers and regulatory agencies. In order to assess immunogenicity of these molecules, appropriate detection, quantitation and characterization of antibody responses are necessary. Inadequately designed antibody assays have led to the hampering of product development or, during licensure, post-marketing commitments (Mire-Sluis et al., 2004).

Antibody-based microarray techniques for the multiplexed detection of cholera toxin beta-subunit, diphtheria toxin, anthrax lethal factor and protective antigen, *Staphylococcus aureus* enterotoxin B, and tetanus toxin C fragment have been developed (Rucker et al., 2005). Two detection schemes were investigated: (i) a direct assay in which fluorescently labeled toxins were captured directly by the antibody array and (ii) a competition assay that employed unlabeled toxins as reporters for the quantification of native toxin in solution. In the direct assay, fluorescence measured at each array element is correlated with labeled toxin concentration to yield baseline binding information (Langmuir isotherms and affinity constants). Extending from the direct assay, the competition assay yields information on the presence, identity, and concentration of toxins. A significant advantage of the competition assay over reported profiling assays is the minimal sample preparation required prior to analysis because the competition assay obviates the need to fluorescently label native proteins in the sample of interest. Although the sensitivity of the direct assay is superior to that of the competition assay, detection limits for unmodified toxins in the competition assay are comparable to values reported previously for sandwich-format immunoassays of antibodies arrayed on planar substrates. As a demonstration of the potential of the competition assay for unlabeled toxin detection, we conclude with a straightforward multiplexed assay for the differentiation and identification of both native *S. aureus* enterotoxin B and tetanus toxin C fragment in spiked dilute serum samples.

The motor neuron degenerative disease spinal muscular atrophy (SMA) is the leading genetic cause of infant mortality and is caused by mutations in the survival of motor neurons (SMN) gene that reduce the expression levels of the SMN protein. A major goal of current therapeutic approaches is to increase SMN levels in SMA patients. Kolb et al., 2006 developed a novel cell immunoassay to quantitatively measure SMN levels from peripheral blood mononuclear cells (PBMCs) using a single anti-SMN antibody. In another study by Hix et al. (2004) a quantitative enzyme immunoassay (EIA) for detection of RBP (retinol-binding protein) was developed. An immunoassay for preß1-HDL (the initial acceptor of cellular cholesterol) using a monoclonal antibody, MAb55201 is useful for clinical measurement of preß1-HDL (Takashi et al., 2003). A newly developed enzyme-linked immunosorbent assay (ELISA) that detects immunoglobulin G antibodies to the 27-kDa *Cryptosporidium parvum* sporozoite surface antigen was used to test many sera collected from pregnant women (Ong et al., 2005). Serological assays may provide more accurate information regarding community levels of *Cryptosporidium* infection.

Recently a commercial enzyme immunoassay PlateliaTM Dengue NS1 AG (Bio-Rad Laboratories) was used to monitor semiquantitatively dengue virus replication in cultured cells (Ludert et al., 2008). These results suggest that the PlateliaTM Dengue NS1 AG kit can be used as a fast and reliable surrogate method for the relative quantitation of dengue virus replication in cultured cells. A immunoassay method may be suitable for determining levels of busulfan in human plasma. It offers the advantages of using a smaller sample size, does not require sample preparation and is less labor intensive than other methods. The ability to make 240 determinations per hour enables effective routine monitoring of busulfan levels in clinical practice.

Immunoassays are also frequently applied in such important areas as the quantitation of biomarker molecules which indicate disease progression or regression, and antibodies elicited in response to treatment with macromolecular therapeutic drug candidates (Findlay et al., 2000). The double antigen bridging immunoassay has been used extensively for detection of immunogenicity responses to therapeutic monoclonal antibodies. A two-step format requires very low coating concentrations and higher conjugate concentrations to achieve maximal sensitivity and suffers from significantly reduced sensitivity at higher coating concentrations. A one-step assay format can greatly reduce the effect of coating concentration variation on assay performance (Bourdage et al., 2005).

Protein microchips are used in immunoassays for detection of antigens or antibodies, as well as to carry out enzymatic reactions and to measure their kinetics in the absence or presence of an inhibitor. A protein microchip can be used several times in different immunoassays and enzymatic kinetic measurements.

A novel point-of-care platform to quantify micro-organisms causing dental infections and/or inflammatory markers reflecting an oral disease status has been studied by some workers. This system is based on a sandwich immunoassay technology known as ABICAP (Antibody Immuno Column for Analytical Processes) using poly-horseradish peroxidase conjugates. This assay enabled to quantify 500 colony-forming units of *Streptococcus sobrinus* per milliliter of saliva. The platform allows rapid and convenient performance chair side of such tests by a dentist or dental hygienist within 20 min at the dental office (Munial et al., 2007).

An indirect enzyme-linked immunosorbent assay (ELISA) for the detection of *Trichomonas vaginalis* (a common sexually transmitted disease) which is both rapid and sensitive (detection limit of approximately 100 trichomonads per ml) has been developed. This assay employs affinity-purified rabbit anti-*T. vaginalis* antibodies in a "sandwich" configuration. It is simple to perform and is neither interfered with nor appears to cross-react with other microorganisms which are common inhabitants of the urogenital tract. In addition to exhibiting a sensitivity of 77%, the specificity of the ELISA was 100%. These results demonstrate that the ELISA is a significant improvement over the wet mount method for the diagnosis of trichomoniasis (Watt et al., 1986). Chronic fatigue syndrome in patients can be detected by immunoassay with cytomegalovirus early antigens from gene products p52 and CM_2 (UL44 and UL57) (Beqaj et al., 2008).

11.6 Utilization and Interpretation of Immunological Tests

Investigation plays an important role in diagnosis of clinical conditions. However, indiscriminate use of tests due to lack of knowledge of false positive and false negative test results, predictive value of a test etc. can cause more problem than benefit. It is imperative for the clinician to know, when to order a test, how to interpret it and what can be the methodological problems with the test. Cystic hydatid disease (hydatidosis) is one of the most important zoonosis that is caused by the larval stage of *Echinococcus granulosus*. As its diagnosis by clinical symptoms alone is difficult and confusing, serologic diagnostic techniques are used to confirm the disease. These techniques can also be used for epidemiologic studies. The commercial human enzyme-linked immunosorbent assay (ELISA) kit for the diagnosis of hydatidosis in sera collected from sheep with hydatidosis concluded that it is possible to use human ELISA kit for the diagnosis of hydatidosis in sheep (Hashemitabar, 2008).

11.7 Conclusion

In conclusion, the number of analytical and clinical investigations relying on these measurement procedures worldwide is exceedingly large. Thus, one can imagine that the numbers of measurements and determinations using immunoassay for routine patient care are astronomical. The impact of diagnostic immunoassays on patients, clinicians, and the healthcare system in general is virtually overwhelming. Given the impact that the immuno assays had on clinical diagnosis and healthcare in general, as well as on the development of a well-established in vitro diagnostic industry, the inventors deserve to be honored again.

References

Arenkov, P., Kukhtin, A., Gemmelll, A., Voloshchuck, S., Chupeeva, V., Mirzabekov, A., 2000, Protein microchips: use for immunoassay and enzymatic reactions. *Ana Biochem* 278(2): 123–131

Beqai, S.H., Lerner, A.M., Fitzgerald, J.T., 2008, Immunoassay with cytomegalovirus early antigens from gene products p52 and CM2 (UL44 and UL57) detects active infection in patients with chronic fatigue syndrome. *J Clin Pathol* 61(5):623–626.

Bosch, A.M.G., Van Hell, H., Brands, J.A.M., Dijkhuizen, D.M., Schuurs, A.H.W.M., 1975, Methods for the determination of total estrogens (TE) and human placental lactogen (HPL) in plasma of pregnant women by enzyme-immunoassay. *Clin Chem* 21:1009.

Bourdage, J.S., Lee, T.N., Taylor, J.M., Willey, M.B., Brandt, J.T., Konrad, R.J., 2005, Effect of double antigen bridging immunoassay format on antigen coating concentration dependence and implications for designing immunogenicity assays for monoclonal antibodies. *J Pharm Biomed Anal* 39(3–4):685–690.

Catt, K., Tregear, G.W., 1967, Solid-phase radioimmunoassay in antibody-coated tubes. *Science* 158:1570–1572.

Engvall, E., Jonsson, K., Perlmann, P., 1971, Enzyme-linked immunosorbent assay. II. Quantitative assay of protein antigen, immunoglobulin G, by means of enzyme-labelled antigen and antibody-coated tubes. *Biochim Biophys Acta* 251(3):427–434.

Engvall, E., Perlmann, P., 1971, Enzyme-linked immunosorbent assay (ELISA). Quantitative assay of immunoglobulin G. *Immunochemistry* 8:871–874.

Engvall, E., Wewer, U.M., 2003 The new frontier in muscular dystrophy research: booster genes. *FASEB J*;17:1579–1584.

Findlay, J.W., Smith, W.C., Lee, J.W., Nordblom, G.D., Das, I., DeSilva, B.S., Khan, M.N., Bowsher, R.R., 2000, Validation of immunoassays for bioanalysis: a pharmaceutical industry perspective *J Pharm Biomed Anal* 21(6):1249–1273.

Geng, D., Shankar, G., Schantz, A., Rajadhyaksha, M., Davis, H., Wagner, C., 2005, Validation of immunoassays used to assess immunogenicity to therapeutic monoclonal antibodies. *J Pharm Biomed Anal* 39(3–4):364–375.

Hashemitabar, G.R., Razmi, G.R., Shahroozian, A., 2008, Application of a modified human enzyme-linked immunosorbent assay kit for diagnosis of hydatidosis in sheep *Iranian J Vet Res* 9(1) serial no. 22:31–35.

Hix, J., Martinez, C., Buchanan, I., Morgan, J., Tam, M., Shankar, A., 2004, Development of a rapid enzyme immunoassay for the detection of retinol-binding protein. *Am J Clin Nutrition* 79(1):93–98.

Howell, E.E., Nasser, J., Schray, K.J., 1981, Coated tube enzyme immunoassay: factors affecting sensitivity and effects of reversible protein binding to polystyrene. *J Immunoassay* 2(3–4): 205–225.

Kolb, S.J., Gubitz, A.K., Olszewski, R.F., Jr., Ottinger, E., Sumner, C.J., Fischbeck, K.H., Dreyfuss, G., 2006, A novel cell immunoassay to measure survival of motor neurons protein in blood cells. *BMC Neurol* 6:6.

Liu, X., Dai, Q., Austin, L., Coutts, J., Knowles, G., Zou, J., Chen, H., Huo, Q., 2008, A one-step homogeneous immunoassay for cancer biomarker detection using glod nanoparticle probes coupled with dynamic light scattering. *J Am Chem Soc* 130(9):2780–2782.

Ludert, J.E., Mosso, C., Ceballos-Olvera, I., del Angel, R.M., 2008, Use of a commercial enzyme immunoassay to monitor dengue virus replication in cultured cells. *Virol J* 5:51.

Mida, T., Miyazaki, O., Nakamura, Y., Hirayama, S., Hanyu, O., Fukamachi, I., Okada, M., 2003, Analytical performance of a sandwich enzyme immunoassay for preβ-HDL in stabilized plasma. *J Lipid Res* 44(3):645–650.

Mire-Sluis, A.R., Barrett, Y.C., Devanarayan, V., Koren, E., Liu, H., Maia, M., Parish, T., Scott, G., Shankar, G., Shores, E., Swanson, S.J., Taniguchi, G., Wierda, D., Zuckerman, L.A., 2004, Recommendations for the design and optimization of immunoassays used in the detection of host antibodies against biotechnology products. *J Immunol Methods* 289(1–2):1–16.

Munial, S., Miethe, P., Netuschil, L., Struck, F., Maier, K., Bauermeister, C., 2007, Immunoassay-based diagnostic point-of-care technology for oral specimen. *Ann N Y Acad Sci* 1098:486–489.

Ong, C.S., Li, A.S., Priest, J.W., Copes, R., Khan, M., Fyfe, M.W., Marion, S.A., Roberts, J.M., Lammie, P.J., Isaac-Renton, J.L., 2005, Enzyme immunoassay of cryptosporidium-specific immunoglobulin G antibodies to assess longitudinal infection trends in six communities in British Columbia, Canada. *Am J Trop Med Hyg* 73(2):288–295.

Rubina, A.Y., Dyukova, V.I., Dementieva, E.I., Stomakhin, A.A., Nesmeyanov, V.A., Grishin, E.V., Zasedateley, A.S., 2005, Quantitative immunoassay of biotoxins on hydrogel-based protein m microchips. *Anal Biochem* 340(2):317–329.

Rucker, V.C., Havenstrite, K.L., Herr, A.E., 2005, Antibody microarrays for native toxin detection. *Anal Biochem* 339(2):262–270.

Schuurs, A.H.W.M., Van Weemen, B.K., 1980, Enzyme-immunoassay: a powerful analytical tool [Review]. *J Immunoassay* 1:229–249.

Schweitzer, B., Wiltshire, S., Lamber, J., O'Malley, S., Kukanskis, K., Zhu, Z., Kingsmore, S.F., Lizardi, P.M., Ward, D.C., 2000, Immunoassays with rolling circle DNA amplification: a versatile platform for ultrasensitive antigen detection. *Proc Natl Acad Sci USA* 97(18):10113–10119.

Seppala, M., Rutanen, E.M., Heikinheimo, M., Jalanko, H., Engvall, E., 1978 Detection of trophoblastic tumour activity by pregnancy-specific ß1 glycoprotein. *Int J Cancer* 21:265–267.

Sipponen, P., Ruoslahti, E., Vuento, M., Engvall, E., Stenman, U., Ihamakit, T., Suirala, M., 1976, CEA and CEA-like activity in gastric cancer. *Acta Hepatogastroenterol* (Stuttg) 13:276–279.

Takashi, M., Miyazaki, O., Nakamura, Y., et al., 2003, Analytical performance of a sandwich enzyme immunoassay for preß1-HDL in stabilized plasma, *J Lipid Res* 44:645–650.

Uotila, M., Ruoslathi, E., Engvall, E., 1981, Two-site sandwich enzyme immunoassay with monoclonal antibodies to human alphafetoprotein. *J Immunol Methods* 42(1):11–15.

Van Cleave, V.H., 2003, Vailidation of immunoassays for anti-drug antibodies. *Dev Biol (Basel)* 112:107–112.

Van Weemen, B.K., Schuurs, A.H.W.M., 1974, Immunoassay using antibody-enzyme conjugates. *FEBS Lett* 43:215–218.

Von der Waat, M., Snelting, A., Cichy, J., Wolters, G., Schuurs, A., Schuurs, A.H.W.M., 1978, Enzyme immunoassay in diagnosis of hepatitis with emphasis on the detection of "e" antigen (HbeAg). *J Med Virol* 3(1):43–49.

Watt, R.M., Philip, A., Wos, S.M., Sam, G.J., 1986, Rapid assay for immunological detection of *Trichomonas vaginalis*. *J Clin Microbiol* 24(4):551–555.
Wolters, G., Kuijpers, L.P.C., Kacaki, J., Schuurs, A.H.W.M., 1976, Enzyme-immunoassay for HbsAg. *Lancet* II:690.

Chapter 12
Phage Display

Abstract Phage display is a method of generating antibody reagents, and phage-antibody libraries to provide a rich source of antibody diversity, potentially providing hundreds of unique antibodies per target. The antibody gene, once isolated, can be conveniently shuttled into a variety of expression formats, providing a renewable resource of antibody protein.

Keywords Antibodies · Combinatorial libraries · Protein–protein interaction · Virions · Phage display · Epitope mapping · Cloning allergens · Drug discovery

12.1 Prologue

The selection of antibodies from combinatorial libraries displayed on the surface of filamentous phage has become an important methodology for the generation of reagent, diagnostic, and therapeutic molecules and for the study of natural immune responses. Using this technique, antibody genes have been cloned from multiple species or expressed directly from large man-made repertoires of antibody-encoding genes.

12.2 Concept

Phage display is the development of phage-displayed peptides and antibody fragments into viable diagnostic reagents (Jayanna et al., 2009). Phage display is a molecular diversity technology that allows the presentation of large peptide and protein libraries on the surface of filamentous phage. Phage display libraries permit the selection of peptides and proteins, including antibodies, with high affinity and specificity for almost any target (Fig. 12.1). A crucial advantage of this technology is the direct link that exists between the experimental phenotype and its encapsulated genotype, which allows the evolution of the selected binders into optimized molecules. Phage display facilitates engineering of antibodies with regard to their size, valency, affinity, and effector functions. The selection of antibodies and peptides from libraries displayed on the surface of filamentous phage has

M. Debnath et al., *Molecular Diagnostics: Promises and Possibilities*,
DOI 10.1007/978-90-481-3261-4_12,

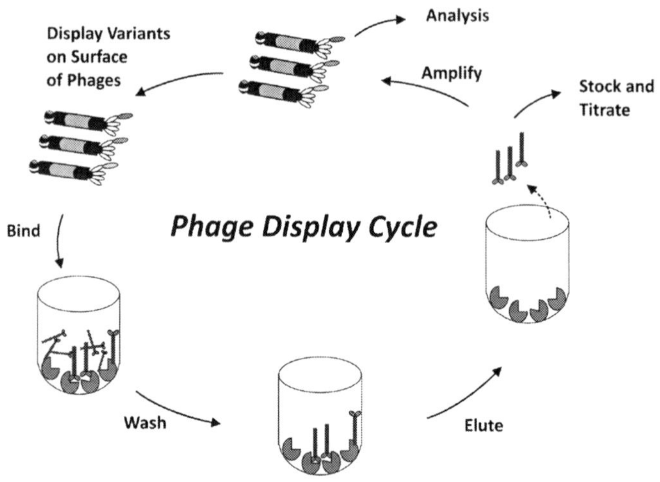

Fig. 12.1 General concept of phage display

proven significant for routine isolation of peptides and antibodies for diagnostic and therapeutic applications (Azzazy and Highsmith, 2000).

Phage display is an advanced technology that can be used to characterize the interactions of antibody with antigen at the molecular level. Phage display is a practical realization of the artificial chemical evolution. Using standard recombinant DNA technology, peptides are associated with replicating viral DNAs that include the peptides' coding sequences. The peptide populations so created are managed by simple microbiological methods. Phage display is an exponentially growing research area, and numerous reviews covering different aspects have been published in recent years.

12.3 Phage-Display Libraries as Populations of Replicable, Mutable Chemicals

Phages are viruses that infect bacterial cells, and many of the vectors used in recombinant DNA research are phages that infect the standard recombinant DNA host: the bacterium *Escherichia coli*. The key feature of recombinant DNA vectors, including phages, is that they accommodate segments of "foreign" DNA-pieces of human DNA, for instance, or even stretches of chemically synthesized DNA. As vector DNA replicates in its *E. coli* host, then, the foreign "insert" replicates along with it as a sort of passenger.

An "expression vector," including a phage-display vector, has an additional feature compared to vectors in general: the foreign DNA is "expressed" as a protein. That is, it programs machinery of the *E. coli* host cell to synthesize a foreign peptide whose amino acid sequence is determined (*via* the genetic code) by the nucleotide sequence of the insert. Phage display differs from conventional expression systems, however, in that the foreign gene sequence is spliced into the gene for one of the phage coat proteins, so that the foreign amino acid sequence is genetically fused to the endogenous amino acids of the coat protein to make a hybrid "fusion" protein. The hybrid coat protein is incorporated into phage particles ("virions") as they are released from the cell, so that the foreign peptide or protein domain is displayed on the outer surface.

A phage-display "library" is a heterogeneous mixture of such phage clones, each carrying a different foreign DNA insert and therefore displaying a different peptide on its surface. Different types of libraries will be formed. Each peptide in the library can replicate. The phage to which the peptide is attached infects a fresh bacterial host cell and it multiplies to produce a huge crop of identical progeny phages displaying the same peptide. And if the phage's DNA suffers a mutation in the peptide coding sequence that mutation is passed on to the phage's progeny and can affect the structure of the peptide. In short, the peptides in a phage-display library have the two key characteristics required for chemical evolution: replicability and mutability.

Because of its accessibility to solvent, a displayed peptide often behaves essentially as it would if it were not attached to the virion surface. This means that many techniques that chemists or biochemists apply to compounds free in solution can be applied more or less unaltered to peptides tethered to a phage. The captured phages are "amplified" by infecting them *en masse* into fresh cells and culturing the cells to yield a large crop of progeny phages, which can serve as the input for another round of affinity purification. Moreover, by periodically introducing mutations into the phage population, the experimenter widens the search for effective ligands by exploring peptide sequences that are not present in the initial phage-display library. Eventually, captured phages are cloned so that the displayed peptides responsible for binding can be studied individually. The amino acid sequence of the peptide is easily obtained by determining the corresponding coding sequence in the viral DNA. This so-called "affinity selection" is the premier example of artificial selection imposed on populations of phage-displayed peptides. Since there is no need to process clones one by one until the final stage, enormous libraries displaying billions of different structures can be easily surveyed for exceedingly rare binding clones.

Phage display is achieved by fusing polypeptide libraries to phage coat proteins. The resulting phage particles display the polypeptides on their surfaces and they also contain the encoding DNA. Library members with particular functions can be isolated with simple selections and polypeptide sequences can be decoded from the encapsulated DNA. The technology's success depends on the efficiency with which polypeptides can be displayed on the phage surface, and significant progress has been made in engineering M13 bacteriophage coat proteins as improved phage display platforms. Functional display has been achieved with all five M13 coat proteins, with both N- and C-terminal fusions. Also, coat protein mutants have been designed and selected to improve the efficiency of heterologous protein display, and in the extreme case, completely artificial coat proteins have been evolved specifically as display platforms. The utility of phage display as a powerful tool in modern biotechnology is well known (Sindhu, 2001).

The selection of peptides and proteins from libraries expressed on the surface of filamentous phage is becoming an important tool in biotechnology. Recent developments have shown that peptides can be selected to bind receptors and antibodies, while semi synthetic antibodies can be selected to bind almost any target. Phage display has allowed the routine isolation of therapeutically interesting human antibodies. Phages are also being utilized to examine the specificities of natural enzymes as well as to evolve novel enzymes de novo (Barbas, 1993).

Phage display technology presents a rapid means by which proteins and peptides that bind specifically to predefined molecular targets can be isolated from extremely complex combinatorial libraries. There are several important ways by which phage display can provide impetus to receptor-based research. Firstly, phage display can be applied, alongside transcriptome and proteome expression profiling techniques, to the identification and characterisation of receptors whose expression is specific to a cell lineage, a tissue or a disease state. Secondly, specific monoclonal antibodies that enable researchers to identify, localize and quantify receptors can be produced very rapidly (weeks). Thirdly, it should be possible to apply phage display to the matching of orphan ligands and receptors. Finally, phage display can be used to identify proteins and peptides that modulate receptor activity. As well as being useful in the study of receptor function, biologically active proteins and peptides could also be used therapeutically, or as leads for drug design. Phage display technologies are powerful tools for selecting binding ligands against purified molecular targets, live cells, and organ vasculature. However, the selection of natural legends using phage display has been limited because of significant problems associated with the display of complex cDNA repertoires (Faix et al., 2004). Hence phage display is ready to play a central role in the study of receptors in the post-genome era (Hartley, 2002).

12.4 Practical Applications of Phage Display

Phage display is a simple yet powerful technology that is used to rapidly characterize protein-protein interactions from amongst billions of candidates. Usual targets for isolation of phage displayed peptide ligands include antibodies, various receptors, other full size proteins or larger fragments thereof. Smaller protein fragments such as synthetic peptides have not been reported as targets for screening of peptide display libraries (Bremnes et al., 1998). This widely practiced technique is used to map antibody epitopes, create vaccines and to engineer peptides, antibodies and other proteins as both diagnostic tools and as human therapeutics.

12.4.1 Target Receptors Used in Affinity Selection

Over the past decade, powerful technologies devoted to survey large molecular libraries for the presence of specific clones using the discriminative power of affinity selection have been developed. Phage surface display technology is the most established of these methods and has revolutionised our ability to select agonistic and antagonistic peptides, antibodies with desired specificity and other drug targets. Thereby ligands are expressed as fusions to phage coat proteins and their respective genes are packaged within the phage. The basic concept of linking the phenotype, expressed as gene product displayed on the phage coat, to its genetic information integrated into the phage genome, creates fusion proteins covalently associated with the infectious particle itself. Binding of the phage to the target molecule offers a selective system by which rare phage carrying the desired gene product can be selected from large phage populations carrying inappropriate sequences. Phage selected in this fashion can be used for subsequent rounds of selection because they are able to self-replicate in suitable host cells, yielding target-specific phage populations after several consecutive rounds of affinity selection (Rhyner et al., 2004).

12.4.2 Epitope Mapping and Mimicking

An "epitope" is the small determinant on the surface of a ligand with which the receptor makes close, geometrically and chemically specific contact. If the ligand is a protein, the epitope is sometimes "continuous," comprising a few adjacent critical amino acids in the primary sequence. For instance, antibodies specific for continuous epitopes on protein antigens typically contact three to four critical amino acids over a six-residue segment. More often, however, protein epitopes are more complex. Many are "discontinuous" because they comprise critical binding residues that are distant in the primary sequence but close in the folded native conformation. And many epitopes, including discontinuous ones, are "conformation-dependent" because they require the context of the overall protein structure to contrain them in a binding conformation.

In many research contexts, it is highly desirable to "map" the epitope to a confined portion of the natural protein ligand. If the epitope is (or might be) continuous and not conformation dependent, random peptide libraries provide a cheap, easy approach to this goal. The receptor is used to affinity select random peptide ligands, and the sequence motif in the selected peptides is compared to the amino acid sequence of the natural ligand. Often, in these cases, the motif clearly matches critical binding amino acids in the natural protein ligand, thereby mapping the epitope to a very narrow part of the overall natural ligand structure. Since this approach uses replicable, widely available, all-purpose random peptide libraries and simple microbiological procedures, it is generally much cheaper and easier than alternative epitope mapping methods that require chemical synthesis of short peptide segments of the ligand's amino acid sequence.

Only rarely will a random peptide library contain a binding motif extending to more than about six amino acids or adequately represent conformation-dependent or discontinuous epitopes. Although receptors recognizing such epitopes often select ligands from random peptide libraries, these artificial ligands seldom bear a recognizable similarity to any part of the natural protein ligand at the amino acid sequence level. An alternative approach in such circumstances is to construct a gene-specific library displaying 15–100 amino acid segments of the natural amino acid sequence-long enough to occasionally include small elements of secondary structure from the native protein. Such libraries sometimes contain good ligands for receptors that fail to select ligands from random peptide libraries. Because it requires construction of a specific library for each new ligand gene, however, this approach is much more arduous than use of all-purpose random peptide libraries.

Affinity selection from random peptide libraries often reveals entirely unexpected ligands-ligands that do not match any linear epitope and that could not have been anticipated from even extensive knowledge of the receptor and/or its natural ligand. This is especially so when the receptor's natural epitope is nonproteinaceous or is a discontinuous or conformation-dependent protein epitope.

12.4.3 Identifying New Receptors and Natural Ligands

A ligand for a receptor can be used as a "probe" to identify new receptors that bind the same ligand (Faix et al., 2004). Phage display technologies are powerful tools for selecting binding ligands against purified molecular targets, live cells, and organ vasculature. However, the selection of natural ligands using phage display has been limited because of significant problems associated with the display of complex cDNA repertoires. These methods will be useful as functional genomics tools for identifying natural ligands from various source tissues.

The selection of phage displayed cDNA repertoires on an immobilized target has been reported to be an efficient way to rapidly identify interacting partners (Hufton, 1999). To date, however, only a few successful applications have been reported. These vectors may be used for selection of phage displayed cDNA libraries with polyclonal sera from patients. This will allow the identifying antigenic cDNA products in such diseases as cancer, viral/bacterial infections or autoimmune disease. Furthermore, by selections with other specific biomolecules, this display system may aid the identification of interacting partners in functional genomics.

12.4.4 Drug Discovery

Many of the receptors used in affinity selection are targets of drug discovery programs, and the peptide ligands selected by them are therefore potential leads to new drugs. Such peptides might act as receptor agonists or antagonists (for example, of enzymes or hormone receptors) or otherwise modulate the receptor's biological effect.

Affinity selection resembles in essence the traditional approach to drug discovery: screening libraries of synthetic compounds or natural products for substances that bind the target receptor and that might therefore be leads to new agonists, antagonists, or modulators. There are important differences, however. Affinity selection has the key advantage that the scale of the search is many orders of magnitude greater than is feasible when chemical libraries must be screened compound by compound- billions of peptides *versus* tens of thousands of chemicals. On the other hand, for most pharmaceutical applications, peptides have poor pharmacological properties, being generally orally unavailable and subject to rapid degradation in the body by naturally occurring enzymes. There is some precedent for synthesizing peptidomimetic compounds that mimic the essential pharmacological features of bioactive peptides on a nonpeptide scaffold (Section VI). But developing peptidomimetics is an arduous and chancy project in medicinal chemistry, and it seems likely that the most important contribution of phage display to drug discovery will be confined to applications where peptides themselves can serve as plausible therapeutics. For example, Wrighton and colleagues used phage display to identify a small peptide agonist of the receptor for eythropoietin, a protein hormone that is administered parenterally in some circumstances. The small peptide, which bears little resemblance to the natural hormone at the amino acid sequence level, might serve as a superior substitute for the much larger protein. Vaccines are another case in which peptides are eminently usable therapeutics.

Peptides composed of D-amino acids are much less susceptible to degradation in the body than peptides composed of the natural L-amino acids. Schumacher and his colleagues have put forth a clever way of using phage display to identify D-amino acid peptide ligands for target receptors. They synthesized chemically the D form of an SH3 domain and used it to affinity select ligands from a random peptide library, whose amino acids are of course the natural L isomers. The D forms of these peptides are therefore ligands for the natural L form of the receptor-the form that would be the actual target of drug discovery.

12.4.5 Epitope Discovery – A New Route to Vaccines and Diagnostics

In recent years, the use of display vectors and in vitro selection technologies has transformed the way in which we generate ligands, such as antibodies and peptides, for a given target. Using this technology, we are now able to design repertoires of ligands from scratch and use the power of phage selection to select those ligands having the desired (biological) properties. With phage display, tailor-made antibodies may be synthesized and selected to acquire the desired affinity of binding and specificity for in vitro and in vivo diagnosis, or for immunotherapy of human disease. A combination of the design and generation of millions to billions of different ligands, together with phage display for the isolation of binding ligands and with functional assays for identifying (and possibly selecting) bio-active ligands, will open even more challenging applications of this inspiring technology, and provide a powerful tool for drug and target discovery well into the next decade (Hoogenboom et al., 1998).

When the receptor used for affinity selection is an antibody, the peptides it selects from random peptide libraries are called "antigenic mimics" of the corresponding natural epitope-the antigenic determinant that elicited the selector antibody in the first place. When these peptides are used in turn to immunize naive animals, some are able to elicit new antibodies that cross-react with the natural epitope, even though the naive animals have never been directly exposed to it. Such peptides are "immunogenic mimics" as well as antigenic mimics. Antigenic and immunogenic mimicry are the basis of "epitope discovery", a new approach to disease diagnosis and vaccine development. Most diseases-particularly infectious diseases-leave their imprint on the complex mixture of antibody specificities that comprises the total serum immunoglobulin population.

Included in this population are disease-specific antibodies-some elicited directly by antigens on a pathogen, others possibly recognizing antigens that reflect the disease process more indirectly. When total serum antibody from a patient is used to affinity select clones from a random peptide library, therefore, some of the selected ligands will correspond to disease-specific antibodies. Of course the patient's pool of antibodies will contain myriad non-disease-specific antibodies, too, so it may require extensive counterselection or screening with antibodies from control subjects (not suffering from the disease) to identify those peptides that correspond to authentic disease-related antibody specificities and that therefore can be considered diagnostic for the disease. This is an eminently "portable" program of discovery, using the same procedure and the same "all-purpose" random peptide libraries regardless of the particular disease. Even in the most difficult cases, it nets a rich diversity of diagnostic peptides with far less work than is required to identify antigenic peptides by direct study of a pathogen's antigenic makeup.

12.4.6 Selection of DNA-Binding Proteins

Phage display may help molecular biologists realize a long-standing goal: to design proteins that specifically bind a given target DNA sequence (O'Neil and Hoess, 1995). Rational design has poor prospects in this field, since there do not seem to be simple rules of complementarily-comparable to those governing base-pairing between complementary single-stranded nucleic acids-by which the sequence specificity of a DNA-binding protein can be predicted from the amino acids at critical positions in its structure. A much more promising approach is to construct a library of randomized variants of a parent DNA binding domain (e.g., one of the zinc-finger domains, a common DNA-binding motif in eukaryotic nuclei) displayed on a filamentous phage; randomization is concentrated on positions that are thought to make sequence-specific contacts with the target DNA in the parent domain. From this library, clones that bind a new target DNA sequence, different from that recognized by the parent domain, are then affinity-selected.

In an experiment analogous to epitope mapping, phage display has been used to map the DNA binding site of SATB1, a nuclear matrix protein that specifically binds the minor groove of a DNA sequence motif called MAR. Using an MAR DNA sequence as the immobilized receptor, affinity-selected peptides from a random peptide library; the predominant peptide shared 50% sequence identity with a nine-residue segment of the SATB1 sequence-a segment that was subsequently shown on independent grounds to be critical for DNA recognition. Phage display has also been used to affinity-select a hexapeptide with some binding preference for the single-stranded heptadeoxycytidilate $(dC)_7$, although in this case no mapping purpose was in view.

12.4.7 Landscape Libraries as a Source of New Materials

The surface landscape of a filamentous virion is a cylindrical array of thousands of repeating subunits composed of the exposed parts of the major coat protein pVIII; this exposed shell accounts for about half the weight of the particle. When a random peptide is displayed on every copy of this protein, it subtends a major fraction (20% or more) of the repeating unit and thus of the entire particle surface. Unless the random peptide is loosely tethered to the bulk of the major coat protein, it is forced to interact with residues in its immediate neighbourhood, and may therefore be constrained in a definite three-dimensional conformation that differs markedly from the surface conformation of wild-type particles and of clones displaying other random peptides. A large population of such clones can therefore be regarded as a library of "organic landscapes".

The ensemble of a random peptide in a landscape library with its surrounding wild-type residues may have emergent properties that are lost when the peptide is excised from its context. Such peptides are analogous to the complementarily-determining regions of antibodies-oligopeptide loops that in the context of the intact protein make most of the specific contacts with antigen but as free peptides seldom have appreciable antigen-binding propensities. Localizable emergent properties are present even when the foreign peptide is displayed on only an occasional pVIII molecule, as in type 88 and 8+8 systems. Nevertheless, the high-density display in landscape phage may greatly enhance overall effectiveness in some applications. For instance, if a single target receptor complex can bind two or more neighbouring peptides on the phage surface, the overall effective affinity may be enhanced many orders of magnitude compared to monovalent binding.

Some emergent properties are not localizable to a single subunit but seem instead to be a global property of the entire surface landscape. Thus, for instance, phage clones that are highly resistant to chloroform were selected from a landscape library; their entire surface is composed of hybrid pVIII subunits displaying a peptide motif that confers resistance to the solvent. In contrast, mosaic phage coated with roughly equal numbers of such hybrid subunits and wild-type subunits showed almost no resistance, indicating that resistance is not an additive property to which each hybrid subunit contributes independently.

Landscape phage might be looked on as a new kind of submicroscopic "fiber." Each phage clone is a type of fiber with unique surface properties. These fibers are not synthesized one by one with some use in mind. Instead, billions of fibers are constructed, propagated all at once in a single vessel and portions of this enormous population are distributed to multiple end-users with many different goals. Each user must devise a method of selecting from this population those fibers that might be suitable for his or her particular application-by affinity selection or whatever other selection principle ingenuity can conjure up.

Localizable or global emergent properties cannot be transferred from the virion surface to another medium; any application that depends on such properties must therefore use phages themselves as the new material. This undoubtedly precludes some applications of phage "fibers": it is doubtful we will be wearing clothes made of them, for instance. Still, filamentous phages are essentially proteins manufactured by a fermentation process and as such are potentially usable in any of the myriad of applications that might be contemplated for such proteins.

12.4.8 Phage Display-Combinatorial Chemistry on the Cheap

The selection of antibodies from combinatorial libraries displayed on the surface of filamentous phage has become an important methodology for the generation of reagent, diagnostic, and therapeutic molecules and for the study of natural immune responses. Using this technique, antibody genes have been cloned from multiple species or expressed directly from large man-made repertoires of antibody-encoding genes. Recent studies demonstrate that the technique allows for the in vitro evolution of antibodies to create molecules whose affinity for antigen exceeds that observed in nature (Rader and Barbas, 1997).

For drug discovery and a handful of other high-profile applications with high commercial stakes, phage display is perhaps not an optimal technology. For the ordinary research user, however, it has the overwhelming advantage that it is cheap and easy. It uses standard microbiological techniques that are familiar to all molecular biologists, and its key resources-phage libraries and clones-are replicable and therefore nearly cost-free after their initial construction or selection. It is astonishing to contemplate that within a single 1.5-ml micro centrifuge tube we can fit a few hundred trillion phage particles displaying billions of different peptide structures-abundance and diversity from which hundreds of different users with altogether different purposes in mind can select clones of great value.

12.4.9 Cloning Allergens by Phage Display

Allergic diseases represent an immune disorder associated with the production of immunoglobulin E (IgE) against normally innocuous antigens (allergens). Almost 20% of the population in industrialized countries suffers from type I allergic symptoms such as allergic rhinitis, conjunctivitis, urticaria or asthma. Although the mechanisms responsible for these allergic reactions are quite well understood, knowledge about the repertoire of molecules able to elicit type I symptoms is still limited. To clone and characterize entire allergen repertoires from complex allergenic sources in a fast and efficient way, new technologies are required (Rhyner et al., 2004). Selective enrichment of cDNA libraries displayed on phage surface with serum IgE from allergic individuals combined with robotic-based high-throughput screening technology has proved to be extremely successful for the rapid isolation of allergens. The basic concept of linking the phenotype, expressed as gene product displayed on the phage coat, to its genetic information integrated into the phage genome, creates fusion proteins covalently associated with the infectious particle itself. Therefore, cDNA libraries displayed on phage surface can be screened for the presence of specific clones using the discriminative power of affinity purification. The selection of IgE-binding clones involves the enrichment of phage binding to serum IgE immobilised to a solid phase during consecutive rounds of affinity selection. As a consequence of the physical linkage between genotype and phenotype, sequencing of the DNA of the integrated section of the phage genome can readily elucidate the amino acid sequence of the surface-displayed allergen. In spite of some biological limitations imposed by *E. coli* as expression host, phage surface display technology has strongly contributed to the rapid isolation of a vast variety of IgE-binding structures (Appenzeller et al., 2001). The screening of cDNA libraries displayed on phage surfaces with immobilized serum IgE from allergic patients reduces the time required for the selection of candidate clones to a few weeks. Robot-assisted high-throughput screening of the enriched library provides a fast and cost-effective way to isolate complete allergen repertoires. The biotechnological production of recombinant allergens derived from these sequences bears a high potential for the improvement of the diagnosis of allergic diseases. It provides valuable data when applied to the investigation of IgE interaction with allergens. Phage display has also been used to characterize novel peanut and fungal allergens. The method has been used to increase our understanding of the molecular basis of allergen-IgE interactions and to develop clinically relevant reagents with the pharmacologic potential to block the effector phase of allergic reactions (Davies et al., 2000).

Phage display technology presents a rapid means by which proteins and peptides that bind specifically to predefined molecular targets can be isolated from extremely complex combinatorial libraries. There are several important ways by which phage display can provide impetus to receptor-based research. Firstly, phage display can be applied, alongside transcriptome and proteome expression profiling techniques, to the identification and characterization of receptors whose expression is specific to a cell lineage, a tissue or a disease state. Secondly, specific monoclonal antibodies that enable researchers to identify, localize and quantify receptors can be produced very rapidly (weeks). Thirdly, it should be possible to apply phage display to the matching of orphan ligands and receptors. Finally, phage display can be used to identify proteins and peptides that modulate receptor activity. As well as being useful in the study of receptor function, biologically active proteins and peptides could also be used therapeutically, or as leads for drug design. Hence phage display is ready to play a central role in the study of receptors in the post-genome era.

12.5 Conclusion

Screening of phage-displayed libraries of proteins and peptides has, for nearly a decade, proven to be a highly effective method for finding much needed "needles" in a vast molecular "haystack". Phage display of proteins has become an important tool for protein engineering. This technique has been used to solve an increasing diversity of problems, including identification of binding motifs for much smaller targets and the use of novel screening (Rodi and Makowski, 1999). New phage derived bio recognition nanomaterials have emerged recently as a result of the in depth study of the genetics and structure of filamentous phage and evolution of phage display technology (Petrenko, 2008). Recently a novel approach of nanocarrier targeting through their fusion with target specific phage coat proteins was reported by Jayanna et al. (2009).

Over the past year, the versatility of the technology has expanded to include the development of DNA-binding proteins with novel specificities, energetics of protein folding and directed evolution of antibodies. In addition, display of expressed cDNA libraries opens an exciting opportunity for studying protein-protein interactions (O'Neil and Hoess, 1995).

References

Appenzeller, U., Blaser, K., Crameri, R., 2001, Phage display as a tool for rapid cloning of allergenic proteins. *Archivum immunologiae et therapiae experimentalis* 49(1):19–25.

Azzazy, H.M., Highsmith, W.E., Jr., 2000, Phage display technology: clinical applications and recent innovations. *Clin Biochem* 35(6):425–445.

Barbas, C.F., III, 1993, Recent advances in phage display. *Curr Opin Biotechnol* 4(5):526–530.

Bremnes, T., Lauvark, V., Lindqvist, B., Blakke, O., 1998, Selection of phage displayed peptides from a random 10-mer library recognising a peptide target. *Immunotechnology* 4:21–28.

Davies, J.M., O'hehir, R.E., Suphioglu, C., 2000, Use of phage display technology to investigate allergen-antibody interactions. *J Allergy Clin Immunol* 105(6 Pt 1):1085–1092.

Faix, P.H., Burg, M.A., Gonzales, M., Ravey, E.P., Baird, A., Larocca, D., 2004, Phage display of cDNA libraries: enrichment of cDNA expression using open reading frame selection. *Biotechniques* 36:1018–1029.

Hartley, O., 2002, The use of phage display in the study of receptors and their ligands. *J Recept Signal Transduct Res* 22(1–4):373–392.

Hoogenboom, H.R., de Bruine, A.P., Hufton, S.E., Hoet, R.M., Arends, J.W., Rooyers, R.C., 1998, Antibody phage display technology and its applications. *Immunotechnology* 4(1):1–20.

Hufton, S.E., Moerkerk, P.T., Meulemans, E.V., de Bruïne, A., Arends, J.W., Hoogenboom, H.R., 1999, Phage display of cDNA repertoires: the pVI display system and its applications for the selection of immunogenic ligands. *J Immunol Methods* 231(1–2):39–51.

Jayanna, P.L., Torchilin, V.P., Petrenko, V.A., 2009, Liposomes targeted by fusion phage proteins. *Nanomedicine* 5(1):83–89.

O'Neil, K.T., Hoess, R.H., 1995, Phage display: protein engineering by directed evolution. *Curr Opin Struct Biol* 5(4):443–449.

Petrenko, V., 2008, Evolution of phage display: from bioactive peptides to bioselective nanomaterials. *Expert Opin Drug Deliv* 5(8):825–836.

Rader, C., Barbas, C.F., III, 1997, Phage display of combinatorial antibody libraries. *Curr Opin Biotechnol* 8(4):503–508.

Rhyner, C., Weichel, M., Fluckiger, S., Hemmann, S., Kleber-Janke, T., Crameri, R., 2004, Cloning allergens via phage display. *Methods* 32(3):212–218.

Rodi, D.J., Makowski, L., 1999, Phage-display technology – finding a needle in a vast molecular haystack. *Curr Opin Biotechnol* 10(1):87–93.

Sidhu, S.S., 2001, Engineering M13 for phage display. *Biomol Eng* 18:57–63.

Chapter 13
Microarray

Abstract DNA microarrays are the modern, parallel version of classic molecular biology hybridization techniques allowing the monitoring the expression of thousands of genes simultaneously in a single run. Microarray allows rapid measurement and visualisation of differential expression of the whole genome scale. Using microarrays, RNA sample from any cell or tissue type can be analysed for changes in transcript levels. Such changes are characteristics of developmental processes, cancer, infection, responses to drugs, sequencing and genotyping. The technique permits the complete analysis of genetic material and the monitoring of expression changes occurring in a biological sample under various conditions. The application of microarrays will have a significant role in clinical and basic science. Microarrays have been used successfully in various research areas including sequencing, single nucleotide polymorphism detection, characterization of protein-DNA interactions, DNA computing, mRNA profiling, and many more. Applications of microarrays also include gene expression studies, disease diagnosis, pharmacogenomics, drug screening, pathogen detection, and genotyping. Improved understanding of the gene changes occurring in human diseases will lead to the development of new diagnostic tools as well as more effective drugs.

Keywords Gene expression · Profiling · Microarray · Molecular medicine · Drug screening · Hybridization · Single nucleotide polymorphism · Transcriptional factors · Expression of genes · SNP · Genotyping

13.1 Prologue

It is widely believed that thousands of genes and their products in a given living organism function in a complicated and orchestrated way that creates the mystery of life. However, traditional methods in molecular biology generally work on a "one gene one experiment" basis which means that the throughput is very limited and the whole picture of gene function is hard to obtain. Another important and interesting question in biology is how gene expression is switched on and off and how genes are regulated. Since almost all cells in a particular organism have an identical

M. Debnath et al., *Molecular Diagnostics: Promises and Possibilities*,

DOI 10.1007/978-90-481-3261-4_13,

genome, differences in gene expression and not the genome content are responsible for cell differentiation during the life of the organism. Gene regulation in eukaryotes is not well understood, but there is evidence that an important role is played by a type of proteins called *transcription factors*. The transcription factors can attach (bind) to specific parts of the DNA, called transcription factor *binding sites* (i.e., specific, relatively short combinations of A, T, C or G), which are located in so-called *promoter* regions. Specific promoters are associated with particular genes and are generally not too far from the respective genes, though some regulatory effects can be located as far as 30,000 bases away, which make the definition of the promoter difficult. Transcription factors control gene expression by binding the gene's promoter and either activating (switching on) the gene's transcription, or repressing it (switching it off). Transcription factors are gene products themselves, and therefore in turn can be controlled by other transcription factors. Transcription factors can control many genes, and some genes are controlled by combinations of transcription factors. Understanding, describing and modelling such gene regulation networks are one of the most challenging problems in functional genomics.

There are over 200 known transcription factors in yeast, over 600 in worm and fly, and over 1,500 in weed, but the real numbers are probably higher, since more than half of the predicted genes in these organism have unknown function and there are transcription factors likely to be found among the unknown genes. In addition, transcription factors are not the only proteins participating in gene regulation. Proteins affecting chromatin structure are another example, and it is known that at least some regulation is happening in the translation stage.

The complete sequencing of several genomes, including that of the human, has signaled the beginning of a new era in which scientists are becoming increasingly interested in functional genomics; that is, uncovering both the functional roles of different genes, and how these genes interact with, and/or influence, each other. Increasingly, this question is being addressed through the simultaneous analysis of hundreds to thousands of unique genetic elements with microarrays. Microarrays and computational methods are playing a major role in attempts to reverse engineer gene networks from various observations (Kothapalli et al., 2002). Note that in reality the gene regulation is likely to be a stochastic and not a deterministic process. Already, analytical strategies have subdivided into distinct "omic" domains, such as genomics, proteomics, and metabolomics. This enables researchers to examine not only genetic elements, but also the corresponding proteins and metabolites derived from these genes. All "omic" technologies share the need for fresh, innovative looks at data analysis. To date, transcriptomics is the most widely studied molecular approach, enabling researchers to examine subtle differences in thousands of mRNA levels between experimental samples, medical biopsies, etc. Although mRNA is not the end product of a gene, the transcription of a gene is both critical and highly regulated, thereby providing an ideal point of investigation (Brazma and Vilo, 2000). Development of microarrays has permitted global measurement of gene expression at the transcript level and provided a glimpse into the coordinated control and interactions between genes.

13.2 The Concept of Microarray

The concept of using microarrays can be traced back 25 years to the introduction of the Southern blot. Modern microarrays analysis was introduced in 1995 by a Stanford University research team led by Pat Brown and Ron Davis. A microarray is an analytical device that comprises an array of molecules (oligonucleotides, cDNAs, clones, PCR products, polypeptides, antibodies, and others) or tissue sections immobilized at discrete ordered or nonordered micrometer-to-millimeter-sized locations on the surface of a porous or nonporous insoluble solid support. Microarrays exploit the preferential binding of complementary single-stranded nucleic acid sequences (Barrett and Kawasakii, 2003). They are miniaturized biological devices consisting of molecules, for examples DNA or protein, named probes, that are orderly arranged at a microscopic scale (Fig. 13.1) onto a solid support such as a membrane or a glass microscopic slide. A microarray is typically a glass (or some other material) slide, on to which DNA molecules are attached at fixed locations (spots). There may be tens of thousands of spots on an array, each containing a huge number of identical DNA molecules (or fragments of identical molecules), of lengths from twenty to hundreds of nucleotides. For gene expression studies, each of these molecules ideally should identify one gene or one exon in the genome; however, in practice this is not always so simple and may not even be generally possible due to families of similar genes in a genome. Microarrays that contain all of the about 6,000 genes of the yeast genome have been available since 1997. The spots are either printed on the microarrays by a robot, or synthesized by photo-lithography (similarly as

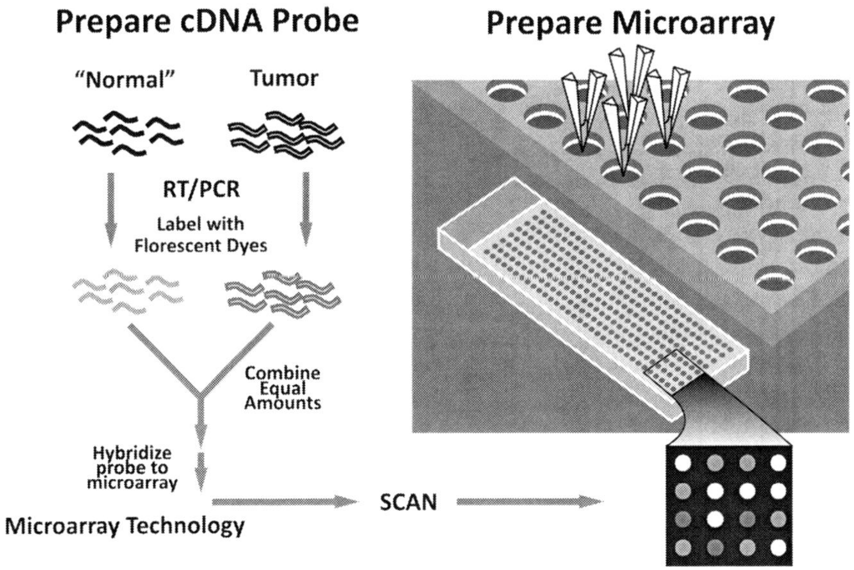

Fig. 13.1 Concept of Microarray

in computer chip productions) or by ink-jet printing (Hughes et al., 2001). The light-generated oligonucleotide array, developed by Affymetrix, Inc. (Santa Clara, CA), involves synthesizing short 25-mer oligonucleotide probes directly onto a glass slide using photolithographic masks. Sample processing includes the production of labeled cRNA, hybridization to a microarray, and quantification of the obtained signal after laser scanning. Regardless of the array used, the output can be readily transferred to commercially available data analysis programs for the selection and clustering of significantly modified genes (Culhane et al., 2003).

DNA microarrays are widely used to analyze genome-wide gene expression patterns and to study genotypic variations. Before a nucleic acid (target) can be used for hybridization to the probes of a microarray, it needs to be extracted from the tissue and labeled. The total mRNA from the cells in two different conditions (Kuo et al., 2002) is extracted and labeled with two different fluorescent labels: for example a green dye for cells at condition 1 and a red dye for cells at condition 2 (to be more accurate, the labeling is typically done by synthesizing single stranded DNA's that are complementary to the extracted mRNA by a enzyme called reverse transcriptase). Frequently, it also needs to be amplified to increase detection sensitivity. Both extracts are washed over the microarray. Labeled gene products from the extracts hybridize to their complementary sequences in the spots due to the preferential binding – complementary single stranded nucleic acid sequences tend to attract to each other and the longer the complimentary parts, the stronger the attraction. To increase throughput and minimize costs without reducing gene expression different fluorescent dyes namely Alexa 488, Alexa 594, Cyanine 3 and Cyanine 5 can also be used for hybridization.

During a hybridization process, labeled target molecules with sequences complementary to the probes are captured quantitatively. Subsequently, a reader measures the amount of label on each probe. To generate accurate and informative data, one of the most critical aspects of these experiments is the quality of both the isolated and the labeled nucleic acid samples (Lockhart et al., 1996).

The dyes enable the amount of sample bound to a spot to be measured by the level of fluorescence emitted when it is excited by a laser. If the RNA from the sample in condition 1 is in abundance, the spot will be green, if the RNA from the sample in condition 2 is in abundance, it will be red. If both are equal, the spot will be yellow, while if neither is present it will not fluoresce and appear black. Thus, from the fluorescence intensities and colours for each spot, the relative expression levels of the genes in both samples can be estimated.

The raw data that are produced from microarray experiments are the hybridized microarray images. To obtain information about gene expression levels, these images should be analysed, each spot on the array identified, its intensity measured and compared to the background. This is called image quantitation.

Microarrays are miniature arrays of gene fragments attached to glass slides. Monitoring the expression of thousands of genes simultaneously is possible with DNA microarray analysis. These devices have been highly effective for the simultaneous detection of large numbers of analytes in a sample, and microarrays have quickly emerged as important analytical tools in many branches of the biological sciences. A microarray-based analytical strategy is quicker and more convenient than serial testing for each analyte, and it has been successfully applied to both immunoassays and DNA-based assays. The current scope of microarray applications includes sequencing by hybridization, resequencing, mutation detection, assessment of gene copy number, comparative genome hybridization, drug discovery, expression analysis, and immunoassay (protein microarrays). In addition, oligonucleotide microarrays have been used for a nonbiological application computing (Hacia, 1999).

The presentation of thousands of gene fragments in a single array allows investigators to define changes in gene expression within a significant fraction of the total genome. For example, cancers will no longer be classified on the basis of a few proteins, but will be characterized by the change in hundreds of specific genes. Animal models will be related to human diseases not by appearance alone but by the changes that occur in thousands of genes. By coupling microarray analysis with the results from genome sequencing projects, researchers can analyze the global patterns of gene expression of an organism during specific physiological responses or developmental processes.

Presently, two technologies dominate the field of high-density microarrays: cDNA arrays and oligonucleotide arrays (Kane et al., 2000). The cDNA array has a long history of development stemming from immunodiagnostic work in the 1980s; however, it has been most widely developed in recent years by Stanford University (California) researchers depositing cDNA tags onto glass slides, or chips, with precise robotic printers (Lipshutz et al., 1999). Labeled cDNA fragments are then hybridized to the tags on the chip, scanned, and differences in mRNA between samples identified and visualized using a variation of the red/green matrix. A tissue microarray using paraffin embedded formalin fixed tissues can be used to evaluate the immunoprofiles of a large set of uterine adenocarcinomas with an extended panel of antibodies, comparing the profile of primary cervical and endometrial adenocarcinomas.

Differentially expressed genes will be defined herein as gene data determined to be statistical outliers from some standard state, and which can not be ascribed to chance or natural variability (Li and Wong, 2001). Various creative techniques have been proposed and implemented for the selection of differentially expressed genes; however, none have yet gained widespread acceptance for microarray analysis (Kane et al., 2000). Despite this, there remains a great impulse to develop new data analysis techniques, partly driven by the obvious need to move beyond setting simple fold change cut-offs which are out of context with the rest of the experimental and biological data at hand. This has been the case for many studies, where the selection of differential gene expression is performed through a simple fold change cutoff, typically between 1.8 and 3.0. There is an inherent problem with this selection criterion, as genes of low absolute expression have a greater inherent error in their measured levels. These genes will then tend to numerically meet any given fold change cut-off even if the gene is not truly differentially expressed. The inverse also holds true, where highly expressed genes, having less error in their measured levels, may not meet an arbitrary fold-change cut-off of 2.0 even when they are truly differentially expressed. Therefore, selecting differentially regulated genes based only on a single fold change across the entire range of experimental data preferentially selects lowly expressed genes. This commonly used approach does not accommodate for background noise, variability, non-specific binding, or low copy numbers, characteristics typical of microarray data which may not be homogeneously distributed. Other approaches entail the use of standard statistical measures such as a student's *t*-test or ANOVA for every individual gene (Johnson and Wichern, 2002). However, due to the cost of repeating microarray experiments, the number of replicates usually remains low, leading to inaccurate estimates of variance (Baldi and Long, 2001). Furthermore, due to the low number of replicates, the power of these "gene-by-gene" statistical tests to differentiate between regulated and non-regulated genes also remains very low.

Protein microarrays are versatile tools for parallel, miniaturized screening of binding events involving large numbers of immobilized proteins in a time-and cost-effective manner. They are increasingly applied for high-throughput protein analyses in many research areas, such as protein interactions, expression profiling and target discovery. While conventionally made by the spotting of purified proteins, recent advances in technology have made it possible to produce protein microarrays through in situ cell-free synthesis directly from corresponding DNA arrays (Stoevesandt et al., 2009).

13.3 Current Challenges of Microarrays

DNA microarray technology is currently an area of great interest (Fig. 13.2). Also called "genechip" technology, it incorporates molecular genetics and computer science on a massive scale. This technology can rapidly provide a detailed view of the simultaneous expression of entire genomes and provide new insights into gene function, disease pathophysiology, disease classification, and drug development (Steven

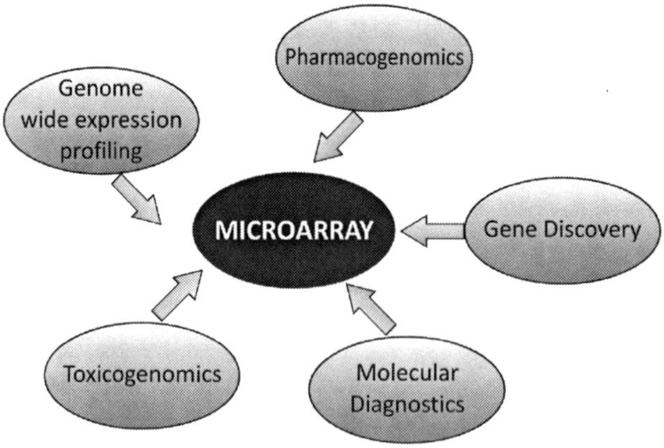

Fig. 13.2 The wide applications of microarrays

and Greenberg, 2001). The application of microarrays will have a significant role in clinical and basic science. Improved understanding of the gene changes occurring in human diseases will lead to the development of new diagnostic tools as well as more effective drugs (Hacia, 1999). In the basic sciences, microarray technology will permit researchers to elucidate the genetic pathways that are followed in the development, aging, and pathology of specific cells or tissues of the body. Microarray technology will help researchers to learn more about many different diseases, including heart disease, mental illness and infectious diseases, to name only a few. One intense area of microarray research at the National Institutes of Health (NIH) is the study of cancer. In the past, scientists have classified different types of cancers based on the organs in which the tumors developed. With the help of microarray technology, however, they will be able to further classify these types of cancers based on the patterns of gene activity in the tumor cells.

Researchers will then be able to design treatment strategies targeted directly to each specific type of cancer. Additionally, by examining the differences in gene activity between untreated and treated tumor cells – for example those that are radiated or oxygen-starved – scientists will understand exactly how different therapies affect tumors and be able to develop more effective treatments.

Within the last decade protein microarray technology has been successfully applied for the simultaneous identification, quantification and functional analysis of proteins in basic and applied proteome research. These miniaturized and parallelized assay systems have the potential to replace state-of-the-art singleplex analysis systems. However, prior to their general application in robust, reliable, routine and high-throughput applications it is mandatory that they demonstrate robustness, sensitivity, automation and appropriate pricing. The microarray analysis provides interesting leads in almost all fields of biology, offers a genome wide glimpse into genetic "terra incogniat," and challenges scientists to explore this unknown world. The biomedical community is developing new methods of data analysis to more efficiently process the massive data sets produced by microarray experiments. Systematic and global mathematical approaches that can be readily applied to a large number of experimental designs become fundamental to correctly handle the otherwise overwhelming data sets. The various applications of DNA microarrays are discussed below in details.

13.3.1 Gene Discovery

The use of high density DNA microarrays to identify sets of genes with similar expression pattern is rapidly becoming a wide spread approach for understanding biological processes. Typically, microarray data is analyzed by hierarchical clustering, self-organizing maps, *K*-means clustering or principle component analysis. Most of these approaches readily identify clusters of tens to hundreds of genes that demonstrate similar expression patterns. One logical systematic approach to study a cluster of genes with similar expression profiles is to analyze the promoter sequences for each member of the gene clusters and attempt to identify transcription factors that might be crucial for regulating their expression.

Detecting and determining the relative abundance of diverse individual sequences in complex DNA samples is a recurring experimental challenge in genomes. DNA microarrays are utilized in the analysis of complex DNA samples. With the objective of discovering novel putative interventions sites for anticancer therapy, transcriptional profiles of breast cancer, lung squamous cell cancer (LSCC), lung adenocarcinoma (LAC), and renal cell cancer (RCC) were compared. They wanted to search for genes not only highly expressed in the majority of patient samples but which also exhibit very low or even absence of expression in a comprehensive panel of 16 critical (vital) normal tissues. In addition to already known tumor-associated genes, the majority of identified genes have not yet been brought into context with tumorigenesis such as genes involved in bone matrix mineralization (OSN, OPN, and OSF-2) in lung, breast, and kidney cancer or genes controlling $Ca(2^{+})$ homeostasis (RCN1,CALCA, S100 protein family). EGLN3, which recently has been shown to be involved in regulation of hypoxia-inducible factor, was found to be highly up-regulated in all RCCs and in half of the LSCCs analyzed. Furthermore, 42 genes, the expression level of which correlated with the overall survival of breast cancer patients, were identified (Favis and Barany 2000; Favia et al., 2000; Brenton et al., 2001).

DNA microarray analysis helped to identify several of those genes involved in pathological processes relating to arthritis, including apoptosis, inflammation, and cellular proliferation. One interesting gene, follistatin-like gene, is highly expressed along the margin of contact between inflammatory synovial pannus and eroding bone, suggesting a role in joint destruction. The identification of genes predisposing to human diseases is of paramount importance. Numerous genes may play a role in drug response and toxicity, introducing a daunting level of complexity into the search for candidate genes. Gene discovery in bladder cancer progression, gastrointestinal tract origin cancer cells have also been identified (Golub et al., 1999). This technology has provided insights into gene function, disease pathophysiology, disease classification and drug development. This technology is also called "genechip" technology as it incorporates molecular genetics and computer science on a massive scale (Steven and Greenberg, 2001).

13.3.2 Gene Expression Profiling

DNA microarray is widely used in gene expression analysis. Expression profiling by this technology permits the identification of genes underlying clinical heterogeneity of many diseases which might contribute to disease progression, thereby improving assessment of treatment and prediction of patient outcome. The gene expression profiles provide a molecular fingerprint of the transcriptome.

Microarray analysis has become a widely used tool for the generation of gene expression data on a genomic scale. The last several years have witnessed remarkable development of new microarray-based automated techniques allowing parallel analysis of multiple DNA samples. Two major current applications of DNA microarrays are gene expression profiling (Lockhart et al., 1996; Mecham et al., 2004) and gene mutation analysis (Hacia et al., 1996; Hacia, 1999). Gene expression profiling using DNA microarrays has already provided results that were not achievable a few years ago. For example, it allowed the molecular classification of leukemias (Golub et al., 1999) and other types of cancers (Brenton et al., 2001) and revelations about genetic network architecture (Tavazoie et al., 1999).

Gene expression monitoring is not the only microarray application, another one is SNP detection. Microarray's are already producing massive amounts of data. These data, like genome sequence data, can help us to gain insights into underlying biological processes only if they are carefully recorded and stored in databases, where they can be queried, compared and analysed by different computer software programs.

Microarray expression analysis has become one of the most widely used functional genomics tools. Efficient application of this technique requires the development of robust and reproducible protocols. Lately optimisation of all aspects of the process, including PCR amplification of target cDNA clones, microarray printing, probe labeling, hybridization, including strategies for data normalization and analysis have been achieved. The 0.1% difference is exploited in DNA fingerprinting used in criminology. Variations in non-coding parts of the genome are analyzed to produce patterns that can reliably distinguish individuals; an exception here is identical twins which are much harder to distinguish with DNA fingerprinting. Particularly important variations in individual genomes are the *single nucleotide polymorphisms* or SNPs, which can occur both in coding and non-coding parts of the genome. SNPs are DNA sequence variations which occur when a single base (A, C, G, or T) is altered so that different individuals may have different letters in these positions. Particular nucleotides in SNP positions within or close to genes can influence the gene's protein product. Some "abnormal" protein variants (mutant variants) are the cause of genetic diseases (Cargill et al., 1999). SNPs may be responsible for many inheritable differences between individuals, and SNP variation may indicate the predisposition to a genetic disease. For example there is evidence that certain combinations of SNPs occur in individuals with Alzheimers disease. SNP analysis is therefore important for diagnostics and a SNP database is being built with the participation of the EBI. SNP analysis can also be used in population genetics, as some SNPs vary in frequency between populations. There are about 3 million SNPs collected in public SNP databases.

The sequencing projects have revealed that the genomes of what look like very different organisms may be quite similar. An example of this is similarity between yeast and worm described in the next section. It is estimated that the difference between human and chimpanzee genomes is only 1–3%, while between human and mouse only about 5–15%, depending on how the similarity is defined and measured (before the sequencing of chimp and mouse are well under way, these are only estimates). These similarities indicate close evolutionary relationships between these mammalian organisms. It is possible to build a *phylogenetic tree* of the evolution of proteins, genes and organisms based on sequence comparisons.

Gene annotation can be taken care to some extent of by links to sequence databases, unfortunately complicated many-to-many relationships between genes in the gene expression matrix and the features (spots) on the array makes it necessary to provide a full and detailed description of each feature on the array, as one gene can relate to several features on the array. The lack of standards in gene naming is another difficulty – a table relating each array feature present in the database to the list of all synonymous names of the respective gene is an essential part of a gene expression database. The ultimate goal of this work is to establish a standard for recording and reporting microarray-based gene expression data, which will in turn facilitate the establishment of databases and public repositories and enable the development of data analysis tools. Detecting and determining the relative abundance of diverse individual sequences in complex DNA samples is a recurring experimental challenge in analyzing genomes. This system should find numerous applications in genome-wide genetic mapping, physical mapping, and gene expression studies.

13.3.3 Pharmacogenomics and Microarray

Pharmacogenomics deals with the genetic basis underlying variable drug response in individual patients. It encompasses sum of all genes. Numerous genes may play a role in drug response and toxicity. The newly emerging genomic technologies enable the search for relevant genes and their variants to include the entire genome and help in the search for candidate genes. Moreover pharmacogenomic analysis can identify the disease susceptibility genes representing potential new drug targets. DNA microarrays has already revolutionised significant part of drug discovery and developmental process. It is rapidly becoming a fundamental tool in discovery based genomic and biomedical research (Wang et al., 2003). DNA microarrays can be used to measure the expression patterns of thousands of genes in parallel, generating clues to gene function that can help to identify appropriate targets for therapeutic intervention. They can also be used to monitor changes in gene expression in response to drug treatments. The DNA microarrays have proven to be a state of the art technique for high throughput comprehensive analysis of thousand of genes in parallel. The application of a DNA microarray to compare normal and pathological cells, tissues or organs may allow, along with classical positional cloning techniques, to speed up the discovery of genes and gene pathways implicated in several diseases. This in turn will result in further application of the DNA microarray technique in the field of pharmacogenomics in order to characterize and validate new therapeutic targets, their mechanism of action, metabolic pathways and unwanted secondary effects. However, the lack of standardized criteria for the analysis and interpretation of the huge amount of data generated by DNA microarrays may hamper its utilization on a routine basis in the human health domain.

In a recent study by Schmoock and his collegues (2009) a rapid oligonucleotide microarray assay based on genetic markers was successfully performed for the accurate identification and differentiation of the agents of glanders and melioidosis *Burkholderia (B.) mallei* and *B. Pseudomallei*. Both the strains are potential agents of bioterrorism and cause infectious diseases of man and animals. The existing bacteriological and immunological methods of identification of *B. mallei* and *B. pseudomallei* are not efficient enough for the rapid diagnosis and typing of strains.

The current concept of drug therapy is to apply treatment for a large population group whereas application of pharmacogenomics will help individual application of drug therapy or on smaller patient subpopulation but whether this individualized medicine will lead to improved and economically feasible therapy is yet to be seen. This is again complicated by various ethical and legal issues.

With the advent of DNA microarray analysis, it is now possible to examine the response of virtually the entire human genome to cellular drug exposure and to uncover a wide variety of genes correlating with the establishment of drug resistance. Genechip arrays are now being used by pharmaceutical companies to improve nearly every aspect of the traditional drug discovery and developmental processes, including target identification, target validation, compound screening, lead optimization and clinical trials. The field of "pharmacogenomics" is likely to vastly increase our understanding of the mechanisms of drug action and how cells respond and adapt to drug exposure. The scientist hope to identify genes influencing drug response and ultimately tailor antihypersensitive therapy for individual patients. Microarrays and DNA analysis enable scientists to explore the whole genome and identify predictive markers of disease to drug response that may ultimately provide more tailored, effective, and safer courses of treatment to help avoid some of over 100,000 annual fatalities from adverse reactions in US alone. Many researchers agree that DNA microarrays will be used as the most important diagnostic tools in near future.

However, DNA microarray studies typically result in the identification of hundreds of genes that may or may not be of relevance in vivo-particularly when large, genetically diverse study populations are used. The challenge to the researcher is to design experimental systems and approaches which minimize variability in the data, increase the reproducibility amongst experiments, allow array data from multiple experiments to be assessed by a variety of statistical, supervised learning, and data clustering approaches, and provide a clear link between drug response and the expression of specific genes (Yuen et al., 2002). Now the possibility to simultaneously analyze the expression of thousands of genes using DNA microarrays has allowed exploring the relationships between gene expression and sensitivity to several anticancer drugs. A number of studies using microarrays for identifying genes governing tumor chemosensitivity focused on tumor cell lines. Some clinical studies have also been carried out to investigate whether tumor gene expression patterns could predict clinical response to chemotherapy. Results of these studies are encouraging, indicating that individualization of drug treatment based on multigenic response predictive markers is feasible. Faced with the skyrocketing expenses of modern healthcare, and microarrays offer a solution with the prospects for more efficient, cost effective, and personalized approaches to patient care.

13.3.4 Molecular Diagnostic Research

Molecular diagnostics is a rapidly advancing field in which insights into disease mechanisms are being elucidated by use of new gene-based biomarkers. Until recently, diagnostic and prognostic assessment of diseased tissues and tumors relied heavily on indirect indicators that permitted only general classifications into broad histologic or morphologic subtypes and did not take into account the alterations in individual gene expression. Global expression analysis using microarrays now allows for simultaneous interrogation of the expression of thousands of genes in a high-throughput fashion and offers unprecedented opportunities to obtain molecular signatures of the state of activity of diseased cells and patient samples. Microarray analysis may provide invaluable information on disease pathology, progression, resistance to treatment, and response to cellular microenvironments and ultimately may lead to improved early diagnosis and innovative therapeutic approaches. Microarray technology has been widely used in the past 3 years to investigate tumor classification, cancer progression, and chemotherapy resistance and sensitivity.

The numbers of microarray based studies identifying new genes or molecular pathways involved in tumor classification, cancer progression, or patient outcome are growing exponentially. We are now in the "postgenomic era", during which the diagnostic, prognostic, and treatment response biomarker genes identified by microarray screening will be interrogated to provide personalized management of patients. Clinicians will be able to use microarrays during early clinical trials to confirm the mechanisms of action of drugs and to assess drug sensitivity and toxicity. Coupled with more conventional biochemical analysis such as IHC and ELISA, microarrays will be used for diagnostic and prognostic purposes.

Array-based comparative genomic hybridization (CGH) measures copy-number variations at multiple loci simultaneously, providing an important tool for studying cancer and developmental disorders and for developing diagnostic and therapeutic targets (Barrett et al., 2004). The results demonstrate that oligonucleotide arrays designed for CGH provide a robust and precise platform for detecting chromosomal alterations throughout a genome with high sensitivity even when using full-complexity genomic samples. Using cDNA microarrays can acquire gene expression profiles that characterize anticancer drug sensitivity at the same time evaluating the genome-wide gene expression profiles of various cancer cell lines to identify the gastrointestinal tract cancer cell-related genes.

Microarray technology is extensively used in biological research. DNA microarrays are an integral part of the process for therapeutic discovery, optimization and clinical validation. At an early stage, investigators use arrays to prioritize a few genes as potential therapeutic targets on the basis of various criteria. Subsequently, gene expression analysis assists in drug discovery and toxicology by eliminating poor compounds and optimizing the selection of promising leads. Integral to this process is the use of sophisticated statistics, mathematics and bioinformatics to define statistically valid observations and to deduce complex patterns of phenotypes and biological pathways. In short, microarrays are redefining the drug discovery process by providing greater knowledge at each step and by illuminating the complex workings of biological systems. cDNA microarrays have been used to acquire gene expression profiles that characterise anticancer drug sensitivity. With the development of microarray based tests for diagnosing a wide range of diseases, the ability to gain more information from a limited clinical sample by using highly parallel expression techniques has emerged. Several research groups have focused on identifying subsets of genes that show differential expression (Favis et al., 2000) between healthy tissues or cell lines and their tumor counterparts to identify biomarkers for several solid tumors, including ovarian carcinomas, oral cancer, melanoma, colorectal cancer, and prostate cancer (Wright et al., 2003). SNP analysis raises the possibility of individual tumor genome-wide allelotyping with potential prognostic and diagnostic applications (Hoque et al., 2003). They provide a basis for DNA-based cancer classification and help to define the genes being modulated, improving understanding of cancer genesis and potential therapeutic targets (Dutt and Beroukhim, 2007). Although the major limiting factors for routine use in a clinical setting at present are cost and access to the microarray technology, it is likely that costs will decrease in the near future and that the technology will become increasingly user friendly and automated.

13.4 Conclusion

DNA microarray analysis is a fast and versatile approach to achieve high throughput exploration of genome structure, gene expression programme and of gene function at both cellular and organism levels. It is a complex multistep process involving numerous specific equipments and requiring a strong expertise in various areas, including molecular biology, image analysis, computing, and statistics.

Protein arrays are emerging to follow DNA chips as possible screening tools. In today's proteome era, efforts are undertaken to adapt microarray technology in order to analyse the expression of a large number of proteins simultaneously and screen entire genomes for proteins that interact with particular factors, catalyse particular reactions, acts as substrates for protein-modifying enzymes and/or as targets of autoimmune responses. Protein microarray technologies, such as reverse-phase protein arrays, provide the unique opportunity to profile tissues and assess the activity of signalling pathways within isolated cell populations. This technology can be used to identify patients likely to benefit from specific treatment modalities and also to monitor therapeutic response in samples obtained during and after treatment. Routine application of genomic and proteomic microarray technologies in clinical practice will require significant efforts to standardise the techniques, controls and reference standards, and analytical tools used. Extensive, independent validation using large, statistically-powered datasets will also be necessary. Inclusion of concomitant genomic and proteomic-based molecular profiling techniques into clinical trial protocols will bring us closer to the reality of patient-tailored therapy. Protein microarray technology allows the simultaneous determination of a large variety of parameters from a minute amount of sample within a single experiment. Assay systems based on this technology are currently used for the identification, quantitation and functional analysis of proteins that are of interest for proteomic research in basic and applied biology and for diagnostic applications. Such novel assays are also of major interest for the pharmaceutical industry, focusing on the identification of biomarkers and the validation of potential target molecules. Sensitivity, reproducibility, robustness and automation have to be demonstrated before this technology will be suitable for high-throughput applications. A rapidly expanding use of this technology is the acquisition of information about the posttranslational modifications of proteins reflecting the activity state of signal pathways and networks, and is now employed for the analysis of biopsy samples in clinical trial research.

The examination of gene expression using microarrays holds tremendous promise for the identification of candidate genes involved in a variety of processes (Tan et al., 2003). The range of applications of microarray technology is enormous. Recent studies in human cancer have demonstrated that microarrays can be used to develop a new molecular taxonomy of cancer, including clustering of cancers according to prognostic groups on the basis of gene expression profiles. The list of potential uses of this technique is not limited to cancer research. For example, the temporal impact on gene expression by drugs, environmental toxins, or oncogenes may be elucidated and regulatory networks and coexpression patterns can then be deciphered. In the last few years since its inception, microarray technology has become a major tool for the investigation of global gene expression of all aspects of human disease and in biomedical research.

Prediction of the future of the gene chip market has been all over the map. Only a few years ago, the market was described as exploding and surging. Indeed the sky seemed to be the limit. Drug discovery had a virtually unlimited appetite for the enabling technologies and the possibilities of expanding chip technology beyond DNA chips to protein, tissue and cell microarrays and eventually lab-on chip technologies looked like a road without end, one that would lead directly into clinical diagnostics and promised land of personalised medicine. In the future, technologies such as gene chip array will enhance genetic medicine and provide novel insights into a patient's susceptibility to disease, enabling a better assessment of prognostic risk factors, quicker diagnosis, and accurate prediction of individual responsiveness to drugs. The predictable consequences of such an approach on the prevention and treatment of diseases could revolutionize medicine. Microarrays are on their way to revolutionize and to allow us to understand how life really works.

References

Baldi, P., Long, A.D., 2001, A Bayesian framework for the analysis of microarray expression data: regularized t-test and statistical inferences of gene changes. *Bioinformatics* 17:509–519.

Barrett, M.T., Scheffer, A., Ben-Dor, A., Sampas, N., Lipson, D., Kincaid, R., Tsang, P., Curry, B., Baird, K., Meltzer, P., Zohar, Y., Bruhn, L., Laderman, S., 2004, Comparative

genomic hybridization using oligonucleotide microarrays and total genomic DNA. *PNAS* 101(51):17765–17770.
Barrett, J.C., Kawasaki, E.S., 2003, Microarrays: the use of oligonucleotides and cDNA for the analysis of gene expression. *Drug Discov Today* 8:134–141.
Brazma, A., Vilo, J., 2000, Gene expression data analysis. *FEBS Lett* 480:17–24.
Brenton, J.D., Aparicio, S.A., Caldas, C., 2001, Molecular profiling of breast cancer: portraits but not physiognomy. *Breast Cancer Res* 3:77–80.
Cargill, M., Altshuler, D., Ireland, J., Sklar, P., Ardlie, K., Patil, N., Shaw, N., Lane, C.R., Lim, E.P., Kalyanaraman, N., Nemesh, J., Ziaugra, L., Friedland, L., Rolfe, A., Warrington, J., Lipshutz, R., Daley, G.Q., Lander, E.S., 1999, Characterization of single-nucleotide polymorphisms in coding regions of human genes. *Nature Genet* 22:231–238.
Culhane, A.C., Perriere, G., Higgins, D.G., 2003, Cross-platform comparison and visualisation of gene expression data using co-inertia analysis. *BMC Bioinform* 4:59.
Dutt, A., Beroukhim, R., 2007, Single nucleotide polymorphism array analysis of cancer. *Curr Opin Oncol* 19:43–49.
Favis, R., Barany, F., 2000, Mutation detection in K-ras, BRCA1, BRCA2, and p53 using PCR/LDR and a universal DNA microarray. *Ann NY Acad Sci* 906:39–43.
Favis, R., Day, J.P., Gerry, N.P., Phelan, C., Narod, S., Barany, F., 2000, Universal DNA array detection of small insertions and deletions in BRCA1 and BRCA2. *Nat Biotechnol* 18: 561–564.
Golub, T.R., Slonim, D.K., Tamayo, P., Huard, C., Gaasenbeek, M., Mesirov, J.P., Coller, H., Loh, M.L., Downing, J.R., Caligiuri, M.A., Bloomfield, C.D., Lander, E.S., 1999, Molecular classification of cancer: class discovery and class prediction by gene expression monitoring. *Science* 286:531–537.
Hacia, J.G., 1999, Resequencing and mutational analysis using oligonucleotide microarrays. *Nat Genet* 21:42–47.
Hacia, J.G., Brody, L.C., Chee, M.S., Fodor, S.P., Collins, F.S., 1996, Detection of heterozygous mutations in BRCA1 using high density oligonucleotide arrays and two-color fluorescence analysis. *Nat Genet* 14:441–447.
Hoque, M.O., Lee, C.C., Cairns, P., Schoenberg, M., Sidransky, D., 2003, Genome-wide genetic characterization of bladder cancer: a comparison of high-density single-nucleotide polymorphism arrays and PCR-based microsatellite analysis. *Cancer Res* 63: 2216–2222.
Hughes, T.R., Mao, M., Jones, A.R., Burchard, J., Marton, M.J., Shannon, K.W., Lefkowitz, S.M., Ziman, M., Schelter, J.M., Meyer, M.R., et al., 2001, Expression profiling using microarrays fabricated by an ink-jet oligonucleotide synthesizer. *Nat Biotechnol* 19:342–347.
Johnson, R.A., Wichern, D.W., 2002, Applied Multivariate Statistical Analysis, 5th edn. Prentice Hall, Upper Saddle River, NJ.
Kane, M.D., Jatkoe, T.A., Stumpf, C.R., Lu, J., Thomas, J.D., Madore, S.J., 2000, Assessment of the sensitivity and specificity of oligonucleotide (50mer) microarrays. *Nucl Acids Res* 28: 4552–4557.
Kothapalli, R., Yoder, S.J., Mane, S., 2002, Microarray results: how accurate are they? *BMC Bioinform* 3:22.
Kuo, W.P., Jenssen, T.K., Butte, A.J., Ohno-Machado, L., Kohane, I.S., 2002, Analysis of matched mRNA measurements from two different microarray technologies. *Bioinformatics* 18: 405–412.
Li, C., Hung Wong, W., 2001, Model-based analysis of oligonucleotide arrays: model validation, design issues and standard error application. *Genome Biol* 2, research/0032.1-0032.11.
Lipshutz, R.J., Fodor, S.P, Gingeras, T.R., Lockhart, D.J., 1999, High density synthetic oligonucleotide arrays. *Nat Genet* 21:20–24.
Lockhart, D.J., Dong, H., Byrne, M.C., Follettie, M.T., Gallo, M.V., Chee, M.S., Mittmann, M., Wang, C., Kobayashi, M., Horton, H., Brown, E.L., 1996, Expression monitoring by hybridization to high-density oligonucleotide arrays. *Nat Biotechnol* 14:1675–1680.

Mecham, B.H., Klus, G.T., Strovel, J., Augustus, M., Byrne, D., Bozso, P., Wetmore, D.Z., Mariani, T.J., Kohane, I.S., Szallasi, Z., 2004, Sequence-matched probes produce increased cross-platform consistency and more reproducible biological results in microarray-based gene expression measurements. *Nucl Acids Res* 32:e74.

Schmoock, G., Ehricht, R., Melzer, F., Rassbach, A., Scholz, H.C., Neubauer, H., Sachse, K., Mota, R.A., Saqib, M., Elschner, M., 2009, DNA microarray-based detection and identification of Burkholderia mallei, Burkholderia pseudomallei and *Burkholderia* spp. *Mol Cell Probes* 23:178–187 (PMID: 19366627).

Steven, A., Greenberg, M.D., 2001, DNA microarray gene expression analysis technology and its application to neurological disorders. *Neurology* 57:755–761.

Stoevesandt, O., Taussig, M.J., He, M., 2009, Protein microarrays: high-throughput tools for proteomics. *Expert Rev Proteomics* 6:145–157.

Tan, P.K., Downey, T.J., Spitznagel, E.L., Jr., Xu, P., Fu, D., Dimitrov, D.S., Lempicki, R.A., Raaka, B.M., Cam, M.C., 2003, Evaluation of gene expression measurements from commercial microarray platforms. *Nucl Acids Res* 31:5676–5684.

Tavazoie, S., Hughes, J.D., Campbell, M.J., Cho, R.J., Church, G.M., 1999, Systematic determination of genetic network architecture. *Nat Genet* 22:281–285.

Wang, H,Y., Malek, R.L., Kwitek, A.E., Greene, A.S., Luu, T.V., Behbahani, B., Frank, B., Quackenbush, J., Lee, N.H., 2003, Assessing unmodified 70-mer oligonucleotide probe performance on glass-slide microarrays. *Genome Biol* 4:R5.

Wright, G., Tan, B., Rosenwald, A., Hurt, E.H., Wiestner, A., Staudt, L.M., 2003, A gene expression-based method to diagnose clinically distinct subgroups of diffuse large B cell lymphoma. *Proc Natl Acad Sci USA* 100:9991–9996.

Yuen, T., Wurmbach, E., Pfeffer, R.L., Ebersole, B.J., Sealfon, S.C., 2002, Accuracy and calibration of commercial oligonucleotide and custom cDNA microarrays. *Nucl Acids Res* 30:e48.

Chapter 14
DNA Biosensors

Abstract As we enter into the twenty-first century, advances in medical technology help in the discovery of more genetic diseases. DNA sensing has become increasingly important for rapid genetic screening and detection. DNA biosensors offer considerable promise for obtaining the sequence specific information in human, viral and bacterial nucleic acid in a faster, simpler and cheaper manner as compared to traditional hybridization assays. The development of a DNA hybridization biosensor helps in wide scale ability of pathogen detection and molecular diagnostics.

Keywords Biosensor · Food diagnostics · Environment monitoring · DNA biosensor · Diagnostics · Nanosensor · Immunosensor · Immobilisation · Bioreceptors · Analytes · DNA diagnostics · Electrochemical biosensors · Nanoparticles · Genosensors · Peptide nucleic acid · PNA · Miniaturization · Nanofabrication

14.1 Prologue

Few scientific areas have witnessed dramatic changes of the magnitude observed recently in DNA diagnostics. With the completion of the human genome project (HGP), we are at the just beginning of a revolution in genetic analysis. The information obtained from the HGP opens a door to tremendous analytical opportunities ranging from diagnostics tests for mutations to the assessment of medical treatment. To continue these advances to exploit these opportunities and to address the growing market needs in the twenty-first century, future devices must link high performances, with speed, simplicity and low cost (Wang, 2002a). With advancement of miniaturization technologies, commercialization of such instruments for molecular diagnostics are now available for genetic testing. Biosensors have several potential advantages over other methods of analysis (Alocilja and Radke, 2003), including sensitivity in the range of ng/mL for microbial toxins and <100 colony-forming units/ml for bacteria.

Nucleic acids have become the ultimate tools in the recognition and monitoring of many important compounds. There is a great demand for detection systems which

M. Debnath et al., *Molecular Diagnostics: Promises and Possibilities*,
DOI 10.1007/978-90-481-3261-4_14,

can not only determine specific DNA fragments, but can also determine the exact total nucleic acid content of a sample. For more than a decade, DNA biosensor technologies are under intense investigation owing to their great promise for rapid and low-cost detection of specific DNA sequences in human, viral and bacterial nucleic acids (Kerman et al., 2001a). As the sequencing of the human genome continues, the mutations responsible for numerous inherited human disorders are now mapped. Pathogens responsible for disease states, bacteria and viruses are also detectable via their unique nucleic acid sequences and interest in their detection continues to grow. The analysis of nucleic acids has gained broad acceptance in diagnostic testing, pharmacological research, and numerous other fields including animal husbandry and detection of transgenes. The growing number of nucleic acid-based tests has stimulated a demand for automated, inexpensive testing devices that also afford miniaturization of the test platform and, ideally, the associated instrumentation.

14.2 The Concept of DNA Biosensor

A biosensor is made from a biological sensing element attached to a signal transducer. The sensing element can be enzymes, antibodies (as in immunosensors), DNA, or microorganisms; and the transducer may be electrochemical, optical, or acoustic in nature. Electrochemical transducers measure changes in current or voltage; optical transducers measure changes in fluorescence, absorbance or reflectance; and acoustic transducers measure changes in frequency resulting from small changes in mass bound to their surface. A signal transducer is an essential component of a biosensor. It converts the recognition event into a measurable signal (Fig. 14.1). The transducer can take many forms depending upon the parameters being measured. The most well developed classes of transducers are potentiometric,

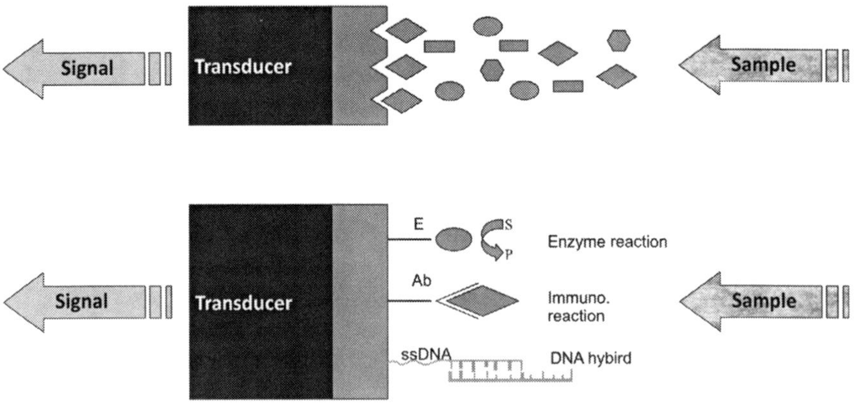

Fig. 14.1 Concept of biosensors

amperometric, conductometric, optical, acoustic or piezoelectric etc. These utilize various electrochemical responses to measure changes in the electrical properties of the biological recognition element. Amperometric biosensors are based on monitoring the current associated with oxidation or reduction of an electroactive species involved in the recognition process. The current produced is linearly proportional to the concentration of the electroactive product, which in turn is proportional to the non-electroactive enzyme substrate (D'Souza et al., 2005).

Transducers based on optical detection techniques have also been used in the field of biosensors. These may employ linear optical phenomenon, including fluorescence, phosphorescence, polarization, rotation, interference, surface plasmon resonance (SPR), total internal reflection fluorescence (TIRF), etc. or non-linear phenomena, such as second harmonic generation. Advantages of optical techniques involve the speed and reproducibility of the measurement.

Most commercial biosensors developed are focused in clinical applications, such as for glucose and lactate. Prospective biosensor market for food, pharmaceutical, agriculture, military, veterinary and environment are still to be explored. The present scenario demands for increased range of detectable analytes with portable device structure. Solving the resulting integration issues will require further convergence with associated technologies such as biochemistry, polymer chemistry, electronics, micro-fluidics and separation technology.

The basic requirement of a biosensor is that the biological material should bring the physico-chemical changes in close proximity of a transducer. In this direction immobilization technology has played a major role. Immobilization not only helps in forming the required close proximity between the biomaterial and the transducer, but also helps in stabilizing it for reuse. The biological material is immobilized directly on the transducer or in most cases, in membranes, which can subsequently be mounted on the transducer. Selection of a technique and/or support would depend on the nature of the biomaterial and the substrate and configuration of the transducer used.

Some of the widely used immobilization techniques include adsorption, entrapment, covalent binding and cross-linking. Immobilization of enzymes and whole cells through adsorption perhaps is the simplest of all the techniques. Enzymes have been immobilized through adsorption on a variety of ion exchange, hydrophobic and affinity surfaces. Most of these techniques have the drawbacks of weak adhesion as well as complexity of the process. Novel techniques have been developed for immobilizing viable or non-viable cells through adhesion on a variety of polymeric surfaces including glass, cotton fabric and synthetic polymeric membranes using polyethylenimine (PEI). This technique is gaining importance in the introduction of enzymes and microbes on transducer surfaces.

14.3 Applications of Biosensors

Biosensors incorporate a biological sensing element that converts a change in an immediate environment to signals conducive for processing. Biosensors have been implemented for a number of applications ranging from environmental pollutant detection to defense monitoring. Biosensors have two intriguing characteristics: (1) they have a naturally evolved selectivity to biological or biologically active analytes; and (2) biosensors have the capacity to respond to analytes in a physiologically relevant manner. Although most biosensors systems have been tested only on non-real samples (such as in distilled water or buffer solutions), a few biosensors applied to real samples have appeared in recent years. Some representative examples of their application to the determination of different classes of key pollutants and environmental quality parameters, such as biological oxygen demand (BOD), toxicity or endocrine effects. The application of biosensors to real samples must be a necessary step before their commercialization, which is, in general, the aim of the device development.

A variety of whole-cell-based biosensors has been developed using numerous native and recombinant biosensing cells. All biosensing cells in use can be classified into two groups in terms of their biosensing mechanisms, constitutive expression and stress- or chemical-specific inducible expression. The use of recombinant whole-cell biosensors in the field requires three components, biosensing cells, a measurement device, and a signal-transducing apparatus, the last two depending on the first and the final applications of the system. The whole-cell-based biosensors are used to monitor different environmental media, such as water, soil, and atmospheric monitoring (Rodriguez-Mozaz and Lopez de Alda, 2006) and for detection of various stressors, including dioxins, endocrine-disrupting chemicals, and ionizing radiation.

Biosensor mechanisms are based on enzymes, whereas bioaffinity biosensors are primarily based on immunochemicals (antibodies). Other biosensors use genetically engineered microorganisms (GEMs). Optical biosensors were identified in 1995 by the IUPAC Commission V-4 as a "new topic of interest." The IUPAC Analytical Chemistry Division noted that these biosensors "combine the exquisite selectivity of molecular recognition of bioreceptors (e.g. antibody, enzyme, nucleic acid probes) and the exceptional sensitivity of spectrochemical detection technologies" for "environmental and biomedical applications".

14.4 Advantages of Biosensors

Biosensors are centrally located in a continuum of analytical technologies ranging from chemical sensors to bioanalytical assays. Although strict definitions are difficult in these often overlapping specialties, the International Union of Pure and Applied Chemistry (IUPAC) is defining biosensors as a subgroup of chemical sensors in which a biologically based mechanism is used for analyte detection. One characteristic of biosensors that distinguishes them from other bioanalytical methods, such as immunosassays and enzyme assays, is that the analyte tracers or catalytic products can be directly and instantaneously measured. For antibody-based biosensors, analyte tracers or unlabeled antibodies are directly detected in a single step, whereas for most immunoassays, an enzyme is attached to the analyte of interest and measurement of the binding of the antibody to the antigen is a multistep process.

Another advantage that biosensors have over bioanalytical assays is that they can regenerate and reuse the immobilized biological recognition element. For enzyme-based biosensors, an immobilized enzyme can be used for repeated assays rather than being discarded after each measurement; this feature allows these devices to be used for continuous or multiple assays. For antibody-based biosensors, chemical immobilization of the antibody to the signal transducer can be beneficial. In some cases, after the analyte has been measured (i.e., as a result of the antibody-analyte binding), the analyte can be stripped from the immobilized antibody and another assay done. In other cases, antibody-based biosensors have been shown to reversibly respond to chemical compounds within seconds or minutes. By contrast, immunoassays, including enzyme-linked immunosorbent assay (ELISA), are typically based on irreversible binding and are thus used only once and discarded.

14.5 Development of DNA Hybridization Biosensor

The first biosensors were reported in the early 1960s and comprised enzymes immobilized to oxygen electrodes. Continued development of this kind of biosensor led to the commercialization of various devices for such applications as the measurement of glucose in blood and the detection of glutamate, aspartame, sulfite, lactose, and ethanol in food products.

DNA biosensors have become an important tool in molecular biology and biotechnology in recent years (Mir and Katakis, 2005). Biosensors are powerful tools aimed at providing selective identification of toxic chemical compounds at ultra levels in industrial products, chemical substances, environmental samples (e.g., air, soil, and water) or biological systems (e.g., bacteria, virus, or tissue components) for biomedical diagnosis. Combining the exquisite specificity of biological recognition probes and the excellent sensitivity of laser-based optical detection, biosensors are capable of detecting and differentiating big/chemical constituents of complex systems in order to provide unambiguous identification and accurate quantification. A new generation of biosensors use antibody and DNA probes (Albers et al., 2003).

Biosensor devices are based on the conversion of nucleic acid recognisation reaction into useful electrical signals and offer considerable promise for DNA diagnostics. By combining the sample handling and measurement steps, biosensors eliminate the need for the sample preparation and hence offer great promise for numerous on-site analytical applications (for which rapid, low-cost measurements are highly desired). A high level of sophistication is commonly employed to mass produce easy-to-use, miniaturized, and often disposable devices (Wang, 2002b). Depending on the nature of the biological process, two general categories of biosensing devices may be distinguished: biocatalytic sensors (based primarily on immobilized enzymes) and affinity devices (utilizing antibodies, receptors and nucleic acids). An important effort has been devoted to develop techniques for immobilizing oligonucleotides on suitable supports. In this context, modified electrodes prepared by electropolymerisation of different monomers aiming the immobilisation of oligonucleotides have had many advances during last decade (Tenreiro et al., 2003). Immobilization of DNA onto carbon nanotubes plays an important role in the development of new types of DNA biosensors (Xian et al., 2005).

The detection technique of the sequence-selective DNA hybridization has been developed through various methods such as optical (Ferguson et al., 1996) electrochemical (Mikkelson, 1996) and piezoelectric transduction modes (Palecek and Jelen, 2002). A great interest in electrochemical method has been attracted because of the high sensitivity, low cost, and compatibility with microfabrication technology (Eunkyung et al., 2003). Electrochemical DNA biosensors exploit the affinity of single-stranded DNA for complementary strands of DNA and are used in the detection of specific sequences of DNA with a view towards developing portable analytical devices. Electrochemical biosensors are normally based on enzymatic catalysis of a reaction that produces ions. The sensor substrate contains three electrodes, a reference electrode, an active electrode and a sink electrode. A counter electrode may also be present as an ion source. The target analyte is involved in the reaction that takes place on the active electrode surface, and the ions produced create a potential which is subtracted from that of the reference electrode to give a signal.

In a recent study, a novel and sensitive assay for the sequence-specific DNA detection has been developed by using bio-bar code techniques. These were based on the electrochemical detection of cadmium ions dissolved from CdS nanoparticles. The "sandwich-type" DNA complexes were fabricated with the thiol-functionalized capture DNA firstly immobilized on the modified electrode (nanoAu-GCE) and hybridized with one end of target DNA, the other end of which was recognized with signal DNA labeled on the surface of Au NP. The experiments indicated that two-base mismatched sequences showed weaker peak current and non-complementary sequences gave no response at all (Ding et al., 2009). In another recent study reported by Du and Collegues (2009), colloidal gold nanoparticles (Au NPs) and carboxyl group-functionalized CdS Nanoparticles (CdS NPs) were immobilized on the Au electrode surface to fabricate a novel electrochemical DNA biosensor. The biosensor exhibited a good repeatability and stability for the determination of DNA sequences. Great progress has been made in this field but there are still numerous challenges to overcome.

The basis for these nucleic-acid hybridization devices is the DNA base pairing. Accordingly, they usually rely on the immobilization of a 25–40 mer short synthetic oligonucleotide (the "probe"), whose sequence is complementary to sought-for targets, onto the physical transducer (Fig. 14.2). Exposure of the sensor to a sample containing the target results in the formation of the hybrid on the surface. Electrochemical, optical or frequency monitoring of this duplex formation can thus result in a useful analytical signal. As with other types of biosensors, high selectivity is crucial for the success of DNA hybridization devices (Williams et al., 2003). The specificity of these genosensors (i.e., their ability to respond to the target

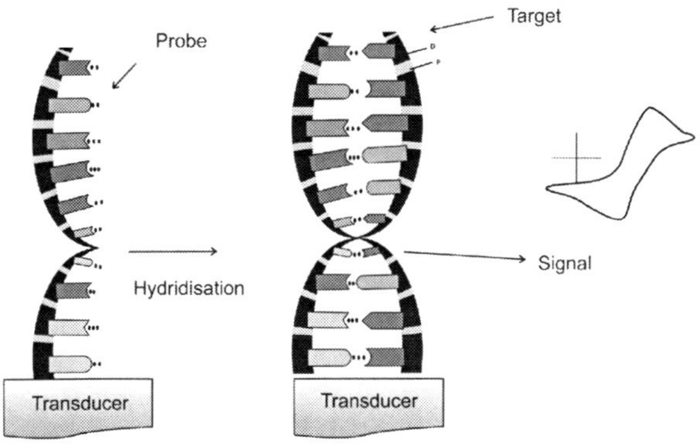

Fig. 14.2 Electrochemical DNA biosensor

sequence in the presence of non-complementary strands) depends primarily on the selection of the probe, and secondarily upon the hybridization conditions (mainly the temperature). To date, however, most DNA biosensors are not capable of selectively discriminating against single-base mismatches, as desired for example, for the detection of disease-related point mutations. The ability to recognize a change in a single nucleotide thus represents a major challenge for DNA biosensor technology. Accordingly, a drastically different approach (relying on the use of new probes) is desired to impart higher selectivity onto nucleic-acid biosensors.

The introduction of peptide nucleic acid (PNA) has opened up exciting opportunities for DNA biosensors (Wang et al., 1997). The unique structural, hybridization, and recognition features of solution-phase PNA (Weiler et al., 1997) can be readily extrapolated onto transducer surfaces in connection with the design of DNA biosensors. Such use of surface-confined PNA recognition layers imparts remarkable sequence specificity onto DNA biosensors and offers other attractive advantages (including greater latitude in the selection of experimental conditions).

Peptide nucleic acids (PNAs) (Nielsen and Egholm, 1999) are molecules of great interest, because they offer great advantages, in comparison with oligonucleotide probes, in molecular diagnosis (Wang et al., 1998). PNAs are DNA mimics in which the sugar-phosphate backbone is replaced by N-(2-aminoethyl) glycine units and hybridize with complementary DNA, forming Watson-Crick double helices. A probe that has recently gained an increasing application is peptide nucleic acid (PNA). Probes made of peptide nucleic acid, which have very strong affinity for complementary DNA sequence, can further improve the specificity. Such probes can more effectively discriminate the pathogenic organisms at the level of single-base mismatches. Additionally, PNA/DNA hybrids are resistant to nuclease attack, due to inability of nucleases and proteases to recognize the peptide backbone, have higher thermal stability and their melting temperature is higher than the corresponding DNA/DNA duplex. They are relatively insensitive to ionic strength due to the neutral charge of PNA.

The efficiency of PNA-DNA hybridization is very high. The stability of PNA-DNA hybrids is greatly affected by the presence of a single base mismatch. This property has been used for the detection of point mutations in advanced diagnostic methods, by means of PCR clamping, affinity electrophoresis, matrix-assisted laser desorption/ionization time-of-flight mass spectrometry, electrochemical biosensors, quartz crystal microbalance (QCM), and microarrays (Weiler et al., 1997).

The recent development of biosensor technologies for biospecific interaction analysis enables the monitoring of a variety of molecular reactions like human immunodeficiency virus type I (HIV-1 in real time by surface plasmon resonance (SPR). If the ligand is a biotinylated single stranded DNA, this technology could monitor DNA-DNA hybridization. This approach could be of great interest in virology, since the hybridization step is often required to confirm specificity of molecular diagnosis (Bianchi et al., 1997).

The development of biosensor technologies for biospecific interaction analysis (BIA) enables us to monitor DNA-DNA and DNA-RNA hybridization in real time by surface plasmon resonance (SPR). This optical technique detects and quantifies changes in refractive index in the vicinity of the surface of sensor chips to which either target DNA or DNA probes could be immobilized. Because the changes in refractive index are proportional to the changes in absorbed mass, the SPR technology allows the monitoring of DNA-DNA hybridization while it is occurring. Therefore SPR-based BIA has been applied to detect HIV-1 infection and genetic mutations affecting the p53 onco-suppressor gene. Despite the fact that this approach could be of great interest in clinical genetics, few data in the literature are available on the use of SPR and biosensor technologies to discriminate between the homozygous and heterozygous states in the case of hereditary diseases caused by genetic mutations. Peptide nucleic acids (PNAs) are excellent probes able to detect the W1282X point mutation of the cystic fibrosis (CF) gene when biospecific interaction analysis (BIA) by surface plasmon resonance (SPR) and biosensor technologies is performed (Feriotto et al., 2001).

Nanoparticles are increasingly finding a wide application in the biological studies due to their unique physical and chemical properties. However, biological and medical applications would require nanoparticles to be conjugated to biomolecules. The combination of nanotechnology, biology, advanced materials and photonics opens the possibility of detecting and manipulating atoms and molecules using nanodevices, which have the potential for a wide variety of medical uses at the cellular level. Nano-biosensors and in situ intracellular measurements of single cells using antibody-based nanoprobes have been developed (Tuan Vo-Dinh 2002). The nanoscale size of this new class of sensors also allows for measurements in the smallest of environments. One such environment that has evoked a great deal of interest is that of individual cells. Using these nanosensors, it is possible to probe individual chemical species and molecular signaling processes in specific locations within a cell. The possibilities to monitor in vivo processes within living cells could dramatically improve our understanding of cellular function, thereby revolutionizing cell biology. Optical nanobiosensors are integrated nanoscale devices consisting of a biological recognition molecule coupled to the optical transducing element such as an optical nanofiber interfaced to a photometric detection system. They are capable of providing specific quantitative, semi-quantitative or qualitative analytical information using biological recognition elements (e.g., DNA, protein) in direct spatial contact with a solid-state optical transducer element. This nanobiotechnology-based devices are being developed in our laboratory and could provide unprecedented insights into intact cell function, allowing, for the first time, studies of molecular functions (such as apoptosis, DNA-protein interactions, protein-protein interaction, functioning of nanomachines, etc.) in the context of the functional cell architecture in a systems biology approach. These devices will lead to novel and powerful tools for fundamental biological research, ultra-high throughput drug screening drug metabolism (Iwuoha et al., 2007) and medical diagnostics applications (Kubik et al., 2005).

DNA hybridization biosensor offers considerable promise for obtaining the sequence-specific information in a simpler, faster and cheaper manner compared to traditional hybridization assays. Such devices thus hold an enormous potential for decentralized clinical diagnosis of genetic or infectious diseases (Basu et al., 2004), for the rapid detection of food-contaminating organisms, and for on-site forensic investigations or environmental testing.

14.6 DNA Biosensor for Molecular Detection of Pathogens

The role of biosensors towards the detection of infectious bacteria (Ivnitski et al., 2000) can assist in achieving the goals for detection of organisms involved in bioterrorism, food poisoning, and clinical problems such as antibiotic resistance, and sensors using several of the different types of transduction modes (Deisingh and Thompson, 2004). Recent engineering advances have enabled the development of electrochemical DNA biosensors with molecular diagnostic capabilities (Liao et al., 2006; Drummond et al., 2003). Electrochemical DNA biosensors offer several advantages compared to alternative molecular detection approaches, including the ability to analyze complex body fluids, high sensitivity, compatibility with microfabrication technology, a low power requirement, and compact instrumentation compatible with portable devices (Gooding, 2002).

Electrochemical DNA sensors consist of a recognition layer containing oligonucleotide probes and an electrochemical signal transducer (Wang, 2002). A well-established electrochemical DNA sensor strategy involves "sandwich" hybridization of target nucleic acids by capture and detector probes. In this strategy, the target is anchored to the sensor surface by the capture probe and detected by hybridization with a detector probe linked to a reporter function. Detector probes coupled to oxidoreductase reporter enzymes allow amperometric detection of redox signals by the sensor electrodes. When a fixed potential is applied between the working and reference electrodes, enzyme-catalyzed redox activity is detected as a measurable electrical current. The current amplitude is a direct reflection of the number of target-probe-reporter enzyme complexes anchored to the sensor surface. Because the initial step in the electrochemical detection strategy is nucleic acid hybridization rather than enzyme-based target amplification, electrochemical sensors are able to directly detect target nucleic acids in clinical specimens, an advantage over nucleic acid amplification techniques, and chylomicrons.

Electrochemical DNA biosensors exploit the affinity of single stranded DNA for complementary strands of DNA and are used in detection of specific sequences of DNA with a view towards developing portable analytic device. Electrochemical DNA biosensor is proposed as a fast and easy screening method of specific DNA sequences in human, viral and bacterial nucleic acids (Ercole et al., 2003). Electrochemical biosensor can be used to detect pathogenic bacteria (Kim et al., 2005). Biosensors for detection of short DNA sequences of chronic myelogenous leukemia, sequence-specific electrochemical biosensing of *M. tuberculosis* DNA (Wang et al., 1997), for detection and discrimination of herpes simplex I and type II viruses from PCR amplified samples (Kerman et al., 2001b) are reported.

The electrochemical biosensors have the capacity for rapid and accurate detection of a wide variety of target molecules in the biological fluids (Liao et al., 2007). These biosensors allow the feasibility of direct detection and species-specific identification of bacterial pathogens in clinical specimens. Genetic polymorphism of human apolipoprotein E in human blood samples was also detected by this biosensor. An electrochemical detection scheme for identification of single nucleotide polymorphisms using hairpin-forming probes (Huang et al., 2002) is reported. A rapid molecular approach for accurate and reliable detection and identification of clinically significant concentrations of bacterial pathogens in clinical urine specimens would be of considerable benefit. Recent engineering advances have enabled the development of electrochemical DNA biosensors with molecular diagnostic capabilities. A very good example is the *C. parvum* biosensor which can detect as few as five oocysts in only 4 h. Current technology requires about 7 days of detection time, since water treatment plants have to send their sample to a testing lab and pay about $400 per analysis. This can be compared to an estimated $25 per analysis with the biosensor if carried out in the water treatment plant.

Electrochemical DNA biosensors offer several advantages compared to alternative molecular detection approaches, including the ability to analyze complex body fluids, high sensitivity, compatibility with microfabrication technology, a low power requirement, and compact instrumentation compatible with portable devices. Because the initial step in the electrochemical detection strategy is nucleic acid hybridization rather than enzyme-based target amplification, electrochemical sensors are able to directly detect target nucleic acids in clinical specimens, an advantage over nucleic acid amplification techniques, such as PCR.

14.7 Biosensors as Analytical Tools in the Food and Drink Industries

The detection of pathogenic bacteria is key to the prevention and identification of problems related to health and safety. Biosensors offer advantages as alternatives to conventional methods due to their inherent specificity, simplicity and quick response (Mello and Kubota, 2002).

Legislation is particularly tough in areas such as the food industry, where failure to detect an infection may have terrible consequences. Miniaturization of biosensors enables biosensor integration into various food production equipment and machinery. Potential uses of biosensors for food microbiology include online process microbial monitoring to provide real-time information in food production and analysis of microbial pathogens and their toxins in finished food. Biosensors can also be integrated into Hazard Analysis and Critical Control Point programs, enabling critical microbial analysis of the entire food manufacturing process (Rasooly and Herold, 2006). Application of the biosensor technique in the field of food processing and quality control is promising.

Recently a nano-biosensor device has been reported (Liu et al., 2007) that works as a molecular transistor, triggered by the presence of specific pathogens on an immunosensor. This device can also measure the amount of pathogen contamination on a particular food or machine, giving processors more data to determine the extent of a problem. The transistor works by processing data through fundamental logic gates. The logic gates operate by converting binding events between an antigen and an antibody into a measurable electrical signal using polyaniline nanowires as the transducer. The logic gates are created by patterning antibodies at different spatial locations in an immunosensor assay. Immunosensors are biosensors that use antibodies to recognize the presence of a pathogen.

Graphite electrodes fabricated by screen-printing have also been used as amperometric detectors in biosensors based on $NAD^{(+)}$-dependent dehydrogenases, tyrosinase, or genetically modified acetylcholinesterases. The mono-enzyme sensors have been optimized as disposable or reusable devices for detection of a variety of substrates important in the food industry (D-lactic acid, L-lactic acid, and acetaldehyde) or in environmental pollution control (phenols and dithiocarbamate, carbamate and organophosphorus pesticides). Tests on real samples have been performed with the biosensors; D-lactic acid and acetaldehyde have been detected in wine and phenols in air. The affinity biosensors field has expanded significantly over the past decade, with a projected global biosensors market growth from $6.1 billion in 2004 to $8.2 billion in 2009, representing major industrial sectors like Pharma, Medicare, and Food (Patel, 2006). Potential markets include the medical, military, food, and environmental industries. Those industries combined have a market size of $563 million for pathogen detecting biosensors and are expected to grow at a compounded annual growth rate of 4.5%. The food market is further segmented into different food product industries. The overall food-pathogen testing market is expected to grow to $192 million and 34 million tests by 2005. The trend in pathogen testing emphasizes the need to commercialize biosensors for the food safety industry as legislation creates new standards for microbial monitoring. With quicker detection time and reusable features, biosensors will be important to those interested in real time diagnostics of disease causing pathogens. As the world becomes more concerned with safe food and water supply, the demand for rapid detecting biosensors will only increase (Evangelyn et al., 2003).

14.8 Potential of Biosensor for Environmental Monitoring

The development of cost-effective on-site methods for environmental monitoring is instrumental to managing risks posed by environmental contamination. Biosensors show the potential to complement both laboratory-based and field analytical methods for environmental monitoring Rodriguez-Mozaz and Lopez de Alda (2004). Although a wide range of biosensors have been reported for potential environmental applications, relatively few of these have progressed into commercial markets. Advances in areas such as toxicity-, bioavailability-, and multianalyte-screening, and incorporation as detectors in chromatographic systems could possibly widen the market and allow these techniques to be more competitive. Stricter regulations and a greater public awareness of environmental issues have necessitated the need to monitor wider range of analytes in air, water and soil, and to do so with greater frequency and accuracy. Analysts currently have a range of portable analytical techniques at their disposal for monitoring across a variety of environmental analytes. The environmental analytical community continues to search for portable analytical techniques that can give reliable, on-site results for a variety of matrices and a host of analytes. More recently, biosensors have emerged as another promising technology in the analyst's armory, especially for applications requiring continuous monitoring (D'Souza et al., 2005).

A variety of biosensors have been reported which measure compounds of environmental interest that are toxic to microorganisms. Major metabolic mechanisms monitored include: the consumption of oxygen, the evolution of protons, and the synthesis of bioluminescence enzymes genetically tied to the expression of metabolic indicators.

The identification and enumeration of microorganisms that can pose human or environmental health problems is another area where biosensor technology may increase the speed and reduce expense associated with specific bioanalytical assays. These biosensors have focused on two mechanisms; immunochemical recognition of surface antigens and identification of DNA sequences that are unique to the organism of interest. In a recent example of an immunochemical approach, antibodies directed toward the red tide-causing plankton *Alexandrium affine* were immobilized to a piezoelectric sensor and used to detect this organism at concentrations as low as 102 cells/ml in sea water. Biosensor methods based on the identification of microorganisms through the detection of unique DNA sequences show significant promise; however, less progress toward practical methods has been achieved. Challenges for these techniques principally involve the isolation and processing of the microorganisms in the environmental sample to isolate and amplify selected DNA sequences prior to the hybridization assay. In this respect, technology borrowed from the intense efforts in micro-scale and automated gene sequencing research has application in the environmental monitoring area.

Biosensors are beginning to move from the proof-of-concept stage to field testing and commercialization in the United States, Europe, and Japan. Several US federal agencies are evaluating the technology for studies of ecological and human exposure. Biosensors have potential for continuous and in situ applications, such as downhole or perimeter groundwater surveillance, and they are suitable for a variety of matrices including soil extracts, groundwater, blood, and urine. Some biosensors can operate in high concentrations of organic solvents (e.g., methanol and acetonitrile) and can be used for in situ monitoring of contaminated organic media or process streams that contain mixed organic wastes. They can be constructed from a wide array of immunochemicals and even genetically engineered microorganisms, and they can be configured to be reversible. The potential for environmental applications lies in the ability of biosensors to measure the interaction of pollutants with biological systems through a biomolecular recognition capability. Biosensors have been used in analysis and monitoring of endocrine-disrupting compounds (EDCs) in the environment (Rodriguez-Mozaz et al., 2004).

An optical whole-cell biosensor based on a genetically engineered bioluminescent catabolic reporter bacterium was developed for continuous on-line monitoring of naphthalene and salicylate bioavailability and microbial catabolic activity potential in waste streams. The bioluminescent reporter bacterium, *Pseudomonas fluorescens* HK44, carries a transcriptional nahG-luxCDABE fusion for naphthalene and salicylate catabolism. Exposure to either compound resulted in inducible bioluminescence. The reporter culture was immobilized onto the surface of an optical light guide by using strontium alginate. This biosensor probe was then inserted into a measurement cell which simultaneously received the waste stream solution and a maintenance medium. Exposure under defined conditions to both naphthalene and salicylate resulted in a rapid increase in bioluminescence. A specific bioluminescence response was obtained after exposure to pollutant mixtures (Heitzer et al., 1994).

Biosensors and biosensor-related techniques that show potential for environmental applications must overcome a number of obstacles to become commercially viable in the highly competitive area of field analytical methods. Some of the obstacles common to all field analytical methods include: the diversity of compounds and the complexity of matrices in environmental samples, the variability in data quality requirements among environmental programs, and the broad range of possible environmental monitoring applications. More specific to biosensor technology, these hurdles include: relatively high development costs for single analyte systems, limited shelf and operational lifetimes for pre-manufactured biorecognition components and relative assay format complexity for many potentially portable biosensor systems.

14.9 Conclusions & Future Challenges

With the advent of personalized medicine, there is a compelling need for rapid and accurate methods for detection of nucleic acid sequence changes in clinical specimens (Krypuy et al., 2006). Micro-Electro-Mechanical Systems or MEMS technology is one of the promising areas that may be going to fulfill these demands in future. The technology is an integration of mechanical elements, sensors, actuators, and electronics on a common silicon substrate through micro fabrication technology. Biochips and sensor arrays for detection of a wide range of hazardous chemical and biological agents can be made out of these MEMS based devices, making it feasible for simultaneous detection of multiple analytes. However, immobilization and stabilization of biomolecules on these nanodevices may be a greater challenge. Utilization of molecular recognition ability of biomolecules like avidin-biotin or streptavidin-biotin in conjunction with a lithographic technique is being investigated for the micro immobilization of enzymes on silicon wafers for biosensor applications. Immobilization of enzymes on silicon supports has attracted attention in biosensor chip technology and a variety of classical techniques have been proposed (Mello and Kubota, 2002).

The opportunity to use any of the biosensors listed above is indeed great. The possibility for miniaturization is also apparent, but the commercial application of these for clinical gene-based diagnostics remains to be realized. By design, many of these approaches can be adapted to a variety of test types, and each has the promise of high sensitivity and specificity. Developers of DNA biosensors believe that in the near future these technologies will enable clinicians, and in some cases patients themselves, to quickly and inexpensively detect a wide variety of genetic-based diseases and conditions, including AIDS, Alzheimer disease, cystic fibrosis, and several forms of cancer. In other arenas, this technology will also make it possible to develop inexpensive strategies for screening of new pharmaceutical agents as well as new genes associated with hitherto uncharacterized diseases. These biosensors will be applied for on-site diagnosis, screening, and surveillance, in order to instantaneously recognize, and consequently reduce the ill effects of, potentially catastrophic biohazardous events.

The application of a portable, easy-to-use and highly sensitive biosensor for the real-time analysis of tumour markers offers significant advantages over current methods. To summarize, the future requirements matches with a biosensor microsystem easy to use, fast (sec), which will allow measurements in real-time with very low volumes (nl) of samples and reagents and which will allow the identification and quantification of biomarkers (at femtomolar level) without using fluorescent or radioactive labels.

We believe the next generation of technology should integrate biology with nanofabrication. Those technologies that directly convert a biological component to an electrical signal are extensively to be explored. In doing so we may be able to avoid the expense and time usually involved in biological amplification of the signal and the use of special readout material such as fluorescent dyes and then the corresponding complexity of the detection device. The envisioned instruments could be battery powered and hand-held, probably very inexpensive and thus could have not only an impact in clinical medicine in the developed countries, but could also be used in the developing world. The technologies under development must either allow researchers to address biological questions that were previously considered impossible or impractical or they must make significant improvements in existing technology for speed, accuracy, throughput or cost. Most commercial biosensors developed are focused in clinical applications, such as for glucose and lactate.

Prospective biosensor market for food, pharmaceutical, agriculture, military, veterinary and environment are still to be explored. Given the existing advances in biological sciences, coupled with advances in various other scientific and engineering disciplines, it is imminent that many analytical applications will be replaced by biosensors. A fruitful fusion between biological sciences and other disciplines will help to realize the full potential of this technology in the future.

References

Albers, J., Grunwald, T., Nebling, E., 2003, Electrical biochip technology—a tool for microarrays and continuous monitoring. *Anal Bioanal Chem* 377:521–527.

Alocilja, E.C., Radke, S.M., 2003, Market analysis of biosensors for food safety. *Biosens Bioelectron* 18(5–6):841–846.

Basu, M., Seggerson, S., Henshaw, J., Jiang, J., del A Cordona, R., Lefave, C., Boyle, P.J., Miller, A., Pugia, M., Basu, S., 2004, Nano-biosensor development for bacterial detection during human kidney infection: use of glycoconjugate-specific antibody-bound gold NanoWire arrays (GNWA). *Glycoconj J* 21:487–496.

Bianchi, N., Rutigliano, C., Tomossetti, M., Feriotto, G., Zorzato, F., Gambari, R., 1997, Biosensor technology and surface plasmon resonance for real-time detection of HIV-1 genomic sequences amplified by polymerase chain reaction. *Clin Diagn Virol* 8(3):199–208.

Deisingh, A.K., Thompson, M., 2004, Biosensors for the detection of bacteria. *Can J Microbiol* 50:69–77.

Ding, C., Zhang, Q., Lin, J.M., Zhang, S.S., 2009, Electrochemical detection of DNA hybridization based on bio-bar code method. *Biosens Bioelectron* 24:3140–3143.

Drummond, T.G., Hill, M.G., Barton, J.K., 2003, Electrochemical DNA sensors. *Nat Biotechnol* 21(10):1192–1199.

D'Souza, S.F., Sandeep, K., Jha Jitendra K., 2005, Environmental biosensors. *IANCS Bull, special issue on Environ Biotech* 4(1):54–59.

Du, P., Li, H., Mei, Z., Liu, S., 2009, Electrochemical DNA biosensor for the detection of DNA hybridization with the amplification of Au nanoparticles and CdS nanoparticles. *Bioelectrochemistry* 75(1):37–43.

Ercole, C., Del Gallo, M., Pantalone, M., Santucci, S., Mosiello, L., Laconi, C., 2003, Rapid identification of viable pathogens *Escherichia coli* subspecies with an electrochemical screen-printed biosensor array. *Biosens Bioelectron* 18:907–916.

Eunkyung, K., Kyuwon, K., Haesik, Y., Youn, T.K., Johyoun, K., 2003, Enzyme-Linked Electrochemical DNA Biosensor using Electrocatalysis of Ferrocenyl-Tethered DendrimerAbs. 1192, 204th Meeting, The Electrochemical Society, Inc.

Evangelyn, C., Alocilja, S., Radke, M., 2003, Market analysis of biosensors for food safety. *Biosens Bioelectron* 18(5–6):841–846.

Ferguson, J.A., Boles, T.C., Adams, C.P., Walt, D.R., 1996, A fiber-optic DNA biosensor microarray for the analysis of gene expression. *Nat Biotechnol* 14(13):1681.

Feriotto, G., Corradini, R., Sforza, S., Bianchi, N., Mischiati, C., Marchelli, R., Gambari, R., 2001, Peptide nucleic acids and biosensor technology for real-time detection of the cystic fibrosis W1282X mutation by surface plasmon resonance. *Lab Invest* 81(10):1415–1427.

Gooding, J.J., 2002, Electrochemical DNA hybridization biosensor. *Electroanalysis* 14: 1149–1156.

Heitzer, A., Malachowsky, K., Thonnard, J.E., Bienkowski, P.R., White, D.C., Sayler, G.S., 1994, Optical biosensor for environmental on-line monitoring of naphthalene and salicylate bioavailability with an immobilized bioluminescent catabolic reporter bacterium. *Appl Environ Microbiol* 60(5):1487–1494.

Huang, T.J., Liu, M., Knight, L.D., Grody, W.W., Miller, J.F., Ho, C.M., 2002, An electrochemical detection scheme for identification of single nucleotide polymorphisms using hairpin-forming probes. *Nucl Acids Res* 30(12):e55.

Ivnitski, D., Wilkins, E., Tien H.T., Ottova, A., 2000, Electrochemical biosensor based on supported planar lipid bilayers for fast detection of pathogenic bacteria. *Electrochem Commun* 2:457–460.

Iwuoha, E., Ngece, R., Klink, M., Baker, P., 2007, Amperometric responses of CYP2D6 drug metabolism nanbiosensor for sertraline: a selective serotonin receptor inhibitor. *IET Nanobiotechnol* 1 (4):62–67.

Kerman, K., Kobayashi, M., Tamiya, E., 2001a, Recent trends in electrochemical DNA biosensor technology. *Meas Sci Technol* 15(2004):R1–R11.

Kerman, K., Meric, B., Ozkan, D., Kara, P., Ozsoz, D., 2001b, Indicator based and indicator-free electrochemical DNA biosensors. *Engineering in Medicine and Biology Society, 2001. Proceedings of the 23rd Annual International Conference of the IEEE* 4:3240–3243.

Kim, S.K., Gopi, K.V., Harada, R., Hiroyuki, T., 2005, Miniaturized portable surface plasmon resonance immunosensor applicable for on-site detection of low-molecular-weight analytes. *Biosen Bioelect* 20(11):2276–2282.

Krypuy, M., Newnham, G.M., Thomas, D.M., Conron, M., Dobrovic, A., 2006, High resolution melting analysis for the rapid and sensitive detection of mutations in clinical samples: KRAS codon 12 and 13 mutations in non-small cell lung cancer. *BMC Cancer* 6:295.

Kubik, T., Bogunia-Kubik, K., Sugisaka, M., 2005, Nanotechnology on Duty in Medical Applications. *Curr Pharma Biotech* 6(1):17–33.

Liao, J.C., Mastali, M., Gau, V., Suchard, M.A., Moller, A.K., Bruckner, D.A., Babbitt, J.T., Li, Y., Gornbein, J., Landaw, E.M., McCabe, E.R., Churchill, B.M., Haake, D.A., 2006, Use of electrochemical DNA biosensors for rapid molecular identification of uropathogens in clinical urine specimens. *J Clin Microbiol* 44(2):561–570.

Liao, J.C., Mastali, M., Li, Y., Gau, V., Suchard, M.A., Babbitt, J., Gornbein, J., Landaw, E.M., McCabe, E.R.B., Churchill, B.M., Haake, D.A., 2007, Development of an advanced electrochemical DNA biosensor for bacterial pathogen detection. *J Mol Diagn* 9:158–168.

Liu, Y., Chakrabartty, S., Alocilja, E., 2007, Fundamental building blocks for molecular biowire based forward error-correcting biosensors. *Nanotechnol* 18:424017 (6 pp).

Mello, L.D., Kubota, L.T., 2002, Review of the use of biosensors as analytical tools in the food and drink industries. *Food Chem* 77(2):237–256.

Mikkelson, S.R., 1996, Electrochemical biosensors for DNA sequence detection. *Electroanalysis* 8:15.

Mir, M., Katakis, I., 2005, Towards a fast-responding, label-free electrochemical DNA biosensor. *Anal Bioanaly Chem* 381(5):1033–1035.

Nielsen, P.E., Egholm, M., 1999, An introduction to peptide nucleic acid. Joseph Wang PNA Biosensors for Nucleic Acid Detection. *Curr Iss Molec Biol* 1:89–104.

Palecek, E., Jelen, F., 2002, Electrochemistry of nucleic acids and development of DNA sensors. *Crit Rev Anal Chem* 3:261–270.

Patel, P.D., 2006, Overview of affinity biosensors in food analysis. *J AOAC Int* 89(3): 805–818.

Rasooly, A., Herold, K.E., 2006, Biosensors for the analysis of food- and waterborne pathogens and their toxins. *J AOAC Int* 89(3):873–883.

Rodriguez-Mozaz, S., Lopez de Alda, M.J., 2006, Biosensors as useful tools for environmental analysis and monitoring. *Anal Bioanal Chem* 386(4):1025–1041.

Rodriguez-Mozaz, S., Marco, M.P., Lopez de Alda, M.J., Barcelo, D., 2004, Biosensors for environmental monitoring of endocrine disruptors: a review article. *Anal Bioanal Chem* 378(3):588–598.

Tenreiro, A., Cordas, C.M., Abrantes L.M., 2003, Oligonucleotide Immobilisation on Polytyramine-Modified Electrodes Suitable for Electrochemical DNA Biosensors. *Portugaliae Electrochimica Acta* 21:361–370.

Tuan, V.-D., 2002, Nanobiosensors: probing the sanctuary of individual living cells *J Cell Biochem* 87(S39):154–161.
Wang, J., 2002a, PNA Biosensors for Nucleic Acid Detection Current Issues *Molec Biol* 1:89–104.
Wang, J., 2002b, Electrochemical nucleic acid biosensors. *Anal Chim Acta* 469:63–71.
Wang, J., Fernandes, J., Kubota, L., 1998, Polishable and renewable DNA hybridization biosensors. *Anal Chem* 70:3699.
Wang, J., Rivas, G., Cai, X., Dontha, R., Shiraishi, H., Farias, P.A.M., Valera, F.S., 1997, Sequence-specific electrochemical biosensing of M. tuberculosis DNA. *Anal Chim Acta* 337:41–48.
Weiler, J., Gausepohl, H., Hauser, N., Jensen, O.N., Hoheisel, J.D., 1997, Hybridization based DNA screening on peptide nucleic acid (PNA) oligomer arrays. *Nucl Acids Res* 25(14):2792.
Williams, E., Pividori, M.I., Merkoci, A., Forster, R.J., Alegret, S., 2003, Rapid electrochemical genosensor assay using a streptovidin carbon-polymer biocomposite electrode. *Biosens Bioelectron* 19(3):165–175.
Xian, Y.Z., Hu, Y., Liu, F., Xian, Y., Wang, H.T., Jin, L.T., 2005, Glucose biosensor based on Au nanoparticles-conductive polyaniline nanocomposite. *Biosens Bioelectron* 21(10):1996–2000.

Chapter 15
Molecular Microbiological Testing

Abstract The application of molecular testing methods in the clinical laboratory has dramatically improved our ability to diagnose infectious diseases. Nucleic acid techniques, such as plasmid profiling, various methods for generating restriction fragment length polymorphisms and the polymerase chain reaction (PCR) are making increasing inroads into clinical laboratories. However, the clinical usefulness of molecular testing will only be maximized to its fullest benefit by appropriate and careful studies correlating clinical findings with assay results.

Keywords Molecular diagnostics · Infectious disease · Diagnosis · PCR · Microbiological testing · Molecular tests · Biotyping · Antibiogram · Resistogram · Bacteriocin typing · Ribotyping · Molecular techniques

15.1 Prologue

Over the past century microbiologists have searched for more rapid and efficient means of microbial identification. Clinical microbiologists and microbiology laboratories are experiencing changes due to evolving views on "healthcare delivery" as an economic activity, due to changes in the medical environment and the demographics of the workforce, and technical evolution. The identification and differentiation of microorganisms has principally relied on microbial morphology and growth variables. Cost-effectiveness of laboratory procedures has been achieved through consolidation and integration of laboratories. Consolidation offers economy of scale and reduction in numbers of on-site staff, but also leads to separation of microbiologists from their clinical colleagues. Integration puts different laboratory disciplines under a single management, and leads to reorganisation of laboratories along common work-lines. Cost-savings combined with on-site availability of laboratories are achieved at the expense of a reduction in the influence of microbiologists in the daily running of the laboratory.

Medically, there is growing emphasis on evidence-based diagnostics. Because of time-delays inherent in culturing, microbiology has a limited impact on patient outcomes. Increased clinical relevance of microbiological testing through rapid testing is mandatory. There is an increasing shortage in Europe and the USA of trained microbiology laboratory technicians and microbiologists. This reinforces the trend

M. Debnath et al., *Molecular Diagnostics: Promises and Possibilities*,
DOI 10.1007/978-90-481-3261-4_15,

towards more automation and integration. Technological advances, particularly in molecular diagnostics, offer the possibility of rapid reporting and improvement of the impact of clinical microbiology on patient management. Molecular tests, however, fit perfectly the concept of an integrated laboratory and may further loosen the link between microbiologist and microbiology tests. The challenge for clinical microbiology will be to use new techniques to improve its cost-effectiveness and impact on infectious disease management (Pfaller, 2001). The future organisation of microbiology laboratories must support this but is itself of secondary importance.

15.2 Concept

Molecular diagnostics of microbial pathogens is an integral part of modern medicine. The development and application of molecular diagnostic techniques has initiated a revolution in the diagnosis and monitoring of infectious diseases. The traditional methods of microbial identification rely solely on the phenotypic characteristics of the organism. Bacterial fermentation, fungal conidiogenesis, parasitic morphology, and viral cytopathic effects are a few phenotypic characteristics commonly used. When methods for microbial genome analysis became available, a new frontier in microbial identification and characterization was opened.

Microbial phenotypic characteristics, such as protein, bacteriophage, and chromatographic profiles, as well as biotyping and susceptibility testing, are used in most routine laboratories for identification and differentiation. At this point, in the twenty-first century, we are on the threshold of another era of discovery, that of molecular diagnostics. Advances in molecular biology over the past 10 years have opened new avenues for microbial identification and characterization (Mullis, 1990). The development and use of new molecular microbiological testing, coupled with an ever-improving understanding of how best to use these precious drugs in the treatment of infection, offers the greatest hope yet for physician prescribing that can retard, or perhaps even reduce, the development of drug resistance in many microbial species.

Nucleic acid techniques, such as plasmid profiling, various methods for generating restriction fragment length polymorphisms, and the polymerase chain reaction (PCR), are making increasing inroads into clinical laboratories. PCR-based systems to detect the etiologic agents of disease directly from clinical samples, without the need for culture, have been useful in rapid detection of uncultivable or fastidious microorganisms. These technologies, however, like the determinations of phenotypic variables, are limited by microbial recovery and growth. Rapid techniques of nucleic acid amplification and characterization have significantly broadened the microbiologists′ diagnostic arsenal. Additionally, sequence analysis of amplified microbial DNA allows for identification and better characterization of the pathogen. Subspecies variation, identified by various techniques, has been shown to be important in the prognosis of certain diseases. Other important advances include the determination of viral load and the direct detection of genes or gene mutations responsible for drug resistance. Increased use of automation and user-friendly software makes these technologies more widely available. In all, the detection of infectious agents at the nucleic acid level represents a true synthesis of clinical chemistry and clinical microbiology techniques.

15.3 Advent of Improved Diagnostics

Diagnostic Microbiology probably started in the late seventeenth century when the Dutch scientist Antoni van Leeuwenhoek made microorganisms visible for the first time. Since then, 3 major revolutions have taken place, all of which had a major impact on the field of clinical microbiology. The first revolution took place at the end of the nineteenth century after the development of solid culture media by Robert Koch. Bacterial culture, nowadays, still remains the cornerstone of clinical microbiology, although technical refinements have made available automated identification and antimicrobial susceptibility testing in most modern laboratories. The second revolution was the development of antigen/antibody detection in the sixties and seventies of the twentieth century.

Early antigen/antibody testing tended to lack some sensitivity and specificity, a problem which is nowadays largely resolved; however, most of these tests still show a window period before detectable antigen/antibody levels appear in the patient. The third and final revolution, the nucleic acid (NA)-based detection of microorganisms started some 25 year ago and is still ongoing. Molecular diagnostic testing has made its way from basic research to become a permanent asset for the clinical microbiology laboratory. In fact, in clinical virology PCR has nearly completely replaced traditional methods like immunofluorescence to detect active viral infection (Tiveljung-Lindell et al., 2009). However, in other fields in clinical microbiology, the use of molecular diagnostics has remained more limited to those microorganisms that are either uncultivable or hard to culture, or those where conventional diagnostic procedures lack sensitivity (Fig. 15.1).

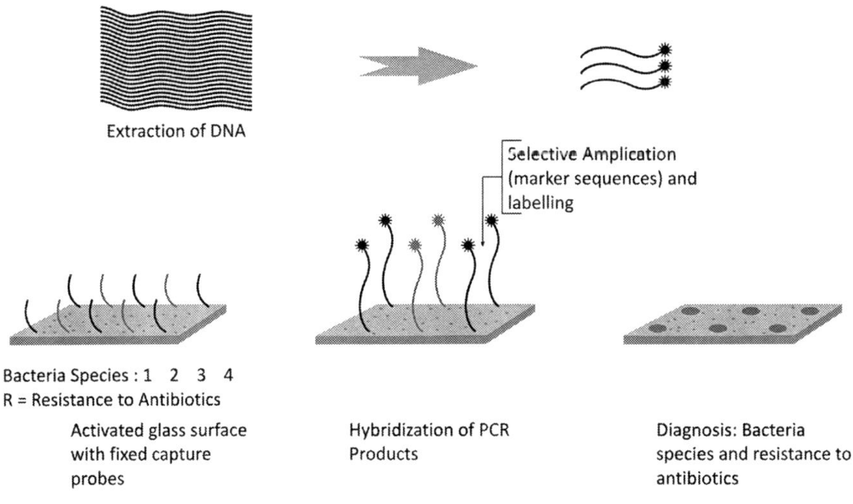

Fig. 15.1 Molecular diagnosis of bacteria using molecular diagnostic techniques

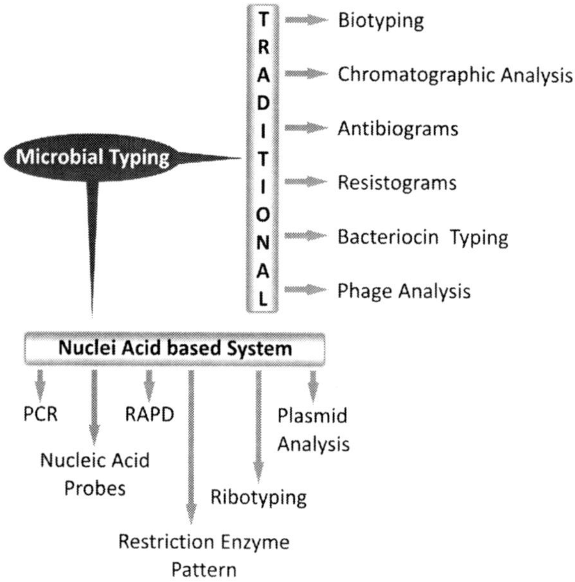

Fig. 15.2 Molecular diagnostic techniques for microorganisms

There are various diagnostic techniques for molecular screening of microorganisms (Fig. 15.2).

15.4 Traditional Microbial Typing

Molecular screening of particular at-risk populations for a group of possible pathogens is an exciting area of development in molecular microbiology. For example, numerous etiologic agents cause debilitating gastroenteritis in immunosuppressed patient populations, including mycobacteria (i.e., *M. avium* complex and *M. genevense*), parasites (i.e., *Cryptosporidum*, *Microsporidum*), viruses (i.e., rotovirus, Norwalk agent), and typical bacterial pathogens (*E. coli* variants, *Salmonella*, *Shigella*, and *Campylobacter*). Traditionally, different methods of detection are used for each group of intestinal pathogens. This requires special media, equipment, and expensive facilities for the culture of mycobacteria; expertise in the identification of parasites in ova and parasite stool preparations; virology facilities; and special media for the workup of bacterial enteric pathogens. Although these tests may be relatively inexpensive individually, an adequate workup for enteric pathogens can be quite costly. Some of the methods are discussed below:

15.4.1 Biotyping

Traditional microbial identification methods typically rely on phenotypes, such as morphologic features, growth variables, and biochemical utilization of organic substrates. The biological profile of an organism is termed a biogram. The determination of relatedness of different organisms on the basis of their biograms is termed biotyping. In many instances, biotyping is used in conjunction with other methods to more accurately profile microorganisms (Tenover et al., 1994).

15.4.2 Antibiograms, Resistograms, and Bacteriocin Typing

The susceptibility or resistance of an organism to a possibly toxic agent forms the basis of the following typing techniques. The antibiogram is the susceptibility profile of an organism to a variety of antimicrobial agents, whereas the resistogram is the susceptibility profile to dyes and heavy metals. Bacteriocin typing is the susceptibility of the isolate to various bacteriocins, i.e., toxins that are produced by a collected set of producer strains. By far, the antibiogram is the most commonly used susceptibility/resistance typing technique, most probably because the data required for antibiogram analysis are available routinely from the antimicrobial susceptibility testing laboratory (Dijkshoorn et al., 1996).

15.4.3 Protein Analysis

Commercially available antibodies are routinely used to specifically identify antigenic proteins from a wide variety of organisms. In some instances, the test may be used only to identify the genus and species of an organism. Monoclonal antibodies directed against the major subtypes of the influenza virus, as well as the various serotypes of *Salmonella*, are commonly used in speciation. Specific antigenic proteins may be detected by antibodies directed against these proteins in immunoblot methods (Mulligan et al., 1988).

15.4.4 Phage Analysis

Bacteriophages, viruses that infect and lyse bacteria, are often specific for strains within a species. The bacteriophage profile may be used to type bacterial strains within a given species. Bacteriophage profiles have been used successfully to type various organisms associated with epidemic outbreaks. However, this typing method is labor-intensive and requires the maintenance of bacteriophage panels for a wide variety of bacteria. Additionally, bacteriophage profiles may fail to identify isolates, are often difficult to interpret, and may give poor reproducibility (Hickman et al., 1991).

15.4.5 Chromatographic Analysis

Chromatographic analysis of short-chain fatty acid production is a routine method used to aid in the identification of anaerobic bacteria. Computer-aided gas–liquid chromatography is commercially available and is a means of microbial identification. This identification system utilizes the type and amount of cellular fatty acids present in the lysate of an organism. Many species have unique cellular fatty acid chromatographic profiles (Stoakes et al., 1994).

15.5 Nucleic Acid-Based Typing Systems

Molecular techniques exist and are being developed that may be used to screen individuals within a particular patient population for the most probable etiologic agents of disease. Nucleic acids extracted from the stool of patients with gastroenteritis may be examined with organism- or group-specific nucleic acid primers and probes. In this manner, one single test may be used to single out the etiologic agent of disease among numerous possibilities.

15.5.1 Plasmid Analysis

Plasmid profile analysis was among the earliest nucleic acid-based techniques applied to the diagnosis of infectious diseases and has proven useful in numerous investigations (Wachsmuth, 1985). This method has also been widely utilized for tracking antimicrobial resistance during nosocomial outbreaks.

15.5.2 Restriction Enzyme Pattern

The relative simplicity of the RFLP profiles generated by PFGE facilitates application of the procedure in identification and epidemiological survey of bacterial pathogens· Fingerprinting, which combines PFGE with Southern transfer and hybridization, has been widely used in studying the tuberculosis nosocomial outbreak in human immunodeficiency virus (HIV)-positive populations (Kristjansson et al., 1994).

15.5.3 Ribotyping

Restriction patterns can be obtained by hybridizing Southern-transferred DNA fragments with labelled bacterial ribosomal operon(s), which encode for 16S and (or) 23S rRNA. This method, called ribotyping, has been shown to have both taxonomic and epidemiological value. Ribotyping assays have been used to differentiate bacterial strains in different serotypes and to determine the serotype(s) most frequently involved in outbreaks (Tenover et al., 1994).

15.5.4 Random Amplified Polymorphic DNA (RAPD)

RAPD has been used to differentiate strains of various species, various serotypes within species, and various subtypes within a serotype (MacGowan et al., 1993). It is, therefore, useful for determining whether two isolates of same species are epidemiologically related.

15.5.5 Nucleic Acid Probes

Nucleic acid probes allow the diagnosis of infections in which the organisms are not easily cultured or cannot be cultured at all. Detection of DNA with direct or culture-amplified gene probe technology has been applied to several organisms, including bacteria, viruses, mycobacteria, fungi, and even certain parasites. The technique has been also used to monitor growth as an indicator of drug resistance or to directly detect genes associated with antibiotic resistance. DNA probes have shortened the time required for probe assay.

15.5.6 Polymerase Chain Reaction

Nucleic acid amplification technology is examined from the critical viewpoint of a clinical microbiologist working in a routine diagnostic bacteriology laboratory. Widely recognised limitations of amplification technology include those of false-positive and false-negative results, the difficulty of obtaining quantitative results, the problem of using this technology for susceptibility testing, and the difficulty of detecting routinely the wide range of possible pathogens contained in a clinical sample. On the positive side, amplification technology brings welcome new possibilities for rapid detection of specific pathogens in a sample, including viruses, slowly growing bacteria, fastidious or uncultivable bacteria, fungi and protozoa. Other possible applications include screening normally sterile clinical samples for non-specific bacterial contamination and the use of amplification-based DNA fingerprinting methods for identification and typing of microorganisms. Nevertheless, it is predicted that – in contrast to research and reference facilities – routine bacteriology laboratories will continue to rely on culture as the preferred "amplification method" for most diagnostic applications.

Nucleic acid amplification techniques increase sensitivity dramatically while still retaining a high specificity. Commercial systems for PCR detection of DNA targets of *Chlamydia trachomatis* and *Mycobacterium tuberculosis* are manufactured by Roche Molecular Systems (Loeffelholz et al., 1992). Numerous modifications of the standard PCR procedure have been developed since its inception. RT-PCR has played an important role in diagnosing RNA-containing virus infections, detecting viable *Mycobacteria* species, and monitoring the effectiveness of antimicrobial therapy. For diagnostic uses, multiplex PCR can be set up to detect internal controls or to detect multiple pathogens from a single specimen (Roberts and Storch, 1997). Another important technical modification is the development of broad-range PCR, in which conserved sequences within phylogenetically informative genetic targets are used to diagnose microbial infection. A broad-range PCR approach has identified several novel, fastidious, or uncultivated bacterial pathogens directly from infected human tissue or blood (Relman et al., 1996). In a recent report, Bannoehr et al. (2009) reported the first diagnostic test for the identification of *Staphylococcus pseudintermedius* involving a simple PCR-restriction fragment length polymorphism approach. The method allows discrimination of *S. pseudintermedius* from the closely related members of the *Staphylococcus intermedius* group and other important staphylococcal pathogens of humans and dogs.

15.6 Current Application of Molecular Diagnostics

In recent years, the demand for quantification of nucleic acid targets has been growing (Crotty et al., 1994). By use of molecular methods, the microbial load of an infecting pathogen may be determined and its genotype may also be evaluated. Viral load data are used to monitor therapeutic responsiveness and may yield prognostic information regarding the progression of disease. Until recently, however, the task of quantitative nucleic acid amplification has been very difficult to accomplish. Because the amplification techniques yielded products in an exponential manner until a plateau was reached, any factor interfering with the exponential nature of the amplification process would therefore affect the result of the quantitative assay. In practice, many factors can affect the efficiency of the PCR reaction throughout the amplification procedures and result in the differences between theoretical and actual yields of the reaction. Now, however, kit-based technologies make it possible for many laboratories to carry out quantitative determinations.

15.6.1 Clinical Microbiology

Traditionally, the clinical medical microbiology laboratory has functioned to identify the etiologic agents of infectious diseases through the direct examination and culture of clinical specimens. Direct examination is limited by the number of organisms present and by the ability of the laboratorian to successfully recognize the pathogen. Similarly, the culture of the etiologic agent depends on the ability of the microbe to propagate on artificial media and the laboratorian's choice of appropriate media for the culture. When a sample of limited volume is submitted, it is often not possible to culture for all pathogens. In such instances, close clinical correlation is essential for the judicious use of the specimen available.

Some microorganisms are either unculturable at present, extremely fastidious, or hazardous to laboratory personnel. In these instances, the diagnosis often depends on the serologic detection of a humoral response or culture in an expensive biosafety level II–IV facility. In community medical microbiology laboratories, these facilities may not be available, or it may not be economically feasible to maintain the special media required for culture of all of the rarely encountered pathogens. Thus, cultures are often sent to referral laboratories. During transit, fragile microbes may lose viability or become overgrown by contaminating organisms or competing normal flora.

The addition of molecular detection methods to the microbiology laboratory has resolved many of these problems. The exquisite sensitivity and specificity of many molecular methods allow the accurate detection of very small numbers of organisms. The direct detection of *M. tuberculosis* nucleic acid from the sputa of smear-negative patients with tuberculosis clearly illustrates this point (Whelen et al., 1995). The technology allows for the rapid and accurate identification of the etiologic agent in a time substantially shorter than traditional methods. This allows for earlier initiation of a focused antimicrobial regimen and decreases the likelihood of disease progression.

In selected situations, the limitations imposed by the ability of an organism to be cultured and the selection of appropriate media and culture conditions may be replaced by the use of molecular microbiology. Microbial DNA/RNA extracted from a clinical specimen may be analyzed for the presence of various organism-specific nucleic acid sequences regardless of the physiological requirements or viability of the organism (Monstein et al., 1996). For example, the inability to culture and analyze the principal etiologic agent of non-A, non-B hepatitis limited medical advances in this area. Using various molecular methods, however, investigators have been able to isolate hepatitis C virus (HCV) nucleic acid (Choo et al., 1989). Analysis and cloning of the HCV genome has provided the viral antigens necessary for the development of specific serologic tests. Currently, RT-PCR allows for the identification, quantification, and sequence analysis of the HCV genome in infected individuals (Hammerle et al., 1996).

Another unculturable microbe that has been specifically detected by PCR and probe analysis is *Tropheryma whippelii*, the causative agent of Whipple disease (Ramzan et al., 1997). Because of the inability of this organism to grow on conventional media and the lack of a serologic test, diagnosis of Whipple disease is usually based on clinical and specific biopsy findings. Patients with Whipple disease often have gastrointestinal manifestations and undergo endoscopy. Small bowel biopsies reveal foamy histiocytes filling the lamina propria. The definitive diagnosis is made with the identification of non-acid-fast, periodic acid-shift-positive, diastase-resistant bacillary forms within the histiocytes. Extraintestinal Whipple disease, principally seen as arthritis and central nervous system involvement, may be missed entirely unless the clinician and pathologist have a high index of suspicion. Even so, the diagnosis in such instances may prove difficult. Advances in the molecular detection of *T. whippelii* have resolved this dilemma. On the basis of bacterial 16S rRNA gene sequence analysis, an emerging pathogen, *Bordetella holmesii*, has been successfully identified in the immunocompromised hosts. Additionally, the DNA from a single clinical specimen, such as a knee fluid aspirate, may be tested for several etiologic agents in a differential diagnosis. In such instances, the specimen may also be analyzed for other fastidious and difficult-to-culture agents of infectious arthritis, such as *N. gonorrhea* or *Borrelia burgdorferi*.

As explained earlier, molecular methods may also be useful in instances of limited specimen volume (Buck, 1996). Even in low-volume specimens, enough DNA/RNA can often be extracted to allow performance of numerous molecular assays. However, though molecular methods are very sensitive, clinically relevant results are ultimately reliant on the submission of quality specimens.

Some organisms, although not difficult to culture, are encountered infrequently and require special media for isolation. In these instances, culturing may not be cost-effective for smaller laboratories because the reagents may expire before usage; these samples may also be sent to reference laboratories for culturing, for the sake of economy. Again, fragile organisms may die in transit or become overgrown by contaminating bacteria, thereby making the subsequent culture useless. If molecular microbiology facilities are not available in community laboratories, nucleic acids extracted by the use of commercially available kits may be sent frozen to molecular reference facilities. Alternatively, if molecular facilities are available, PCR primers and probes for relatively rare microorganisms may be maintained frozen at –70°C for extended periods and used when needed. This may eliminate the need for special culture media and circumvent problems related to specimen transit. As molecular techniques become more widely available, the spectrum of rapid and cost-effective clinical microbiology testing available to smaller laboratories can be extended.

Molecular methods of detection may also play a role in laboratory safety. Organisms such as *Coxiella burnetti*, *M. tuberculosis*, *Coccidioides immitis*, and several viruses causing severe hemorrhagic fevers are laboratory hazards. These organisms are easily cultured, but may infect laboratory personnel and cause serious illness or death. The handling of these organisms requires specially trained personnel, special equipment, and expensive ventilated facilities. All of these increase laboratory costs. Molecular methods may be used to detect such organisms directly from clinical specimens, without exposing laboratory personnel to biologically amplified organisms. After the initial extraction procedure, only non-infectious materials are handled.

The molecular detection of microbes with a known susceptibility profile is an effective replacement of the traditional culture. An excellent example is the molecular detection of *Bordetella pertussis*. This organism is a relatively slow grower, requires specially supplemented and more costly media, and has a known susceptibility profile. The molecular detection of *Bordetella pertussis* can save time, lower laboratory costs with regard to special media, and allow for the more rapid initiation of effective therapy. If variable antimicrobial susceptibility profiles exist, culture for susceptibility testing is still necessary. Molecular methods for the detection of antimicrobial-resistant strains are in development and in the future may replace traditional susceptibility testing. Until then, molecular screening may be used to determine which patients should be cultured for subsequent susceptibility testing.

At the beginning of this century we were just beginning to recognize the potential for use of molecular diagnostics in the rapid and accurate detection of infectious pathogens (Araj, 2000; Louie et al., 2000). There was a general acceptance that laboratories have started molecular testing in pathogen detection as well as resistance determination for individual patients. Perhaps the first molecular test that was successful in reducing antibiotic use was the application of PCR to the detection of enteroviral meningoencephalitis (Dumler and Valsamakis, 1999). In 2003, several reports appeared assessing the impact of early detection of infectious agents (Hallin et al., 2003), determination of specific antimicrobial agent resistance and the positive impact of rapid results on the prescribing of vancomycin. An important innovation by Lapierre et al. (2003) was the development and testing of a novel real-time PCR assay for direct detection of fluoroquinolone resistance in staphylococci, finding a 99.8% correlation between PCR results and MIC measurements, and thus demonstrating the potential to increase dramatically the speed of resistance detection using this new technology. Stefanelli et al. (2003) in another excellent technological advance, were able to predict diminished penicillin susceptibility for all *Neisseria meningitidis* tested within a few hours using real-time PCR.

In a recent report (Ramond et al., 2009) for diagnosis of respiratory virus infections in young children, modern molecular diagnostics tools like quantitative real-time PCR assays, multiplex PCR, microarray hybridization in an integrated molecular diagnostic device, the Infiniti analyzer were successfully used for detection and analysis. Detection of adenoviruses of groups A, B, C, and E; coronaviruses HKU1, 229E, NL63, and OC43; enteroviruses A, B, C, and D; rhinoviruses of genotypes A and B; influenza viruses A and B; human metapneumoviruses (HMPV) A and B, human respiratory syncytial viruses (HRSV) A and B; and parainfluenza viruses of types 1, 2, and 3 allowed a thorough etiological assessment of respiratory viruses infecting children.

15.6.2 Clinical Epidemiology and Infection Control

The investigation and control of nosocomial infections is a complex issue that involves clinical, infection-control, and laboratory personnel. The efforts of both the microbiologist and the hospital epidemiologist are facilitated greatly by the availability of the newer molecular epidemiological typing techniques. Molecular diagnostic techniques have been successfully used in the investigation and control of classical and emerging nosocomial pathogens, such as the enterobacteriaceae, *Pseudomonas aeruginosa*, *Staphylococcus aureus*, coagulase-negative staphylococci, enterococci, *Candida albicans*, *M. tuberculosis*, and *Chlamydia pneumoniae*. Application of DNA probe-based assays allows the diagnosis of other nosocomial infections caused by respiratory syncytial virus, varicella-zoster virus, herpes simplex virus, and legionella to be made in only a few hours. Several putative outbreaks of infections have been investigated by molecular techniques. Such examples include investigation of several temporally clustered cases of *Streptococcus pyogenes* invasive disease in Air Force recruits (Musser et al., 1994), a case cluster of lymphogranuloma venereum caused by *Chlamydia trachomatis* serovar L1 in homosexual men, and an outbreak of *E. coli* O157:H7 infection from contaminated deer.

Molecular techniques are being used increasingly in epidemiological and clinical investigations. Use of DNA hybridization assays in cervical swabs or fresh cervical biopsy specimens to determine HPV infection and viral types has provided helpful information for clinical assessment and treatment of patients (Morrison et al., 1991). By using PCR followed by automated direct sequencing, several studies have revealed that the most common genotypes of HCV. A new PCR-based HCV genotyping system has been recently developed to identify various HCV genotypes in clinical studies. Molecular techniques have been used to directly detect resistance genes or mutations that result in resistance in organisms. The *mecA* gene that codes for resistance to methicillin in *Staphylococci* has been detected by PCR and multiplex PCR assays (Ubukat et al., 1992). Defining the mutations responsible for resistance to microbial agents has led to new methods for monitoring efficacy of antimicrobial therapy. Successful investigations have been carried out on both bacterial and viral resistance mechanisms.

15.7 Promise of Molecular Testing

New diagnostics will also need to sense those settings where infection is actually occurring. To avoid unnecessary prescribing of antimicrobial agents, rapid tests will be particularly useful when they can indicate what type of antibacterial, antiviral or symptomatic therapy (in the absence of infection) is most beneficial. The newest and most exciting possibility for specific outpatient diagnosis is that based upon amplification technology such as PCR. Rapidity of testing and cost are always key issues. With the advent of real-time PCR we finally have as a reality access to a reliable diagnostic test with same-day results. The cost of this testing also now approximates to that of conventional culture-based processing. Thus, we are on the threshold of microbiology diagnostics that, when carried to its expected conclusion, can provide sufficient information to permit specific application or avoidance of antimicrobial chemotherapy.

The newer applications of molecular diagnostics known as gene "chip" and "microarray" and "nanoparticle" technology offer the potential to solve many remaining impediments to rapid detection of important infectious agents in health care. Since these technologies do not require organism viability, and thus avoid any adverse effect of longer specimen transport, they can be successfully applied to both the in- and outpatient settings. Also, the resulting test rapidity theoretically will provide relevant information within a few hours, which would limit any necessary empirical treatment to one or two doses. Several companies currently possess the technical expertise and laboratory research infrastructure to bring a useful diagnostic testing approach to the clinical trial stage very shortly. One example is the new technology company Nanosphere, Inc. (Northbrook, IL, USA), which is developing gold nanoparticle technology to detect molecular DNA, RNA and protein biomarker targets using automated instrumentation, without the need for prior amplification (Ehghanian et al., 1997).This testing could detect likely pathogens responsible for important clinical scenarios, such as respiratory disease symptom complexes, implicating the key bacterial, viral or atypical microbial pathogens responsible. Simplified automation opens the potential for testing to be done near the patient at a peripheral site. If molecular testing methods were widely applied, it could be done at a very manageable cost and would have a significant impact on lowering unnecessary antibiotic prescribing.

Thus, the application of molecular testing methods in the clinical laboratory has drastically improved our ability to diagnose infectious diseases. However, the clinical usefulness of molecular testing will only be maximized to its fullest benefit by appropriate studies correlating clinical findings with assay results. As methods become more refined, automated, and standardized, the use of amplification methods to detect infectious agents will become more valuable. Because amplification methods and the interpretation of their results are continuously evolving and becoming more refined, it is imperative for clinical microbiologists as well as clinicians to remain current and knowledgeable in all aspects, including the chemistry and microbiology of the infectious organism as well as in general medicine. Questions regarding basic laboratory issues including the clinical need, the cost benefit for testing the patent, and other issues as well as direct and indirect costs must be addressed and answered. Of great import, the specific clinical niche for the amplification test (i.e., the clinical question that a PCR result will answer) must also be delineated. And finally, whether an amplification assay is and remains the best approach must be addressed during as well as after assay development. On the short term, there has been an increase in molecular diagnostic testing within the clinical microbiology laboratory. The role of culture will therefore move from a primary detection tool to an epidemiological tool. On the long run, developments will be technology driven, eventually leading to "real-time" point of care diagnostics.

However, some additional hurdles remain other than the technological issues. For example, in the United States the estimated cost of regulatory requirements for approval and ongoing modification of multitargeted molecular tests, added to the expense of first developing them (in excess of $100 million) are considered prohibitive by the industry. This is due primarily to the fact that regulations have not changed to encompass test platforms having the capacity to detect a variety of genetic targets from multiple microbial species, with the aim of providing a unified diagnosis based on interpretation of a complex pattern using multiple results. Because of what is at stake – managing antimicrobial agent resistance – it clearly appears in the best interest of all to develop a modernized set of regulations that facilitate development and application of this testing.

15.8 Assay Validation-Analytic Sensitivity and Specificity

Once optimized and its specificity confirmed yet prior to its validation in a clinical setting, an amplification assay must be evaluated for its analytic sensitivity and specificity using a background milieu of clinical material. However, it is essential that key decisions be made prior to further assay development and evaluation employing clinical material. First, the type of specimen (or specimens) acceptable for assaying by PCR must be determined as well as the physical criteria for the specimen's (or specimens') suitability for analysis, including optimum source and volume, appropriate collection method, transport and storage conditions, and specimen longevity. The choice of specimen(s) plays a key role in the performance and interpretation of test results because if any of the criteria are not fulfilled, the sensitivity and specificity of the assay will vary accordingly (Noordhoek et al., 1996). This becomes of paramount importance in quantitative amplification assays used to determine viral clearance such as for HCV (Van Vliet et al., 2001). In general, plasma or serum is preferred over whole blood or leukocytes when the target organism is predominantly extracellular (e.g., HCV RNA, human immunodeficiency virus-1 RNA).

Once the acceptable specimen type (or types) has been delineated, the method for extraction must be selected. There are numerous methods in the literature for extracting either DNA or RNA from clinical specimens; however, they are beyond the scope of this discussion. A host of commercially available kits are available from which to choose as well as a number of automated methods for specimen preparation. These commercially available kits and automated systems are advantageous over manual methods of extraction in that they are more rapid and cost-effective and provide more consistent results. Nevertheless, the most optimal method for extracting nucleic acid must also be determined for each specimen type. The volume of specimen to be used for extraction must be taken under consideration; sample volumes vary considerably depending on the amplification method, but most assays use between 100 and 250 [μ]l. Lastly, the method for detection of the target must be selected. Again, numerous options are available, such as detection by agarose gel electrophoresis or hybridization capture assays employing the capture of a biotinylated amplicon by a target-specific probe and detection by streptavadin-horseradish peroxidase etc.

In the mid-1980s, the polymerase chain reaction (PCR) methodology for the amplification of minute amounts of target DNA was successfully developed and then introduced into clinical use; such technology has led to a revolution in diagnostic testing. Nucleic acid amplifications have become crucial in the diagnosis of selected infectious disease agents. Initially, molecular diagnostic tests were performed only in highly specialized or research laboratories; however, these assays are now more widely performed in all sections of the clinical laboratory. This has been largely a result of commercial assays, such as those advocated by the Food and Drug Administration, having been introduced into the market. Clearly, as the technology advances, additional commercial, Food and Drug Administration-approved methods will become available and enable even the smallest laboratory to employ amplification technologies for the detection of microorganisms. Nevertheless, in the meantime, there is a significant demand for amplification assays to detect the presence of a variety of microorganisms in clinical specimens for which there are no commercially available kits. Thus, clinical laboratories, particularly those associated with academic medical centers, must frequently develop their own "in-house" assays to accommodate the demand for the laboratory diagnosis of infectious diseases by amplification methods.

Despite the significant advantages and strengths that amplification methods offer in terms of the rapid and sensitive detection of infectious agents as well as the quantification of pathogens such as human immunodeficiency virus-1 to monitor therapy or disease outcome, there are limitations and caveats to these assays that must be understood. For example, studies have demonstrated that there is significant variation in the ability of "home-brew" or in-house assays among clinical laboratories to reliably detect infectious agents. Moreover, as studies are published that correlate clinical findings with results of amplification tests, it has become exceedingly evident that although molecular diagnostic assays enhance diagnostic capabilities, their results must be clearly interpreted within the clinical context and performance of the laboratory assay. In other words, a thorough understanding of the parameters of the molecular assay, including the respective procedural limitations and the target organism's microbiology and pathogenesis, is critical for the proper interpretation of results. Clearly, ongoing clinical research to correlate amplification results with clinical findings as well as strict adherence to guidelines for method validation for in-house PCR assays are prerequisite.

In general, PCR and other amplification tests are perceived as expensive. Not only are there costs for reagents and equipment, but there are also training costs associated with teaching microbiologists to perform these molecular diagnostic assays. Through rapid diagnosis, more expensive and/or invasive procedures may be prevented, hospital stays shortened, or the unnecessary administration of antibiotics prevented. Presently, in light of the expense of these assays coupled with an ever-increasing demand for testing as the list of clinical syndromes increases for some infectious agents, the appropriate use of these assays is imperative. Studies centering on the use of molecular assays in terms of cost and other benefits are now being published. More of these types of studies need to be performed so that the appropriate use of molecular assays is accomplished.

15.9 Conclusion

New pathogens will continue to emerge, and as soon as they are recognized will need to be incorporated into existing diagnostic test menus. Mixed viral and bacterial infections can also be daunting tasks for diagnosis and therapy, but if the molecular tests developed not only delineate all that is present (signals) but also differentiate what is responsible for illness in the host (sensors), then we will be at a true crossroads in the treatment of infectious diseases. Dunne et al. (2002) described a future scenario where this testing develops as we envisage here, and postulated that by the year 2025 sophisticated samplers will painlessly obtain the necessary material, followed by automated analysers to process simultaneously DNA, RNA, protein, glycopeptides and exopolysaccharides to detect any of a possible 168 pathogenic microbes as well as toxins and resistance genotypes, all completed and yielding a diagnosis within 15 min. We believe this future is attainable even earlier, and that bringing these advances to fruition is critical for mastering antimicrobial agent prescribing in order to manage their use in a rational way and finally reverse the increasing resistance that threatens these precious agents we have as necessary aids to all our health. The technology and vision are here, we just need the will to do it.

References

Araj, G.F., 2000, Available laboratory tests to guide antimicrobial therapy. *Leb Med J* 48:199–202.

Bannoehr, J., Franco, A., Iurescia, M., Battisti, A., Fitzgerald, J.R., 2009, Molecular diagnostic identification of *Staphylococcus pseudintermedius*. *J Clin Microbiol* 47(2):469–471

Beekmann, S.E., Dickema D.J., Chapin K.C., Doern, G.V., 2003, Effects of rapid detection of bloodstream infections on length of hospitalization and hospital charges. *J Clin Microbiol* 41, 3119–3125.

Buck, G.E., 1996, Detection of Bordetella pertussis by rapid-cycle PCR and colorimetric microwell hybridization. *J Clin Microbiol* 34:1355–1358.

Choo, Q.L., Kuo, G., Weiner, A.J., Overby, L.R., Bradley, D.W., Houghton, M., 1989, Isolation of a cDNA clone derived from a blood-borne non-A, non-B viral hepatitis genome. *Science* 244(4902):359–362.

Crotty, P.L., Staggs, R.A., Porter, P.T., Killeen, A.A., McGlennen, R.C., 1994, Quantitative analysis in molecular diagnostics. *Hum Pathol* 25(6):572–579.

Dijkshoorn, L., Aucken, H., Gerner-Smidt, P., Janssen, P., Kaufmann, M.E., Garaizar, J., Ursing, J., Pitt, T.L., 1996, Comparison of outbreak and nonoubreak *Aacinetobacter baumannii* strains by genotypic and phenotypic methods. *J Clin Microbiol* 34(6):1519–1525.

Dumler, J.S., Valsamakis, A., 1999, Molecular diagnostics for existing and emerging infections. *Am J Clin Pathol* 112 (1): S33–S39.

Dunne, W.M., Pinckard, J.K., Jr., Hooper, L.V., 2002, Clinical microbiology in the year 2025. *J Clin Microbiol* 40:3889–3893.

Elghanian, R., Storhoff, J.J., Mucic, R.C., Letsinger, R.L., Mirkin, C.A., 1997, Selective colorimetric detection of polynucleotides based on the distance-dependent optical properties of gold nanoparticles. *Science* 277(5329):1078–1081.

Hallin, M., Maes, N., Byl, B., Jacobs, F., De Gheldre, Y., Struelens, M.J., 2003, Clinical impact of a PCR assay for identification of *Staphylococcus aureus* and determination of methicillin resistance directly from blood cultures. *J Clin Microbiol* 41(8):3942–3944.

Hammerle, T., Falkner, F.G., Dorner, F., 1996, A sensitive PCR assay system for the quantitation of viral genome equivalents: hepatitis C virus (HCV). *Arch Virol* 141:2103–2114.

Hickman-Brenner, F.W., Stubbs, A.D., Farmer, J.J., 1991, Phage typing of Salmonella enteritidis in the United States. *J Clin Microbiol* 29:2817–2823.

Isenberg, H.D., 2003, Clinical microbiology: past, present, and future. *J Clin Microbiol* 41: 917–918.

Kristjansson, M., Samore, M.H., Gerding, D.N., DeGirolami, P.C., Bettin, K.M., Karchmer, A.W., Arbeit, R.D., 1994, Comparison of restriction endonuclease analysis, ribotyping and pulsed-field gel electrophoresis for molecular differentiation of *Clostridium difficicile* strains. *J Clin Microbiol* 32(8):1963–1969.

Lapierre, P., Huletsky, A., Fortin, V., Picard, F.J., Roy, P.H., Ouellette, M., Bergeron, M.G., 2003, Real-time PCR assay for detection of fluoroquinolone resistance associated with grlA mutations in *Staphylococcus aureus*. *J Clin Microbiol* 41(7):3246–3251.

Loeffelholz, M.J., Lewinski, C.A., Silver, S.R., Purohit, A.P., Herman, S.A., Buonagurio, D.A., Dragon, E.A.,1992, Detection of *Chlamydia trachomatis* in endocervical specimens by polymerase chain reaction. *J Clin Microbiol* 30(11):2847–2851.

Louie, M., Louie, L., Simor A.E., 2000, The role of DNA amplification technology in the diagnosis of infectious diseases. *Can Med Assoc J* 163:301–309.

MacGowan, A.P., O'Donaghue, K., Nicholls, S., McLauchlin, J., Bennett, P.M., Reeves, D.S., 1993, Typing of *Listeria* spp. By random amplified polymorphic DNA (RAPD) anakysis. *J Clin Microbiol* 38(5):322–327.

Monstein, H.J., Kihlstrom, E., Tiveljung, A., 1996, Detection and identification of bacteria using in-house broad range 16S rDNA PCR amplification and genus-specific DNA hybridization probes, located within variable regions of 16S rRNA genes. *APMIS* 104: 451–458

Morrison, E.A., Ho, G.Y., Vermund, S.H., Goldberg, G.L., Kadish, A.S., Kelley, K.F., Burk, R.D., 1991, Human papillomavirus infection and other risk factors for cervical neoplasia: a case-control study. *Int J Cancer* 49(1):6–13.

Mulligan, M.E., Kwok, R.Y., Citron, D.M., John, J.F., Jr., Smith, P.B., 1988, Immunoblots, antimicrobial resistace and bacteriophage typing of oxacillin-resistant *Staphylococcus aureus*. *J Clin Micribiol* 26:2395–2401.

Mullis, K.B., 1990, The unusual origin of the polymerase chain reaction. *Sci Am* 262:56–65.

Musser, J.M., Kapur, V., Peters, J.E., Hendrix, C.W., Drehner, D., Gackstetter, G.D., Skalka, D.R., Fort, P.L., Maffei, J.T., Li, L.L., 1994, Real-time molecular epidemiologic analysis of an outbreak of *Streptococcus pyogenes* invasive disease in US Air Force trainees. *Arch Pathol Lab Med* 118(2):128–133.

Noordhoek, G.T., Embden, J.D.A., Van Kolk, A.H., 1996, Reliability of nucleic acid amplification for detection of Mycobacterium tuberculosis: an international collaborative quality control study among 30 laboratories. *J Clin Microbiol* 34:2522–2525.

Pfaller, M.A., 2001, Molecular approaches to diagnosing and managing infectious diseases: practicality and costs. *Emer Infect Dis* 7:312–318.

Ramzan, N.N., Loftus, E., Jr., Burgart, L.J., Rooney, M., Batts, K.P., Wiesner, R.H., Fredricks, D.N., Relman, D.A., Persing, D.H., 1997, Diagnosis and monitoring of Whipple disease by polymerase chain reaction. *Ann Intern Med* 126(7):520–527.

Raymond, F., Carbonneau, J., Boucher, N., Robitaille, L., Boisvert, S., Wu, W.K., De Serres, G., Boivin, G., Corbeil, J., 2009, Comparison of automated microarray detection with real-time PCR assays for detection of respiratory viruses in specimens obtained from children. *J Clin Microbiol* 47(3):743–750.

Relman, D.A., Schmidt, T.M., Gajadhar, A., Sogin, M., Cross, J., Yoder, K., Sethabutr, O., Echeverria, P., 1996, Molecular phylogenetic analysis of the human intestinal pathogen, *Cyclospora*, suggests that it is closely related to *Eimentia*. *J Infect Dis* 173(2):440–445.

Roberts, T.C., Storch, G.A., 1997, Multiple PCR for diagnosis of AIDS-related central nervous system lymphoma and toxoplasmosis. *J Clin Microbiol* 35:268–269.

Stefanelli, P., Carattoli, A., Neri, A., Fazio, C., Mastrantonio, P., 2003, Prediction of decreased susceptibility to penicillin of *Neisseria meningitides* strains by real-time PCR. *J Clin Microbiol* 41(10):4666–4670

Stoakes, L., John, M.A., Lannigan, R., Schieven, B.C., Ramos, M., Harley, D., Hussain, Z., 1994, Gas-liquid chromatography of cellular fatty acids for identification of *staphylococci*. *J Clin Microbiol* 32(8):1908–1910.

Tenover, F.C., Arbeit, R., Archer, G., Biddle, J., Byrne, S., Goering, R., Hancock, G., Hebert, GA, Hills, B., Hollis, R., 1994, Comparison of traditional and molecular methods of typing isolates of *Staphylococcus aureus*. *J Clin Microbiol* 32(2):407–415.

Tiveljung-Lindell, A., Rotzen-Ostlund, M., Gupta, S., Ullstrand, R., Grillner, L., Zweygberg-Wirgart, B., Allander, T., 2009, Development and implementation of molecular diagnostic platform for daily rapid detection of 15 respiratory viruses. *J Med Virol* 81(1):167–175.

Ubukata, K., Nakagami, S., Nitta, A., Yamane, A., Kawakami, S., Sugiura, M., Konno, M., 1992, Rapid detection of the mecA gene in methicillin-resistant *staphylococci* by enzymatic detection of polymerase chain reaction products. *J Clin Microbiol* 30(7):1728–1733.

Van Vliet, K.E., Muir, P., Echevarria, J.M., Klapper, P.E., Cleator, G.M., Van Loon, A.M., 2001, Multicenter proficiency testing of nucleic acid amplification methods for the detection of enteroviruses. *J Clin Microbiol* 39(9):3390–3392.

Wachsmuth, K., 1985, Genotypic approaches to the diagnosis of bacterial infections: plasmid analyses and gene probes. *Infect Control* 6:100–109.

Whelen, A.C., Felmlee, T.A., Hunt, J.M., Williams, D.L., Roberts, G.D., Stockman, L., Persing, D.H., 1995, Direct genotypic detection of *Mycobacterium tuberculosis* reifampicin resistance in clinical specimens by using single-tube heminested PCR. *J Clin Microbiol* 33(3):556–561.

Chapter 16
Proteomic Technology

Abstract Fuelled by ever-growing DNA sequence information, proteomics, the large scale analysis of proteins, has become one of the most important disciplines for characterizing gene function, for building functional linkages between protein molecules, and for providing insight into the mechanisms of biological processes in a high-throughput mode. The long-term challenge of proteomics is enormous: to define the identities, quantities, structures and functions of complete complements of proteins, and to characterize how these properties vary in different cellular contexts. Functional proteomics is a promising technique for the rational identification of novel therapeutic targets by elucidation of the function of newly identified proteins in disease-relevant cellular pathways.

Keywords Post genomic · Proteome · Proteomics · Mass spectrometry · Protein profiling

16.1 Prologue

The dawn of the post-genome era is leading to extraordinary opportunities in biomedicine. With recent innovative approaches resulting in the development of new therapies such as small molecular inhibitors, therapeutic antibodies, recombinant proteins and gene therapy, there is increasing need for improved understanding of the basic molecular mechanisms that are exploited by such treatments. The term functional genomics has been used to describe analysis of changes in gene expression in response to various experimental conditions. However, there are some limitations of genomic analysis that it cannot provide complete information of cellular, sub cellular and supracellular functions, in which proteins, not genes, govern the functions. Simply put, differential developmental stages of the animals are resulted from changes in global protein expression or the proteome, while gene expression remains static. This can be explained by dynamic processes, especially posttranslational modifications that modify gene products and cause functional changes in the proteins. Indeed, studies of protein expression, rather than gene expression, have generally been used for several decades for biological study. Advances in molecular

M. Debnath et al., *Molecular Diagnostics: Promises and Possibilities*,

DOI 10.1007/978-90-481-3261-4_16,

biology techniques during the past two decades have led to better understanding the genetic and molecular bases of a wide variety of human diseases.

16.2 Concept

Proteomics is one of the most important of the so-called "post-genomic" approaches to understanding gene function because it is the proteins expressed by genes that are ultimately responsible for all processes that take place within the cell (Zhu et al., 2003). But, while proteins may yield the most important clues to cellular function, they are also the most difficult of the cell's components to detect on a large scale. A universal definition of proteomics remains elusive. However, the word "proteomics" derives from a Greek word "proteus" meaning an ancient Greek god, regarded by some as a symbol of the original matter from which we are created. Proteomics refers to the study of proteins at both the structural and functional levels. Distinct changes occur in proteins during the transformation of a normal cell into a neoplastic cell that range from altered expression, differential protein modification, changes in specific activity, to aberrant localizations all of which affect cellular function (Tyer and Mann, 2003).

In 1975, two-dimensional polyacrylamide gel electrophoresis (2-DE) was simultaneously described by O'Farrell and by Klose. Their techniques allowed, for the first time, the separation of complex mixtures of proteins into individual components. Using this technique, initially 1,100 protein components were resolved from *Escherichia coli* on a 2-D gel. Because techniques for protein identification were limited at that time and due to a lack of standardized reagents, 2-DE was not extensively used. However, recent advances in mass spectrometry (MS), in parallel with successes in genomic analysis, have made it possible to identify proteins separated by 2-DE (Cunnigham, 2000).

The high-throughput capability of MS, coupled with 2-DE separations of proteins, has made proteomic analysis practical. This field of research has acquired the name "Proteomics". "PROTEins expressed by the genOME" have been termed "Proteome", a term firstly used in the public by Marc Wilkins at the first Siena proteomic conference in 1994. Proteomics is a field that Anderson and Anderson have defined as "the use of quantitative protein-level measurement of gene expression to characterize biological processes (e.g., disease processes and drug effects) and decipher the mechanisms of gene expression control". It is a field of study that is separated from, but complementary to genomics. The field has expanded rapidly as the number of published proteomic articles has been increasing rapidly.

Within the last 5 years the field of proteomics has changed the understanding of molecular biology. Proteins manifest physiological as well as pathophysiological processes in a cell or an organism, and proteomics describe the complete protein inventory in dependence on in vivo parameters. Disease mechanism or drug effects both affect a protein profile and, vice versa, characterising protein profiles reveals information for the understanding of disease and therapy. Analytical methods for proteomics are based on conventional tools for protein characterization. The technical challenge is the complete coverage of physico-chemical properties for thousands of proteins. Nucleic acids display a relative chemical homogeneity and therefore genomics was considered more promising in the past than proteomics. Further improvements in proteomics technologies will likely change this course with proteomics complementing genomics as a tool to study life sciences (Kellner, 2000; Huang et al., 2001).

16.3 Methods to Perform Proteomic Analysis

Proteomic analysis can be simply classified as three main categories (Fig. 16.1):

(a) Expression proteomics (b) bioinformatic analysis (c) functional proteomics. While the most important part of proteomic analysis is the functional study, expression proteomics may be a necessary initial step. Unless a hypothesis has been generated by prior study, expression proteomics can be used to screen biological samples to identify candidate proteins for further functional studies. To date, most proteomic studies have compared protein expression in normal and disease states in cells and tissues.

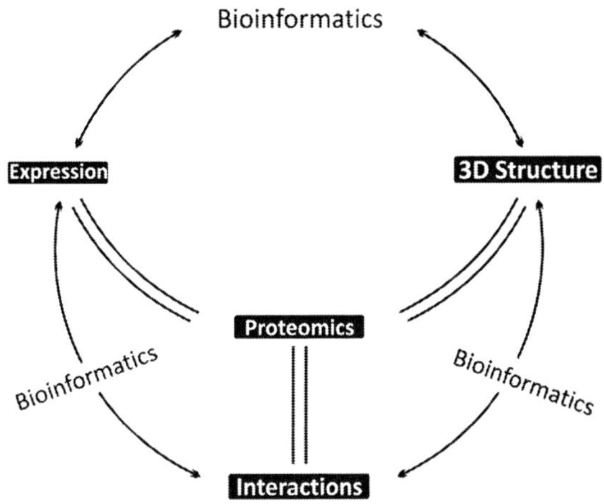

Fig. 16.1 Proteomics analysis showing the major components

Expression proteomics alone, however, does not provide any functional or physiological significance. Functional proteomics and other functional studies will take role after candidates or hypotheses have been generated from initial expression studies.

A typical process in expression proteomics can be explained as below: Protein extraction and separation, protein identification, and bioinformatics based analyses. Protein extraction is to isolate the proteins from tissues. Protein separation can be performed either by 2-DE or by LC techniques. The proteins are then identified by various MS methods. Bioinformatics-based analyses are used to obtain protein information to guide further functional proteomic studies. The final step in proteomic study is to generate a new hypothesis from both expression and functional proteomic data. The novel hypothesis is then examined by various methods.

(a) Expression Proteomics

The initial process of expression proteomics is to prepare the samples to be ready for proteomic analysis, either by gel-based or gel-free methods. Because of differences in protein physical and chemical properties, different extraction protocols may favour proteins from differential compartments of the entire proteome.

For example, most of the extraction protocols for 2-DE separation have a limitation in solubilizing membrane proteins. After extraction or isolation of proteins from cells and tissues, the proteins can be separated either by 2-DE or by liquid chromatography (LC) that requires specific conditioning for each procedure.

Thereafter, those proteins are identified, mostly by MS-based methods. Expression proteomics can be performed as to identify proteins expressed in a specific cell or tissue. However, comparing protein expression between two (or more) differential experimental conditions or disease states will provide more information to facilitate understanding physiology and pathogenic mechanisms. *Quantitative proteomics* takes role to determine differential protein abundance in various conditions.

Quantitation can be performed by indirect method (intensity analysis) or by direct measurement using isotope labelling. While the intensity analysis is simple and widely used, it is limited by sensitivity and accuracy, which allow new technologies of MS-based quantitative techniques to be developed.

(b) Bioinformatics Analysis

After the proteins are identified, bioinformatics analysis will take role in expanding the initial protein information. Additionally, bioinformatics analysis can be used to guide the future direction of functional proteomic studies. The following are additional information that can be obtained by bioinformatics analysis:

(1) Primary, secondary, and tertiary structures;
(2) Sequence alignment and homology;
(3) Motifs and domains;
(4) Protein interactions and networks;
(5) Potential PTMs;
(6) Potential transmembrane regions;
(7) Subcellular locations;
(8) Miscellaneous.

Several bioinformatics analytical tools are freely accessible at http://us.expasy.org/tools/. Additional information obtained from bioinformatics analysis may facilitate and make functional analysis be more focused.

16.4 Functional Proteomics

To understand the role of targeted proteins in cellular functions, high throughput comprehensive analyses of protein–protein interactions, protein complexes, and PTMs are necessary. Protein interactions and complexes can be analyzed by co-immunoprecipitation techniques followed by 2-DE or by affinity chromatography, protein chip, tandem (sequential) affinity tag purification (TAP), biomolecular interaction analysis-mass spectrometry (BIA-MS), etc. All of the applications are based on "ligand fishing" by using the specific antibody or tag-specific receptor as a "bait" and the "prey" or interacting proteins from ligand (analyte)-containing elution can be further analyzed by MS techniques (Colland and Daviet, 2004).

In addition to the aforementioned screening steps for determination of PTMs, a confirmation method may be needed in some circumstances. The methods employed depend on the chemistry involved in the PTMs. For example, phosphoproteins can be detected by MS-based methods or by Western blot analysis using a phosphoprotein-specific antibody that can be performed in parallel and compared with stained 2-D gel. An alternative approach to phosphoproteomics is using enzymatic radiolabeling, using the protein of interest. The protein sample is treated with the appropriate substrate containing 32P, which is incorporated into the protein. Subsequent separation with 2-DE allows identification and quantitation by autoradiography. Fluorescence is another means to detect protein-bound phosphate group by staining with phosphate-specific fluorescence reagents. BO-INI, a fluorescent label, provides a rapid screening method for phosphoproteins without the use of more dangerous radioactive materials.

Protein kinases and phosphatases can be compared as the "yin and yang" of protein phosphorylation and cell signaling. Significant functional roles of phosphorylation can be determined by exposing the cells to various conditioning of protein kinases, phosphatases, and phosphatase inhibitors.

16.5 Implications of the Human Genome Project

After the completion of the Human Genome Project, it has become evident that the complexity of organisms is only to a small part the result of direct gene expression from the genome, and it is clear that the simple concept of one gene-one protein is incorrect. One important reason for this is that the product of one gene can be transformed to a whole family of gene products, i.e., one gene can produce multiple mature mRNAs via alternative splicing and other mechanisms. Furthermore, the correlation between mRNA and protein concentrations has been demonstrated to be insufficient to predict protein expression levels from quantitative mRNA data, since protein levels are regulated by degradation as well.

Although the correlation between mRNA levels and protein abundance was very good for a limited number of highly abundant proteins, it was poor for proteins with lower expression levels. In this group, 30-fold differences in protein

Fig. 16.2 The concept of proteomics

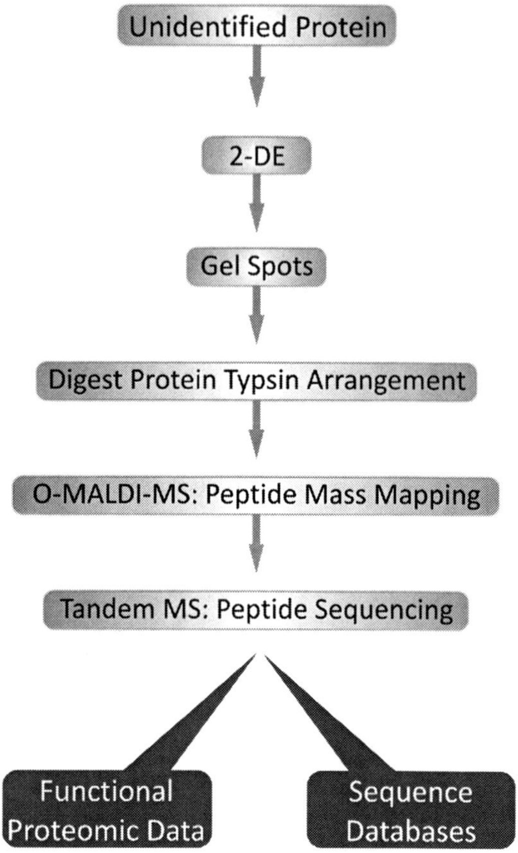

abundance were found for proteins with the same mRNA levels. Proteins from genes with very low expression levels could not be detected at all in this study with a current two-dimensional (2D) electrophoresis mass spectrometry approach (Fig. 16.2). Posttranslational modifications such as glycosylation, phosphorylation, and ubiquitination produce further variations by increasing the number of components from the standard 20 amino acid to more than 140 possible amino acid forms. These modifications undergo rapid changes and usually are not mutually exclusive. Therefore, the study of the genome or even mRNA levels (the transcriptome) will reveal only a small spectrum of the response to a particular stimulus. Even among the diseases known to be based on specific genetic defects, only a very small number are likely to be monogenic, since cellular systems complex interactions with a high level of redundancy.

Conversely, the function of a large number of the protein products that are encoded by these genes is still unclear. Direct investigation of the proteome provides a more complete representation of changes in the status of an organism.

However, there exist several impediments to such an approach, the sheer complexity of the proteome being the most important one. Diversity is another issue, since there are at least 250 different types of human cells, each of which contains at least 2,000–6,000 different primary proteins, and posttranslational modifications will multiply this number. It has been estimated that the different types of human cells may differ from each other in 400 unique proteins. Another important factor is the dynamic range of concentrations of proteins, since one cell can contain between one and more than 100,000 copies of a single protein. Finally, the proteome of organisms is dynamic and changes with environment and with time.

16.6 Measurement Using a Proteomic Approach

Proteomic investigation of a given cell or other biological system should ideally detect all proteins and their functional responses to a stimulus. Given this goal, it is not surprising that no approaches currently come close to achieving this. However, despite the present limitations, a proteomics based approach has the unique advantage to identify changes in protein patterns (clusters) between different states of the organism.

Consequently, the screening for markers of disease has been one of the principal objectives in a large number of proteomics studies. This application uses a limited segment of the potential power of proteomics, which should be able to evaluate coherently the complex changes in the proteome (or a significant segment of it) in multifactorial diseases. This implies not only a gain in knowledge due to the massive increase in the data acquired from one set of experiments at one time point, but also provides additional information compared with conventional approaches by yielding insights into the complex interactions among different proteins and pathways.

This type of discovery is difficult to accomplish with reductionist methods and should improve our understanding of complex pathologies, like sepsis or acute lung injury, which involve multiple and constantly interacting components of the immune system and signalling pathways. There has been an expansion of proteomics into "functional proteomics," the correlation of changes in the proteome with different states of the organism. This field is currently expanding in several different dimensions. "Protein profiling techniques" take a global view at complex protein samples, such as plasma. Given the complexity of these samples, these techniques need to be streamlined to achieve high throughput.

The resulting protein patterns have diagnostic value as biomarkers on their own and indicate directions for more specific investigations. The application of protein profiling to tissue samples provides a combination of spatial information and protein profiles. The current results clearly indicate that these techniques are a valuable complement to histology.

The continuing improvement in protein identification will provide further insights into pathological processes and will most likely be especially valuable in cancer research. The application of mass spectrometry technology to the evaluation of "protein modifications" further extends the scope of proteomic analysis in depth. The physiological responses of an organism are only to a small part represented by changes in protein concentrations; especially, rapid responses to stimuli are transmitted by the modification of existing proteins. In spite of this complexity, this emerging field has, therefore, a large potential for clinically relevant research. The development of quantitative proteomics has widened the applicability of these techniques beyond a purely descriptive study design. Novel techniques in this field, namely differential gel electrophoresis (DIGE) and isotope-coded affinity tagging (ICAT), allow the direct comparison of samples, e.g., of different disease states.

16.7 Use of Mass Spectrometry in Proteomics

It is now possible to examine the expression of more than 1,000 proteins using mass spectrometry technology coupled with various separation methods. The typical proteomics experiment consists of five stages (Fig. 16.3).

In stage 1, the proteins to be analysed are isolated from cell lysate or tissues by biochemical fractionation or affinity selection. This often includes a final step of one-dimensional gel electrophoresis, and defines the "sub-proteome" to be analysed.

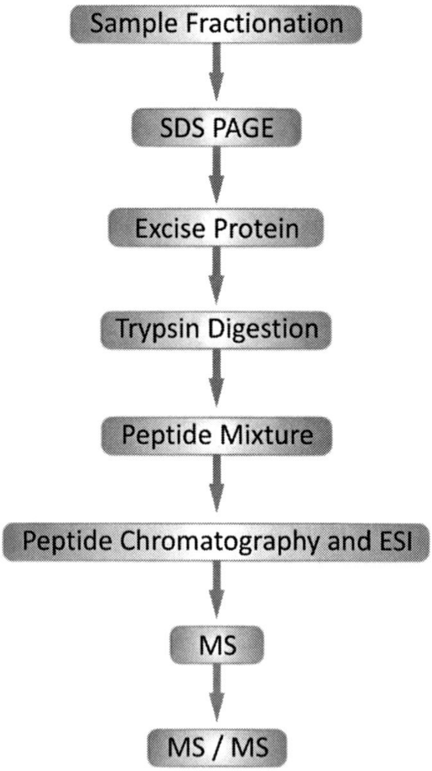

Fig. 16.3 The five steps of proteomics

MS of whole proteins is less sensitive than peptide MS and the mass of the intact protein by itself is insufficient for identification.

Therefore, proteins are degraded enzymatically to peptides in stage 2, usually by trypsin, leading to peptides with C-terminally protonated amino acids, providing an advantage in subsequent peptide sequencing. In stage 3, the peptides are separated by one or more steps of high-pressure liquid chromatography in very fine capillaries and eluted into an electrospray ion source where they are nebulized in small, highly charged droplets. After evaporation, multiple protonated peptides enter the mass spectrometer and, in stage 4, a mass spectrum of the peptides eluting at this time point is taken (MS1 spectrum, or "normal mass spectrum").

The computer generates a prioritized list of these peptides for fragmentation and a series of tandem mass spectrometric or "MS/MS" experiments follows (stage 5). These consist of isolation of a given peptide ion, fragmentation by energetic collision with gas, and recording of the tandem or MS/MS spectrum. The MS and MS/MS spectra are typically acquired for about one second each and stored for matching against protein sequence databases. The outcome of the experiment is the identity of the peptides and therefore the proteins making up the purified protein population.

16.8 Proteomics: From Basic Research to Diagnostic Application

The long-term challenge of proteomics is enormous: to define the identities, quantities, structures and functions of complete complements of proteins, and to characterize how these properties vary in different cellular contexts. One critical step in tackling this goal is the generation of sets of clones that express a representative of each protein of a proteome in a useful format, followed by the analysis of these sets on a genome-wide basis. Such studies enable genetic, biochemical and cell biological technologies to be applied on a systematic level, leading to the assignment of biochemical activities, the construction of protein arrays, the identification of interactions, and the localization of proteins within cellular compartments (Phizicky et al., 2003).

Within the last 5 years the field of proteomics has changed the understanding of molecular biology. Proteins manifest physiological as well as pathophysiological processes in a cell or an organism, and proteomics describe the complete protein inventory in dependence on in vivo parameters. Disease mechanism or drug effects both affect a protein profile and, vice versa, characterizing protein profiles reveals information for the understanding of disease and therapy. Analytical methods for proteomics are based on conventional tools for protein characterization. The technical challenge is the complete coverage of physico-chemical properties for thousands of proteins. Nucleic acids display a relative chemical homogeneity and therefore genomics was considered more promising in the past than proteomics. Further improvements in proteomics technologies will likely change this course with proteomics complementing genomics as a tool to study life sciences (Kellner, 2000).

Proteomics opens new horizons in many research areas of life sciences. This is particularly true for research efforts in the field of medicine. Clinical proteomics may be defined as a subset of proteomics activities in the field of medicine, which promises to accelerate the discovery of new drug targets and protein disease markers useful for in vitro diagnostics (IVD). IVD is based on the extracorporeal analysis of tissues and body fluids. Thus, it is expected that new pharmaceutical treatment opportunities will emerge and that the number and value of protein diagnostics will increase (Table 16.1).

The latter is of special interest, because reliable diagnostic information, in particular IVD data, is essential for choosing the appropriate intervention. Technologies applied in proteomics research, in particular SELDI-TOF MS and protein array techniques, are thought to be moving from research-focused applications to clinical laboratories as routine instruments for protein analysis. In conjunction with the routine implementation of such technologies in the clinical laboratory it has been argued that multimarker profiling approaches or pattern signatures will be the next generation of protein IVD's and shift paradigms in IVD. These goals and expectations for clinical proteomics should be assessed critically in view of the role of IVD in health care. The success of clinical proteomic marker searches and of the technologies applied in proteomics and their ability to enter routine IVD testing depends on the ability to fulfil IVD requirements and to adequately address IVD needs.

16.9 Role of In Vitro Diagnostics (IVD) in Health Care

IVD plays an essential role in health care (Fig. 16.4) by (1) providing reliable information about a person's condition and (2) supporting treatment and decision making, both occurring in conjunction with a physicians' examination, which may include anamnesis, physical examination, and in vivo diagnostics (e.g., computed tomography and nuclear magnetic resonance imaging).

A third purpose, which has been proposed by McNeil and Adelstein, is to better understand disease mechanisms and evolution. This last purpose is outside the realm of routine IVD, but is nevertheless important, since it should provide better IVDs that will hopefully enter that realm. IVD tests are estimated to contribute up to 94% of the objective data in clinical records and may influence 60–70% of critical decision making. In addition, major cost reductions within the worldwide health expenditures could be achieved. All the applications of IVD is given in Table 16.1.

Discovery of novel protein biomarkers is essential for successful drug discovery and development.

These novel protein biomarkers may aid accelerated drug efficacy, response, or toxicity decision making based on their enhanced sensitivity and/or specificity. These biomarkers, if necessary, could eventually be converted into novel diagnostic marker assays. Proteomic platforms developed over the past few years have

Table 16.1 Purposes for the application of IVD tests or respective marker types

Marker type	Examples
Acute markers are used when an acute disease event occurs and should help in the process of differential diagnosis to provide the information necessary for a specific treatment	cTns that are used to rule-in AMI and B-type natriuretic peptide (BNP) and N-terminal pro BNP (NT-proBNP) to rule-out heart failure in patients with acute dyspnoe
Screening markers identify diseased preferentially in an asymptomatic stage within a population to start treatment as soon as possible which usually goes along with a high treatment success. Generally, screening markers are applied in population subgroups with increased risk and disease frequency to achieve adequate PVs	The deletion of protein in urine, in particular albumin, is a screening tool for renal diseases
Primary risk assessment markers are used to assess the risk of a healthy individual to suffer from a disease in the future	Total cholesterol, triglycerides, LDL cholesterol, HDL cholesterol, high sensitivity C-reactive protein (hsCRP), among others are markers used for primary cardiovascular risk assessment
Secondary risk assessment, prognostic or progression markers are used to determine how the disease may potentially develop, the risk of a diseased patient to suffer recurrent or other disease events, etc.	Cardiac troponins are used in the secondary risk assessment for AMI patients as an indicator for the risk of adverse outcomes, i.e., morbidity and mortality. As primary risk assessment markers, secondary risk assessment markers help to determine aggressiveness of a treatment and balance the benefits and side effects
Disease staging or classifications markers help to classify different disease states	Determination of certain proteins in urine, kidney and urinary tract diseases can be classified into prerenal, renal, and postrenal
Treatment response stratification markers are used to predict the likeliness of a response to a pharmacon before its application to prevent adverse effects and to initiate the most effective therapeutic treatment	Hemostasis markers may be determined before a distinct anticoagulation therapy is started. Viral resistance markers and antibiotics resistance testing in bacteria are further examples
Treatment of therapeutic monitoring markers are used to determine and monitor the effectiveness of a treatment	Blood lipids are applied to follow the impact of exercise, nutrition, and eventually therapeutic interventions
Therapeutic drug monitoring is based on the determination of pharmaca administered to treat a patient. In this sense, it is not a marker type, but listed for the sake of completeness	Immnosuppressive drugs, like sirolimus, everolimus, tacrolimus, and cyclosporine A have to be monitored closely to prevent graft rejection and to minimize adverse therapy effects
Compliance markers provide information on treatment compliance	Glycosylated hemoglobin A (HbAlc) can be used to monitorinsulin therapy compliance80 and carbohydrate deficient tranferrin (CDT) may be applied to control compliance with alcohol withdrawal therapy

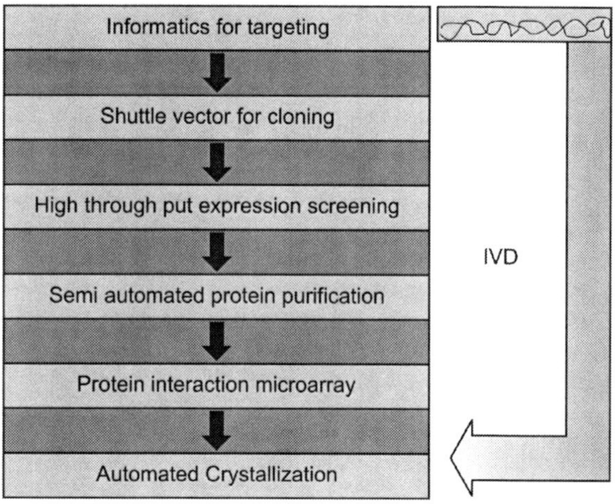

Fig. 16.4 The various applications of in vitro diagnostics

given us the ability to rapidly identify novel protein biomarkers in various biological matrices from cell cultures (lysates, supernatants) to human clinical samples (serum, plasma, and urine; Gao et al., 2005). The discovery of new highly sensitive and specific biomarkers for early disease detection and risk stratification coupled with the development of personalized "designer" therapies holds the key to future treatment of complex diseases such as cancer. Mounting evidence confirms that the low molecular weight (LMW) range of the circulatory proteome contains a rich source of information that may be able to detect early stage disease and stratify risk. Current mass spectrometry (MS) platforms can generate a rapid and high resolution portrait of the LMW proteome. Emerging novel nanotechnology strategies to amplify and harvest these LMW biomarkers in vivo or ex vivo will greatly enhance our ability to discover and characterize molecules for early disease detection, sub classification and prognostic capability of current proteomics modalities. Ultimately genetic mutations giving rise to disease are played out and manifested on a protein level, involving derangements in protein function and information flow within diseased cells and the interconnected tissue microenvironment. Newly developed highly sensitive, specific and linearly dynamic reverse phase protein microarray systems are now able to generate circuit maps of information flow through phosphoprotein networks of pure populations of micro dissected tumour cells obtained from patient biopsies. This type of technology will provide the foundation for the development of individualized combinatorial therapies of molecular inhibitors to target tumor-specific deranged pathways regulating key biologic processes including proliferation, differentiation, apoptosis, immunity and metastasis. Hence future therapies will be tailored to the specific deranged molecular circuitry of an individual patient's disease. The successful transition of these groundbreaking proteomic technologies from research tools to integrated clinical diagnostic platforms will require ongoing continued development, and optimization with rigorous standardization development and quality control procedures (Calvo et al., 2005).

Proteomic approaches to the identification of novel biomarkers for cancer diagnosis and staging have traditionally relied on the identification of differentially expressed proteins between tumor cells and their normal counterparts based on the patterns of protein expression observed by two-dimensional gel electrophoresis (2D-PAGE). Recent advances in mass spectrometry and in the informatics and statistical tools necessary to interpret mass spectrometric data have revolutionized the approach to defining new tumor markers. The combinations of SELDI mass spectrometry, retentate affinity chromatography, and statistical algorithms for pattern recognition have engendered a great deal of interest in "proteomic profiling" as a diagnostic too (Rodland, 2004). Recent advances in proteomic instrumentation and computational methodologies offer unique chance to rapidly identify these new candidate markers or pattern of markers. The combination of retentate affinity chromatography and surfaced-enhanced laser desorption/ionization time-of-flight (SELDI-TOF) mass spectrometry is one of the most interesting new approaches for cancer diagnostic using proteomic profiling (Solassol et al., 2005).

16.10 Goals of Proteomics

Functional proteomics is a promising technique for the rational identification of novel therapeutic targets by elucidation of the function of newly identified proteins in disease-relevant cellular pathways. Of the recently described high-throughput approaches for analyzing protein–protein interactions, the yeast two-hybrid (Y2H) system has turned out to be one of the most suitable for genome-wide analysis. However, this system presents a challenging technical problem: the high prevalence of false positives and false negatives in datasets due to intrinsic limitations of the technology and the use of a high-throughput, genetic assay (Colland and Daviet, 2004). The sequencing of the human genome and that of numerous pathogens has opened the door for proteomics by providing a sequence-based framework for mining proteomes. As a result, there is intense interest in applying proteomics to foster a better understanding of disease processes, develop new biomarkers for diagnosis and early detection of disease, and accelerate drug development. This interest creates numerous opportunities as well as challenges to meet the needs for high sensitivity and high throughput required for disease-related investigations (Hanash, 2003).

The goal of proteomics is a comprehensive, quantitative description of protein expression and its changes under the influence of biological perturbations such as disease or drug treatment. Quantitative analysis of protein expression data obtained by high-throughput methods has led us to define the concept of "regulatory homology" and use it to begin to elucidate the basic structure of gene expression control in vivo. Such investigations lay the groundwork for construction of comprehensive databases of mechanisms (cataloguing possible biological outcomes), the next logical step after the soon to be completed cataloguing of genes and gene products.

Mechanism databases provide a roadmap towards effective therapeutic intervention that is more direct than that offered by conventional genomics approaches (Anderson and Anderson, 1998).

The goal of interaction proteomics that studies the protein-protein interactions of all expressed proteins is to understand biological processes that are strictly regulated by these interactions. The availability of entire genome sequences of many organisms and high-throughput analysis tools has led scientists to study the entire proteome (Hanash et al., 2002). There are various high-throughput methods for detecting protein interactions such as yeast two-hybrid approach and mass spectrometry to produce vast amounts of data that can be utilized to decipher protein functions in complicated biological networks (Cho et al., 2004).

16.11 Conclusion

Complete description of the complex network of cellular mechanisms and use of the network to predict the full range of cellular behaviour are major goals of systems biology. A key role in contemporary biology will focus on the elucidation of protein functions and the definition of cellular mechanisms at the molecular level. The attainment of these targets is strictly dependent on the identification of individual proteins within functional complexes in vivo and can be played by functional proteomics (Monti et al., 2009).

Isolation of interacting proteins relies on either affinity-based or immunoprecipitation procedures in which the protein bait and its specific partners can be fished out by their specific binding to ligand molecules immobilized on insoluble supports. These approaches led to the final identification of several proteins belonging to distinct complexes endowed with different biological functions. Assignment of each protein to a specific complex constitutes a tremendous problem that can only be partially solved using protein-protein interaction databases and literature information. The development of prefractionation methodologies to separate individual protein complexes while preserving their native interactions might then represent an essential tool for the future of functional proteomics. Following prefractionation, the complex associated to a specific biological function can be isolated using affinity purification techniques. Functional proteomics approaches able to describe individual proteins belonging to complexes involved in specific cellular functions will have a terrific impact on future systems biology studies.

One of the most pressing issues facing the pharmaceutical and biotechnology industry is the tremendous dropout rate of lead drug candidates. Over the last two decades, several new genomic technologies have been developed in hopes of addressing the issues of target identification and lead candidate optimization. Tremendous progress has been made in the past few years in generating large-scale data sets for protein–protein interactions, organelle composition, protein activity patterns and protein profiles in cancer patients. But further technological improvements, organization of international proteomics projects and open access to results are needed for proteomics to fulfil its potential (Tyrers and Mann, 2003).

References

Anderson, N.L., Anderson, N.G., 1998, Proteome and proteomics: new technologies, new concepts, and new words. *Electrophoresis* 19(11):1853–1861.
Calvo, K.R., Liotta, L.A., Petricoin, E.F., 2005, Clinical proteomics: from biomarker discovery and cell signalling profiles to individualized personal therapy. *Biosci Rep* 25(1–2):107–125.
Cho, S., Park, S.G., Lee, D.H., Park, B.C., 2004, Protein–protein interaction networks: from interactions to networks. *J Biochem Mol Biol* 37(1):45–52.
Colland, F., Daviet, L., 2004, Integrating a functional proteomic approach into the target discovery process. *Biochimie* 86(9–10):625–632.
Cunningham, M.J., 2000, Genomics and proteomics: the new millennium of drug discovery and development. *J Pharmacol Toxicol Meth* 44(1):291–300.
Gao, J., Garulacan, L.A., Storm, S.M., Opiteck, G.J., Dubaquie, Y., Hefta, S.A., Dambach, D.M., Dongre, A.R., 2005, Biomarker discovery in biological fluids. *Methods* 35(3):291–302. Epub 12.
Hanash, S., 2003, Disease proteomics. *Nature* 422(6928):226–232.
Hanash, S.M., Bobek, M.P., Rickman, D.S., Williams, T., Rouillard, J.M., Kuick, R., Puravs, E., 2002, Integrating cancer genomics and proteomics in the post-genome era. *Proteomics* 2(1): 69–75.
Huang, J.X., Mehrens, D., Wiese, R., Lee, S., Tam, S.W., Daniel, S., Gilmore, J., Shi, M., Lashkari, D., 2001, High-throughput genomic and proteomics analysis using microarray technology. *Clin Chem* 47(10):1912–1916.
Kellner, R., 2000, Proteomics. Concepts and perspectives. *Fresenius J Anal Chem* 366(6–7): 517–524.
Monti, M., Cozzolino, M., Cozzolino, F., Vitiello, G., Tedesco, R., Flagiello, A., Pucci, P., 2009, Puzzle of protein complexes in vivo: a present and future challenge for functional proteomics. *Expert Rev Proteomics* 6(2):159–169.
Phizicky, E., Bastiaens, P.I., Zhu, H., Snyder, M., Fields, S., 2003, Protein analysis on a proteomics scale. *Nature* 422(6928):208–215.
Rodland, K.D., 2004, Proteomics and cancer diagnosis: the potential of mass spectrometry. *Clin Biochem* 37(7):579–583.
Solassol, J., Boulle, N., Maudelonde, T., Mange, A., 2005, Clinical proteomics: towards early detection of cancers. *Med Sci (Paris)* 21(8–9):722–729.
Tyers, M., Mann, M., 2003, From genomics to proteomics. *Nature* 422(6928):193–197.
Zhu, H., Bilgin, M., Snyder, M., 2003, Proteomics. *Ann Rev Biochem* 72:783–812.

Chapter 17
Nanomedicine

Abstract Nanomedicine is the medical use of molecular-sized particles to deliver drugs, heat, light or other substances to specific cells in the human body. Engineering particles to be used in this way allows detection and/or treatment of diseases or injuries within the targeted cells, thereby minimizing the damage to healthy cells in the body. Nanoparticles can attach to cells infected with various diseases and allow a lab to identify, in a blood sample, the particular disease.

Keywords Nanotechnology · Nanodiagnostics · Medicine · Nanometer · Nanotube · Nanoparticles · Quantum dots · Drug delivery · Nanospheres · Dendrimers · Therapeutics · Nanochips · Nanocarrier

17.1 Prologue

Scientists are working now to create novel nanostructures that serve as new kinds of drugs for treating cancer, Parkinson's and cardiovascular disease; to engineer nanomaterials for use as artificial tissues that would replace diseased kidneys and livers, and even repair nerve damage; and to integrate nanodevices with the nervous system to create implants that restore vision and hearing, and build new prosthetic limbs. Nanotechnology can help in the creation of functional materials, devices and systems through control of matter on the nanometer length scale (1–100 nm) exploiting novel phenomena and properties (physical, chemical, biological, mechanical, electrical...) at that length scale. The term "nanotechnology" is now referred to a future manufacturing technology based on molecular machine systems.

Nanotechnology has the potential to play a great impact on healthcare and agriculture. Nanomanufacturing integrates science and engineering knowledge and develops new processes and systems to assure quality nanomaterials, to control the assembly of molecular-scale elements, and to predictably incorporate nanoscale elements into nano-, micro-, and macroscale products utilizing new design methods and tools. Efforts in this area are directed toward enabling the mass production of reliable and affordable nanoscale materials, structures, devices, and systems for healthcare.

M. Debnath et al., *Molecular Diagnostics: Promises and Possibilities*,
DOI 10.1007/978-90-481-3261-4_17,

The challenge for nanotechnology is whether these principles can be used to engineer novel constructs in addition to natural ones. Examples of nanotechnology in modern use are the manufacture of polymers based on molecular structure, and the design of computer chip layouts based on surface science. Despite the great promise of numerous nanotechnologies such as quantum dots and nanotubes, real commercial applications have mainly used the advantages of colloidal nanoparticles in bulk form, such as suntan lotion, cosmetics, protective coatings, drug delivery, and stain resistant clothing.

17.2 Concept

Nanotechnology is the way discoveries made at the nanoscale is put to work. Nanotechnology will soon allow many diseases to be monitored, diagnosed and treated in a minimally invasive way and it thus holds great promise of improving health and prolonging life. Whereas molecular or personalized medicine will bring better diagnosis and prevention of disease, nanomedicine might very well be the next breakthrough in the treatment of diseases.

This scenario challenges our current concept of treatment, whereby physicians identify a disease and prescribe a drug to treat it. Quite probably, the concept of a drug targeted against a single illness will no longer be accurate. Many diseases induce changes to cells and tissue, such as those that occur in tumor development, which involve numerous genes and proteins. Even a simple infection triggers a complex cascade of gene and protein actions both in the infected cells and in the cells of the immune system. Instead of using a therapeutic that tackles rapidly dividing cells to kill tumor cells – which in the course of treatment affects normal body cells – or an antibiotic that kills beneficial bacteria as well, it would be much safer and more efficient to use therapeutics that target only the molecules that are involved in pathogenesis. It is logical to assume that, as soon as we know the precise sequence and pattern of molecular events for a specific disease, we can devise such drugs to interfere with pathological processes and correct what has gone wrong.

Some of these new, so-called "smart"; drugs are already available or are under development. They include small peptides that target specific surface receptors on cancer cells and also gene therapeutics – gene constructs antisense oligonucleotides and small interfering RNAs (siRNAs) – that induce changes specifically in metabolism or gene expression. Another example is the protease inhibitor that blocks HIV protease and thus prevents the virus from maturing (Friedman et al., 1993). On the diagnostic side, there are further advances: new diagnostic tools based on gene or gene expression analyses allow physicians to select the most effective drugs to treat various cancers or to circumvent drug resistance in anti-HIV therapy. These are the first harbingers of this new concept of medicine, and the next few years will probably see the arrival of many more peptides, antibodies and other molecules.

However, no matter how good and "smart" these therapeutics are, the problem remains of how to target these drugs to the appropriate cells or tissues. Furthermore, many of these drugs require carrier systems as they are either hydrophobic or need to be protected from enzyme destruction while they are on their way to the target cells. Even existing therapeutics of organic and inorganic origin could be vastly improved in terms of safety and efficacy using methods that deliver them safely to their designated area. Regardless of whether a "smart" drug is new or old, one of the main challenges for biomedical research is the refinement of drug delivery (Hong and Claymann, 2000). The efficiency of chitosan encapsulated DNA based RSV vaccine was studied by Boyoglu et al. (2009). Antigenic regions of RSV F, M2, and G genes were cloned into the phCMV1 vector resulting in a DNA vaccine vector named DR-FM2G. This vector was used to formulate DNA/Chitosan nanoparticles (DCNP). Differential Scanning Calorimetry (DSC) showed DCNP to be more stable than naked DNA or chitosan, offering protection of DNA degradation by nucleases.

It is therefore not surprising that targeted drug delivery has become a leading area in medical research. However, many technologies that are being exploited have various limitations, and it seems that appropriate solutions could be found in nanotechnology. So, what is it that makes nanotechnology appealing for drug delivery? First, it makes it possible to produce devices and constructs smaller than 1 μm, which enhances function and creates completely new material properties. Second, it fits neatly in the biological setting, as biological molecules and cellular structures are about 100 nm or smaller – the size scale of this technology. Nanotechnology also creates new possibilities (Fig. 17.1) in engineering devices with enormous precision through self-assembly and directed assembly (Hilt, 2004). In terms of drug delivery, it offers various advantages over traditional systems. Nanoscale drug-delivery

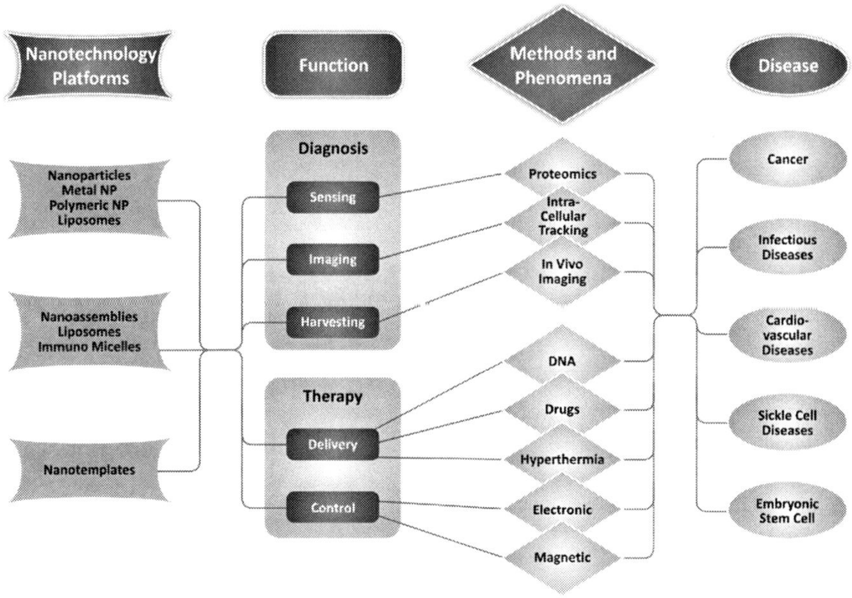

Fig. 17.1 Nanotechnology has diverse functions in detection and diagnosis of diseases

devices – nanocarriers – are able to overcome biological barriers and may provide targeted delivery of drugs because of their small size. They may also be able to solubilize drugs for intravascular delivery, improve the stability of therapeutic agents and control drug release in target tissues for optimal therapeutic efficacy. In a recent study polymeric micelles were used as core-shell nanoparticles that offer considerable advantages for cancer diagnosis and therapy (Le Garrec et al., 2004). Their relatively small size (10–100 nm), ability to solubilize hydrophobic drugs as well as imaging agents, and improved pharmacokinetics provide a useful bioengineering platform for cancer applications. Several similar polymeric micelle formulations are currently undergoing phase I/II clinical trials, which have shown improved antitumor efficacy and reduced systemic toxicity (Blanco, 2009).

17.3 Nanotechnology in Medicine

The rapid progress of nanotechnology-based applications in medicine is due to the huge interest from therapeutic and diagnostic companies. So far, nanotechnology has been an area of "technology push", with substantial investments in both generic research and technology development, in the absence of clear market demand. Worldwide, private investments in research have recently overtaken public funding, but the potential advantages of nanotechnology-based products compared to other alternatives are not yet clear. Also, there is a growing global debate on the ethical, legal and social aspects of nanotechnology, in particular the potential risks to human health and the environment posed by engineered nanomaterials. Nanotechnology is still mainly a solution looking for problems to solve, including sustainable development issues. These nanoparticles act as nanoscopic medicine (Fig. 17.2). Qdots identify the location of cancer cells in the body. Nanoparticles deliver chemotherapy drugs directly to cancer cells to minimize damage to healthy cells. Nanoshells

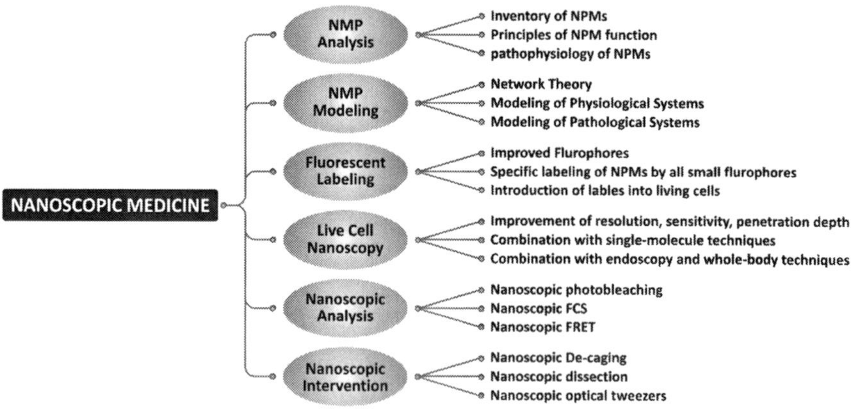

Fig. 17.2 Nanoscopic medicine and its applications

concentrate the heat from infrared light to destroy cancer cells with minimal damage to surrounding healthy cells. Nanotubes are used in broken bones to provide a structure for new bone material to grow.

Worldwide, governments and private companies invested about €4 billion (approx. $5.6 billion) in nanotechnology research and development in 2004, and this figure is likely to rise dramatically in the coming years. Hundreds of nanotechnology-based products are already on sale, and many more are in the pipeline. About 65 countries, including EU member states, Japan and the United States, as well as developing countries and emerging economies, are currently funding nanotechnology research (Table 17.1). Let us discuss some of the latest work in the field of nanomedicine.

The main interests currently lie in improving diagnostic methods and in developing better drug delivery systems to improve disease therapy. More generally, the scientific community is increasingly focusing its attention on the novel chemical and physical properties of nano-sized materials to develop new applications in regard to human health.

Table 17.1 Nanotechnology in medicine: company directory

Company	Product
CytImmune	Gold nanoparticles for targeted delivery of drugs to tumors
Nucryst	Antimicrobial wound dressings using silver nanocrystals
Nanobiotix	Nanoparticles for targeted delivery of drugs to diseased cells
Oxonica	Disease identification using gold nanoparticles (nanotags)
Nanotherapeutics	Nanoparticles for improving the performance of drug delivery by oral, inhaled or nasal methods
NanoBio	Antimicrobial nanoemulsions for nasal delivery to fight viruses (such as flu and colds) and bacteria
Novavax	Drug delivery through the skin using micellar nanoparticles and oral drug delivery with propriety structures called Novasomes
BioDelivery Sciences	Targeted oral drug delivery to diseased cells using a nanocrystalline structure called a cochleate

17.3.1 Use of Quantum Dots

Nanotechnology today deals mainly with two rather different but complementary types of material: nano-sized structures (or nanoparticles) and nanoporous materials. There are already some exciting developments in the field of diagnostics based on the use of nanoparticles; in particular fluorescent semiconductor quantum dots (QDs). QDs are monodisperse inorganic nanocrystalline particles made from semiconducting material and are typically 2–10 nm in size – about the size of a protein or a short sequence of DNA. They can be linked to biomolecules to form sensitive long-lived probes that target and identify specific cellular compounds. As fluorescent probes, QDs have several advantages over conventional organic dyes: their emission spectra are narrow and symmetrical on the basis of their size and material composition, and they exhibit excellent photostability. In addition, they display broad absorption spectra, which make it possible to excite many QDs to different colours with a single excitation light source (Wu et al., 2003). This is certainly an advantage in studying multiple biological targets simultaneously in the cell. The high photo stability of QDs also allows real-time monitoring or tracking of intracellular processes in vivo over extended periods. Dubertret et al. (2002) showed this in a breakthrough experiment by labeling a living frog embryo with more than a billion individual QD particles encapsulated in phospholipids copolymer micelles. Their experiments showed that these QD micelles are non-toxic and stable in biological environments, which suggests the possibility of tracing cell lineages in embryogenesis experiments. Bioconjugated QDs have also been used for DNA hybridization and high-throughput genotyping of single-nucleotide polymorphisms (SNPs), the most common type of genetic variation between individuals. SNPs are very good markers for disease-causing genes, and they hold further potential for personalized medicine as markers for differential drug responses (Xu, 2003). As the labeling of individual molecules or cell structures in living cells or tissues is becoming an increasingly important tool in diagnostics, QDs, because of their many advantages over organic dyes, have a large potential for new and improved diagnostic tests in medicine.

The small size of nanoparticles endows them with properties that can be very useful in oncology, particularly in imaging. Quantum dots (nanoparticles with quantum confinement properties, such as size-tunable light emission), when used in conjunction with MRI (magnetic resonance imaging), can produce exceptional images of tumor sites. These nanoparticles are much brighter than organic dyes and only need one light source for excitation. This means that the use of fluorescent quantum dots could produce a higher contrast image and at a lower cost than today's organic dyes. Another nanoproperty, high surface area to volume ratio, allows many functional groups to be attached to a nanoparticle, which can seek out and bind to certain tumor cells. Additionally, the small size of nanoparticles (10–100 nm), allows them to preferentially accumulate at tumor sites (because tumors lack an effective lymphatic drainage system). A very exciting research question is how to make these imaging nanoparticles do more things for cancer. For instance, is it possible to manufacture multifunctional nanoparticles that would detect, image, and then proceed to treat a tumor? This question is currently under vigorous investigation; the answer to which could shape the future of cancer treatment.

17.3.2 Cell Targeting

One of the most important problems in medicine is the proper distribution and targeting of drugs and other therapeutic agents within the patient's body. Increasing the efficiency of delivering therapeutic molecules to their final target cells is therefore a priority for the pharmaceutical industry. A large amount of research is being undertaken to develop new "advanced" materials that incorporate a biologically active substance, transport it to the right place and release it there in a controlled way. This concept is particularly interesting for potent but non-specifically acting drugs, such as anticancer drugs, that cause several severe side effects in healthy tissues or cells. Cell targeting can be achieved either by "passive" methods, in which a polymer conjugate or particle system is captured by a physiological uptake mechanism of the target cell, or by "active" targeting – the attachment of a homing moiety, such as monoclonal antibodies or ligands in the form of sugars or lectins – to deliver the drug to the right cell by attaching it to specific receptors on cell surfaces (Davis, 1997).

It is just a little step further to combine these two characteristics – attaching or encapsulating the therapeutic agent within a carrier material and providing this material with the means to home in on target cells – to create a drug-delivery vessel that will increase drug efficiency while circumventing side effects in non-targeted tissues. One possibility is to use homing peptides with high selectivity in vivo. Blood vessels, for instance, express molecular markers that distinguish between the vasculatures of individual organs, tissues and tumors, and unique sets of peptides with high selectivity for these markers have been identified. Akerman et al. (2002) showed the effective targeting of the vasculatures of normal lungs and tumors in a living mammalian organism by using peptide-coated QDs. These results opened up the possibility of using other nanomaterials with more far-reaching properties, for example a material that could target tumor cells through the folate receptor, while also having a fluorescent label for imaging and tracking of the drug. Quintana et al. (2002) developed dendrimer-based nanodevices optimized for the intracellular targeting of drugs, imaging agents and other materials.

17.3.3 New Nanocarriers for Drug Delivery

In the good old days of nanotechnology, some 2–3 years ago, it became clear that researchers were making fascinating progress in revolutionizing drug delivery by using nanoparticles as delivery agents. Previously, the ideal drug carrier was something out of science fiction when injected into the body it transports itself to the correct target, such as a tumor, and delivers the required dose at this target. With the advent of nanomedicine, this idea, nicknamed the "magic bullet" concept, is rapidly becoming a reality. Currently used pharmaceutical nanocarriers like liposomes, micelles, nanoemulsions, polymeric nanoparticles and many others demonstrate a broad variety of useful properties, such as for instance increased longevity in the blood, specific targeting to certain disease sites, or enhanced intracellular penetration. Researchers have also begun to combine several properties in order to develop multifunctional nanocarriers. Overall, these nanocarriers already have proven quite successful in practice with great potential for the development of novel molecular-based diagnostics and therapeutics. On the other hand, they could also be useful in understanding nanotoxicity. The scientists found that the binding and activation of membrane receptors and subsequent protein expression strongly depend on nanoparticle size. Although all nanoparticles within the 2–100 nm size range were found to alter signaling processes essential for basic cell functions – including cell death – 40- and 50-nm nanoparticles demonstrated the greatest effect.

Not surprisingly, drug-delivery systems based on nanotechnology have attracted much commercial interest in the pharmaceutical industry (Oh et al., 2005). According to a report, the global market for drug-delivery products and services is projected to exceed US$67 billion in 2009 (NanoMarkets, 2005). The market value for delivery technologies in cancer treatment alone was US$2.6 billion in 2002, and is expected to grow to US$15.4 billion. As a part of this, nanotechnology-based drug-delivery systems are expected to create a market of about US$1.7 billion by 2009 and more than US$4.8 billion by 2012.

Some of the systems already used in drug delivery are nanospheres and nanoparticles – both composed of oligomeric or polymeric units – and nanocapsules consisting of a hydrophobic core surrounded by a polymer wall. These drug carriers can entrap, dissolve or encapsulate therapeutic molecules, chemically attach them or adsorb them at the surface. Many polymers, such as polyvinyl alcohol, polyethylene glycol (PEG), poly-*N*-vinyl pyrolidone, polyethyleneimine (PEI), polyethylene oxide (PEO), phosphatidylethanolamine (PE), polylactides (PLA) and polylactide-co-glycolides (PLGA) are already in use. For example, two newly synthesized nanoparticles, methoxypolyethylene glycol(MPEG)–PLA copolymer and pluronic PEO–polypropylene oxide–PEO triblock copolymer, are now being investigated for the delivery and controlled release of paclitaxel, which could reduce unwanted side effects and improve therapy (Dong and Feng, 2004).

However, a major drawback of polymer carrier systems is their low drug-carrying capacity. Each molecule of PEG, for example, can carry only two drug molecules. Thus, polymeric micelles are spherical, colloidal particles spontaneously formed by amphiphilic molecules in water and are progressively being exploited, as their inner cores have a higher drug-loading capacity. Micelles also increase the solubility and stability of hydrophobic agents, improve their pharmacokinetics and decrease the toxicity of the drug. The first generation of drug-delivery devices for anticancer therapeutics were liposomes – colloidal, vesicular structures based on phospholipid bilayers of different sizes, some as small as 20 nm. But, as liposomes are made of phospholipids, they are taken up by macrophages. Additionally, they move from blood vessels into other tissues, such as the liver and spleen. Several researchers have therefore tried to increase the circulation time of liposomes in the bloodstream by adding different molecules that reduce their uptake by macrophages. However, lipid carriers, even if easy to produce, often display toxicity that limits their use. Poly *N*-vinyl pyrrolidone block poly D,L-lactide and has been suggested as an alternative drug carrier system. In water, it self-assembles into micelles that efficiently solubilize hydrophobic anticancer drugs, such as paclitaxel and etoposide. In vitro and in vivo studies showed that the micelles release paclitaxel in a controlled manner and are stable in aqueous media for several months.

Platins are widely used cancer drugs that could benefit from improved drug navigation to reduce their severe side effects. Several carrier systems, such as micelles and liposomes, have been used to deliver platin drugs, but these formulations have shown a higher selectivity for the liver and spleen than for tumour tissue. A recent in vivo study used PEG–poly(glutamic acid) block copolymer micelles as carriers, which accumulated significantly and selectively in solid tumours. Administration of these cisplatin-incorporated micelles caused total eradication of primary tumours.

As drugs become larger and less water-soluble, new delivery systems such as nanoporous materials are needed. These have defined pore sizes of 1–100 nm or larger, and are being studied particularly for controlled drug release. By making the pores only slightly larger than the drug molecule, the rate of diffusion of the molecules can be controlled, regardless of the amount of the drug inside the capsule. Among various approaches, dendrimers are polymers with a branching structure that are synthesized one step at a time to add a new monomer layer with each synthesis – have recently gained in popularity due to their low systemic toxicity and high drug-load ability. Dendrimers have a central core, an interior branched structure and external reactive groups, and move through vascular pores and into tissue more efficiently than larger carriers. Several researchers are therefore studying the possibility of using them as drug-delivery devices, including methotrexate (Quintana et al., 2002), indomethacin (Chauhan et al., 2004), paclitaxel and doxorubicin.

However, the synthesis of dendrimers is a long, difficult and expensive process, and the yield of functional particles decline successively with each synthesis step. A new, less time-consuming and more efficient method for dendrimer synthesis is consequently a promising development. Choi et al. (2005) prepared two types of dendrimer that carry complementary single stranded, non-coding DNA. The DNA molecules bound to each other in solution, thus forming two-dendrimer complexes. One dendrimer was prepared for imaging applications and the second was specifically designed to target cancer cells through the folate receptor. The self-assembled dendrimer clusters were shown to be well formed and functional.

Some of the most advanced cutting-edge technologies to synthesize such complex drug-delivery vessels involve the use of dendrimers and fullerenes when used in combination with peptides (Akerman et al., 2002). Dendrimers are polymeric monodisperse macromolecules with a precisely defined chemical structure. These branching synthetic molecules can be grown nanometre by nanometre to reach the desired size. They have a spherical shape and sufficiently large openings and cavities to carry small molecules. Dendrimers can be used as potential drug-delivery agents in at least two ways: drug molecules can be either physically trapped inside the molecule or covalently attached to the surface. In both cases, the dendrimer backbone serves just as a scaffold to arrange the functional units in space, without having any other inherent function. At present, the most active research area in dendrimer-based therapeutics is the entrapment of drugs for subsequent and controlled release, as has been shown with the cytostatic agents adriamycin and methotrexate. Another potential application could be the use of dendrimers coated with sialic acid – crucial for the influenza virus to attach to the cell surface – to "fool" the virus into attaching itself to these dendrimers instead of to the cell. Dendrimers have also been used commercially as non-viral vectors for DNA delivery. In addition, dendrimers are non-immunogenic and are thus uniquely suited as carrier structures for drugs or bioactive molecules without falling prey to the immune system (Freitas, 2002).

Several research groups have been testing another, completely different, class of molecules for drug delivery that has also raised huge interest among chemists and physicists: fullerenes. Fullerenes and their relatives – endohedral and metal-coated fullerenes, carbon nanotubes, carbon nanoparticles and porous carbons – are carbon-based nano-sized structures that represent the third allotropic form of elemental carbon after graphite and diamond. By far the most common is C_{60}, also called a "buckyball". Foley et al. (2002) proposed using fullerenes as drug-delivery agents because their structure mimics the clathrin scaffolds that mediate endocytosis. They showed that a fullerene derivative is able to cross the external cellular membrane and that it localizes preferentially to mitochondria. However, drug delivery by fullerenes is still confined to the research laboratory and it will take many years of research to develop the first clinical applications. Nevertheless, fullerenes have been shown to hold promise as radical scavengers for the protection of neurons, for the inhibition of HIV protease and as an antibacterial agent. Another possible application is to be found in nuclear medicine, in which fullerenes could be used as an alternative to chelating compounds that prevent the direct binding of toxic metal ions to serum components. This could increase the therapeutic potency of radiation treatments and decrease their adverse effect profile, because fullerenes are resistant to biochemical degradation within the body (Cagle et al., 1999).

17.3.4 Gene Therapy

Gene therapy is another application that might benefit from nanotechnology, given that it faces the same problems of delivering a therapeutic agent. Although the classic use of gene therapy is to treat monogenetic diseases by supplying a correct form of the defective gene, most of the approved and tested gene therapy protocols available at present aim at treating cancer patients. Nevertheless, the main problem now in gene therapy is to find safe methods for delivering genes to human cells. Three main types of gene-delivery systems have been described: viral vectors, non-viral vectors (Nakanishi, 2003) and the direct injection of corrective DNA or RNA into tissues. There are several serious questions about the safety of viral vectors, such as immunogenicity and the reversion of the engineered form to the wild type. Non-viral vectors and direct injection seem in many ways advantageous with regard to safety, but they are less effective (Davis, 1997). This is where nanotechnology could provide a solution to this problem.

Another important research field is nucleic-acid delivery for gene therapy. With the development of genomics and proteomics, exciting new possibilities for these therapeutic approaches have arisen (Kraljevic et al., 2004). But the clinical use of gene therapy will depend strongly on the development of safe and efficacious delivery systems to protect and deliver the drug to the target cell, as the carrier system has to overcome many natural barriers. When arriving at the target cell, it must be taken up through endocytosis and escape the endosome by disrupting the membrane; only then can it deliver the nucleic acid to the nucleus. The clinical use of gene therapy will depend strongly on the development of safe and efficacious delivery systems to protect and deliver the drug to the target cell.

Two groups of vectors have been used for gene delivery: viruses and non-viral vectors. During recent years, it has become obvious that viruses harbour the risk of unwanted immune responses and have inadequate specificity for cell targeting. Non-viral systems could overcome some of these problems. Most importantly, they can be used with a higher degree of safety, as they do not integrate into the host chromosome and cause fewer problems with immune response. They can also carry more nucleic acids than viruses can, which allows the delivery of larger genes. Additionally, they are suitable for attaching various types of targeting molecule and provide better cell-type specificity. Finally, synthetic systems are much easier to characterize and control than viral vectors.

Some promising non-viral delivery systems include cationic lipids and polymers, which bind negatively charged DNA to form stable complexes due to their charge. Csaba et al. (2005) combined PLGA with polyoxyethylene derivatives (poloxamers and poloxamines) to create a carrier system that was able to encapsulate plasmid DNA and was well tolerated in cell culture. In addition, PEG–PEI nanoplexes were successfully tested as delivery vehicles for siRNA to tumour neovasculature. The siRNA nanocomplex self-assembled by exploiting the PEI's cationic domain, which binds to negatively charged nucleic acids. PEG was used for steric stabilization and further ligation of the targeting peptide sequence. Similarly, PLGA nanospheres containing phosphothionated antisense DNA were shown in animal experiments to be an effective gene-delivery system for the treatment of restenosis – the narrowing of coronary arteries after angioplasty. Another group prepared a protein-based nanocarrier using albumin matrices to entrap different antisense oligonucleotides. The results were promising, as human albumin molecules showed good biocompatibility and transfection efficacy. However, a basic problem is the limited knowledge of the biological effects of polymeric complexes formed with DNA. For example, complexes formed by combining PEI and DNA have a higher transfection efficiency if the polymer is added to the DNA during preparation and not vice versa (Merdan et al., 2002).

Dendrimers are also appropriate vehicles for gene delivery (Cohen-Sacks et al., 2002; Ooya et al. 2004). Kim et al. (2004) have synthesized a new triblock polyamidoamine (PAMAM) dendrimer copolymer with PEG, which formed highly watersoluble polyplexes with plasmid DNA. It revealed little cytotoxicity despite its poor degradability and achieved high transfection efficiency. Cationic liposomes, such as Lipofectin®, are already being tested in clinical trials of non-viral gene therapy because of their low immunogenic response. These lipid-gene complexes have the potential to transfer large pieces of DNA – up to 1 million base pairs – into cells. However, their transfection efficiencies are still low when compared with viral vectors, and problems with cytotoxicity remain. According to a recent study, it seems that, at present, the limiting factor for cationic lipid carriers seems to be the tight association of a fraction of the delivered exogenous DNA with cationic cellular molecules, which may prevent optimal transcriptional activity (Ewert et al., 2004).

Antisense DNA is a potential gene-specific therapeutic agent. These short oligodeoxynucleotide fragments are complementary in sequence to a portion of the targeted mRNA. The aim of antisense therapy is to use them to interfere with gene expression, thereby preventing the translation of mRNA into protein (Lambert et al., 2001). There are some problems in antisense therapy, mainly the poor stability of oligonucleotides against nuclease activity and their low intracellular penetration, which limits their therapeutic use. To solve these problems, particulate nano-sized carriers have been proposed as a delivery method. Cationic polymerized lipid-based nanoparticles, for instance, when coupled to the $\alpha_V\beta_3$-integrin-targeting ligand, can deliver genes selectively to angiogenic blood vessels in tumour-bearing mice. The $\alpha_V\beta_3$-integrin receptor is strongly expressed in endothelial cells and has a key function in cell survival during angiogenesis in vivo. It might be particularly useful for gene-delivery strategies to endothelial cells because it potentiates the internalization of several viruses, thereby facilitating gene transfer. Hood et al. (2002) tested its therapeutic efficacy by generating $\alpha_V\beta_3$-ligand nanoparticles conjugated to a mutant *Raf* gene, which blocks endothelial signalling and angiogenesis in vivo. *Raf* is a component of the signalling pathway that is important in neovascularization, and blocking this pathway suppresses angiogenesis. The delivery of a mutant *Raf* to angiogenic blood vessels blocks endothelial *Raf* activity and induces apoptosis. Systemic injection of the $\alpha_V\beta_3$-nanoparticle/*Raf* into mice resulted in the regression of established primary and metastatic tumours.

Choy et al. (2000) used inorganic layered double hydroxides (LDH) that consist of cationic brucite-like layers and exchange-able interlayer anions. The unique ion-exchange capability of LDHs enables these inorganic matrices to encapsulate functional biomolecules, such as antisense DNA, that are negatively charged in aqueous media, by a simple ion-exchange reaction to form bio-LDH nanohybrids. These hydroxide layers protect the intercalated antisense molecule from degradation by DNase. Once LDH-antisense hybrids have entered the cell, the hydroxide layers are removed by dissolution in the lysosome, where the pH is slightly acidic, and the encapsulated biomolecules are released inside the cell.

17.3.5 Molecular Medicine

Nanoporous materials, another class of inorganic molecules, have recently attracted interest for use in molecular medicine. Such materials, with pore sizes ranging from less than 2–50 nm, include zeolites and related molecular sieves. Zeolites are naturally occurring or synthetic crystalline aluminosilicates. The pore network of a typical zeolite, which is confined by the framework, consists of cavities and connecting windows of uniform size. As a result of these unique properties, zeolites can absorb gas and water molecules, facilitate ion exchange and act as "molecular sieves" with long-term chemical and biological stability. Zeolites have therefore already become interesting subjects in different areas of chemical research and are now widely used in industrial, agricultural, environmental and biological technology (Mumpton, 1999). There are obvious structural similarities between the cages of zeolites and the substrate-binding sites of enzymes, which have led to the development of zeolite structures that mimic enzyme functions. Another approach would be to incorporate key features of selected enzymes, such as metal complexes, in zeolites – so-called "ship-in-a-bottle" complexes – that could be used as oxygen carriers, mimicking haemoglobin, cytochrome P450 and iron–sulphur proteins (Bedioui, 1995).

Several toxicological studies of a natural zeolite named clinoptilolite proved that this compound is non-toxic and safe for use in human and veterinary medicine. Recent experiments in vitro and in vivo have suggested that clinoptilolite could be used as an adjuvant in anticancer therapy (Pavelic, 2001). So far, zeolites have been successfully used as detoxicants and decontaminants – when added to animal diets they reduce the level of heavy metals and organopoisoning, radionuclides and ammonia – as well as antibacterial and antidiarrhoeal agents. Zeolites have been used for haemodialysis, and also in cartridges in haemoperfusions. Zeolite powder has been shown to be effective in the treatment of athlete's foot and in reducing the healing time of wounds and surgical incisions. Tissue conditioners containing silver-exchanging zeolites showed a strong antimicrobial effect. On the basis of these findings and positive double-blind clinical studies, the drug Zeomic (Sinamen-Zeomic, Nagoya, Japan) was recently approved in Japan as an antimicrobial agent for dental treatments (Pavelic and Hadzija, 2003).

17.3.6 Construction of Nanostructure Template

In the long term, nanotechnologies will further enable the construction of clearly defined two-dimensional and three-dimensional structures on inorganic, organic or biological templates. The fabrication of such nanostructures on various substrates can be achieved by exploiting a variety of procedures based on either molecular self-assembly or micropositioning technology. Particularly interesting is the process of molecular self-assembly, in which molecules – or parts of molecules – spontaneously form ordered aggregates without any human intervention. Such self-assembly is centrally important in many processes in life. The formation of molecular crystals, colloids, lipid bilayers and phase-separated polymers, the folding of nucleic acids into their functional forms, and the association of a ligand with its receptor are all examples of self-assembly (Whitesides and Boncheva, 2002; Zhang et al., 2002). Consequently, the term "self-assembly" was, until recently, applied almost exclusively to biological structures. However, self-assembling materials are becoming a significant topic in current research and it is assumed that in the twenty-first century they might become new building blocks comparable to those of alloys, plastics and semiconductors in the twentieth century.

Numerous self-assembling systems have already been developed: block copolymers, surfactant-like materials, scaffolds for three-dimensional cell culture, DNA-based structures and models with which to study protein folding and protein conformational diseases, such as Alzheimer's disease. A research team at the Scripps Research Institute (La Jolla, CA, USA) has developed antibiotic agents based on self-assembling cyclic peptide nanotubes that attach to, and poke holes through, bacterial cell membranes, thus killing the cell. These self-assembling peptide nanotubes cleared infections of antibiotic-resistant bacteria in mice, even when injected far from the site of infection (Fernandez-Lopez et al., 2001). Another promising example is a vaccine consisting of self-assembling virus-like particles for the prevention of infection of the genital tract by human papilloma virus, which can cause cervical cancer. Such particles are now being developed by MedImmune (Gaithersburg, MD, USA) and GlaxoSmithKline (Uxbridge, UK). Moreover, self-assembling biological materials are of great interest in advanced medicine because they can serve as bioactive extracellular matrices (ECMs). They provide cells and tissues with the appropriate three-dimensional architecture for normal growth and development, which is crucial for the proper understanding required for any successful clinically relevant therapies and for targeted drug delivery. PuraMatrix (3DM, Cambridge, Massachusetts, USA), for example, is one commercially available line of synthetic ECMs (Holmes, 2002).

17.3.7 Targeting Drugs to Cells

A further challenge for improving drug delivery is creating carriers that specifically home in on target cells to deliver the drug while leaving normal cells untouched (Ravi et al., 2004). There are two possible strategies to achieve this goal, namely passive and active targeting. Passive targeting requires an increase in the stability of the drug carrier in the body. As nanoparticles tend to accumulate in and around cancer tissues, the longer the circulation time, the more drug carriers accumulate at the tumour. Additionally, nanoparticles can be coated with PEG to reduce non-specific interactions and absorption by endothelial cells (Otsuka et al., 2003).

Active targeting is achieved by incorporating homing moieties, which facilitate uptake by certain tissues or cell types into the drug-delivery vehicle. Such vehicles concentrate the drug at the target site, thus increasing efficacy and decreasing side effects. Much emphasis has been placed on finding adhesion molecules – antibodies and so-called small "homing peptides" – that are specific for different tissues, particularly for the vasculature. To find such markers, phage display libraries of small peptides were screened in live mice, and peptides that directed phages to a specific target in the body were selected. Interestingly, most organ- and tumourspecific receptors found so far have been peptidases. For example, dipeptidyl peptidase IV and membrane dipeptidase are selectively expressed in lung vessels, whereas aminopeptidase P was found to be expressed in breast gland vasculature. Essler and Ruoslahti (2002) also identified a nonapeptide that targets aminopeptidase P in the breast vasculature in mice. Their results showed that this peptide was able to target not only normal breast tissue but also premalignant tissue and primary breast tumours.

Targeting drugs to blood vessels holds great promise (Orosz et al., 2004). Because tumour vessels express specific markers of angiogenesis, cancer chemotherapy directed against angiogenic vessels induces tumour regression while avoiding drug resistance. Some tumour-homing peptides that have already been successfully tested are the RGD-motif tripeptide that binds to integrins $\alpha_V\beta_3$ and $\alpha_V\beta_5$, both overexpressed in tumour vasculature; the NGR-motif tripeptide that binds to aminopeptidase N, a membrane protein that is specific for angiogenesis if expressed in the vasculature; and peptides directed to nucleolin, a new marker for angiogenic vessels. Additionally, peptides that target the vasculature of various tumours have been recently discovered, such as markers for pancreatic islet cell tumours and skin tumours (Chen et al., 2001). Some authors have already successfully used NGR-motif peptides to target drug-delivery vehicles loaded with tumour necrosis factor and tachyplesin, which are both anticancer compounds (Curnis et al., 2000). Another homing device that is similarly interesting is a new pentapeptide; liposomes carrying this peptide and adriamycin effectively suppressed tumour growth and were able to target angiogenic vessels of glioblastomas.

Another promising approach is used to attack lymph-node tumour cells. These cells carry specific markers, which can be targeted specifically by a cyclic amino-acid peptide (LyP-1) that does not bind to normal lymphatic tissues (Laakkonen et al., 2002). The authors reported that LyP-1 had a proapoptotic and cytotoxic effect on tumour cells, and that systemic administration of the LyP-1 peptide inhibited breast cancer xenograft growth in mice. It was later found that LyP-1 also binds to cells in other kinds of tumours, including prostate cancer in mice (Laakkonen et al., 2004). LyP-1 offers even more clinical value than simply being a homing device for lymph-node tumours: as cancer often spreads through lymphatic vessels, it could be used to target drugs to lymphatic vessels and to block metastases by destroying tumour cells close to the lymphatic system.

In addition to antibodies and small peptides, inorganic compounds are being investigated for their ability to specifically target tumours. In an experiment, Akerman et al. (2002) showed that quantum dots (qdots) coated with PEG were able to avoid the reticuloendothelial system (RES), which reduced their accumulation in the liver and spleen but did not affect their accumulation in tumour tissue. The qdots were further coupled with homing peptides and showed excellent abilities for "homing" to the vasculature of normal lungs and to tumours. Qdots have traditionally been used for molecular imaging, but their application could be expanded for drug-delivery purposes as well.

Colloidal silica nanoparticles with cationic surfaces are interesting vehicles for targeted gene delivery as they bind to plasmid DNA and transfect cells in vitro. The silica particles have been studied in the mouse, where they increased gene expression in the lung while causing either low or no cell toxicity. Layered double hydroxide (LDH) also performed well, both in vitro and in vivo. It was first described by Choy et al. (2000) as a nanosized inorganic "clay" with the ability to intercalate biomolecules such as DNA, ATP and nucleosides. The inorganic lattice of LDH provides biochemical properties that could be exploited for gene delivery and for the prolonged release of drugs (Kwak et al., 2002). Kwak et al. (2004) prepared uniformly sized LDHs in the range 100–200 nm and injected them into adult male rats to investigate their safety. No serious systemic effects were observed for LDHs at concentrations below 200 mg/kg, but if deposited extravascularly, LDHs were locally irritating. Recently, Tyner et al. (2004) used LDHs as delivery vehicles for camptothecin. Their idea was to encapsulate camptothecin in an anionic micelle derived from a biocompatible surfactant with a negative charge, which allowed the intercalation of camptothecin-loaded micelles into LDH layers. These complexes inhibited the growth of 9L glioma cells in vitro similar to the "naked" drug. However, the real advantage lies in the fact that these nanohybrids can be administered in a dose-controlled fashion to increase solubility. Thus it successfully attached a few homing molecules to the outside surface of LDHs, which showed the good targeting abilities of the hybrids. In general, LDH-micelle constructs could become excellent drug-carrier systems as they are biocompatible, show good targeting properties and successfully avoid the RES. However, they are not biodegradable and further studies are needed to assess their safety.

17.4 Nanodiagnostics

What if there was a way doctors could determine your likelihood of contracting a disease without you actually acquiring the symptoms of that disease? Current diagnostic methods focus on technologies that detect the symptoms of disease. This may include the measurement of a particular antibody produced by the body in response to infection or the observation of a specific bacterium that is known to cause the disease. Methods such as these are slow and inefficient because they involve recognizing a disease based on the patient contracting the disease first. The applications of nanotechnology in molecular diagnostics, indicates, researchers within the field of nanodiagnostics are currently studying the ways nanotechnology will be able to extend the limits of current molecular diagnostic techniques (1). Nanotechnology implemented within current diagnostic equipment has the potential of analyzing entire genomes in minutes instead of hours. Based on which DNA sequences are deviated from the normal, doctors will be able to determine an individual's predisposition to either cancer or a specific disease. Current research into microfluidic technology is making it possible to integrate a number of complex diagnostic procedures into one simple device that will be able to give on the spot diagnosis.

One of the most common techniques used today to analyze DNA sequences is hybridization, or the pairing of separated strands of DNA with complementary DNA strands of known sequence that act as probes. Currently, DNA chips called DNA microarray assays are used to analyze DNA. A company called Nanogen has developed a product called the "Nanochip" that employs the power of an electronic current that separates DNA probes to specific sites on the array based on charge and size. Once these probes are on specific sites of the nanochip, the test sample (blood) can then be analyzed for target DNA sequences by hybridization with these probes. The DNA molecules that hybridize with target DNA sequences fluoresce, which is detected and relayed back to an onboard system through platinum wiring that is present within the chip. The secret behind this nano-chip is that each test site can be controlled electronically from the system's onboard computer. In other words, the chip can place different probes in different sites according to what DNA sequence is of interest.

The field of nanodiagnostics is currently seeing a trend towards hand-held devices that are easy to use and are marketable to customers. One company has created a product called "Gluco-watch" which permeates your skin with fluidic nanochip biosensors that sense the level of blood sugar and then relate this read out to a wrist-watch. Nanodiagnostic technology such as this have the potential to free up large proportions of the health care industry that are devoted to monitoring patients' (especially the elderly) blood composition. Routine appointments with the hospital for blood tests would no longer be needed if such devices became available to the public.

Nanodiagnostic devices would also make the job of a physician easier. As of now, doctors order tests to confirm their hypotheses on what they think a particular disease may be. With the use of nanodevices, doctors would be able to differentiate between diseases based on mutated DNA sequences. The field of nano diagnostics raises certain ethical concerns related with the testing of blood. For example, if a nanochip were to be able to analyze our entire DNA sequence from a drop of blood, would it be morally correct for hospitals to know an individual's entire genetic makeup? Shouldn't individuals have some say in whether or not hospitals have access to these records? Another area of concern is with the use of MEMS devices within the body. If the capsule device breaks down in one's stomach, harmful metal oxide particles could cause the introduction of free radicals that are harmful to cells. Researchers still do not know if the introduction of these devices within one's body would leave residual nanoparticles that would be harmful to the digestive system. It is clear that the field of nanodiagnostics still has a long way to go before diagnostic equipment will be available to consumers.

Another amazing discovery is a biosensing nanodevice may revolutionize health screenings for diseases like anthrax, cancer and antibiotic resistant *Staphylococcus aureus* (MRSA). Even more incredible than the device itself, is that it is based on the world's tiniest rotary motor: a biological engine measured on the order of molecules. The enzyme F1-adenosine triphosphatase, better known as F1-ATPase. This enzyme, only 10–12 nm in diameter, has an axle that spins and produces torque.

It was reported that the enzyme can be armed with an optical probe (gold nanorod) and manipulated to emit a signal when it detects a single molecule of target DNA. This is achieved by anchoring a quiescent F1-ATPase motor to a surface. A single strand of a reference biotinylated DNA molecule is then attached to its axle. The marker protein, biotin, on the DNA is known to bind specifically and tightly to the glycoprotein avidin, so an avidin-coated gold nanorod is then added. The avidin-nanorod attaches to the biotinylated DNA strand and forms a stable complex. When a test solution containing a target piece of DNA is added, this DNA binds to the single complementary reference strand attached to the F1-ATPase. The

DNA complex, suspended between the nanorod and the axle, forms a stiff bridge. Once ATP is added to the test solution, the F1-ATPase axle spins, and with it, the attached (now double-stranded) DNA and nanorod. The whirling nano-sized device emits a pulsing red signal that can then be detected with a microscope. Studies with the F1-ATPase in the laboratory show that since it can detect single DNA molecules, it far exceeds the detection limits of conventional PCR technology. Such a detection instrument based on the F1-ATPase enzyme would also be faster and more portable. Thus a nano-sized F1-ATPase is utilized to produce a DNA detection instrument. It is roughly the size of a small tissue. Rapid and sensitive biosensing of nucleic acids and proteins is vital for the identification of pathogenic agents of biomedical and bioterrorist importance. This provides a new avenue through which to analyze genotypes and forensic evidence. In a recent study, use of antimicrobial peptides (AMPs) immobilized in electroactive nanostructured films was reported for applications in the pharmaceutical industry and diagnosis.

17.5 Molecular Electronics

One of the many fascinating concepts in nanotechnology is the vision of molecular electronics. Molecular electronics engineers of tomorrow might use individual molecules to perform the functions in an electronic circuit that are performed by semiconductor devices today.

DNA, the blueprint of life, and electronics seem to be two completely different things but it appears that DNA could offer a solution to many of the hurdles that need to be overcome in further scaling down electronic circuits beyond a certain point. The reason why DNA could be useful in nanotechnology for the design of electric circuits is the fact that it actually is the best nanowire in existence – it self-assembles, it self-replicates and it can adopt various states and conformations. Not surprisingly, performing reliable experiments on a single oligo-DNA molecule is an extremely delicate task as partly contradicting research reports demonstrate. Different DNA transport experiments have shown that DNA may be insulating, semiconducting, or metallic. Among the numerous factors that could impact the results are the quality of the DNA-electrode interface, the base pair, the charge injection into the molecule, or environmental effects such as humidity or temperature.

DNA-based nanostructuring is one approach that could lead to promising results. It has already been shown that DNA could be used to structure nanoscale surfaces. Nanoscale objects of very different size can be deposited simultaneously and site-selectively onto DNA-displaying surfaces, based on sequence-specific DNA-DNA duplex formation.

Already widely used today, DNA microarrays (DNA chips) that can contain hundreds of thousands of distinct probes, each bearing a reagent whose molecular recognition of a complementary molecule can lead to a signal that is detected by an imaging technology, most often fluorescence. The traditional solid-phase array is a collection of DNA molecules attached to a solid surface, such as glass, plastic or silicon chip. The ability of DNA sequences to bind to surfaces site-selectively has intrigued scientists who are excited by the vision of molecular electronics and who struggle with finding robust techniques that would allow the structuring of surfaces with nanometer precision.

Researchers have now demonstrated a novel carbon nanotube-based nanoelectronic platform as proof of concept that single DNA molecules can be detected. This novel detection technique is based on change in electrical conductance upon selective hybridization of the complementary target DNA with the single stranded probe attached to the system. The single-stranded sequence-specific probe DNA whose ends are modified with amine is attached between two carbon nanotubes/nanowires using dielectrophoresis (DEP). This platform can be used for understanding how electrical charge moves through DNA which could help researchers understand and perhaps develop a technique for reversing the damage of DNA done by oxidation and mutation.

One of the potential applications would be identification of specific genes based on the hybridization-induced change in electrical signal. The researchers explain that their current detection platform could be used in an application for the electrical detection of several other gene sequences on a single chip, although such simultaneous detection of several different gene sequences using an array of nanoelectrodes is a major challenge which currently is under investigation by the research team.

This platform could also be used in fundamental research studies understanding the properties of DNA at single molecular level. The proposed technology has a wide-range application that includes but not limited to revealing the presence of disease related genes, bacterial and viral.

The human body so far is the ultimate "wet computer", a highly efficient, biomolecule-based information processor that relies on chemical, optical and electrical signals to operate. Researchers are trying various routes to mimic some of the body's approaches to computing. Prominent among them is DNA computing, a form of computing which uses DNA and molecular biology instead of the traditional silicon-based computer technologies. The success of nanorobotics requires the precise placement and subsequent operation of specific nanomechanical devices at particular locations, thereby leading to a diversity of structural states. The structural programmability of DNA makes it a particularly attractive system for nanorobotics. A large number of DNA-based nanomechanical devices have been described, controlled by a variety of methods. These include pH changes and the addition of other molecular components, such as small molecule effectors, proteins and DNA strands. The most versatile of these devices are those that are controlled by DNA strands.

The key problems on the nanoscale is that the multivalent character of the molecular interactions also strengthens undesirable forces that are weak on the level of small molecules, such as unspecific adsorption, or formation of small partially matched duplexes between DNA strands.

17.6 Nanotechnology Biosensor for Pathogen Detection

Three pathogens, *Salmonella*, *Listeria*, and *Toxoplasma*, are responsible for 1,500 deaths each year. *Salmonella* is the most common cause of foodborne deaths and responsible for millions of cases of foodborne illness a year. Sources are raw and undercooked eggs, undercooked poultry and meat, dairy products, seafood, fruits and vegetables so basically more or less everything you eat. Early detection of foodborne pathogenic bacteria, especially *Salmonella*, is therefore an important task in microbiological analysis to control food safety. Several methods have been developed in order to detect this pathogen; however, the biggest challenges remain detection speed and sensitivity. A novel nanotechnology-based biosensor is showing great potential for foodborne pathogenic bacteria detection with high accuracy.

In principle, the nano-technology based biosensor could be used for detecting other foodborne pathogenic bacteria such as *E. coli*, *Staphylococcus*, *Campylobacter* and food toxins such as Ricin, Abrin, or *Clostridium botulinum* if the proper antibody is selected for the conjugation with nanorod substrates. Additionally, the fluorescent detection dye can also be replaced by other types of dyes or potentially quantum dots that may allow for multiplex detection. This novel nanobiosensor could have broad appeal to the food industry, food safety inspection agencies, government agencies overseeing food safety, and researchers focused on safety and biosecurity research.

17.7 Nanotechnology Composite Materials for Next Generation Biomedical Applications

Polymethylmethacrylate (PMMA), a clear plastic, is a pretty versatile material. Plexiglas windows are made from PMMA. Acrylic paints contain PMMA. It also remains one of the most enduring materials in orthopedic surgery where it has a central role in the success of total joint replacement. Being part of a group of medical materials called "bone cement", its use includes the fixation of biomaterials such as artificial joints to bone, the filling of bone defects and, also, as a drug-delivery system. Beginning in the 1970s, many successful results have been reported for total hip replacement using PMMA cement; however, failures of fixation have also occurred. The fixation strength of PMMA cement to bone is mainly dependent on mechanical interlocking, but it is known that a fibrous tissue layer intervenes between cement and bone – PMMA cement never bonds directly to the bone.

One of the problems associated with the conventional types of bone cement used is their unsatisfactory mechanical and exothermic reaction properties. Other problems with PMMA cement include the biological response, leakage of the monomer of methylmethacrylate and a high curing temperature, which can damage cell activity. Ideally, a bone cement material must functionally match the mechanical behavior of the tissue to be replaced, it must be able to form a stable interface with the surrounding natural tissue and be effective in guided tissue regenerative procedures, it should be easy to handle, biologically compatible, non-supporting of microbial growth, and non-allergenic.

Hydroxyapatite (HA) is another key constituent of bone cements because of its ability to bond chemically with living bone tissues; this is due to its similar chemical composition and crystal structure to apatite in the human skeletal system. However, the intrinsic brittleness and poor strength of sintered HA restricts its clinical applications under load-bearing conditions.

A novel nanocomposite of carbon-nanotube-reinforced PMMA/HA is a demonstration of how nanomaterials will play an increasing role in the synthesis of next-generation biomedical applications.

The combination of PMMA and hydroxyapatite with multi-walled carbon nanotubes (MWCNTs) seems to be a path of great promise to be explored – carbon nanotubes, due to their small dimensions and high aspect ratio, exhibit exceptional physical and chemical properties. No other material can compete with their outstanding combination of mechanical, thermal and electronic properties.

Carbon nanotubes (CNTs) have shown promise as an important new class of multifunctional building blocks and innovative tools in a large variety of nanotechnology applications, ranging from nanocomposite materials through nanoelectronics to biomedical applications. The exploration of CNTs in biomedical applications is well underway and exploratory uses have included CNT-coated implants, drug delivery and CNTs as components of biosensors. Notwithstanding the still not satisfactorily addressed issue of toxicity, CNTs' properties such as high strength, high electrical and thermal conductivities, and high specific surface area render them particularly useful in the fabrication of nanocomposite-derived biomedical devices.

In one particular area, biomaterials applied to bone – CNTs are anticipated to improve the overall mechanical properties for applications such as high-strength arthroplasty prostheses expected to remain in the body for a long time, or fixation plates and screws that will not fail or impede healing of bone. In addition, CNTs are expected to be of use as local drug delivery systems or scaffolds to promote and guide bone tissue regeneration. A new study by Japanese scientists clearly demonstrates that multi-walled CNTs (MWCNTs) have good bone-tissue compatibility, permitting bone repair and becoming closely integrated with bone tissue. Furthermore, under certain circumstances, their results indicate that MWCNTs accelerate bone formation.

17.8 A Carbon Nanotube Biosensor

A carbon nanotube electronic sensor was developed to detect prostate cancer cells based on the unique properties of carbon nanotubes, such as nanometer diameter, high aspect ratio, high electrical conductivity and the fact that they are chemically inert. It is clear that detecting cancer at the earliest stage provides the greatest chance of survival. Currently, detection techniques are improving. More patients are being diagnosed with localized prostate cancer and fewer patients are being diagnosed with disseminated disease. The sensor might also be used in the bloodstream to detect the concentration of cancer cells and hence infer the metastatic potential of the cancer. Initial testing showed that a carbon nanotube array electrode could quickly detect prostate cancer cells in solution. This nanotube electrode could be used to form the needle sensor.

The carbon nanotube electronic sensor was developed starting from the synthesis of long, highly aligned carbon nanotubes in the form of towers. The nanotube towers were embedded in a microfluidic channel and LNCaP prostate cancer cells were flowed through the channel. Electrochemical impedance measurements were performed and were able to distinguish different concentrations of cancer cells in the flowing solution. Based on these results, it is expected that the carbon nanotube electronic sensor could be used to develop generic cell-based biosensors to detect different types of cancer and disease.

Development of the nanotube biosensor is part of an overall trend towards miniaturization in medicine. Electronic biosensing is attractive because it offers the advantages of a quick response, low cost and ease of use as an additional diagnostic tool to supplement traditional immunoassay procedures. In particular, carbon nanotube electronic sensors offer advantages such as enhanced current density, low detection limits and improved signal-to-noise ratios. The preliminary electrochemical impedance results reported here using deionized water, buffer solution and LNCaP prostate cells indicate that nanotube electrodes and electrochemical analysis can be portable, have disposable electrodes and provide real-time measurements as a lab on a chip. The next step in the research is to functionalize the electrode to capture cancer cells. The sensor can then be tested to detect cancer cells in blood and to predict the metastatic potential of diagnosed prostate tumors. If successful, this sensor could improve the efficiency of point-of-care and clinical testing, and might be used as a remote portable device to screen for different diseases.

The potential applications of biosensors in daily life are tremendous: from environmental and water analysis to personalized healthcare, from quality food control to home patient monitoring, in addition, other aspects with strong impact on public healthcare – such as drug screening and genetic disease control – will certainly benefit from biosensors. These electrodes can be designed in a microarray structure in order to achieve parallel detection of several metabolites, for instance glucose, urea, lactate and cholesterol. Besides these catalytic biosensors, the researchers are also developing affinity biosensors that use impedimetric detection to study substrate-protein binding, as well as DNA hybridization. Affinity interaction can be used for the design of immunosensors by immobilization of enzymes on the surface of diamond ion field-effect transistors (ISFETs) in order to develop enzyme field-effect transistors (ENFETs).

17.9 Conclusion & Future Directions

Nanotechnology supports and enhances advances in genomics and proteomics. Specifically, it offers many interesting possibilities for improving drug delivery. The rapid progress of nanotechnology-based applications in medicine is due to the huge interest from therapeutic and diagnostic companies, which have already introduced many nanotechnology-based devices to everyday practice, among them new drug-delivery systems. Future research should further improve their targeting properties, given that one of the most important problems in medicine is the controlled and exact delivery of a drug to diseased tissue (Liu and Frechet, 1999).

However, the promising solutions offered by nanotechnology pose some inevitable questions. We still do not know whether organic and inorganic nanoparticles, once they have entered the body, can cause damage to other tissues. Additionally, we need new safety standards as the novel and unexplored characteristics of nanoscale materials demand appropriate quality-control measures. The further development of new strategies in drug delivery depends largely on the establishment of such standards and their subsequent implementation in research and manufacturing.

The abilities discussed here might well take years or decades to develop. It is quite natural to ask: "When might we see these systems actually used?" The scientifically correct answer is, of course, "We don't know." That said, it is worth noting that if progress in computer hardware continues as the trend lines of the last 50 years suggest, we should have some form of molecular manufacturing in the 2010–2020 time frame. After this, the medical applications will require some additional time to develop.

The remarkably steady trend lines in computer hardware, however, give a false sense that there is a "schedule" and that developments will spontaneously happen at their appointed time. This is incorrect. How long it will take to develop these systems depends very much on what we do. If focused efforts to develop molecular manufacturing and its medical applications are pursued, we will have such systems well within our lifetimes. If we make no special efforts the schedule will slip, possibly by a great deal.

As might be appreciated, developing these systems within our lifetimes would be advantageous for a variety of reasons. The fact that biological processes are in a way dependent on molecular machines and clearly defined structures shows that building new nanomachines is physically possible (Freitas, 2002). It also means that we will not necessarily need fundamental new technologies to make further advances in nanotechnology if we can make proper use of those found in nature – which renders the future of nanotechnology foreseeable. Although many of the ideas developed in nanomedicine might seem to be in the realm of science fiction, only a few more steps are needed to make them come true, so the "time-to-market" of these technologies will not be as long as it seems today.

References

Akerman, M.E., Chan, W.C., Laakkonen, P., Bhatia, S.N., Ruoslahti, E., 2002, Nanocrystal targeting in vivo. *Proc Natl Acad Sci USA*, 99:12617–12621.

Bedioui, F., 1995, Zeolite-encapsuled and clay-intercalated metal porphyrin, phthalocyanine and Schiff-base complexes as models for biomimetic oxidation catalysts: an overview. *Coord Chem Rev* 144:39–68.

Blanco, E., 2009, Multifunctional micellar nanomedicine for cancer therapy. *Exp Biol Med* (Maywood) 234(2):123–131.

Boyoglu, S., Vig, K., Pillai, S., Rangari, V., Dennis, V.A., Khazi, F., Singh, S.R., 2009, Enhanced delivery and expression of a nanoencapsulated DNA vaccine vector for RSV. *Nanomedicine* (Article in Press, Corrected Proof) doi:10.1016/j.nano.2009.02.004

Cagle, D.W., Kennel, S.J., Mirzadeh, S., Alford, J.M., Wilson, L.J., 1999, In vivo studies of fullerene-based materials using endohedral metallofullerence radiotracers. *Proc Natl Acad Sci USA*, 96:5182–5187.

Chauhan, A.S., Jain, N.K., Diwan, P.V., Khopade, A.J., 2004, Solubility enhancement of indomethacin with poly(amidoamine) dendrimers and targeting to inflammatory regions of arthritic rats. *J Drug Target* 12(9–10):575–583.

Chen, Y., Xu, X., Hong, S., Chen, J., Lu, N., Underhill, C.B., Creswell, K., Zhang, L., 2001, RGD-tachyplesin inhibits tumor growth. *Cancer Res* 61(15):2434–2438.

Choi, Y., Thomas, T., Kotlyar, A., Islam, M.T., Baker, J.R., Jr., 2005, Synthesis and functional evaluation of DNA-assembled polyamidoamine dendrimer clusters for cancer cell-specific targeting. *Chem Biol* 12(1):35–43.

Choy, J.H., Kwak, S.Y., Jeong, Y.J., Park, J.S., 2000, Inorganic layered double hydroxides as nonviral vectors. *Angew Chem Int Edn* 39:4041–4045.

Cohen-Sacks, H., Najajreh, Y., Tchaikovski, V., Gao, G., Elazer, V., Dahan, R., Gati, I., Kanaan, M., Waltenberger, J., Golomb, G., 2002, Novel PDGF/R antisense encapsulated in polymeric nanospheres for the treatment of restenosis. *Gene Ther* 9(23):1607–1616.

Csaba, N., Caamano, P., Sanchez, A., Dominguez, F., Alonso, M.J., 2005, PLGA: poloxamer and PLGA: poloxamine blend nanoparticles: new carriers for gene delivery. *Biomacromolecules* 6(1):271–278.

Curnis, F., Sacchi, A., Borgna, L., Magni, F., Gasparri, A., Corti, A., 2000, Enhancement of tumor necrosis factor/antitumor immunotherapeutic properties by targeted delivery to aminopeptidase N (CD13). *Nat Biotechnol* 18(11):1185–1190.

Davis, S.S., 1997, Biomedical applications of nanotechnology – implications for drug targeting and gene therapy. *Trends Biotechnol* 15:217–224.

Dong, Y., Feng, S.S., 2004, Methoxy poly(ethylene glycol)-poly(lactide) (MPEG-PLA) nanoparticles for controlled delivery of anticancer drugs. *Biomaterials* 25:2843–2849.

Dubertret, B., Skourides, P., Norris, D.J, Noireaux, V., Brivanlou, A.H., Libchaber, A., 2002, In vivo imaging of quantum dots encapsulated in phospholipid micelles. *Science* 298: 1759–1762.

Essler, M., Ruoslahti, E., 2002, Molecular specialization of breast vasculature: a breast-homing phage-displayed peptide binds to aminopeptidase P in breast vasculature. *Proc Natl Acad Sci USA* 99:2252–2257.

Ewert, K., Slack, N.L., Ahmad, A., Evans, H.M., Lin, A.J., Samuel, C.E., Safinya, C.R., 2004, Cationic lipid-DNA complexes for gene therapy: understanding the relationship between complex structure and gene delivery pathways at the molecular level. *Curr Med Chem* 11(2):133–149.

Fernandez-Lopez, S., Kim, H.S., Choi, E.C., Delgado, M., Granja, J.R., Khasanov, A., Kraehenbuehl, K., Long, G., Weinberger, D.A., Wilcoxen, K.M., Ghadiri, M.R., 2001, Antibacterial agents based on the cyclic D,L – peptide architecture. *Nature* 412(6845):452–455.

Foley, S., Crowley, C., Smaihi, M., Bonfils, C., Erlanger, B.F., Seta, P., Larroque, C., 2002, Cellular localization of water-soluble fullerene derivative. *Biochem Biophys Res Comm* 294(1): 116–119.

Freitas, R.A., Jr., 2002, The future of nanofabrication and molecular scale devices in nanomedicine. *Stud Health Technol Inform* 80:45–59.

Friedman, S.H., Decamp, D.L., Sijbesma, R.P., Srdanov, G., Wudl, F., Kenyon, G.L., 1993, Inhibition of the HIV-1 protease by fullerene derivatives – model building studies and experimental verification. *J Am Chem Soc*, 115:6506–6509.

Hilt, J.Z., 2004, Nanotechnology and biomimetic methods in therapeutics: molecular scale control with some help from nature. *Adv Drug Deliv Rev* 56:1533–1536.

Holmes, T.C., 2002, Novel peptide-based biomaterial scaffolds for tissue engineering. *Trends Biotechnol* 20:16–21.

Hong, F.D., Clayman, G.L., 2000, Isolation of a peptide for targeted drug delivery into human head and neck solid tumors. *Cancer Res* 60:6551–6556.

Hood, J.D., Bednarski, M., Frausto, R., Guccione, S., Reisfeld, R.A., Xiang, R., Cheresh, D.A., 2002, Tumor regression by targeted gene delivery to the neovasculature. *Science* 296(5577):2404–2407.

Kim, T.I., Seo, H.J., Choi, J.S., Jang, H.S., Baek, J.U., Kim, K., Park, J.S., 2004, PAMAM–PEG–PAMAM: novel triblock copolymer as a biocompatible and efficient gene delivery carrier. *Biomacromolecules* 5(6):2487–2492.

Kraljevic, S., Stambrook, P.J, Pavelic, K., 2004, Accelerating drug discovery. *EMBO Rep* 5: 837–842.

Kwak, S.Y., Jeong, Y.J., Park, J.S., Choy, J.H., 2002, Bio-LDH nanohybrid for gene therapy. *Solid State Ionics* 151:229–234.

Kwak, S.Y., Kriven, W.M., Wallig, M.A., Choy, J.H., 2004, Inorganic delivery vector for intravenous injection. *Biomaterials* 25:5995–6001.

Laakkonen, P., Akerman, M.E., Biliran, H., Yang, M., Ferrer, F., Karpanen, T., Hoffman, R.M., Ruoslahti, E., 2004, Antitumor activity of a homing peptide that targets tumor lymphatics and tumor cells. *Proc Natl Acad Sci USA* 101(25):9381–9386.

Laakkonen, P., Porkka, K., Hoffman, J.A., Ruoslahti, E., 2002, A tumor-homing peptide with a targeting specificity related to lymphatic vessels. *Nat Med* 8:751–755.

Lambert, G., Fattal, E., Couvreur, P., 2001, Nanoparticulate systems for the delivery of antisense oligonucleotides. *Adv Drug Deliv Rev* 47:99–112.

Le Garrec, D., Gori, S., Luo, L., Lessard, D., Smith, D.C., Yessine, M.A., Ranger, M., Leroux, J.C., 2004, Poly(N-vinylpyrrolidone)-block-poly(D,L-lactide) as a new polymeric solubilizer for hydrophobic anticancer drugs: in vitro and in vivo evaluation. *J Control Release* 99(1): 83–101.

Liu, M., Frechet, M.J., 1999, Designing dendrimers for drug delivery. *Pharm Sci Technol Today* 2:393–401.

Merdan, T., Kopecek, J., Kissel, T., 2002, Prospects for cationic polymers in gene and oligonucleotide therapy against cancer. *Adv Drug Deliv Rev* 54:715–758.

Mumpton, F.A., 1999. La roca majica: Uses of natural zeolites in agriculture and industry. *Proc Natl Acad Sci USA* 96:3463–3470.

Nakanishi, H., 2003, Nonviral genetic transfer of Fas ligand induced significant growth suppression and apoptotic tumor cell death in prostate cancer in vivo. *Gene Ther* 10:434–442.

NanoMarkets, 2005, Nano Drug Delivery. NanoMarkets LC, Sterling, VA, USA.

Nishiyama, N., Okazaki, S., Cabral, H., Miyamoto, M., Kato, Y., Sugiyama, Y., Nishio, K., Matsumura, Y., Kataoka, K., 2003, Novel cisplatin-incorporated polymeric micelles can eradicate solid tumors in mice. *Cancer Res* 63(24):8977–8983.

Oh, K.S., Lee, K.E., Han, S.S., Cho, S.H., Kim, D., Yuk, S.H., 2005, Formation of core/shell nanoparticles with a lipid core and their application as a drug delivery system. *Biomacromolecules* 6(2):1062–1067.

Ooya, T., Lee, J., Park, K., 2004, Hydrotropic dendrimers of generations 4 and 5: synthesis, characterization, and hydrotropic solubilization of paclitaxel. *Bioconjug Chem* 15: 1221–1229.

Orosz, K.E., Gupta, S., Hassink, M., Abdel-Rahman, M., Moldovan, L., Davidorf, F.H., Moldovan, N.I., 2004, Delivery of antiangiogenic and antioxidant drugs of ophthalmic interest through a nanoporous inorganic filter. *Mol Vis* 10:555–565.

Otsuka, H., Nagasaki, Y., Kataoka, K., 2003, PEGylated nanoparticles for biological and pharmaceutical applications. *Adv Drug Deliv Rev* 55:403–419.

Pavelic, K., 2001, Natural zeolite clinoptilolite: new adjuvant in anticancer therapy. *J Mol Med* 78:708–720.

Pavelic, K., Hadzija, M., 2003, Handbook of Zeolite Science and Technology, Auerbach, S.M., Carrado, K.A., Dutta, P.K., eds., Marcel Dekker, New York, USA, pp. 1143–1174.

Quintana, A., Raczka, E., Piehler, L., Lee, I., Myc, A., Majoros, I., Patri, A.K., Thomas, T., Mulé, J., Baker, J.R., Jr., 2002, Design and function of a dendrimer-based therapeutic nanodevice targeted to tumor cells through the folate receptor. *Pharma Res* 19(9):1310–1316.

Ravi, K.M., Hellermann, G., Lockery, R.F., Mohapatra, S.S., 2004, Nanoparticle-mediated gene delivery: state of the art. *Expert Opin Biol Ther* 4(8):1213–1224.

Tyner, K.M., Schiffman, S.R., Giannelis, E.P., 2004, Nanobiohybrids as delivery vehicles for camptothecin. *J Control Release* 95:501–514.

Whitesides, G.M., Boncheva, M., 2002, Beyond molecules: self-assembly of mesoscopic and macroscopic components. *Proc Natl Acad Sci USA* 99:4769–4774.

Wu, X., Liu, H., Liu, J., Haley, K.N., Treadway, J.A., Larson, J.P., Ge, N., Peale, F., Bruchez, M.P., 2003, Immunofluorescent labeling of cancer marker Her2 and other cellular targets with semiconductor quantum dots. *Nat Biotechnol* 21(1):41–46.

Xu, H., 2003, Multiplexed SNP genotyping using the QbeadTM system: a quantum dot-encoded microsphere-based assay. *Nucl Acids Res* 31:E43.

Zhang, S., Marini, D.M., Hwang, W., Santoso, S., 2002, Design of nanostructured biological materials through self-assembly of peptides and proteins. *Curr Opin Chem Biol* 6(6):865–871.

Chapter 18
Biomarkers

Abstract In the post-genomic age, the means to search the entire genome for biomarkers has become available, but the conventional approaches to biomarker discovery are entirely inadequate to yield results with the new technology. Finding clinically useful biomarker panels with sensitivity and specificity equal to that of cystoscopy is a problem of systems biology. Biomarkers are capable of bridging the distance between basic research and late-stage clinical trials and can even shorten it. Many pharmaceuticals and biotech companies are investing additional resources to bring biomarkers into the clinic. They hope that these efforts to practice pharmacogenomic research both in the preclinical and clinical stages will increase the odds of finding a successful drug. Most of the biomarker discovery utilize the microarrays for RNA analysis.

Keywords Toxicity · Biomarkers · Microarray · Proteomics · Genomics · Metabonomics · Polymorphism · Haplotype · Validation · Cancer · Genetic markers · Cancer biomarkers · Candidate markers

18.1 Prologue

The molecular biology revolution and the advent of genomic and proteomic technologies are facilitating rapid advances in our understanding of the molecular details of cell and tissue function. These advances have the potential to transform toxicological and clinical practice, and are likely to lead to the supplementation or replacement of traditional biomarkers of cellular integrity, cell and tissue homeostasis, and morphological alterations that result from cell damage or death. New technologies that permit simultaneous monitoring of many hundreds, or thousands, of macro- and small molecules ("omics" technologies) promise to allow functional monitoring of multiple (or perhaps all) key cellular pathways simultaneously. Elucidation of cellular responses to molecular damage, including evolutionarily conserved inducible molecular defense systems, suggests the possibility of new biomarkers based on molecular responses to functional perturbations and cellular damage (MacGregor, 2003). Our improved understanding of the molecular basis of

M. Debnath et al., *Molecular Diagnostics: Promises and Possibilities*,
DOI 10.1007/978-90-481-3261-4_18,

various pathologies suggests that monitoring specific molecular responses may provide improved prediction of human outcomes. Responses that can be monitored directly in the human should provide "bridging biomarkers" that may eliminate much of the current uncertainty in extrapolating from laboratory models to human outcome. Another aspect of genomics is our enhanced ability to associate DNA sequence variations with biological outcomes and individual sensitivity. The human genome sequence has revealed that sequence variations are very common, and may be an important determinant of variation in biological outcomes. The impending availability of a complete human haplotype map linked to standard genetic markers greatly facilitates identification of genetic variations that convey sensitivity or resistance to chemical exposures. Genetic approaches have already linked a large number of genetic variants (polymorphisms) with human diseases and adverse reactions from exposure to drugs or toxicants, suggesting an important role in sensitivity to drugs and environmental agents, disease susceptibilities, and therapeutic responses. As these opportunities are transformed into reality, regulatory toxicological practice is likely to be shaped in the future by the combination of conventional pathology, toxicology, molecular genetics, biochemistry, cell biology, and computational bio-informatics – resulting in the broad application of molecular approaches to monitoring functional disturbances.

18.2 Concept of Biomarkers

Biomarkers are defined as endogenous or injected molecules whose presence or metabolism correlates with important disease related physiological processes and disease outcomes (Ferber, 2002). They should be identified or defined molecular entities to facilitate comparison across laboratories and technology platforms. A biomarker is a characteristic that is objectively measured and evaluated as an indicator of normal biologic processes, pathogenic processes, or pharmacologic responses to a therapeutic intervention.

In recent years, progress has been strong in the discovery and implementation of nucleic acid biomarkers, but the identification of protein biomarkers is as yet less advanced. Biomarkers for recurrence and progression could make a great contribution, but in spite of decades of research, no biomarkers are commercially available with the requisite sensitivity and specificity (Hurst, 2009). Biomarkers have many uses in laboratory and clinical investigations and in drug discovery. Biomarkers are useful for diagnosing, classifying, or grading the severity of disease in both laboratory and clinical settings (Atkinson, 2001). They may be able to supply efficacy, toxicity, and mechanistic information for the preclinical and clinical phases of drug discovery and be applied with therapeutics to produce commercial tests that aid patient selection or drug dosing (personalized medicine). Because biomarkers and surrogate end point markers can accelerate the speed and decrease the risk of drug discovery, they are highly sought after. The development process is complex. Investigators need a complete development plan and, most important, access to sufficient, well-characterized samples. Unfortunately, many promising biomarkers never make it into clinical practice or even broad application in clinical or laboratory research.

A biomarker can serve as a clinical endpoint, surrogate endpoint, or both. Clinical endpoint is defined as a characteristic or variable that reflects how a patient feels, functions or survives. Surrogate endpoint is defined as a biomarker intended to substitute for a clinical endpoint. A surrogate endpoint is expected to predict clinical benefit (or harm, or lack of benefit or harm) based on epidemiologic, therapeutic, pathophysiologic or other scientific evidence. In proteomics, such a biomarker could be a protein or peptide (native or modified) identified through use of various technologies, or a pattern of peaks identified through time-off light based technologies as discussed in the latter part of this article. To this end, biomarkers are proposed to identify unique molecular signatures detrimental to certain disease states. Why proteomics? The reason for the use of protein-based assays is straightforward. Mammalian systems are much more complex than can be deciphered by their genes alone. Expression analysis directly at the protein level is necessary to unravel the critical changes that occur as part of disease pathogenesis. This is because proteins are often expressed at levels and forms that cannot be predicted from DNA or mRNA analysis. Proteins represent the dynamic states of cells and therefore indicate any abnormal activity in real-time. Protein-based assays have remained the mainstay of the diagnostic field for the past several decades and are likely to be expanded by the emergence of newer, high throughput technologies in coming years.

There are more than 200 post-translational modifications that proteins could undergo that affect function, protein-protein and nucleic acid-protein interaction, stability, targeting, half-life etc., all contributing to a potentially large number of protein products from one gene.

18.3 Biomarkers on the Horizon

In recent years, the development of microarrays and proteomics has brought the power of whole genome analysis to the field of biomarkers. Naively, many in the field have assumed that high-dimension studies of patient samples will magically yield robust, sensitive and specific biomarkers. Misconceptions about the technology and misunderstandings of the underlying biology of disease have plagued the field. Instead of there being "pathways" in which signalling is linear and definable, the reality is a large, interconnected network of cooperating proteins that regulate cellular growth, death and differentiation. Alterations in this network tend to ripple outward in unpredictable ways. Moreover, this complex system responds to complex inputs from the local tumour environment as well as all other biological variables affecting the organism. Individual molecular markers tend to lack sensitivity and specificity because, unlike morphologically-based grading, they are imperfectly reflective of the overall phenotype and overly sensitive to the cellular network. At a practical level, this means the probability of finding a single biomarker with the requisite sensitivity and specificity is vanishing small (Hurst, 2009). Genomic and proteomic technologies have significantly increased the number of potential DNA, RNA and protein biomarkers under study.

18.3.1 DNA Biomarkers

Circulating DNA and tumour cells were among the first markers evaluated for cancer staging (Sidransky, 2002). Increased serum DNA concentrations are associated with cancer (principally metastatic cancer) and with other conditions such as sepsis and autoimmune disease. A number of studies suggest circulating tumour cells in the blood or bone marrow as indicators of systemic metastasis, but the clinical sample sizes have been small and the long-term survival benefit remains to be assessed. Mutations in oncogenes, tumour-suppressor genes, and mismatch-repair genes can serve as DNA biomarkers.

For instance, mutations in the oncogene KRAS predict metastatic spread in various tumour types, and there are mutations in the gene that encodes the tumour suppressor p53 in more than half of sporadic cancers. Germline inheritance of a TP53 mutation (Li–Fraumeni syndrome) confers a risk of developing many of the same cancers. Mutations in other cancer-related genes, such as the RAS oncogene or the tumour-suppressor genes CDKN2A (cyclin-dependent kinase inhibitor A, which encodes p16INK4A), APC (the adenomatouspolyposis coli gene) and RB1 (the retinoblastoma gene), also have potential as markers for prognosis or selection of therapy.

Epigenetic regulation of transcription and translation can also be important in many diseases including carcinogenesis. Histone deacetylation, lysine-specific histone-H3 methylation, and promoter region CpG methylation can function through transcriptional abrogation of tumour-suppressor genes (for example, CDKN2A, TP53, APC or the breast cancer 1 gene, BRCA1) or DNA mismatch-repair genes (for example, MLH1 or the O6-methyl-guanine-DNA methyltransferase gene, MGMT). They can also function through effects on apoptosis, invasion and the cell cycle.

Single-nucleotide polymorphisms (SNPs) are associated with increased cancer risk, and HAPLOTYPE assessment can be predictive for several cancers, including those of prostate, breast and lung. Because five well-known cancer susceptibility genes ataxia telangiectasia mutated (ATM), breast cancer 1 (BRCA1, BRCA2, RAD51 and TP53) show low haplotype diversity within ethnicities, perhaps as few as 10% of the SNPs will have to be sequenced for useful haplotype-based risk assessment. Genome-wide SNP analysis has been reported but SNPs are not currently used for formal cancer staging or grading.

18.3.2 RNA Biomarkers

Whereas most DNA markers are evaluated individually, many high – throughput technologies have been developed to assess mRNA expression comprehensively.

Among them are Affymetrix and NimbleGen arrays that are produced by light-directed in situ synthesis of oligonucleotides, Rosetta-Agilent ink-jet-printed arrays, Differential Display, Serial Analysis of Gene Expression (SAGE), and Bead Based Methods. Quantitative real-time RT-PCR is generally considered the "gold standard" against which other methods are validated, and it can now be performed at relatively high-throughput – for example, by using MICRO FLUIDIC cards.

Most RNA-based biomarkers undergoing clinical evaluation consist of multi-gene molecular patterns or "fingerprints". Although such patterns can be more accurate than single-molecule markers, choosing which genes to include in the pattern adds an additional layer of statistical complexity, prompting new developments in biostatistics, bioinformatics and data visualization. Molecular markers and their patterns have been analysed by various Supervised Algorithms, most prominently by double hierarchical clustering methods that lead to colour-coded "clustered image maps" (CIMs).

Pattern-based RNA-expression analysis of clinical breast cancers has identified previously unknown molecular subtypes that are associated with differences in survival. That analysis has also provided increased prognostic capability, predicted response to neoadjuvant therapy, predicted likelihood of metastasis in lymph-node negative patients and correctly predicted tumour grade from Laser Capture Micro dissected specimens. The transcript levels of enzymes important for drug metabolism have been used preclinically to predict the response to chemotherapy in lung and colon cancers. Similar approaches have led to novel discoveries for other cancers, including melanoma, leukaemias, lymphomas, and carcinomas of the lung, prostate and colon. Such cancer "snapshots" taken at the time of diagnosis can be expected to further the goals of cancer staging. Extensive validation studies will be required, however, to move those developments from clinical research to standard practice in staging. Several companies have attempted to do so and have made their RT-PCR based gene signatures available to the public for use in predicting survival. However, these and other RNA-based markers have not yet undergone rigorous, prospective clinical validation, and they have not been approved by the FDA.

18.3.3 Protein Biomarkers

Technology limits our capacity to interrogate the proteome of complex protein mixtures (e.g., blood or tissue) for biomarker discovery. It has been estimated that blood contains more than 100,000 different protein forms with abundances that span 10–12 orders of magnitude. Mass spectrometry (MS), the leading tool in proteomics, currently has a limited capacity to identify and quantitate proteins in complex mixtures (Chaurand et al., 2004). Proteins are often digested to 20 and 50 peptides, further increasing the complexity of species in a mixture. Mass spectrometers that use electrospray ionization provide high resolving power for small mass differences, and tandem instruments (MS/MS) can identify a peptide's protein of origin by examining the peptide pattern following induced fragmentation (Negm et al., 2002; Issaq et al., 2002).

This technology allows researchers to identify hundreds of the most abundant proteins from a complex mixture. Other mass spectrometric techniques, such as matrix-assisted laser desorption ionization/time-of-flight (MALDI-TOF) MS, can display hundreds of the most abundant intact proteins in a mixture, although identification of the proteins is not possible without applying additional laborious technologies (Anderson and Anderson, 1998). Most of the known biomarkers in blood occur at very low abundance and would not be revealed by either of these techniques.

These facts lead to several conclusions. First, simple profiling of body fluids that contain complex mixtures of proteins that span huge dynamic ranges is unlikely to be very effective with present technologies (Caprioli et al., 1997). Therefore, candidate biomarkers will need to be identified prior to an attempt to correlate these biomarkers in body fluids with disease status. A corollary of this conclusion is that the subsequent analysis of these candidates in body fluids will likely require reagents specific to each candidate to detect and quantitate them at low abundance.

18.3.3.1 Identifying Candidate Protein Biomarkers

Any proteins that are differentially expressed in cancer tissue when compared to normal tissue, or any proteins that are known to be involved in the cancer process, are good sources of candidate biomarkers for cancer. One of the standard methods to identify candidate biomarkers from the first category is to compare transcription profiles from disease vs. normal tissue to identify differentially expressed transcripts. Another approach that has received attention is to identify genes amplified in cancer tissues. While these are fruitful approaches, they will not detect the vast majority of protein biomarkers those that arise by post-transcriptional mechanisms.

Direct identification of proteins that are differentially present in cancer tissue due to changes in translation, degradation, or post-translational modification is well within the capacity of MS or even 2D-gel analysis followed by LC-MS (Diamandis, 2004; Fenn et al., 1989). Given the limitations of both methods, it is reasonable to expect to be able to identify hundreds of candidates from any particular fraction of disease tissue. Although these methods will be biased toward the abundant proteins, it is the abundant proteins that are more likely to be present in significant quantities in blood. One of the several approaches that may be useful is the identification of specific protein markers of tumour viability in tumour tissue itself. These markers, or protease fragments of these, could then be specifically targeted for analysis in body fluids. This may lead to "high quality" biomarkers directly associated with the disease tissue (Aebersold and Goodlett, 2001).

For the analysis of body fluids, the question is which fraction-(s) to examine. Here, we are in unknown territory, since we do not know which fraction of cellular proteins is more likely to be enriched for biomarkers that can ultimately be identified in blood. Some fractions worth examining are cell surface proteins, secreted proteins, phosphorylated or glycosylated proteins, products of proteolytic digestion, etc. Analysis of any one of these fractions could be done either following tumour cell enrichment, which has the advantage of amplifying the signals over normal tissue, or in unfractionated tumor tissue, which offers the advantage of preserving stromal cell proteins that may also be valuable markers of disease. No single laboratory can hope to examine all these sources comprehensively, and this consideration suggests that teams will need to collaborate on identification of biomarkers for even a single cancer type. In our search for biomarker candidates, we should not overlook the wealth of accumulated knowledge about cancer and cell biology. During the last forty years, we have achieved an impressive understanding of the molecular fundamentals of cancer.

Cancer arises from a single cell as a result of successive genetic changes that alter a number of cellular processes, including growth control, senescence, apoptosis, angiogenesis, and metastasis. Additionally, many cancers appear to have activated inflammation and wound healing genetic expression programs. These changes are driven by abnormal methylation or mutation. While the proteins that function in each of these cellular circuits are often thought of as drug targets, they also represent strong candidates for biomarkers.

Bioinformatics approaches can easily generate a list of hundreds of proteins associated with each of these processes. Examining the body fluids of cancer patients for all of these proteins will require reagents in the form of antibodies, aptamers, and isotopically labeled peptides. No single laboratory can be expected to produce the reagents for even a single process, and even if the reagents were made available, no single laboratory could examine the proteins in the blood of cancer patients for more than one process. These considerations lead to a need for teams of laboratories concentrating on a single cancer site and for a centralized source of standardized reagents with known performance characteristics. In addition to the large number of protein categories that must be analyzed to discover biomarker candidates, there is also the issue of choice of body fluid to examine as the ultimate source of diagnostic information. Investigators are currently examining plasma, serum, urine, sputum, and other materials, and there is no consensus as to the best site for even one specific cancer. This fact increases the dimensionality of the problem by requiring that biomarker discovery be pursued in as many different body fluids as possible, further reinforcing the need for team science.

A handful of the FDA-approved cancer biomarkers in clinical use are single proteins, and most are serum-derived. AFP, β-HCG and LDH are used in the AJCC system to stage testicular cancer. Other proteins, although not formally used for staging, are important for prognosis and selection of therapy. For example, the expression of HER2/NEU and Cytokeratins can be used to refine the prognosis of breast cancers. HER2/NEU, EGFR and KIT are used clinically to predict if breast cancers, colon cancers or GISTs will respond to trastuzumab, cetuximab or imatinib, respectively. Similarly, expression of ER or PR is necessary for hormonal therapies to be effective against breast cancer. Just as pattern-based RNA biomarkers frequently outperform single RNA markers in tumour classification, prognosis or prediction of response to therapy, protein-based "fingerprints" may outperform individual protein markers. Technologies such as differential in-gel electrophoresis (DIGE), two-dimensional polyacrylamide gel electrophoresis (2D-PAGE) and multidimensional protein identification technology (MudPIT) can be used for higher-throughput profiling with microgram quantities of protein. Other high-throughput technologies, such as the Reverse Phase Microarry and surface-enhanced laser desorption ionization time-of-flight (SELDI-TOF) mass spectrometry, are more sensitive (in the femtomolar range) and can cover more of the 12 orders of magnitude range of serum-protein expression levels.

Emerging nanotechnologies, such as Immuno PCR, Field Effect Transistor (FET) based protein detection and Quantum Dots, promise further increases in the sensitivity of protein markers, but those techniques are currently experimental. Protein quantity by itself might not be the salient marker parameter.

18.4 New Applications for Established Biomarkers

Biomarkers can play roles before cancer diagnosis (in risk assessment and screening), at diagnosis and after diagnosis (in monitoring therapy, selecting additional therapy and detecting recurrence). Markers that are currently considered for risk assessment or screening may also prove useful in cancer staging or prediction of response at the time of diagnosis. For example, the BRCA1 (breast cancer 1) gene can be used in breast cancer, both for risk assessment and as a predictor of 10-year survival. For patients with HIV/AIDS, the viral load and CD4-positive T-cell count predict the likelihood of acquiring an AIDS-related cancer and the probability that such a tumour will respond to highly active anti-retroviral therapy (HAART).

For risk assessment and screening, a marker must generally be inexpensive, highly specific and minimally invasive. Those requirements do not necessarily apply to markers for staging or grading. As noted in the text, markers considered infeasible for screening because they would yield too many false positives may still be useful after diagnosis. For example, serum CEA (carcinoembryonic antigen) is increased in colon, breast and lung cancer, but also in many benign conditions. It is, in that sense, non-specific, but increases in the context of known colon cancer strongly suggest recurrence of that malignancy (Cui et al., 2003). Similarly, although AFP (α-fetoprotein) and β-HCG (human chorionic gonadotropin-β) can be increased for many reasons, their reliability in assessing testicular cancer burden following diagnosis accounts for their integration into staging. PSA (prostate-specific antigen), cancer antigen (CA), CA19-9, and other, similar markers may also prove useful in similar contexts, but they have not been integrated into TNM staging because their expression often fails to correlate with tumour burden (Dhanasekaran et al., 2001).

The following are some of the diseases that have characteristic biomarkers that physicians can use for diagnosis:

- In Alzheimer's disease, elevations in beta secretase creates amyloid/beta-protein, which causes plaque to build up in the patient's brain, which causes dementia. Targeting this enzyme decreases the amyloid/beta-protein and so slows the progression of the disease. A procedure to test for the increase in amyloid/beta-protein is immunohistochemical staining, in which antibodies bind to specific antigens or biological tissue of amyloid/beta-protein.
- Heart disease is commonly assessed using several key protein based biomarkers. Standard protein biomarkers for CVD include interleukin-6, interleukin-8, serum

amyloid A protein, fibrinogen, and troponins. cTnI cardiac troponin I increases in concentration within 3–12 h of initial cardiac injury and can be found elevated days after an acute myocardial infarction. A number of commercial antibody based assays as well as other methods are used in hospitals as primary tests for acute MI.

Proteomic analysis of kidney cells and cancerous kidney cells (Bandara et al., 2003) is producing promising leads for biomarkers for renal cell carcinoma and developing assays to test for this disease. In kidney-related diseases, urine is a potential source for such biomarkers. Recently, it has been shown that the identification of urinary polypeptides as biomarkers of kidney-related diseases allows to diagnose the severity of the disease several months before the appearance of the pathology (Morrow and Braunwald, 2003).

18.4.1 Development Pathways

The discovery phase needs high-quality, well-characterized samples that may be human or from animal models. Once a promising lead is found, the presence of the biomarker should be confirmed in different samples. The next stage is to develop a clinically useful assay (often in serum or urine) and validate if it can detect established disease. The clinical utility of the biomarker is established in a retrospective longitudinal study and a prospective study and finally to determine whether the biomarker screening strategy can reduce the burden of disease. The final stage, often not appreciated, is the commercial development of the assay by industry.

18.4.2 Identification of Biomarkers

Candidate markers can be identified initially from cDNA array data on the NCI-60 cancer cell line panel, and then sequence-verified by re-sequencing of the clones and corroborated using Affymetrix oligonucleotide arrays (Fig. 18.1). Reverse-phase

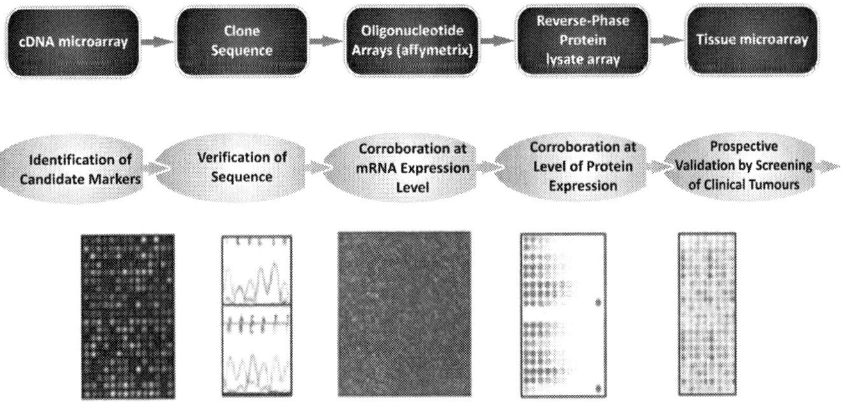

Fig. 18.1 Use of multiple molecular technologies in combination to identify candidate biomarkers

proteomic arrays later showed that the selectivities of the candidate biomarkers held up at the protein level, and tissue arrays indicated the same selectivity at the level of clinical tumour specimens. Candidate biomarkers to distinguish between colon and ovarian cancers of an unknown primary location can be identified and verified in this way.

18.4.3 Chronology of Biomarker Development

A biomarker is first identified, and then evaluated for a particular clinical indication (Fig. 18.2). Analytical and clinical validations must be performed before submission for US Food and Drug Administration approval. Alternatively, the marker might bypass the FDA approval process if it is to be used for "research purposes only".

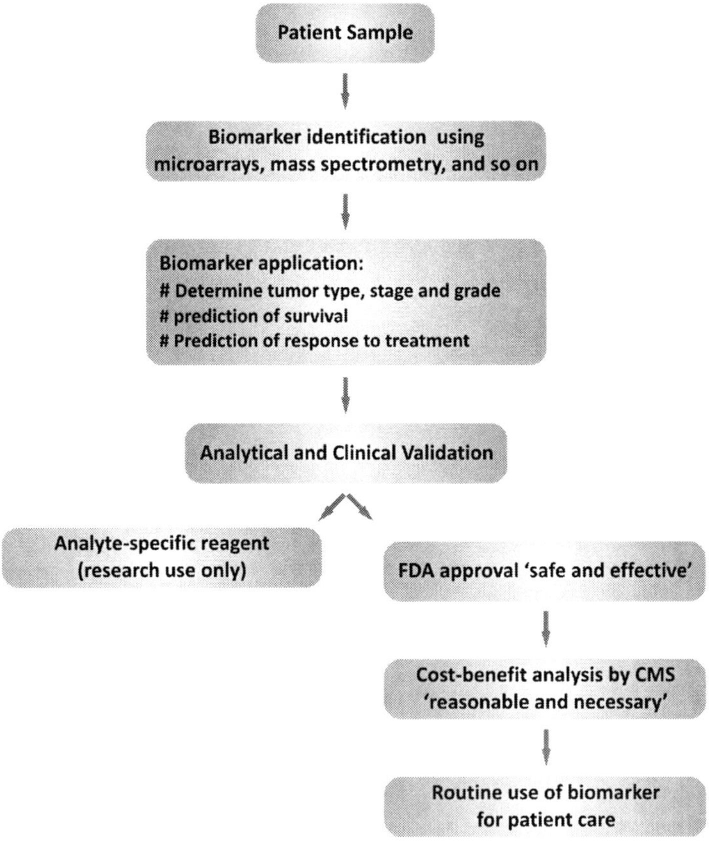

Fig. 18.2 Chronology of biomarker development

18.4.4 Role of Biomarkers in Cancer Detection, Diagnosis, and Prognosis

Biomarkers hold great promise for cancer detection, diagnosis, and prognosis because of their potential ability in identifying unique molecular signatures detrimental to certain pathophysiological states (Hanash, 2003). However, the progress has been slow in bringing biomarkers to clinical fruition. In the last several decades, only a few biomarkers, such as PAP Smears, PSA, and CA125, have found their way to clinical application. Since 1998, only two protein tests have been introduced in Food and Drug Administration-approved clinical tests. It is said that the biomarker pipeline is becoming dry due to the lack of discovery while the validation of existing biomarkers is slow. It seems ironic because most of the high throughput technologies, such as genomics and proteomics were supposed to be transforming the biomarker pipeline and clinical landscape through rapid discovery of biomarkers for clinical application, especially for early detection of cancer.

The reason for such a dismal state-of-the-science for biomarkers in clinical application among many reasons affecting the introduction of protein-based tests are the lack of required specificity, sensitivity, and other performance characteristics for a requested clinical purpose, such as diagnosis, prognosis, or monitoring. Another reason for the lack of clinically relevant biomarkers is the lack of an organized effort to move biomarkers from discovery to development to validation to clinical application.

As a consequence, much work in this area is fragmented into numerous small and disconnected studies without complete evaluation. Usually, the results of these studies cannot even be generalized to the population as a whole. In 1999, realizing the need for such an effort, the National Cancer Institute (NCI) established a consortium, the Early Detection Research Network (EDRN), to accelerate the development of biomarkers for cancer detection and diagnosis with the hope that some of these biomarkers may also serve as predictive markers for treatment and chemoprevention. The EDRN is based on the premise that a vertical approach to biomarker research in an integrated, multidisciplinary environment will facilitate collaboration among technology developers, basic scientists, clinicians, epidemiologists, biostatisticians, and other health professionals, and therefore expedite clinical applications. When this research is conducted within a consortium of collaborating investigators in a systematic and concerted fashion, translations of basic scientific discoveries in genomics, and proteomics into medical benefits are expedited.

Structured around four main components, the Network comprises a group of Biomarkers Developmental Laboratories (BDL), Biomarkers Reference Laboratories (BRL), Clinical Epidemiology and Validation Centers (CEVC), and a single Data Management and Coordinating Center (DMCC). BDLs develop and characterize new biomarkers, or refine existing biomarkers. BRLs serve as a resource for clinical and laboratory validation of biomarkers, including technological development, standardization of assay methods, and refinement. CEVCs conduct the early phases of clinical and epidemiological research on the application

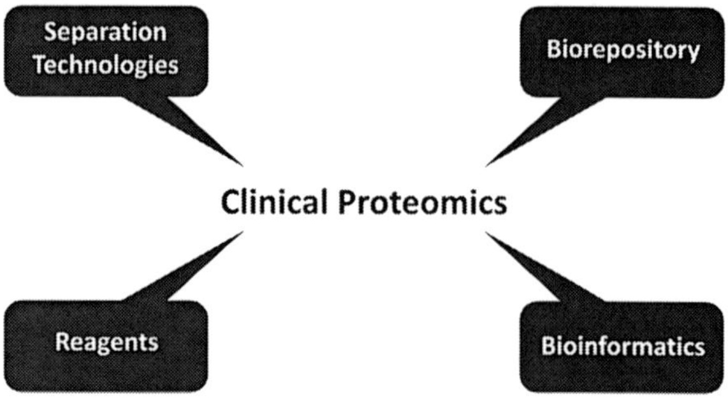

Fig. 18.3 Building blocks of clinical proteomics

of biomarkers. Statistical, logistics and informatics support is provided through a DMCC. This Center develops the theoretical statistical approaches to pattern analysis of multiple markers simultaneously. Also in 1999, a formal definition was given for a biomarker for a variety of applications.

Identifying and understanding these changes is the underlying theme in clinical proteomics with necessary infrastructures and resources (Fig. 18.3). The deliverables include identification of biomarkers that have utility both for early detection and for determining therapy.

These blocks comprise technologies (separation as well as fractionation technologies), analytical specific reagents, biorepository and bioinformatics. Separation technologies refer to instruments or devices that enable separation and identification of proteins and also include various types of fractionation technologies for depleting most abundant proteins or enriching less abundant proteins for their subsequent identification on mass-spectrometer based technologies.

Reagents refer to analytes or affinity reagents, such as antibodies, aptamers, and single chain antibodies to capture molecules of interest. Biorepository refers to the collection of clinically annotated biological specimens, including the serum and plasma. Bioinformatics refers to the collection of statistical and computational tools in preprocessing and postprocessing of data generated from high throughput platforms, such as microarrays, protein profiles (intensity vs m/z) for identification of candidate biomarkers or the development of disease classifiers.

This proposed life cycle depicts the stages in biomarker development and validation (Fig. 18.4). In the Discovery Phase, a promising lead (biomarker) is identified through basic research in animal models, tissue, or cell cultures.

In the Development Phase, a biomarker is identified in human specimens; an assay for the biomarker is developed or refined; analytical sensitivity of the assay for the biomarker is tested; and the clinical sensitivity of the assay measured to meet specific biological questions such as detecting mutations for biological events, or specific clinical questions such as detecting precancerous lesions.

Once the breast cancer is diagnosed, biomarkers are used for sub classification and for prediction of response, particularly to targeted therapies. Cancer antigen

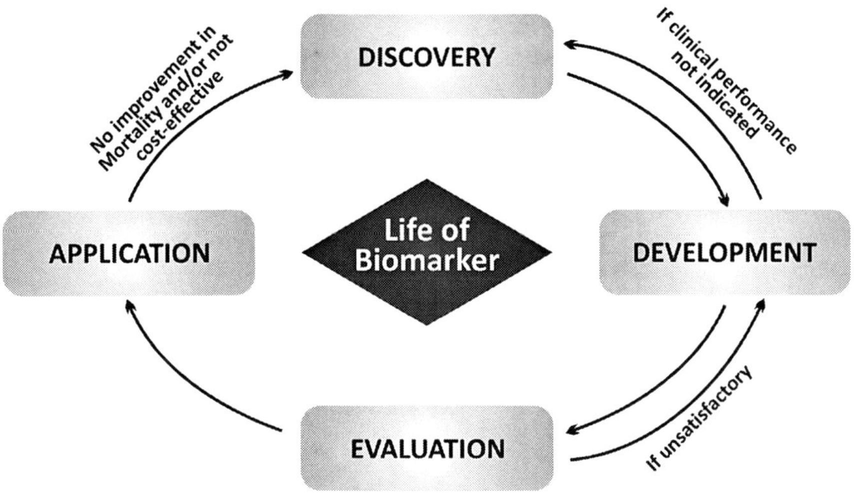

Fig. 18.4 Life cycle of biomarker

(CA) 15-3 might on rare occasions be used to monitor for recurrence. ER, oestrogen receptor; FISH, fluorescence in situ hybridization; IHC, immunohistochemistry; PET, positron-emission tomography; PR, progesterone receptor.

In the Evaluation Phase, a biomarker is subjected to broad questions to verify the intended use, such as whether the biomarker is applicable to biological and clinical settings in a variety of conditions. Biomarker is subjected to rigorous evaluation for precision, reproducibility, accuracy and other performance characteristics, including sensitivity and specificity in controlled study designs, such as a case-control or a large prospective trial. In the Application Phase, a biomarker test is field-tested for a specific clinical use, such as screening of cancer, and for its cost and effectiveness in reducing disease burden and mortality due to the disease. An unsuccessful verification or benefit may lead to return to the discovery phase and to cycle re-entry (Fig. 18.5). The advantages and disadvantages of biomarkers are discussed in Table 18.1.

18.4.5 Biomarkers in Cancer Therapy

The molecular analysis of cancer presents a genetic landscape that is orthogonal to the standard organ-specific view. Within an organ several genetic pathways may lead to cancer, and the same sets of altered genes appear across organs. This situation is leading to an increasing emphasis on the genetic characterization of cancer as biomarkers for prognosis and therapeutic intervention. With this comes the concomitant advantage of measuring proteins, the products of these genes. Since

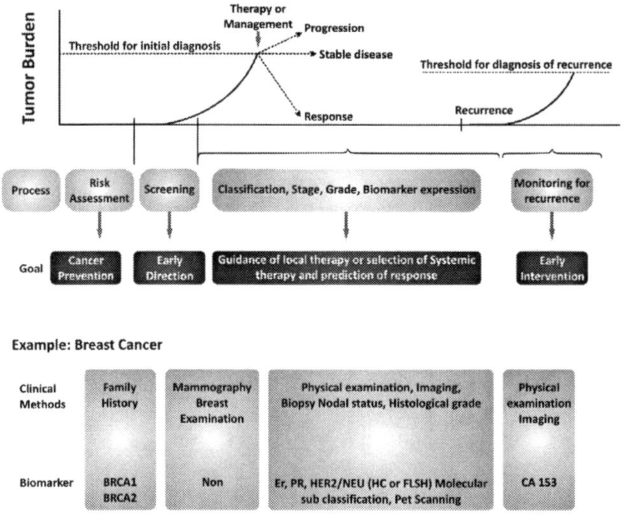

Fig. 18.5 Schematic representations of the uses of biomarkers at different stages in the clinical evolution of cancer, with breast cancer biomarkers as an example. (**a**) Before diagnosis, markers might be used for risk assessment and screening. At diagnosis, markers can assist with staging, grading, and selection of initial therapy. Later, they can be used to monitor therapy, select additional therapy, or monitor for recurrent disease. (**b**) Breast cancer biomarkers as an example. Genetic studies of BRCA1 (breast cancer 1) and BRCA2 are often performed in patients with a high risk of familial breast cancer

functional proteins may change significantly on a temporal and geographical basis, protein biomarkers promise to have a significant impact in cancer detection and therapy.

More accurate diagnosis of the hundreds of different types of cancer will permit more effective choice of therapy and will enhance clinical trials. Indeed, as therapies become more targeted to specific signal transduction and metabolic pathways, it is becoming of paramount importance to document the existence of those pathways in the target cancers. Consider some specific genomic examples. Targeting of breast cancers (Shin et al., 2002) with HERCEPTIN (trastuzumab) is not indicated if the patient's tumor does not overexpress Her-2/neu.1,2. Similarly, GLEEVEC (imatinib) is most effective against cancers that carry the Bcr-Abl translocation. Genetic alterations or transcript array profiles allow the stratification of many organ-specific cancers (e.g., breast, leukemia, lymphoma, sarcoma) into different subtypes that have distinctive therapeutic outcomes. For example, Myc gene amplification status predicts outcome for childhood neuroblastoma. The quantity of Bcr-Abl transcript predicts risk of disease recurrence in chronic myelogenous leukemia (CML) long before clinical symptoms recur. As these examples suggest, it will be important to develop informative biomarkers for virtually all cancer subtypes. Real-time markers of the physiological state of the patient, whether markers of disease process or those indicating efficacy of treatment will bring enormous benefit to patient care (Hammond and Traube, 2002). This would greatly improve individualized

Table 18.1 The various advantages and disadvantages of platforms[a] in biomarker discovery

Platform	Advantages for biomarker discovery	Disadvantages for biomarker discovery
Measurement of mRNA expression (e.g., differential display, SAGE, microarray)	Able to screen large number of "genes" commercially available	RNA levels may not directly relate to protein levels provide no information about posttranslational protein modifications difficult to handle large volume of data
2-D DIGE[b]	Assay of the actual biomarker not mRNA Allows identification of previously unknown biomarkers Can quantify amplitude of change in bio marker well established technique	Poor technique for difficult-to-solubilize proteins (e.g., membrane proteins), low-abundance proteins, and low-molecular weight protein Not high throughput, i.e., labor intensive
SELDI[c]	Well suited to generating a pattern of peptide peaks Corresponding to a disease biomarker high throughput, less labor intensive, and cheaper than 2-D electrophoresis can focus on certain subsets of proteins	Difficult to identify proteins Difficult to measure protein abundance specimen handling can have large impact on quality
LC/MS/MS[d]	Higher throughput than 2D DIGE Can identify protein by amino acid sequencing Increased yield of membrane proteins and low abundance proteins	Need to use ICAT[e] Abundance
Tissue microarray	High-throughput validation and prioritization of tissue Biomarkers (Pepe stage 1b) Obtain protein location by immunohistochemistry	Immunohistochemistry: need antibody; cannot detect "unknown" proteins. In situ hybridization – detects mRNA only quantitation issues Specimen quality issues
SNP	May produce unexpected new leads about pathogenesis of and biomarkers for disease	Only gives information about an individual's risk of disease, not presence of disease per se Provides no information about expression of protein

[a]2-D DIGE, two-dimensional difference gel electrophoresis; TOF, time of flight mass spectrometry; SELDI, surface-enhanced laser desorption ionization; ICAT, isotope-coded affinity tags; SNF, single nucleotide polymorphism.

[b]Proteins from normal and diseased samples are labelled with different flurescent dyes and then separated by two dimensional electrophoresis. Size of peptide (mass to charge ratio) is calculated based on the length of time for the peptide to travel through a vacuum.

[c]Proteins from a sample(s) bind to a chip if the coating of the chip allows an adequate protein-surface affinity. For example, hydrophobic proteins bind to a hydrophobic chip surface. Then the proteins are identified by a TOF mass spectrometer.

[d]Complex peptide mixtures are separated by chromatography (e.g., reverse phase, cation exchange), then the chromatography fractions are analysed by TOF mass spectrometry; When two TOF mass spectrometers are used in "series," this is referred to as MS/MS. This allows actual peptide sequencing.

[e]Proteins from two different sources (e.g., disease versus normal) can be labelled with "light" and "heavy" tags. After LC/MS/MS, the relative abundances of different peptides in the two samples can be calculated.

dosing and agent selection. Indeed, some chemotherapeutic agents are currently individualized by adjusting the dose according to the patient's individual metabolic characteristics. Moreover, a series of agents could be tested on the same patient in a matter of weeks to determine empirically the most effective therapy.

18.4.6 Biomarkers in Early Detection

Short of prevention, improved diagnostics to detect cancer at an early stage, when it is curable with contemporary methods, would provide the greatest benefit for cancer patients. For most cancers, 5- and even 10-year survival often approaches 90% for cancer detected at an early stage, while it may drop to 10% or less for cancer detected at a late stage 6. It is well established that screening to detect cancer earlier saves lives. For example, the PAP smear strongly reduces mortality through early detection of pre-neoplastic cervical cancer lesions, as does colonoscopy for colon cancer. Furthermore, both tests have been widely employed despite their significant inconvenience, unpleasantness, cost, and requirement for clinical expertise. These successful protocols have created a receptive social environment that will encourage the rapid application of new tests. What is needed, however, are affordable and effective diagnostic tests for more types of cancer (Pepe et al., 2001).

The risk of cancer recurrence is high in those patients who have previously had cancer, even for those who have been in remission for 5 or more years. Cancer survivors constitute a high-risk group that would benefit from improved tests for early detection of disease recurrence if effective therapy is available. Monitoring CML patients during GLEEVEC therapy and in the post-transplant setting for the persistence of the Bcr-Abl transcript is already an effective technique for detecting disease recurrence at an early stage.

18.4.7 Biomarkers in Risk Assessment

If we could segment the population into smaller groups at increased risk for specific cancers, then screening individuals for early cancer detection will be more cost-effective and prevent fewer unnecessary interventions. Molecular biomarkers that predict risk are essential. Success in identifying individuals at increased risk has been achieved for many cancers through epidemiological studies that identify strong environmental or behavioral risk factors and by genetic studies that identify mutations underlying rare inherited cancer syndromes. However, the use of biomarkers in the assessment of risk for sporadic disease remains largely unexplored (Paweletz et al., 2001).

Epidemiologic studies indicate that lifestyle, diet, and environmental exposures significantly affect the risk for sporadic disease, but few advances have been made in identifying biomarkers that reflect the stable, cumulative molecular changes associated with disease or mediate this risk. While stochastic genetic alterations occur infrequently and are difficult to detect, there is increasing interest in more common, stable genetic, and epigenetic changes in histologically normal or premalignant tissue, reflective of deleterious exposure and associated with increased risk for malignant progression. For example, DNA mutation, methylation, and ploidy changes are highly correlated with an increased risk for cancer of the esophagus. Similarly, the loss of imprinting of insulin-like growth factor 2 (IGF2) in peripheral blood lymphocytes in subjects at risk for colorectal cancer is a risk marker. Many familial cancer-prone syndromes are due to defects in DNA repair. A study by Scott and Roberts revealed that approximately 40% of breast cancer patients, prior to treatment, exhibit a defect in DNA double strand break repair in their white blood cells. Therefore, biomarkers related to DNA repair capacity might be effective for risk stratification.

For example, 8-oxoguanine DNA N-glycosylase (OGG) activity in peripheral blood monocytes is associated with risk of lung cancer. Noninvasive monitoring of biomarkers in individuals at risk for cancer or with cancer would benefit enormously by better methods for determining cancer risk, detecting and localizing cancer at its earliest stage, profiling for therapeutic decision making, and monitoring response to therapy in real time. For some of these applications, we will not know whether a tumor exists or, if so, its anatomical site. Thus, there is a need for biomarkers that can be monitored noninvasively in readily available bodily fluids. Tumors "leak" DNA and proteins into circulation. Tumors also induce dramatic alterations of surrounding stroma (e.g., alterations in basement membranes, angiogenesis, and lymphogenesis) and release proteases that digest normal tissue and plasma proteins. It is therefore reasonable to expect many biomarkers to be present in blood and other fluids. Indeed, several individual plasma proteins (i.e., prostate-specific antigen [PSA], cancer antigen 125 [CA125], carcinoembryonic antigen [CEA], and alpha fetoprotein [AFP] antigen) are in clinical use as markers of the presence of a tumor, response to therapy or of tumor recurrence. Progress is being made in detecting mutated or methylated tumor DNA in accessible bodily fluids including blood, urine, sputum, and stools.

Academic and commercial entities are currently developing highly sensitive and specific panels of biomarkers for different cancer sites. DNA biomarkers found in bodily fluids may change with tumor progression and correlate with aggressiveness and therapeutic response. In addition to DNA- and RNA-based diagnostics, other molecules offer diagnostic potential; many new contrast agents for molecularly targeted imaging are under development. The technology for discovering and developing nucleic acid biomarkers and new imaging agents is robust and is being actively pursued in academic and commercial laboratories (Reyzer et al., 2004).

18.5 Organization of a Program in Protein Biomarker Discovery

The proposed scheme of organization for biomarker discovery is explained taking 'Cancer' as an example. The proposed initiative is structured into three integrated core components: Informatics, Reagents, and Technology Assessment. In addition to its development function, each core component would support, through services and resources, various satellite projects organized around cancer sites, biomarker mines, and pilot projects. These components do not need to be localized at a single physical site; with the aid of information technology, they can be "virtual" sites.

18.5.1 Cancer Site Teams

The Cancer Site Teams shown in Fig. 18.6 represent groups of investigators dedicated to biomarker discovery at a particular cancer site, e.g., breast, lung, prostate, and colon. Each team would have clinicians with access to human tissue at that site, scientists working on mouse models of the particular cancer, and other researchers who are skilled in protein fractionation and enrichment techniques, antibody preparation, mass spectrometry, and informatics.

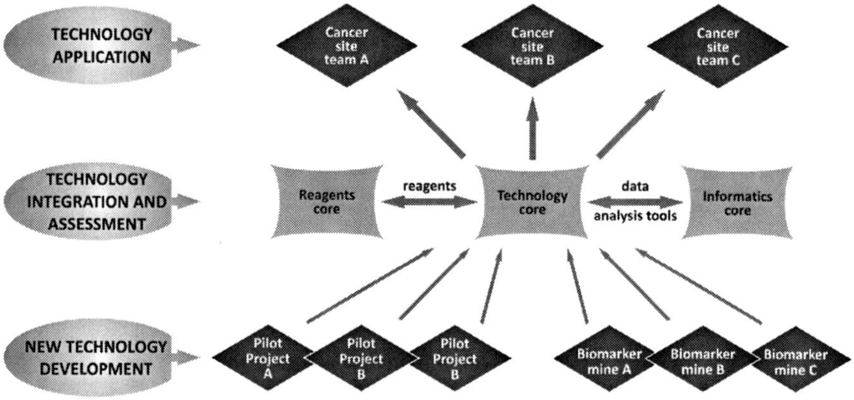

Fig. 18.6 Fully integrated clinical biomarker discovery technology program centers around Core resources for technology assessment, reagents, and informatics

18.5.2 Biomarker Mines

The Biomarker Mine components would comprise single investigators or small groups dedicated to optimizing the methods for discovery in a particular category of biomarkers, such as cell surface or secreted proteins. Optimization of procedures for tissue workup, bioinformatics, and the biochemical fractionations and analysis for a specific category of biomarkers requires special attention that is unlikely to occur in the Cancer Site Teams who are more committed to discovering biomarkers for a particular cancer site.

The Biomarker Mine components would instead develop analytical methods to identify candidate biomarkers in a specific mine, identify and rank prospective candidates, develop reagents for their enrichment and detection, work collaboratively with one or more Cancer Site teams and the Technology Assessment Core to mine for cancer-specific biomarkers, deposit data into the Informatics Core, and share data and reagents with other components.

18.5.3 Informatics Platform

The Informatics Core will develop tools that enable laboratories to communicate efficiently and to compare data. This core would develop a standardized data format to facilitate cross-platform comparisons. It would also provide an opensource suite of analysis tools compatible with this standard data format to facilitate standardization of analysis across laboratories and allow meaningful comparisons of results. Additionally, a central database for storing the data of the programs would be housed in the Informatics Core. Finally, the Informatics Core will assemble and curate data sources for candidate biomarkers.

18.5.4 Reagents Core

The Reagents Core would organize tools for biomarker discovery, maintaining a virtual database of reagents, their characteristics, and their performance data, along with reagent request forms. Reagents as defined here include mice, mouse and human tissues and plasma, antibodies for candidate enrichment, standard batches of bodily fluids, standard protein and/or peptide mixtures for spiking fluids prior to analysis, and other standard reagents developed or discovered by investigators. Reagents, along with data on their performance and quality, will be acquired and dispersed quickly to other core facilities and satellites.

One of the most frequently identified needs in the biomarker discovery field is for high quality tissues from cancer patients and controls. It is unclear at present how best to collect, prepare, and store tissue or bodily fluids. For this reason, tissue collections should probably be coordinated at the inception of a project and occur prospectively. The greatest confounder of effective biomarker discovery is the false positives generated because the control tissue is not matched sufficiently to the disease tissue. There are several possible sources of valid control tissue, including "normal" tissue from the same organ and individual as the cancer, normal tissue from other healthy individuals (including identical twins where possible) and tissue from individuals with disease of the same organ but of a different nature (e.g., inflammation). Ideally, each should be compared with disease tissue. Initial evaluation of biomarkers will require hundreds of clinically annotated tissue and blood samples. To evaluate early detection capability, plasma must be collected from early-stage patients (together with stored tissue) as well as from pre-symptomatic individuals who are later diagnosed with cancer (Pepe et al., 2004).

To evaluate clinical response, plasma obtained from well-controlled clinical trials with clinical outcomes is essential. For protein biomarker discovery, it is essential to have access to many antibodies for detection of candidates that are likely to be present at low concentrations in the blood. One could easily create a list of more than a thousand proteins known to be involved in cancer-related processes, each of which is a potential candidate biomarker. Moreover, as candidate proteins are discovered by proteomic methods that distinguish disease from normal tissue, additional antibodies will be needed for these candidates. Several thousand antibodies will likely be required, as pairs of antibodies for enzyme-linked immunosorbent assays (ELISA) or other tests, such as proximity-based detection.

Many of the antibodies developed for a particular cancer site will be reused at other cancer sites, thereby promoting efficient use of resources. In addition to antibody-based methods, other approaches, such as VICAT, are available to identify and quantitate a candidate peptide at low concentration in complex mixtures such as blood. Anti-peptide antibodies will be useful for enrichment in the SISCAPA approach, and appropriate isotopically labeled peptide reagents will be needed.

New technologies will be evaluated through pilot projects and "biomarker mines" comprised of individual investigators and smaller research teams. Upon standardization by the Cores, new proteomic technologies can be tested in the clinical setting by Cancer Site Teams dedicated to biomarker discovery at a particular cancer site.

18.5.5 Incorporating Technology Improvements

Although it is possible to improve the discovery of protein biomarkers, the performance of technologies for protein separation and identification remains the major limitation to translate biomarker discovery into therapies and diagnostics. Technology improvements are likely to occur over the next few years, and these advances should be rapidly incorporated into the discovery process. Technology improvement should be considered in mass spectrometry, protein arrays, protein fractionation, protein detection, protein quantitation, and other appropriate methodologies. To support such efforts, the Technology Assessment Core will evaluate mature, commercialized technologies central to biomarker discovery to provide laboratories with the best possible techniques and protocols. Initially, this Core would use standard reference plasma to systematically compare existing technologies in each step of the biomarker discovery process. The best-performing technologies would then be integrated into an optimized platform against which new technologies (discovered via pilot projects and biomarker mines) could be tested. The ability of the integrated and optimized platform to identify biomarkers would likely best be assessed using mouse models.

This Core would also provide data to the Informatics Core for algorithm development and would deposit useful reagents (including reference plasma and mouse tissues) into the Reagents Core for dissemination. Finally, this Core would collaborate with Cancer Site components to implement optimized technologies to identify biomarkers in human samples. This Core might be a virtual core that requests performance criteria from investigators with new technology. To encourage new technology development prior to commercialization a pilot project program might be useful. The pilot projects would be single-investigator projects designed to test a new technology, such as protein chips for biomarker discovery. Where appropriate, promising new technologies would be reproduced and tested against current standards by the Technology Assessment Core.

18.6 Conclusion

Understanding of the entire complex biomarker development process and using a team approach are required for a successful biomarker development project. Every step requires validation, of both assay performance and diagnostic utility, as the biomarker moves toward the clinic. Advances in genomics, proteomics and molecular pathology have generated many candidate biomarkers with potential clinical value.

For the future we need to assess technologies central to biomarker discovery in order to provide laboratories with the best possible techniques and protocols; develop an open-source suite of analysis tools compatible with this standard data format to facilitate standardization of analysis across laboratories and allow meaningful comparisons of results; establish a well-structured database for store and organize the data for candidate biomarkers and establish a source for standardized reagents, including human specimens, mouse models, antibodies, and other reagents.

Finally, for an effective diagnosis and tailored therapy, there is a need to integrate data from a variety of platforms, such as genomics, imaging, and proteomics, for a particular cancer site from multiple studies. To date, only a few large-scale integrated molecular profiling efforts have combined data obtained from multiple studies, or combined data obtained through two different global profiling platforms (genomic and transcriptomic, or transcriptomic and proteomic) for the same set of study samples.

After spending huge sums of money over the years on biomarker research there is remarkably little to show for the effort. The needs for clinically effective biomarkers for bladder cancer and other cancers are great, and only a fresh approach based on the powerful new technologies available recently are needed. However, along with new technology must come a new understanding that old paradigms are not adequate in the post-genomic age (Kennedy, 2002). To summarize, the new paradigm should consist of the following: further improvement in understanding of the complex, interacting system of genes and proteins to be able to develop relevant tests for dysregulation of the system, further development of mathematical techniques to analyze data within the system paradigm, and a high quality set of clinical studies using genome-wide techniques that captures the spectrum of disease. The intelligent development of biomarkers truly is a problem in systems biology.

References

Aebersold, R., Goodlett, D.R., 2001, Mass spectrometry in proteomics. *Chem Rev* 101:269–295.

Anderson, N.L., Anderson, N.G., 1998, Proteome and proteomics: New technologies, new concepts and new words. *Electrophoresis* 19:1853–1861.

Atkinson, A.J., 2001, Biomarkers and surrogate endpoints: preferred definitions and conceptual framework. *Clin Pharmacol Ther* 69:89–95.

Bandara, L.R., Kelly, M.D., Lock, E.A., Kennedy, S., 2003, A potential biomarker of kidney damage identified by proteomics: preliminary findings. *Biomarkers* 8(3–4):272–286.

Caprioli, R.M., Farmer, T.B., Gile, J., 1997, Molecular imaging of biological samples: localization of peptides and proteins using MALDI-TOF MS. *Anal Chem* 69(23):4751–4760.
Chaurand, P., Schwartz, S.A., Caprioli, R.M., 2004, Profiling and imaging proteins in tissue sections by MS. *Anal Chem* 76:86A–93A.
Cui, H., Cruz-Correa, M., Giardiello, F.M., Hutcheon, D.F., Kafonek, D.R., Brandenburg, S., Wu, Y., He, X., Powe, N.R., Feinberg, A.P., 2003, Loss of IGF2 imprinting: a potential marker of colorectal cancer risk. *Science* 299(5613):1753–1755.
Dhanasekaran, S.M., Barrette, T.R., Ghosh, D., Shah, R., Varambally, S., Kurachi, K., Pienta, K.J., Rubin, M.A., Chinnaiyan, A.M., 2001, Delineation of prognostic biomarkers in prostate cancer. *Nature* 412(6849):822–826.
Diamandis, E.P., 2004, Mass spectrometry as a diagnostic and a cancer biomarker discovery tool: opportunities and potential limitations. *Mol Cell Proteomics* 3:367–378.
Fenn, J.B., Mann, M., Meng, C.K., Wong, S.F., Whitehouse, C.M., 1989, Electrospray ionization for mass spectrometry of large biomolecules. *Science* 246:64–71.
Ferber, G., 2002, Biomarkers and proof of concept. Methods find. *Exp Clin Pharmacol* 24:35–40.
Hammond, M.E., Taube, S.E., 2002, Issues and barriers to development of clinically useful tumor markers: a development pathway proposal. *Semin Oncol* 29:213–221.
Hanash, S., 2003, Harnessing immunity for cancer marker discovery. *Nat Biotechnol* 21:37–38.
Hurst, R.E., 2009, Does the biomarker search paradigm need re-booting? *BMC Urol* 9:1.
Issaq, H.J., Veenstra, T.D., Conrads, T.P., Felschow, D., 2002, The SELDI TOF MS approach to proteomics: protein profiling and biomarker identification. *Biochem Biophys Res Commun* 292(3):587–592.
Kennedy, S., 2002, The role of proteomics in toxicology: identification of biomarkers of toxicity by protein expression analysis. *Biomarkers* 7:269–290.
MacGregor, J.T., 2003, The future of regulatory toxicology: impact of the biotechnology revolution toxicological. *Sciences* 75:236–248.
Morrow, D.A., Braunwald, E., 2003, Future of biomarkers in acute coronary syndromes: moving toward a multimarker strategy. *Circulation* 108:250–252.
Negm, R.S., Verma, M., Srivastava, S., 2002, The promise of biomarkers in cancer screening and detection. *Trends Mol Med* 8:288–293.
Paweletz, C.P., Trock, B., Pennanen, M., Tsangaris, T., Magnant, C., Liotta, L.A., Petricoin, E.F., 2001, Proteomic patterns of nipple aspirate fluids obtained by SELDI-TOF: potential for new biomarkers to aid in the diagnosis of breast cancer. *Dis Markers* 17(4):301–307.
Pepe, M.S., Etzioni, R., Feng, Z., Potter, J.D., Thompson, M.L., Thornquist, M., Winget, M., Yasui, Y., 2001, Phases of biomarker development for early detection of cancer. *J Natl Cancer Inst* 93(14):1054–1061.
Pepe, M.S., Etzioni, R., Feng, Z., Potter, J.D., Thompson, M.L., Thornquist, M., Winget, M., Yasui, Y., 2004, Early changes in protein expression detected by mass spectrometry predict tumor response to molecular therapeutics. *Cancer Res* 64(24):9093–9100.
Reyzer, M.L., Caldwell, R.L., Dugger, T.C., et al., 2004, Early changes in protein expression detected by mass spectrometry predict tumor response to molecular therapeutics. *Cancer Res* 64:9093–9100.
Shin, B.K., Wang, H., Hanash, S., 2002, Proteomics approaches to uncover the repertoire of circulating biomarkers for breast cancer. *J Mammary Gland Biol Neoplasia* 7:407–413.
Sidransky, D., 2002, Emerging molecular markers of cancer. *Nat Rev Cancer* 2:210–219; doi:10.1038/nrc755.

Chapter 19
Diagnosis and Monitoring of Infections

Abstract As molecular techniques for identifying and detecting microorganisms in the clinical microbiology laboratory have become routine, questions about the cost of these techniques and their contribution to patient care need to be addressed. Molecular diagnosis is most appropriate for infectious agents that are difficult to detect, identify, or test for susceptibility in a timely fashion with conventional methods. During the last 10 years, the detection of infectious disease agents has begun to include the use of nucleic acid-based technologies. Diagnosis of infection caused by parasitic organisms is the last field of clinical microbiology to incorporate these techniques and molecular techniques (e.g., PCR and hybridization assays) have recently been developed for the detection, species differentiation, and phylogenetic analysis.

Keywords PCR · Infections · Detections · Diagnosis · Sexually transmitted diseases · Vector borne diseases · Viral infections · Bacterial infections · Fungal infections · Molecular methods · Molecular epidemiology

19.1 Prologue

The tools of molecular biology have proven readily adaptable for use in the clinical diagnostic laboratory and promise to be extremely useful in diagnosis, therapy, epidemiologic investigations and infection control. Although technical issues such as ease of performance, reproducibility, sensitivity, and specificity of molecular tests are important, cost and potential contribution to patient care are also of concern. Molecular methods may be an improvement over conventional microbiologic testing in many ways. Currently, their most practical and useful application is in detecting and identifying infectious agents for which routine growth-based culture and microscopy methods may not be adequate.

Nucleic acid-based tests used in diagnosing infectious diseases use standard methods for isolating nucleic acids from organisms and clinical material. Because the target DNA or RNA may be present in very small amounts in clinical specimens, various signal amplification and target amplification techniques have been

M. Debnath et al., *Molecular Diagnostics: Promises and Possibilities*,
DOI 10.1007/978-90-481-3261-4_19, © Springer Science+Business Media B.V. 2010

used to detect infectious agents in clinical diagnostic laboratories (Vaneechoutte and Van Eldere 1997). Although mainly a research tool, nucleic acid sequence analysis coupled with target amplification is clinically useful and helps to detect and identify previously uncultivable organisms and characterize antimicrobial resistance gene mutations, thus aiding both diagnosis and treatment of infectious diseases. Automation and high-density oligonucleotide probe arrays (DNA chips) also hold great promise for characterizing microbial pathogens.

Although most clinicians and microbiologists enthusiastically welcome the new molecular tests for diagnosing infectious disease, the high cost of these tests is of concern. Despite the probability that improved patient outcome and reduced cost of antimicrobial agents and length of hospital stay will outweigh the increased laboratory costs incurred through the use of molecular testing, such savings are difficult to document.

19.2 Concept of Diagnosis and Monitoring of Infections

Over the past several years, the development and application of molecular diagnostic techniques has initiated a revolution in the diagnosis and monitoring of infectious diseases. As technology advances, diagnostic tests continue to improve, and each year, we are presented with new alternatives to the standard procedures. Given the plethora of diagnostic alternatives, diagnostic tests must be evaluated to determine their place in the diagnostic armamentarium. Criteria include sensitivity and specificity, positive and negative predictive values, likelihood ratios for positive and negative tests, and receiver operating characteristic curves. The prevalence of disease influences the diagnostic utility of tests, particularly when diseases are rare or very frequent (Fardy, 2009).

Microbial phenotypic characteristics, such as protein, bacteriophage, and chromatographic profiles, as well as biotyping and susceptibility testing, are used in most routine laboratories for identification and differentiation. Nucleic acid techniques, such as plasmid profiling, various methods for generating restriction fragment length polymorphisms, and the polymerase chain reaction, are making increasing inroads into clinical laboratories (Wagar, 1996).

PCR-based systems to detect the etiologic agents of disease directly from clinical samples, without the need for culture, have been useful in rapid detection of unculturable or fastidious microorganisms (Schluger et al., 1994). Additionally, sequence analysis of amplified microbial DNA allows for identification and better characterization of the pathogen. Subspecies variation, identified by various techniques, has been shown to be important in the prognosis of certain diseases. Other important advances include the determination of viral load and the direct detection of genes or gene mutations responsible for drug resistance. Increased use of automation and user-friendly software makes these technologies more widely available. In all, the detection of infectious agents at the nucleic acid level represents a true synthesis of clinical chemistry and clinical microbiology techniques (Tang et al., 1997).

19.2.1 Detection and Identification of Pathogens Without Target Amplification

Commercial kits containing non-isotopically labeled nucleic acid probes are available for direct detection of pathogens in clinical material and identification of organisms after isolation in culture (Table 19.1). Use of solution-phase hybridization has allowed tests to be performed singly or in batches in a familiar microwell format.

Although direct detection of organisms in clinical specimens by nucleic acid probes is rapid and simple, it suffers from lack of sensitivity. Most direct probe detection assays require at least 10^4 copies of nucleic acid per microliter for reliable detection, a requirement rarely met in clinical samples without some form of amplification. Amplification of the detection signal after probe hybridization improves sensitivity to as low as 500 gene copies per microliter and provides quantitative capabilities. This approach has been used extensively for quantitative assays of viral load (HIV, hepatitis B virus [HBV] and hepatitis C virus [HCV]) (Table 19.1) but does not match the analytical sensitivity of target amplification-based methods, such as polymerase chain reaction, for detecting organisms.

Table 19.1 FDA-approved molecular diagnostic tests for infectious diseases

Test	Method	Company
Chlamydia trachomatis detection	PCR[a]; LCR; TMA; Hybrid capture	Roche Abbott Gen-Probe Digene
Neisseria gonorrhoeae detection	LCR; Hybrid capture	Abbott Digene
C. trachomatis/N. gonorrhoeae	Screening/detection; Hybridization	Becton-Dickinson
Mycobacterium tuberculosis detection	PCR TMA	Roche; Gen-Probe HPV screening Hybrid capture Digene
CMV	Hybrid capture	NASBA Digene Organon Teknika
Grp A strep detection	Hybridization HIV quantitation PCR	Gen-Probe Roche
Gardnerella, T. vaginalis, and *Candida*	Hybridization	Becton-Dickinson
Culture confirmation for bacteria and fungi	Hybridization	Gen-Probe

[a]PCR = polymerase chain reaction; LCR = ligase chain reaction; TMA = transcription-mediated amplification.

The commercial probe systems that use solution-phase hybridization and chemiluminescence for direct detection of infectious agents in clinical material include the PACE2 products of Gen-Probe and the hybrid capture assay systems of Digene and Murex (Table 19.1). These systems are user friendly, have a long shelf life, and are adaptable to small or large numbers of specimens. The PACE2 products are designed for direct detection of both *Neisseria gonorrhoeae* and *Chlamydia trachomatis* in a single specimen (one specimen, two separate probes). The hybrid capture systems detect human papilloma virus (HPV) in cervical scrapings (Ciotti et al., 2004), herpes simplex virus (HSV) in vesicle material, and cytomegalovirus (CMV) in blood and other fluids (Cinque et al., 1995). All these tests have demonstrated sensitivity exceeding that of culture or immunologic methods for detecting the respective pathogens but are less sensitive than PCR or other target amplification-based methods.

The signal amplification-based probe methods for detection and quantitation of viruses (HBV, HCV, HIV) are presented in an enzyme immunoassay-like format and include branched chain DNA probes (Chiron) and QB replicase (Gene-Trak) methods (Table 19.1). These methods are not as sensitive as target amplification-based methods for detection of viruses; however, the quantitative results have proven useful for determining viral load and prognosis and for monitoring response to therapy.

Probe hybridization is useful for identifying slow-growing organisms after isolation in culture using either liquid or solid media. Identification of mycobacteria and other slow-growing organisms such as the dimorphic fungi (*Histoplasma capsulatum, Coccidioides immitis,* and *Blastomyces dermatitidis*) has certainly been facilitated by commercially available probes. All commercial probes for identifying organisms are produced by Gen-Probe and use acridinium ester-labeled probes directed at species-specific rRNA sequences (Table 19.1). Gen-Probe products are available for the culture identification of *Mycobacterium tuberculosis, M. avium-intracellulare* complex, *M. gordonae, M. kansasii, Cryptococcus neoformans*, the dimorphic fungi, *N. gonorrhoeae, Staphylococcus aureus, Streptococcus pneumoniae, Escherichia coli, Haemophilus influenzae, Enterococcus* spp., *S. agalactiae*, and *Listeria monocytogenes*. The sensitivity and specificity of these probes are excellent, and they provide species identification within one working day. Because most of the bacteria listed, plus *C. neoformans*, can be easily and efficiently identified by conventional methods within 1–2 days, many of these probes have not been widely used. The mycobacterial probes, on the other hand, are accepted as mainstays for the identification of *M. tuberculosis* and related species.

19.2.2 Nucleic Acid Amplification

Nucleic acid amplification provides the ability to selectively amplify specific targets present in low concentrations to detectable levels; thus, amplification-based methods offer superior performance, in terms of sensitivity, over the direct (non-amplified) probe-based tests. PCR (Roche Molecular Systems, Branchburg, NJ) was the first such technique to be developed and because of its flexibility and ease of performance remains the most widely used molecular diagnostic technique in both research and clinical laboratories. Several different amplification-based strategies have been developed and are available commercially (Table 19.1). Commercial amplification-based molecular diagnostic systems for infectious diseases have focused largely on systems for detecting *N. gonorrhoeae*, *C. trachomatis*, *M. tuberculosis*, and specific viral infections (HBV, HCV, HIV, CMV, and enterovirus) (Table 19.1).

Given the adaptability of PCR, numerous additional infectious pathogens have been detected by investigator-developed PCR assays (Table 19.2). In many instances, such tests provide important and clinically relevant information that would otherwise be unavailable since commercial interests have been slow to expand the line of products available to clinical laboratories. In addition to qualitative detection of viruses, quantitation of viral load in clinical specimens is now recognized to be of great importance for the diagnosis, prognosis, and therapeutic monitoring for HCV, HIV, HBV, and CMV. Both PCR and nucleic acid strand-based amplification systems are available for quantitation of one or more viruses.

The adaptation of amplification-based test methods to commercially available kits has served to optimize user acceptability, prevent contamination, standardize reagents and testing conditions, and make automation a possibility. It is not clear to what extent the levels of detection achievable by the different amplification strategies differ. None of the newer methods provides a level of sensitivity greater than that of PCR. In choosing a molecular diagnostic system, one should consider the range of tests available, suitability of the method to workflow, and

Table 19.2 Noncommercial nucleic acid-based tests for clinically important viral and bacterial pathogens

Organism	Specimen type	Clinical indication
Epstein-Barr virus (EBV)	Cerebrospinal fluid (CSF)	EBV lymphoproliferative disorder
Herpes simplex virus (HSV) types 1 and 2	CSF; Vitreous humor	Encephalitis
Varicella-zoster virus (VZV)	Various tissues	VZV reactivation
JC virus	CSF	Progressive mutifocal leukoencephalopathy
Enterovirus	CSF	Aseptic meningitis
Parvovirus B19	Amniotic fluid; serum	Hydrops fetalis; Anemia
Adenovirus	Urine; tissues; blood;	Immunocompromised patients, transplant recipients
Ehrilichia	Blood; human	Granulocytic and monocytic ehrlichiosis
Bordetella pertussis	Nasopharyngeal aspirate	Whooping cough
Legionella pheumophila	Respiratory	Atypical pneumonia
Chlamydia pneumoniae	Respiratory	Atypical pneumonia
Mycoplasma pneumoniae	Respiratory	Atypical pneumonia
Helicobacter pylori	Gastric fluid; stool	Peptic ulcer disease

All tests use polymerase chain reaction. The list is not all-inclusive.

cost. Choosing one amplification-based method that provides testing capabilities for several pathogens is certainly practical.

Amplification-based methods are also valuable for identifying cultured and noncultivatable organisms. Amplification reactions may be designed to rapidly identify an acid-fast organism as *M. tuberculosis* or may amplify a genus-specific or "universal" target, which then is characterized by using restriction endonuclease digestion, hybridization with multiple probes, or sequence determination to provide species or even subspecies delineation. Although identification was initially applied to slow-growing mycobacteria, it has applications for other pathogens that are difficult or impossible to identify with conventional methods.

19.2.3 Detecting Antimicrobial-Drug Resistance

Molecular methods can rapidly detect antimicrobial-drug resistance in clinical settings and have substantially contributed to our understanding of the spread and genetics of resistance. Conventional broth- and agar-based antimicrobial susceptibility testing methods provide a phenotypic profile of the response of a given microbe to an array of agents. Although useful for selecting potentially useful therapeutic agents, conventional methods are slow and fraught with problems. The most common failing is in the detection of methicillin resistance in staphylococci, which may be expressed in a very heterogeneous fashion, making phenotypic characterization of resistance difficult. Currently, molecular detection of the resistance gene, *mec A*, is the standard against which phenotypic methods for detection of methicillin resistance are judged.

Molecular methods may be used to detect specific antimicrobial-drug resistance genes (resistance genotyping) in many organisms (Table 19.3). Detection of specific point mutations associated with resistance to antiviral agents is also increasingly important. Screening for mutations in an amplified product may be facilitated by the use of high-density probe arrays (Gene chips).

Despite its many potential advantages, genotyping will not likely replace phenotypic methods for detecting antimicrobial-drug resistance in the clinical laboratory in the near future. Molecular methods for resistance detection (Table 19.3) may be applied directly to the clinical specimen, providing simultaneous detection and identification of the pathogen plus resistance characterization. Likewise, they are useful in detecting resistance in viruses, slow-growing or nonviable organisms, or organisms with resistance mechanisms that are not reliably detected by phenotypic methods. However, because of their high specificity, molecular methods will not detect newly emerging resistance mechanisms and are unlikely to be useful in detecting resistance genes in species where the gene has not been observed previously. Furthermore, the presence of a resistance gene does not mean that the gene will be expressed, and the absence of a known resistance gene does not exclude the possibility of resistance from another mechanism. Phenotypic antimicrobial susceptibility testing methods allow laboratories to test many organisms and detect newly emerging as well as established resistance patterns.

Table 19.3 Molecular methods for detecting antimicrobial resistance[a]

Organism(s)	Antimicrobial agent(s)	Gene	Detection method
Staphylococci	Methicillin; Oxacillin	*mec A* [b]	Standard DNA probe; PCR; Branched chain DNA probe
Enterococci	Vancomycin	*van A, B, C, D* [c]	Standard DNA; probe PCR
Enterobacteriaceae *Haemophilus influenzae* *Neisseria gonorrhoeae*	Beta-lactams	*bla* TEM *bla* SHV [d]	Standard probe; PCR and RFLP; PCR; sequencing
Enterobacteriaceae and gram-positive cocci	Quinolones	Point mutations in *gyr A*, *gyr B*, *par C* and *par E*	PCR and sequencing
Mycobacterium tuberculosis [e]	Rifampin Isoniazid Streptomycin	Point mutatins in *rpo B*; *kat G*, *inh A*, and *ahp C*; in *emb B;* *rps L* and *rrs*	PCR and SSCP PCR and sequencing PCR and SSLP PCR and SSCP PCR and RFLP
Herpes viruses[f]	Acyclovir and related drugs	Mutatins in the TK gene	PCR and sequencing
Iv[g]	Nucleoside reverse transcriptase inhibitors	Protease inhibitors Point mutations in RT gene Point mutations in PROT gene	PCR and sequencing PCR and LIPA PCR and sequencing

[a]Adapted from Pfaller and Herwaldt (1997).
[b]*mecA* encodes for the altered penicillin binding protein PBP2a′; phenotypic methods may require 48 h incubation or more to detect resistance and are less than 100% sensitive. Detection of *mecA* has potential for clinical application in specific circumstances.
[c]Vancomycin resistance in enterococci may be related to one of four distinct resistance genotypes of which *vanA* and *vanB* are most important. Genotypic detection of resistance is useful in validation of phenotypic methods.
[d]The genetic basis of resistance to beta-lactam antibiotics is extremely complex. The bla_{TEM} and bla_{SHV} genes are the two most common sets of plasmid encoded beta-lactamases. The presence of either a bla_{TEM} or bla_{SHV} gene implies ampicillin resistance. Variants of the bla_{TEM} and bla_{SHV} genes (extended spectrum beta-lactamases) may also encode for resistance to a range of third-generation cephalosporins and to monobactams.
[e]*M. tuberculos* is very slow growing. Four weeks or more may be required to obtain phenotypic susceptibility test results. Detection of resistance genes in *M. tuberculosis* has potential for clinical application in the short term.
[f]There are no phenotypic methods sufficiently practical for routine clinical detection of resistance to antiviral agents. Genotypic methods represent a practical method for routine detection of antiviral resistance.
[g]Abbreviations not defined in text: RFLP, restriction fragment length polymorphism; SSCP, single-stranded conformational polymorphism; LIPA, line probe assay; TK, thymidine kinase; RT, reverse transcriptase; PROT, protease.

19.3 Molecular Diagnostics of Infections

During the past 10–15 years, we have seen expansive growth of the use of molecular technology in the clinical laboratory for diagnosing infectious diseases (Jeffrey et al., 2001). As a result, many laboratories are able to offer more sensitive testing, faster turnaround times, and ultimately improved patient care. The gold standard in bacteriology largely remains culture, primarily due to cost accounting and the potential complex nature of associated infections (i.e., urine, wound, and respiratory cultures).

However, in circumstances in which there may be minute quantities of a specific pathogen present, the patient may have received antibiotics prior to specimen collection or the etiologic agent may require unusual culture conditions, molecular detection offers a great advantage to culture techniques. In many virology laboratories, molecular detection has supplanted cell culture techniques for the identification of several viral pathogens and in many cases has become the new gold standard. Though molecular techniques can offer an abundance of added benefits when used to augment current gold standards such as culture and/or serology, the optimal use of molecular methodologies in microbiology resides with specimens in which a limited number of pathogenic organisms are sought and in cases where the enhanced sensitivity and faster turnaround time of molecular methods far outweighs the increased cost.

19.3.1 Sexually Transmitted Infections

Sexually transmitted infections (STIs) constitute an important world-wide public health problem (Msuya et al., 2009). The use of sensitive and specific laboratory methods for diagnosing this condition is crucial to reduce the transmission and sequelae of STI. *Neisseria gonorrhoeae* and *Chlamydia trachomatis* are the pathogens most frequently involved in urethral and cervical infection. Culture continues to be the gold standard for diagnosing gonorrhea. Nucleic acid amplification assays are considered the new gold standard for *C. trachomatis*, although culture is till the most specific technique. Genital ulcers due to *Treponema pallidum, Haemophilus ducreyi*, or herpes simplex virus have little clinical and bacteriological correlation; therefore, it is essential to establish the microbiological diagnosis. Lesions present in the primary or secondary period of syphilis can be diagnosed by dark field microscopy. Serologic diagnosis for the remaining periods is based on non-treponemal tests associated with confirmatory treponemal tests. Cell culture is considered the gold standard for herpes simplex virus although molecular methods also have a sensitivity and specificity near 100%. Currently, microbiologic diagnosis of *H. ducreyi* and venereal lymphogranuloma is achieved with the use of molecular methods on samples obtained from the ulceration or lymph adenopathy. The diagnosis of genital warts in immunocompetent patients is based on clinical findings in most cases because the lesions are sufficiently characteristic. Culture is considered the reference method in *Trichomonas vaginalis* infection (Vázquez et al., 2008).

Several new techniques for diagnosis of the sexually transmitted diseases have been developed during the past decade. The rapid, nonculture techniques for direct detection of *Chlamydia trachomatis, Neisseria gonorrhoeae*, and herpes simplex virus are reviewed, and their advantages and disadvantages are compared with conventional diagnostic methods (Woods, 1995). Polymerase chain reaction tests were performed on specimens from consecutive male patients with genital ulcers to detect sexually transmitted pathogens. PCR was also performed for the detection of *Neisseria gonorrhoeae*, *Chlamydia trachomatis*, and *Trichomonas vaginalis* on urethral specimens from consecutive subjects with dysuria or urethral discharge. Antibody tests for syphilis and Herpes simplex virus type-2 (HSV-2) and human immunodeficiency virus antibodies were performed. Men at risk of genital ulcers should be asked about relevant symptoms with and without prompting and examined clinically to maximize the likelihood of correct diagnosis and treatment. The finding of a high prevalence of HSV-2 and associated dysuria cautions against providing empirical treatment for gonorrhoea and chlamydia in ulcer patients with dysuria but without urethral discharge. Innovative strategies to limit the burden of HSV-2 infection in this population are required (O'Farrell et al., 2008).

Even then, microbiological screening is incorporated into STI control strategies; lack of access to appropriate services (especially in rural and remote areas), reluctance of at-risk populations to attend for treatment, fear of invasive genital examinations, and lower sensitivities of conventional diagnostic assays reduces the effectiveness of such programmes. Therefore, accurate, cost-effective, reliable diagnostic assays (preferably those which can be used in the field) are needed to impact on the incidence of the various STIs, as well as HIV. With the advent of molecular technologies, including target and signal amplification methods, diagnoses of STIs have been revolutionised and allow the use of non or minimally invasive sampling techniques, some of which are self-collected by the patient, e.g. first-void urine, cervico-vaginal lavage, low vaginal swabs, and tampons. Most studies evaluating such self-sampling with molecular diagnostic techniques have demonstrated an equivalent or superior detection of STIs as compared to conventional sampling and detection methods. These sampling methods can also be used to determine prevalence of STIs in various populations, but particularly those with difficult access to medical care (Garland and Tabrizi, 2004).

The presence of *Trichomonas vaginalis* (TV) is associated with an increased risk of coinfection with *Chlamydia trachomatis* (CT) and/or *Neisseria gonorrhoeae* (NG) in female patients presenting to the ED with symptoms consistent with a sexually transmitted disease (STD; White et al., 2005). Vaginal swabs are the specimens of choice when screening for *Chlamydia trachomatis* and *Neisseria gonorrhoeae:* results from a multicenter evaluation of the APTIMA assays for both infections (Schachter et al., 2005). The Gen-Probe APTIMA Combo 2 (AC2) is a Food and Drug Administration-cleared nucleic acid amplification test (NAAT) for the detection of *Chlamydia trachomatis* and *Neisseria gonorrhoeae* from urine and urogenital swab specimens. The Centers for Disease Control and Prevention have recommended confirmation of positive NAAT results in low-prevalence populations. APTIMA CT (ACT) and APTIMA GC (AGC) are two discrete NAATs for

C. trachomatis and *N. gonorrhoeae* detection that are suitable for confirming AC2-positive results because they target different nucleic acid sequences. Thus, ACT and AGC relative sensitivity compared to that of AC2 was 100%. All APTIMA assays detected more confirmed positive results than culture, DFA, and LCx. The performance of APTIMA assays was not altered by the use of various swab types and by long-term storage of specimens. All APTIMA assays are highly sensitive and rapid. ACT and AGC can be recommended for confirmation of positive results from other NAATs, such as AC2 and LCx (Boyadzhyan et al., 2004).

19.3.2 Vector Borne Disease

Lyme borreliosis caused by the spirochete *Borrelia burgdorferi* is now the most common vector borne disease in North America, Europe and Asia. It is a multi-systemic infection which may cause skin, neurological, cardiac or rheumatologic disorders. The use of serologic testing and its value in the diagnosis of Lyme disease remain confusing and controversial for physicians, especially concerning persons who are at low risk for the disease. Samples drawn from patients within four weeks of disease onset are tested by Western blot technique for both immunoglobulin M and immunoglobulin G antibodies; samples drawn more than four weeks after disease onset are tested for immunoglobulin G only (DePietropaolo et al., 2006).

Current serologic Lyme disease tests use whole borrelia cells as the source of antigen. These assays are difficult to standardize and to optimize for sensitivity and specificity.

The recently introduced highly sensitive PCR methodology can be used for amplification and subsequent identification of *B. burgdorferi* specific sequences. The diagnostic sensitivity of PCR was comparable to and even higher than in vitro culture. PCR was significantly more sensitive than a histological *B. burgdorferi* specific immunophosphatase-staining method. The sensitivity of PCR was 71%, which was superior to culture and serology. There is clear evidence for PCR being the most sensitive and specific test for detection of *B. burgdorferi* in skin biopsies from patients with both early and late dermatoborreliosis. PCR can be used as a diagnostic aid in these patients. However, in general the measurement of specific intrathecal antibody production in patients with neuroborreliosis was superior to PCR. Furthermore, strain differences were of importance for selection of suitable antigens for diagnostic assays and for vaccine development. Since then, *B. burgdorferi* isolates have been studied by phenotypic and genotypic traits and have been shown to be highly heterogeneous (Lebech, 2002).

19.3.3 Viral Infections

Viral infections are common causes of respiratory tract disease in the outpatient setting. Some viruses, such as influenza, respiratory syncytial virus (RSV), cytomegalovirus (CMV), and varicella-zoster virus (VZV), are relatively common. Others, such as adenovirus, severe acute respiratory syndrome (SARS)-coronavirus,

Hantavirus, and the viral hemorrhagic fevers (VHFs), are rare but have an immense public health impact. In cases of acute LRI, rapid diagnosis of the pathogens may be very important for initiating antiviral therapy, stopping viruses affecting humans throughout the world. Current methods used in the diagnosis of infections with RSV A and B, influenza virus A and B, and HPIV-1, -2, and costs of treating children with LRIs exceed $700 million in the United States. Respiratory syncytial virus (RSV) type A -3 include virus isolation, shell vial assay, antigen detection, and serology. Major limitations of these techniques include and type B, influenza A and B viruses, and human parainfluenza virus (HPIV) types 1, 2, and 3 cause about 80–90% of viral low sensitivity, low specificity, and prolonged time to completion.

Infants, the elderly, and individuals with compromised cardiac, pulmonary, or immune systems are at the greatest risk. The development of rapid molecular techniques such as gene amplification (e.g., PCR) has allowed for the detection of small risk of serious complications from these viruses. A multiplex quantitative reverse transcription-polymerase chain reaction-enzyme hybridization assay (Hexaplex; Prodesse, Milwaukee) was developed and used to rapidly detect and quantitate RNA of respiratory syncytial viruses A and B, influenza viruses A and B, and human parainfluenza virus types 1, 2, and 3 in nasal wash specimens in a single test. Primers and probes originated from highly conserved regions of each viral genome (Fan et al., 1998). PCR will become the first-line diagnostic test for viral meningitis and encephalitis (Jeffery et al., 1997).

Molecular biology-based assays are invaluable tools for the management of chronic viral hepatitis. They can be used to test blood donations, diagnose active infection, help to establish the prognosis, guide treatment decisions, and assess the virological response to therapy (Pawlotsky, 2002).

Human papilloma virus are strictly epitheliotropic and host-specific. Currently, 57 different types based on deoxyribonucleic acid sequence rather than serology have been identified. Some papilloma virus types infect the skin, while others infect the oral or genital mucous membranes. Human Papilloma virus type 16 is often associated with invasive carcinomas of the uterine cervix and of their intra-epithelial precursors. If various techniques for the detection of human papilloma virus may be used (cytology, electron microscopy and immunocytochemistry), as will become apparent, nucleic acid hybridization is the most reliable method for viral diagnosis and genotyping (Morinct, 1991).

Zika virus (ZIKV) is an emerging mosquito-borne flavivirus circulating in Asia and Africa. Human infection induces an influenza-like syndrome that is associated with retro-orbital pain, oedema, lymphadenopathy, or diarrhea. Diagnosis of Zika fever requires virus isolation and serology, which are time consuming or cross-reactive. A one-step RT-PCR assay to detect ZIKV in human serum was successfully reported by Faye et al., (2008).

Detection of virus genomes and their mutants by molecular technology is essential in clinical practice for patients with hepatitis B virus(HBV) and hepatitis C virus (HCV) related liver diseases (Silva et al., 2006). A multiplex PCR (M-PCR) assay with colorimetric detection was devised for the simultaneous amplification of DNA targets from *Haemophilus ducreyi*, *Treponema pallidum*, and herpes simplex virus

(HSV) types 1 and 2. The M-PCR assay is more sensitive than standard diagnostic tests for the detection of HSV, *H. ducreyi*, and *T. pallidum* from genital ulcers (Orle et al., 1996).

Polymerase chain reaction is a highly sensitive technique for the detection of hepatitis B virus-DNA and hepatitis C virus-RNA in serum, liver tissue, and peripheral mononuclear blood cells. Polymerase chain reaction detection of hepatitis C virus-RNA in serum may be the only means of confirming acute hepatitis C infection and also of identifying viraemia in the chronic disease, particularly in anti-hepatitis C virus antibody-negative individuals. It can also be used for direct evaluation of mother to child hepatitis C virus transmission. As in hepatitis B, polymerase chain reaction can be used for monitoring reinfection with hepatitis C virus after liver transplant, and has proved invaluable in identification of different hepatitis C virus genotypes. The efficacy of antiviral treatment can also be monitored using polymerase chain reaction. Polymerase chain reaction has thus shown numerous advantages for disease detection and monitoring despite the limitations imposed, for example, by possible contamination problems and semiquantitative evaluations (Bréchot, 1993).

A nested polymerase chain reaction (PCR) was evaluated for the detection of cytomegalovirus (CMV) DNA in cerebrospinal fluid (CSF). CMV PCR on CSF is highly sensitive and specific. It should be considered a rapid and reliable diagnostic method for CMV infection of the central nervous system (Cinque, 1995).

Molecular detection of viruses has extended beyond the standard therapeutic monitoring of viral loads in specific patient populations. Cell culture techniques are insensitive due to the low viral burden typically found associated with encephalitis and perhaps also the presence of host neutralizing antibodies. CSF culture for HSV detects less than 2% of clinically determined adult HSV encephalitis cases and 40% of neonatal central nervous system (CNS) disease. In contrast, HSV NAA is positive in most adult cases resulting in sensitivity and specificity > 95% and is 75% sensitive and 100% specific for neonatal meningitis. The rapid diagnosis of HSV encephalitis can prevent a brain biopsy and rapidly determine the need for acyclovir therapy. Enterovirus is the most common cause of aseptic meningitis in the summer and fall months in temperate climates and accounts for 10–20% of encephalitis cases. Nucleic acid amplification has also been successfully applied to other etiologies of viral CNS disease, such as CMV and varicella-zoster virus (VZV), but these assays have not been implemented as broadly as those for HSV and EV, so are still transitioning to becoming the method of choice. It should be noted that not all encephalitis viruses are readily detected by NAA. For example, due to the short period of viremia in many carboviral infections (i.e., West Nile Virus), CSF NAA has low sensitivity, and the gold standard remains serology. CSF NAA false negative results can occur due to collection of CSF very early or very late in illness, rapid viral clearance in immunocompetent hosts, and NAA inhibitors. False positive CSF NAA results also occur primarily due to lack of data to suggest the detection of certain viral nucleic acids correlates with clinical CNS disease, but can also be caused by the presence of peripheral blood in the CSF. While CSF NAA is considered by many the diagnostic standard of care as discussed above, the lack of standardized FDA-approved assays has made implementation of CSF NAA difficult in nonacademic settings. While most laboratories offering CSF NAA use qualitative methods, data indicate a role for quantitative CSF NAA in differentiating nonspecific presence of virus and virus-associated disease, to aid in prognosis for improved patient management, and in monitoring antiviral therapy.

19.3.4 Bacterial Infections

The quest for the search of rapid, cheap, easy to perform and highly sensitive and specific tests has resulted into the introduction and application of various new tests for the diagnosis of bacterial meningitis. Most of these tests are based on immunological principles, viz. counter-immunoelectrophoresis, latex agglutination test, co-agglutination, radio-immunoassay, haemagglutination inhibition as well as study of immune profile of cerebrospinal fluid. Apart from these certain non-immunological tests viz. Limulus amoebecyte lysate test, gas liquid chromatography, nitroblue tetrazolium dye reduction test have also been evaluated to make the laboratory diagnosis of this important clinical entity rapid and more reliable (Ichhpujani and Bhatia, 1984)

The tuberculin skin test for immunologic diagnosis of *Mycobacterium tuberculosis* infection has many limitations, including being confounded by Bacillus Calmette-Guérin (BCG) vaccination or exposure to nontuberculous mycobacteria. *M. tuberculosis*-specific antigens that are absent from BCG and most nontuberculous mycobacteria have been identified (Tiwari et al., 2007). Mori et al., (2004) examined the use of two of these antigens, CFP-10 and ESAT-6, in a whole blood IFN-gamma assay as a diagnostic test for tuberculosis in BCG-vaccinated individuals.

An enzyme-linked immunoassay for diagnosis of tuberculosis was devised by using a shotgun immunoexpression library in the _gt11 vector. DNA from a virulent *M. tuberculosis strain* was sheared and expressed to generate shotgun polypeptides.

Promising mycobacterial DNA cassettes were subcloned and expressed into the glutathione *S*-transferase (GST) fusion vector pGEX-5X-1 with a strong *tac* promoter and were expressed in *Escherichia coli* BL21. These recombinant proteins when used in ELISA differentiated BCG-vaccinated healthy subjects and patients with active tuberculosis and proved to be effective in detecting pulmonary as well as extrapulmonary tuberculosis, with an overall sensitivity of 84.33% and an overall specificity of 93.62% (Bisen et al., 2003). A simple, rapid and cost-effective glycolipid antibody based liposome agglutination assay viz., TB Screen Test, for diagnosis of patients with pulmonary and extrapulmonary tuberculosis with 94% sensitivity and 98.3% specificity was developed (Tiwari et al., 2005). The assay clearly differentiated healthy controls and *M. bovis* BCG-vaccinated subjects from those with active tuberculosis. A TB/M card test based on liposomal agglutination test was developed by Tiwari et al. (2007) for detection of glycolipid antigens of *M. tuberculosis*. The test was shown to be effective in detecting glycolipid antigens of *M. tuberculosis* in clinical samples from patients with active TB with as low as 1 ng/ml analytical sensitivity, 97.4% clinical sensitivity and 96.9% specificity. The test was proved to be highly economical, rapid, simple and instrument free. The test seems fairly useful for mass testing of a variety of biological specimens (cerebrospinal, pleural and synovial fluids, serum, tissue biopsy extract) from patients with tuberculous meningitis, pulmonary TB and other extra-pulmonary TB in endemic countries.

The sensitivity of NAA detection of MTB in smear-positive respiratory specimens is 96.9%, and the specificity is 100%, whereas the sensitivity and specificity in smear-negative specimens is 72.0 and 99.3%, respectively. It should be noted that NAA of MTB does not replace the need for routine mycobacterial culture and susceptibility testing. In addition to the direct detection of MTB, techniques such as probe-based technology and sequence analysis can be applied to cultured isolates to decrease the time to identification over routine biochemical analysis. Rapid identification of MTB impacts not only patient care, but also infection control. Due to the increasing frequency of isolation of mycobacterial species associated with immunocompromised hosts and the increased incidence of multi-drug resistant MTB, it has become imperative to offer accurate yet rapid diagnostic tools for the detection and identification of mycobacteria (Tiwari et al., 2007). A debate exists regarding the gold standard for the laboratory diagnosis of *Bordetella pertussis*. Historically, culture plates collected at the patient's bedside (i.e., cough plates) have been considered the reference method. Although culture is very specific, its sensitivity suffers partially due to the organism's fastidious nature, but primarily because the highest sensitivity for culture occurs before patients are symptomatic. NAA remains positive for longer after therapy than culture, and NAA is also positive for a longer period after onset of symptoms. Therefore, NAA is useful for patients presenting later in their illness. NAA testing allows for same-day results and since erythromycin-resistant *B. pertussis* is still rare, a cultured isolate is rarely needed for antimicrobial susceptibility testing. Multiple studies have demonstrated significant increased detection of *B. pertussis* when comparing NAA to culture: reported PCR-positive, culture-negative samples range from 13 to 88%. However, due to potential false positive and false negative results with *B. pertussis* NAA procedures, it is strongly recommended that results be considered in the context of patient clinical presentation, and clinically inconsistent results should be confirmed by a second method.

NAA is also being used in bacteriology to detect antimicrobial resistance. Since antimicrobial resistance can be multi-factorial, this practice is limited to organisms in which the results can be interpreted with confidence in regard to the genotypic relationship to clinical treatment and/or infection control precautions. Such examples are direct detection of vancomycin-resistant *Enterococcus* (VRE) and methicillin-resistant *Staphylococcus aureus* (MRSA) from rectal and nares surveillance cultures, respectively. Screening patients for VRE and MRSA carriage is a key strategy for preventing the spread of these organisms in health care settings. NAA technology reportedly increases VRE detection by up to 120%. In addition, enterococci that confer low-level intrinsic resistance, and thus not considered "true" VRE, are accurately ruled out preventing unnecessary contact precautions and contributing to hospital savings. NAA detection of MRSA has been shown to be equal in sensitivity to culture-based methods, but has the advantage of offering a faster turnaround time, thus impacting hospital cost savings. However, it should be noted that direct specimen testing for MRSA comes with limitations, often including a lower positive predictive value than conventional methods. More recently, new strains of MRSA have appeared that are associated with skin and soft tissue infections in outpatients and are called community-associated MRSA (CA-MRSA). The increasing incidence of CA-MRSA is causing overall rates of MRSA to rise. Therefore, it has become even more important to quickly and accurately identify resistant isolates.

19.3.5 Fungal Infections

The incidence of invasive fungal infections has increased considerably in recent years. Mycetomas are chronic, granulomatous, subcutaneous infections caused by either actinomycetes bacteria or eumycetes fungi. The disease is endemic in the tropics and is characterised by a slow progression with risks of bone and visceral involvement. There have been notable advances in improved molecular techniques for species identification. Carbapenems, oxazolidinones and triazoles have emerged as promising therapeutic options, but access to drug therapies in developing countries remains limited by the poor availability and high costs (Ameen and Arenas, 2008). The molecular assay using PCR-ELISA help in the diagnosis of invasive fungal infections at the early stage of infection, before clinical manifestations (Badiee et al., 2009).

19.3.6 Practical Applications of Molecular Methods in the Clinical Microbiology Laboratory

Commercial kits for the molecular detection and identification of infectious pathogens have provided a degree of standardization and ease of use that has facilitated the introduction of molecular diagnostics into the clinical microbiology laboratory (Table 19.1). The use of nucleic acid probes for identifying cultured organisms and for direct detection of organisms in clinical material was the first exposure that most laboratories had to commercially available molecular tests. Although these probe tests are still widely used, amplification-based methods are increasingly employed for diagnosis, identification and quantitation of pathogens, and characterization of antimicrobial-drug resistance genes. Commercial amplification kits are available for some pathogens (Table 19.1), but some clinically important pathogens require investigator-designed or "home-brew" methods (Table 19.2). In addition, molecular strain typing, or genotyping, has proven useful in guiding therapeutic decisions for certain viral pathogens and for epidemiologic investigation and infection control.

19.3.6.1 Molecular Epidemiology

Laboratory characterization of microbial pathogens as biologically or genetically related is frequently useful in investigations. Several different epidemiologic typing methods have been applied in studies of microbial pathogens (Table 19.4). The phenotypic methods have occasionally been useful in describing the epidemiology of infectious diseases; however, they are too variable, slow, and labor-intensive to be of much use in most epidemiologic investigations. Newer DNA-based typing methods have eliminated most of these limitations and are now the preferred techniques for epidemiologic typing. The most widely used molecular typing methods include plasmid profiling, restriction endonuclease analysis of plasmid and genomic DNA, Southern hybridization analysis using specific DNA probes, and chromosomal DNA profiling using either pulsed-field gel electrophoresis (PFGE) or PCR-based methods. All these methods use electric fields to separate DNA

Table 19.4 Genotypic methods for epidemiologic typing of microorganisms[a]

Method	Examples	Comments
Plasmid analysis	Staphylococci enterobacteriaceae	Plasmids may be digested with restriction endonuclease, only useful when organisms carry plasmids
Restriction endonuclease analysis of chromosomal DNA with electrophoresis	Enterococci; *Staphylococcus aureus Clostridium difficile; Candida* spp	Large number of bands difficult to interpret not amenable to computer analysis
PFGE	Enterobacteriaceae; Staphylococci Enterococci; *Candida* spp.	Fewer bands; Amenable to computer analysis very broad application
Genome restriction fragment length polymorphism analysis; ribotyping, insertion sequence probe fingerprinting	Enterobacteriaceae; Staphylococci *Pseudomonas aeruginosa* *Mycobacterium tuberculosis* *Candida* spp.	Fewer bands; computer analysis Sequence-based profiles Automated
PCR-based methods: repetitive elements PCR spacer typing, selective amplification of genome restriction fragments, multilocus allelic sequence-based typing	Enterobacteriaceae *Acinetobacter* spp. Staphylococci *M. tuberculosis* HCV	Crude extracts and small amounts of DNA may suffice
Library probe genotypic hybridization schemes: multilocus probe dot-blot patterns, high-density oligonucleotide patterns	*Burkholderia cepacia* *S. aureus* *M. tuberculosis* High-density	Unambiguous yes-no result Less discrimination than other methods Couple with DNA chip technology

[a]The table contains examples of available methods and applications and is not intended to be all-inclusive; Adapted from Pfaller.

fragments, whole chromosomes, or plasmids into unique patterns or fingerprints that are visualized by staining with ethidium bromide or by nucleic acid probe hybridization. Molecular typing is performed to determine whether different isolates give the same or different results for one or more tests. Epidemiologically related isolates share the same DNA profile or fingerprint, whereas sporadic or epidemiologically unrelated isolates have distinctly different patterns. If isolates from different patients share the same fingerprint, they probably originated from the same clone and were transmitted from patient to patient by a common source or mechanism.

Molecular typing methods have allowed investigators to study the relationship between colonizing and infecting isolates in individual patients, distinguish contaminating from infecting strains, document nosocomial transmission in hospitalized patients, evaluate reinfection versus relapse in patients being treated for an infection, and follow the spread of antimicrobial-drug resistant strains within and between hospitals over time. Most available DNA-based typing methods may be used in studying nosocomial infections when applied in the context of a careful epidemiologic investigation. In contrast, even the most powerful and sophisticated typing method, if used indiscriminately in the absence of sound epidemiologic data, may provide conflicting and confusing information.

19.3.6.2 Financial Considerations

Molecular testing for infectious diseases includes testing for the host's predisposition to disease, screening for infected or colonized persons, diagnosis of clinically important infections, and monitoring the course of infection or the spread of a specific pathogen in a given population. It is often assumed that in addition to improved patient care, major financial benefits may accrue from molecular testing because the tests reduce the use of less sensitive and specific tests, unnecessary diagnostic procedures and therapies, and nosocomial infections. However, the inherent costs of molecular testing methods, coupled with variable and inadequate reimbursement by third-party payers and managed-care organizations, have limited the introduction of these tests into the clinical diagnostic laboratory.

Not all molecular diagnostic tests are extremely expensive. Direct costs vary widely, depending on the test's complexity and sophistication. Inexpensive molecular tests are generally kit based and use methods that require little instrumentation or technologist experience. DNA probe methods that detect *C. trachomatis* or *N. gonorrhoeae* are examples of low-cost molecular tests. The more complex molecular tests, such as resistance genotyping, often have high labor costs because they require experienced, well-trained technologists. Although the more sophisticated tests may require expensive equipment (e.g., DNA sequencer) and reagents, advances in automation and the production of less-expensive reagents promise to decrease these costs as well as technician time. Major obstacles to establishing a molecular diagnostics laboratory that are often not considered until late in the process are required licenses, existing and pending patents, test selection, and billing and reimbursement.

Molecular screening programs for infectious diseases are developed to detect symptomatic and asymptomatic disease in individuals and groups. Persons at high risk, such as immunocompromised patients or those attending family planning or obstetrical clinics, are screened for CMV and *Chlamydia*, respectively. Likewise, all blood donors are screened for bloodborne pathogens. The financial outcome of such testing is unknown. The cost must be balanced against the benefits of earlier diagnosis and treatment and societal issues such as disease epidemiology and population management.

One of the most highly touted benefits of molecular testing for infectious diseases is the promise of earlier detection of certain pathogens. The rapid detection of *M. tuberculosis* directly in clinical specimens by PCR or other amplification-based methods is quite likely to be cost-effective in the management of tuberculosis. Other examples of infectious disease that are amenable to molecular diagnosis and for which management can be improved by this technology include HSV encephalitis, *Helicobacter pylori* infection, and neuroborreliosis caused by *Borrelia burgdorferi*. For HSV encephalitis, detection of HSV in cerebrospinal fluid (CSF) can direct specific therapy and eliminate other tests including brain biopsy. Likewise, detection of *H. pylori* in gastric fluid can direct therapy and obviate the need for endoscopy and biopsy. PCR detection of *B. burgdorferi* in CSF is helpful in differentiating neuroborreliosis from other chronic neurologic conditions and chronic fatigue syndrome.

As discussed earlier, molecular tests may be used to predict disease response to specific antimicrobial therapy. Detection of specific resistance genes (*mec A, van A*) or point mutations resulting in resistance has proven efficacious in managing disease. Molecular-based viral load testing has become standard practice for patients with chronic hepatitis and AIDS. Viral load testing and genotyping of HCV are useful in determining the use of expensive therapy such as interferon and can be used to justify decisions on extent and duration of therapy. With AIDS, viral load determinations plus resistance genotyping have been used to select among the various protease inhibitor drugs available for treatment, improving patient response and decreasing incidence of opportunistic infections.

Pharmacogenomics is the use of molecular-based tests to predict the response to specific therapies and to monitor the response of the disease to the agents administered. The best examples of pharmacogenomics in infectious diseases are the use of viral load and resistance genotyping to select and monitor antiviral therapy of AIDS and chronic hepatitis. This application improves disease outcome; shortens length of hospital stay; reduces adverse events and toxicity; and facilitates cost-effective therapy by avoiding unnecessary expensive drugs, optimizing doses and timing, and eliminating ineffective drugs.

Molecular strain typing of microorganisms is now well recognized as an essential component of a comprehensive infection control program that also involves the infection control department, the infectious disease division, and pharmacy. Molecular techniques for establishing presence or absence of clonality are effective in tracking the spread of nosocomial infections and streamlining the activities of the infection control program. A comprehensive infection control program uses active surveillance by both infection control practitioners and the clinical microbiology laboratory to identify clusters of infections with a common microbial phenotype (same species and antimicrobial susceptibility profile). The isolates are then characterized in the laboratory by using one of a number of molecular typing methods to confirm or refute clonality.

19.3.6.3 Challenges and Opportunities

The field of molecular infectious disease testing has grown so rapidly that the diagnostic industry has not kept up. To fill this void, independent investigators have turned to the development of user-defined, or "homebrew," molecular detection methods in the clinical laboratory. The implementation of user-defined NAA testing has revolutionized clinical molecular infectious disease testing. In addition, commercially-available non-FDA-approved NAA assays are increasingly becoming available as analyte specific reagents (ASRs). Though all reagents necessary for the amplification reaction can be purchased commercially, assay development and verification studies must be performed by individual laboratories. In many cases, there are no comparative studies between user-defined NAA procedures, including ASRs, limiting the comparative value of assays between institutions (particularly in viral load monitoring) and restricting the application of such procedures to more experienced laboratories. It is not without considerable cost that a molecular infectious disease diagnostic lab is developed. It represents an institutional commitment because the costs may only be offset when analysis of hospital-wide cost savings is employed (i.e. shorter hospital stays, decreased use of unncecessary antibiotics). The costs incurred not only stem from instrumentation purchases, but also from the dedicated, expert staff required for such testing. Since many academic medical centers have resorted to implementing user-defined assays, verification and validation studies are substantial and require extensive resources, including time, staff, and expertise. These studies are crucial to defining the performance of the assay and determining appropriate clinical utilization. The FDA, diagnostic companies, and major molecular infectious disease laboratories need to work together to resolve the poor standardization that exists between laboratories using user-defined assays or ASRs.

19.4 Conclusion

The applications of molecular technology in clinical microbiology are endless, but challenges also abound. We are still learning what many NAA results mean in terms of infectious etiology. With the use of molecular technology to detect potential etiologic agents of disease, we need to remember Koch's postulates. Is the mere presence of an organism's nucleic acid convincing evidence of disease causation? Undoubtedly, additional clinical scientific evidence is needed to make such a claim, and such evidence or lack thereof should be considered when interpreting molecular infectious disease results. Though there is still much to be learned regarding the appropriate application and interpretation of molecular infectious disease testing, there are numerous exciting opportunities on the horizon. User-defined assays and ASRs have allowed experienced laboratories to offer critical diagnostic services that have yet to become available with FDA clearance. As investigators refine molecular applications for infectious disease testing, diagnostic companies market such applications, quality control and government organizations standardize results, and as costs associated with implementation decrease and reimbursement increases, molecular infectious disease testing will not only be available in academic medical centers and reference laboratories, but will also transition to community hospitals, thus more globally impacting patient care.

References

Ameen, M., Arenas, R., 2008, Emerging therapeutic regimes for the management of mycetomas. *Expert Opin Pharmacother* 9(12):2077–2085.

Badiee, P., Kordbacheh, P., Alborzi, A., Malekhoseini, S., Ramzi, M., Mirhendi, H., Mahmoodi, M., Shakiba, E., 2009, Study on invasive fungal infections in immuno compromised patients to present a suitable early diagnostic procedure. *Int J Infect Dis* 13(1):97–102.

Bisen, P.S., Garg, S.K., Tiwari, R.P., Tagore, P.R., Chandra, R., Karnik, R., Thaker, N., Desai, N., Ghosh, P.K., Fraziano, M., Colizzi, V., 2003, Analysis of shotgun expression library of *Mycobacterium tuberculosis* genome for immunodominant polypeptides: Potential use in serodiagnosis. *Clin Diagn Lab Immunol* (Am Soc Microbiol, Washington) 10(6):1051–1058.

Boyadzhyan, B., Yashina, T., Yatabe, J.H., Patnaik, M., Hill, C.S., 2004, Comparison of the APTIMA CT and GC assays with the APTIMA combo 2 assay, the Abbott LCx assay, and direct fluorescent-antibody and culture assays for detection of Chlamydia trachomatis and Neisseria gonorrhoeae. *J Clin Microbiol* 42(7):3089–3093.

Brechot, C., 1993, Polymerase chain reaction for the diagnosis of viral hepatitis B and C. *Gut* 34:39–44.

Cinque, P., Vago, L., Terreni, M.R., Brytting, M., Marenzi, R., Castagna, A., Lazzarin, A., Linde, A., 1995, Diagnosis of cytomegalovirus infection of the nervous system in AIDS by polymerase chain reaction analysis of cerebrospinal fluid. *Scand J Infect Dis Suppl* 1995(99):92–94.

Ciotti, M., Sesti, F., Paba, P., Benedetto, A., Patrizi, L., Criscuolo, A., Piccione, E., Branca, M., Syrjänen, K., Favalli, C., 2004, Human papillomavirus (HPV) testing in the management of women with abnormal Pap smears. Experience of a colposcopy referral clinic. *Eur J Gynaecol Oncol* 25(5):577–584.

DePietropaolo, D.L., Powers, J.H., Gill, J.M., Foy, A.J., 2006, Diagnosis of Lyme disease. *Del Med J* 78(1):11–18.

Fardy, J.M., 2009, Evaluation of diagnostic tests. *Methods Mol Biol* 473:127–136.

Faye, O., Faye, O., Dupressoir, A., Weidmann, M., Ndiaye, M., Alpha Sall, A., 2008, One-step RT-PCR for detection of Zika virus. *J Clin Virol* 43(1):96–101.

Garland, S.M., Tabrizi, S.N., 2004, Diagnosis of sexually transmitted infections (STI) using self-collected non-invasive specimens. *Sex Health* 1(2):121–126.

Ichhpujani, R.L., Bhatia, R., 1984, Immunodiagnosis of bacterial meningitis. *J Commun Dis* 16(1):30–37.

Jeffery, K.J., Read, S.J., Peto, T.E., Mayon-White, R.T., Bangham, C.R., 1997, Diagnosis of viral infections of the central nervous system: clinical interpretation of PCR results. *Lancet* 349(9048):313–317.

Jeffrey, M., Martin, S., González, L., Ryder, S.J., Bellworthy, S.J., Jackman, R., 2001, Differential diagnosis of infections with the bovine spongiform encephalopathy (BSE) and scrapie agents in sheep. *J Comp Pathol* 125(4):271–284.

Lebech, A.M., 2002, Polymerase chain reaction in diagnosis of Borrelia burgdorferi infections and studies on taxonomic classification. *APMIS Suppl* 105:1–40.

Mori, T., Sakatani, M., Yamagishi, F., Takashima, T., Kawabe, Y., Nagao, K., Shigeto, E., Harada, N., Mitarai, S., Okada, M., Suzuki, K., Inoue, Y., Tsuyuguchi, K., Sasaki, Y., Mazurek, G.H., Tsuyuguchi, I., 2004, Specific detection of tuberculosis infection: an interferon-gamma-based assay using new antigens. *Am J Respir Crit Care Med* 170(1):59–64.

Morinet, F., 1991, Virologic diagnosis of papillomavirus infections. *Rev Prat* Feb 41(4): 327–328.

Msuya, S.E., Uriyo, J., Hussain, A., Mbizvo, E.M., Jeansson, S., Sam, N.E., Stray-Pedersen, B., 2009, Prevalence of sexually transmitted infections among pregnant women with known HIV status in northern Tanzania. *Reproductive Health* 6:4.

O'Farrell, N., Morison, L., Moodley, P., Pillay, K., Vanmali, T., Quigley, M., Sturm, A.W., 2008, Genital ulcers and concomitant complaints in men attending a sexually transmitted infections clinic: implications for sexually transmitted infections management. *Sex Transm Dis* 35(6):545–549.

Orle, K.A., Gates, C.A., Martin, D.H., Body, B.A., Weiss, J.B., 1996, Simultaneous PCR detection of *Haemophilus ducreyi*, *Treponema pallidum*, and herpes simplex virus types 1 and 2 from genital ulcers. *J Clin Microbiol* 34(1):49–54.

Paternoster, D.M., Cester, M., Resente, C., Pascoli, I., Nanhorngue, K., Marchini, F., Boccagni, P., Cillo, U., Ribaldone, R., Amoruso, E., Cocca, N., Cuccolo, V., Bertolino, M., Surico, N., Stratta, P., 2008, Human papilloma virus infection and cervical intraepithelial neoplasia in transplanted patients. *Transplant Proc* 40(6):1877–1880.

Pawlotsky, J.M., 2002, Molecular diagnosis of viral hepatitis. *Gastroenterology* 122(6):1554–1568.

Pflaller, M., Herwaldt, H., 1997, Role of molecular epidemiology in infection control. *Infect Dis Clin North Am* 11(2):257–278.

Schachter, J., Chernesky, M.A., Willis, D.E., Fine, P.M., Martin, D.H., Fuller, D., Jordan, J.A., Janda, W., Hook, E.W. 3rd, 2005, Vaginal swabs are the specimens of choice when screening for Chlamydia trachomatis and Neisseria gonorrhoeae: results from a multicenter evaluation of the APTIMA assays for both infections. *Sex Transm Dis* 32(12):725–728.

Schluger, N.W., Kinney, D., Harkin, T.J., Rom, W.N., 1994, Clinical utility of the polymerase chain reaction in the diagnosis of infections due to *Mycobacterium tuberculosis*. *Chest* 105: 1116–1121.

Silva, L.K., Silva, M.B.S., Rodart, I.F., Lopes, G.B., Costa F.Q., Melo, M.E., Gusmão, E., Reis, M.G., 2006, Prevalence of hepatitis C virus (HCV) infection and HCV genotypes of hemodialysis patients in Salvador, Northeastern Brazil. *Braz J Med Biol Res* 39(5): 595–602.

Tang, Y.W., Procop, G.W., Persing, D.H., 1997, Molecular diagnostics of infectious diseases. *Clin Chem* 43:2021–2038.

Tiwari, R.P., Tiwari, D., Garg, S.K., Chandra, R., Bisen, P.S. 2005, Glycolipids of mycobacterium tuberculosis strain H37Rv are potential serological markers for diagnosis of active tuberculosis. *Clin Diagn Lab Immunol* (Am Soc Microbiol, Washington) 12:465–473.

Tiwari R.P., Bharmal, R.N., Kartikeyan, S., Bisen, P.S., 2007, Development of a rapid liposomal agglutination card test for the detection of glycolipid antigen in patients with meningial, pulmonary and other extra-pulmonary tuberculosis. *Int J Tuberculosis Lung Dis* (International Union Against Tuberculosis and Lung Disease, Paris, France) 11(10):1143–1151.

Tiwari, R.P., Hattikudur, N.S., Bharmal, R.N., Kartikeyan, S., Deshmukh, N.N., Bisen, P.S., 2007, Modern approaches to a rapid diagnosis of tuberculosis: Promises and challenges. *Tuberculosis* (Elsevier The Netherland) 87:193–201.

Vázquez, F., Lepe, J.A., Otero, L., Blanco, M.A., Aznar, J., 2008, Microbiological diagnosis of sexually-transmitted infection. *Enferm Infecc Microbiol Clin* 1:32–37.

White, M.J., Sadalla, J.K., Springer, S.R., Counselman, F.L., 2005, Is the presence of Trichomonas vaginalis a reliable predictor of coinfection with Chlamydia trachomatis and/or Neisseria gonorrhoeae in female ED patients? *Am J Emerg Med* 23(2):127–130.

Woods, G.L., 1995, Update on laboratory diagnosis of sexually transmitted diseases. *Clin Lab Med* 15(3):665–684.

Wagar, E.A., 1996, Defining the unknown: molecular methods for finding new microbes. *J Clin Lab Anal* 10:331–334.

Vaneechoutte, M., Van Eldere, J., 1997, The possibilities and limitations of nucleic acid amplification technology in diagnostic microbiology. *J Med Microbiol* 46:188–194.

Chapter 20
Diagnosis of Mutation and Genetic Disorders

Abstract All diseases have a genetic component. However, the extent to which genes contribute to disease varies and much remains to be learned. Mutations may be inherited or developed in response to environmental stresses such as viruses or toxins. The ultimate goal is to use this information to treat, cure, or if possible, prevent the development of disease. Advances in understanding the genetic mechanisms behind these diseases enable the development of early diagnostic tests, new treatments, or interventions to prevent disease onset or minimize disease severity.

Keywords Single gene disorders · Mutations · Genetic disorders · DNA diagnosis · Cystic fibrosis transmembrane conductance regulator gene · *CFTR* · *BRCA1* and *BRCA2* · Retinoblastoma · Inherited disorder · Monogenic disorders · Pharmacogenetics · Duchenne muscular dystrophy · Haemophilia A and B · Phenylketonuria · Thalassaemia · Multiple endocrine neoplasia · Factor V · Hemochomatosis · Colorectal cancer · Comparative genome hybridization

20.1 Prologue

The exponential increase in the discovery of genes over the past few years has transformed the DNA diagnosis of genetic disorders from a minor research-based activity to a major professional operation. For any genetic disease, once the defective gene is identified, knowledge of the pathogenic mutations is indispensable to offer DNA diagnosis. DNA diagnosis is offered at pre- and postnatal levels either by direct or indirect approaches. Direct mutational analysis and linkage studies with highly polymorphic intragenic markers are carried out depending on the feasibility of their detection. This is not as easy as it may sound, because many genes, including the Cystic Fibrosis Transmembrane Conductance Regulator (*CFTR*) gene, the breast cancer genes (*BRCA1* and *BRCA2*), and the retinoblastoma gene (*RB1*), lack mutational hot spots necessitating an exhaustive analysis of coding and flanking intronic and regulatory sequences.

M. Debnath et al., *Molecular Diagnostics: Promises and Possibilities*,
DOI 10.1007/978-90-481-3261-4_20,

20.2 Concept

The analysis of human DNA, RNA, chromosomes, proteins and certain metabolites in order to detect heritable disease-related genotypes, mutations, phenotypes or karyotypes for clinical purposes is essential. In the past decade, with the availability of technology and knowledge fueled by the investment and interest in the Human Genome Project, molecular diagnostics has recently enabled laboratories to offer diagnostic and predictive tests for inherited disorders (Harper et al., 2008). The technology will continue to become easier to apply and more affordable. Large amounts of genetic information can be determined at increasing efficiencies.

Historically, genetic testing focused on single-gene disorders, where a disease is caused by a mutation in one gene. The classic example is sickle cell anemia. A single nucleic acid base change is responsible for the sickle cell trait, when inherited from one parent or sickle cell disease when the nucleic acid base change is inherited from both parents. Although molecular diagnostics is available, sickle cell disease is most often diagnosed with hemoglobin electrophoresis. For single-gene disorders, mutation analysis is useful in diagnosis, confirmation of diagnosis, predisposition, prognosis and family planning. As with Huntington's disease, the identification of the gene and availability of testing opened another door to additional social and medical issues. In contrast, testing an individual with multiple endocrine neoplasia type-2 (MEN-2) for the RET proto-oncogene can avoid medullary carcinoma if the individual is positive and undergoes a prophylactic thyroidectomy.

The pace of the molecular dissection of human disease can be measured by looking at the catalog of human genes and genetic disorders identified so far in *Mendelian Inheritance in Man* (Holzman and Watson, 2002) and in OMIM, its online version, which is updated daily (www.ncbi.nlm.nih.gov/omim). For 1100 genes, at least one disease-related mutation has been identified. Because different mutations in the same gene often result in more or less distinct disorders, the total number of diseases for which OMIM lists mutations approaches 1500. The number of disease genes discovered so far is 1112. This number does not include at least 94 disease-related genes identified as translocation gene-fusion partners in neoplastic disorders.

Apart from the genes that cause monogenic disorders, the discovery of functional sequence variants that confer genetic susceptibility to common multifactorial disorders, such as cardiovascular disease, psychiatric disorders, autoimmune disorders and cancer, also creates demand for high throughput testing for clinically relevant polymorphisms. Single Nucleotide Polymorphisms (SNPs) are the most abundant type of DNA sequence variations in the human genome, though only miniscule fraction causes significant changes in amino acid sequence. One area where SNPs may have an immediate impact on patient care is the individual response to drug therapy, commonly referred to as pharmacogenetics. A genetic variability in *N*-acetyl transferase (NAT-2), for instance, is associated with a high incidence of peripheral neuropathy when taking isoniazid, an antituberculosis drug. A variant in the core promoter of the *ALOX5* gene, on the other hand, is responsible for the failure of some asthma patients to respond to treatment with ALOX5-pathway modifiers. The ultimate method for detection and definition of mutations is direct sequencing. But it comes at significant cost and labour, and the vast majority of sequencing reactions will only exclude the presence of a mutation. This has led to the development of physical, chemical and biological mutation screening methods that exclude the presence of mutations at a fraction of the cost of sequencing, but provide little, if any, information on the location and nature of sequence variation in mutated gene fragments. Most methods do well with regard to specificity, but fail miserably when it comes to sensitivity. They include such popular methods as single-strand conformation analysis, gel electrophoresis-based heteroduplex analysis, and denaturing gradient gel electrophoresis.

The diagnosis of a genetic disease requires a comprehensive clinical examination composed of three major elements:

1. A physical examination
2. A detailed medical family history
3. Clinical and laboratory testing if available.

While primary care providers may not always be able to make a definitive diagnosis of a genetic disease, their role is critical in collecting a detailed family history, considering the possibility of a genetic disease in the differential diagnosis, ordering testing as indicated and when available, appropriately referring patients to genetic specialists.

20.3 Factors Regulating a Genetic Disease

There are several factors that raise the possibility of a genetic disease in a differential diagnosis. One major factor is the occurrence of a condition among family members that is disclosed when the family history is obtained. The occurrence of the same condition in more than one family member (particularly first-degree relatives), multiple miscarriages, stillbirths, and childhood deaths are all suggestive of a genetic disease. Additionally, family history of common adult conditions (heart disease, cancer, and dementia) that occur in two or more relatives at relatively young ages may also suggest a genetic predisposition.

Other clinical symptoms that are suggestive of a genetic disease include developmental delay/mental retardation and congenital abnormalities. Dysmorphologies, often involving the heart and face, as well as growth problems are suggestive of a genetic disorder caused by an inherited mutation, a spontaneous mutation, a teratogen exposure, or unknown factors. While these clinical features may be caused by a number of factors, genetic conditions should also be considered as part of the differential diagnosis, particularly if the patient expresses several clinical features together that might be indicative of a syndrome (for example, mental retardation, distinct facies, and heart defect). Some physical features may appear unique or slightly different than the average such as wide-set or droopy eyes, flat face, short fingers, and tall stature. While these rare and seemingly mild features may not immediately be suggestive of a genetic disease to a primary care provider, an evaluation by a genetics specialist may be helpful in ruling in/out a genetic disease.

While many genetic conditions appear during childhood, a genetic condition should not entirely be ruled out in adolescents or adults. Often a genetic disease can remain undetected for several years until an event such as puberty or pregnancy triggers the onset of symptoms or the accumulation of toxic metabolites manifests in disease. In these cases, a detailed family history and physical examination should be performed and a referral made to a genetics specialist if needed.

20.4 Genetic Testing

Genetic tests can be used for many different purposes (Holtzman and Watson, 2002; Fig. 20.1). *Newborn screening* is the most widespread use of genetic testing. Almost every newborn in the U.S. is screened for several genetic diseases. Early detection of these diseases can lead to interventions to prevent the onset of symptoms or minimize disease severity. *Carrier testing* can be used to help couples to learn if they carry and thus risk passing to their children – an allele for a recessive condition such as cystic fibrosis, sickle cell anaemia, and Tay-Sachs disease. This type of testing is typically offered to individuals who have a family history of a genetic disorder and to people in ethnic groups with an increased risk of specific genetic conditions. If both parents are tested, the test can provide information about a couple's risk of having a child with a genetic condition. *Prenatal diagnostic testing* is used to detect changes in a foetus's genes or chromosomes. This type of testing is offered to couples with an increased risk of having a baby with a genetic or chromosomal disorder.

A tissue sample for testing can be obtained through amniocentesis or chorionic villus sampling. *Predictive or predispositional* genetic testing can identify individuals at risk of getting a disease prior to the onset of symptoms. These tests are particularly useful if an individual has a family history of a specific disease and an intervention is available to prevent the onset of disease or minimize disease severity. Predictive testing can identify mutations that increase a person's risk of developing disorders with a genetic basis, such as certain types of cancer.

Several different methods are currently used in genetic testing laboratories. The type of test will depend on the type of abnormality that is being measured. In general, three major types of genetic testing are available: cytogenetic, biochemical, and molecular testing.

20.4.1 Cytogenetic Testing

Cytogenetics involves the examination of whole chromosomes for abnormalities. Chromosomes of a dividing human cell can be clearly analyzed under a microscope. White blood cells, specifically T lymphocytes, are the most readily accessible cells for cytogenetic analysis since they are easily collected from blood and are capable

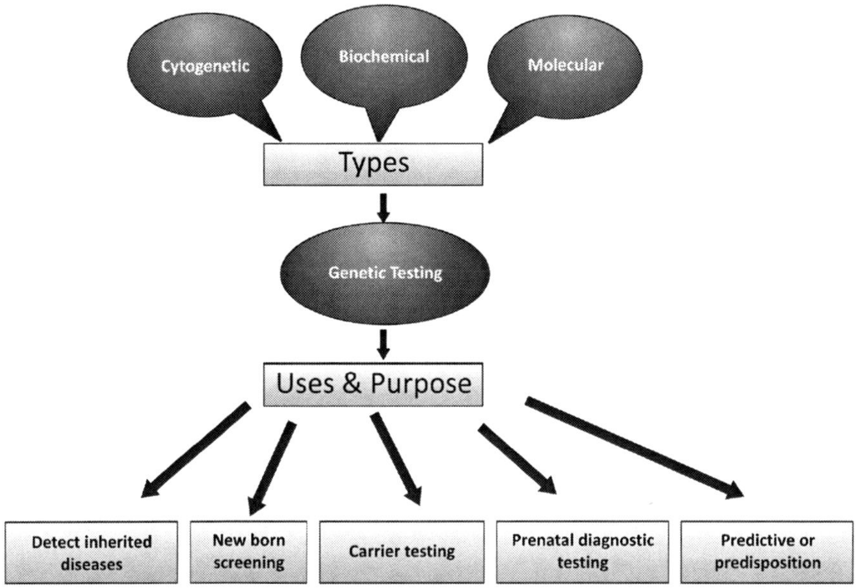

Fig. 20.1 Genetic tests may be used to confirm a diagnosis in a symptomatic individual or used to monitor prognosis of a disease or response to treatment

of rapid division in cell culture. Cells from other tissues such as bone marrow (for leukemia), amniotic fluid (prenatal diagnosis), and other tissue biopsies can also be cultured for cytogenetic analysis Following several days of cell culture, chromosomes are fixed, spread on microscope slides, and then stained. The staining methods for routine analysis allow each of the chromosomes to be individually identified. The distinct bands of each chromosome revealed by staining allow for analysis of chromosome structure.

20.4.2 Biochemical Testing

The enormous numbers of biochemical reactions that routinely occur in cells require different types of proteins. Several classes of proteins exist to fulfill multiple functions, such as enzymes, transporters, structural proteins, regulatory proteins, receptors, and hormones. A mutation in any type of protein can result in disease if the mutation results in failure of the protein to correctly function. Clinical testing for a biochemical disease utilizes techniques that examine the protein instead of the gene. Tests can be developed to directly measure protein activity (enzymes), level of metabolites (indirect measurement of protein activity), and the size or quantity of protein (structural proteins). These tests require a tissue sample in which the protein is present, typically blood, urine, amniotic fluid, or cerebrospinal fluid. Because proteins are more unstable than DNA and can degrade quickly, the sample must be collected, stored properly, and shipped promptly according to the laboratory's specifications.

20.4.3 Molecular Testing

For small DNA mutations, direct DNA testing may be the most effective method, particularly if the function of the protein is not known and a biochemical test cannot be developed. A DNA test can be performed on any tissue sample and requires very small amounts of sample. Some genetic diseases can be caused by many different mutations, making molecular testing challenging. For example, more than 1000 mutations in the cystic fibrosis transmembrane conductance regulator (CFTR) gene can cause cystic fibrosis (CF). It would be impractical to sequence the entire CFTR gene to identify the causative mutation since the gene is quite large. However, since the majority of CF cases are caused by approximately 30 mutations, this smaller group of mutations is first tested before more comprehensive testing is performed (Grody, 1999).

20.5 Current Status of Molecular Diagnosis of Some Common Genetic Diseases

Beyond the complexity of genetic sequencing, with rigorous quality control and assurance, is the difficulty in interpreting the results. As with HIV genotyping, some mutations have no known clinical significance and others impart antiretroviral drug resistance. Sometimes, two mutations act to cancel out the impact of one another. Likewise, some mutations in the gene associated with Gaucher's disease cause neuronopathic disease and other mutations do not. As a final example, some mutations in the CFTR cause cystic fibrosis. Other mutations have low penetrance. That is, these genes seem to have no clinical significance in the majority of individuals, but when combined with other gene expressions, impart classical or nonclassical cystic fibrosis disease.

The current state of molecular diagnosis of some common genetic diseases, including cystic fibrosis, Duchenne muscular dystrophy, haemophilia A and B, phenylketonuria, and thalassaemia, has been reported. Data on carrier detection and prenatal diagnosis are presented and some objective problems and obstacles hampering efficient molecular diagnosis are discussed. The necessity for molecular diagnosis of some other inherited diseases (for example, von Willebrand's disease, Martin-Bell syndrome, polycystic kidney disease, Huntington's disease, and myotonic dystrophy) is stressed. The need for establishing new diagnostic centres dealing with the most common diseases, as well as rare genetic diseases, is substantiated. Perspectives on the implementation of new molecular methods and new technical approaches (preimplantation embryo diagnosis, fetal cells selected from maternal blood) are briefly outlined. It is still too early to estimate the significance of our Human Genome Programme as a whole, but its important impact on medical genetic studies in this country is quite obvious and has been highly appreciated.

There are a myriad of technologies capable of detecting single mutations or SNPs. These platforms are capable of performing medium throughput testing with standard laboratory robotic liquid handlers and at a reasonable hardware cost. These methods include allele-specific oligonucleotide hybridization, PCR RFLP, allele-specific PCR, Line Probe Assays (reverse dot blots), Invader (Third Wave Technologies), ReadIT (Promega), Nanochips (Nanogen), Homogenous PGR (includes TaqMan and Molecular Beacon) and many others. Gene quantitation can be done by several techniques, including fluorescent in situ hybridization (FISH), usually performed in the cytogenetics laboratory with the use of a computerized fluorescent microscope, real-time PCR analysis and comparative genome hybridization.

20.5.1 Cystic Fibrosis

Cystic fibrosis (CF) is one of the most commonly inherited diseases in the Caucasian populations (Bouchara et al., 2009). Those affected have high levels of sodium and chloride (salt) in their sweat. More importantly, a thick, sticky mucous in the lungs causes persistent coughing, wheezing and frequent lung infections, including pneumonia.

In 1989, a research team led by Dr. Francis Collins at the University of Michigan and Lap-Chee Tsui and John Riodan at Toronto's Hospital for Sick Children discovered the gene responsible for cystic fibrosis. The protein is the cystic fibrosis transmembrance conductance regulator (CFTR). Since the discovery of the most common mutation that causes cystic fibrosis, approximately 1000 additional mutations have been identified. Given that approximately 10 million Americans are carriers for cystic fibrosis and 30,000 have the disorder, testing for cystic fibrosis carrier status prior to conception (or if necessary, early after conception) is currently recommended. While cystic fibrosis is not curable, there are some treatments that greatly increase the life span and quality of life for patients with CF. Today, cystic fibrosis mutational analysis is the most commonly performed molecular diagnostic test for inheritable disorders. Patterns of allelic polymorphisms in CF have been analysed and studied. Large scale testing of the AF508 mutation has become possible after adopting the PCR technique for dried blood spots on filter paper. The other major mutations of the CFTR gene known to be quite common in western populations (G551D, R553X, R334W, W1282X, R551X, 1716+ 12T-*C) are detected only occasionally in Russian samples (1–3%). However, CFTR gene mutation 3732delA (exon 19), recently found in southern France, was detected in almost 7% of Russian CF chromosomes. Moreover, 1677delTA, originally discovered turned out to be a major mutation inthe populations of the Black Sea Basin.

Three more new CFTR gene mutations (E504Q exon 10, W1282R exon 19, and S 1196X) have been recently shown by SSCP analysis followed by direct sequencing. At least three different mutations, AF508, 1677delTA, and 3732delA, might be used both for carrier detection and prenatal diagnosis of CF. Identification of new major CFTR gene mutations, specific to our native populations, as well as RFLP analysis of new intragenic polymorphisms, such as the recently discovered highly polymorphic minisatellite DNA sequences in introns 6, 8, and 17b, is helping in more efficient application of molecular analysis in CF patients (Grody, 1999).

20.5.2 Duchenne Muscular Dystrophy

Molecular diagnosis of Duchenne muscular dystrophy (DMD) were carried out with intragenic and flanking DNA probes. They were later supplemented with multiplex polymerase chain reaction (MPCR) for exon deletion detection in the dystrophin gene. Both carrier detection and prenatal diagnosis have been done. A relatively low deletion detection rate with the standard set of exons tested by MPCR, a somewhat unusual pattern of deletion distribution along the DMD cDNA, and a significant proportion of extensive deletions extending through the major part of the gene. One of the possible approaches, not yet tried here so far, concerns RNA amplification supplemented with MPCR analysis of cDNA. Of special diagnostic value for at least some male fetuses at risk might also be the application of western blotting to muscle biopsies or direct immunocytochemical studies of dystrophin in muscle fibres. This approach might be of great benefit for the otherwise uninformative DMD families requesting prenatal diagnosis during the second trimester of pregnancy.

20.5.3 Haemophilia A

Molecular diagnosis of haemophilia A was carried out using DNA probes. Highly polymorphic flanking DNA (probe St14/TaqI) and intragenic polymorphic sites HindIII (intron 19), BclI (intron 18), and XbaI (intron 22) were used both for population studies and diagnostic purposes. Some intragenic polymorphic sites were used for polymerase chain reaction detection. Several prenatal diagnoses by PCR have been carried out. Mutation identification by SSCP analysis followed by direct sequencing of altered exons and amplification-mismatch detection studies of the factor VIII gene are now in progress.

20.5.4 Haemophilia B

Intragenic DNA probes are available for RFLP analysis of the factor IX gene. Southern blot RFLP analysis with these probes was later replaced by PCR for detection of intragenic polymorphic sites TaqI, XmnI (both in intron C), and HinfI/DdeI (intron A). Polymorphism identification was done with original sets of oligoprimers. Thus, direct identification of mutations by means of SSCP analysis or by the amplification-mismatch detection technique is advisable.

20.5.5 Phenylketonuria

Large scale newborn screening programmes for phenylketonuria (PKU), either by the Guthrie test or by an automated fluorescent assay are in practice. According to already available data the frequency of PKU in newborns varies between 1 in 5000 to 1 in 8000 in different regions, with an average of around 1 in 6000. Molecular analysis of PKU was initiated with the cDNA probe of the PAH gene. These studies were substantially helped later by the PCR method. Allele specific hybridisation supplemented with direct sequencing of exon 12 of the PAH gene showed the mutation in codon 408 as well as the exon-intron splicing mutation.

20.5.6 β Thalassaemia

Molecular studies of β thalassaemia was started in 1975 and was initially confined to mRNA analysis and its application to deletion detection in the A globin gene (L Lymborskaya). RFLP analysis in 32 thalassaemia patients with a severe form of the disease showed 11 different haplotypes, one of which was encountered in half of all affected subjects. This particular haplotype was actually in strong linkage disequilibrium with mutations in codon 8.

One of these mutations (deletion of a G nucleotide between codons 82 and 83) was found for the first time. Oligoprimer sets suitable for allele specific amplification and thus for direct identification of globin gene mutations have been suggested.

20.5.7 Wilson's Disease (Hepatolenticular Degeneration)

Identification and biochemical investigations of copper binding protein caeruloplasmin (CP) in Wilson's disease patients were later extended to CP mRNA and CP gene molecular analysis. However, these studies failed to disclose any mutations of the CP gene in patients and thus indicated that the CP gene is not involved by itself in Wilson's disease. Curiously enough, in situ gene mapping with an original fragment of rat CP-DNA as a hybridisation probe disclosed a positive hybridisation signal not only on chromosome 3 (3q23–25, close to the transferrin gene) but also on chromosome 13 (13q23–24), that is, somewhere very close to the still unknown gene responsible for Wilson's disease. Thus the problem of the molecular nature of Wilson's disease is still awaiting a solution. Meanwhile chromosome 13 is one of the genome units selected by our Human Genome Project for detailed molecular analysis. Identification of the gene for Wilson's disease is one of the urgent tasks.

20.5.8 α Antitrypsin Deficiency

DNA analysis of α antitrypsin deficiency (AD) was carried out. RFLP analysis was carried out in 659 patients with chronic non-specific lung disease and a substantial preponderance of the abnormal Z allele was found. An unusual neutral mutation of the MaeIII site in exon 3 of the AT gene in 20–30% of normal subjects with the MI allele, as well as its linkage disequilibrium with the Z allele in AD patients, was discovered. An allele specific amplification system for direct detection of the common mutation in codon 342 of AD patients with the Z haplotype was elaborated and reported.

20.5.9 Familial Hypercholesterolaemia and Other Lipoprotein Disorders

Molecular analysis has been confined to 20 families with familial hypercholesterolemia. No major mutations or rearrangements in the low density lipoprotein receptor gene have been reported so far. RFLP analysis of different polymorphic sites of the receptor gene in normal and affected subjects is in progress. Correlation of particular genotypes of Apo CIII, Apo B 100, and Apo Al genes with blood cholesterol lipoprotein levels in patients with cardiac ischaemic disease and in the general population was reported. A simple approach based on PCR mediated site directed mutagenesis has been suggested for the identification of the common mutation in codon 3500 of the apo Bio gene.

20.5.10 Huntington's Disease Gene

Dr. George Huntington, along with his father and grandfather, published their observations in 1872 on familial cases of chorea near their home on Long Island, NY. The genetic disorder they described is now known as "Huntington's disease" (HD). Nearly 1% of Americans has HD or is at risk of passing along the disease to a child. HD affects as many people as hemophilia, cystic fibrosis or muscular dystrophy combined. In 1993, the HD gene was isolated and, eventually, a direct genetic test was developed that can accurately determine carrier status for the HD gene. The HD gene was found to contain a specific section with a pattern of so-called "trinucleotide repeats" which is expanded in people with HD. In most cases, the repeated pattern occurs 30 times or less. In HD it occurs more than 40 times (Harper et al., 2008).

The test cannot predict when symptoms will begin, and therapy is palliative. In the absence of effective treatment and a cure, most individuals at risk elect not to take the test. HD is one of many trinucleotide expansion diseases characterized by genetic anticipation. This phenomenon manifests when a genetic disease either presents earlier, or presents with more severe symptoms in subsequent generations. For example, myotonic dystrophy has a broad spectrum of presentation varying from congenital myotonia to the seventh decade of life. Because of genetic anticipation, early and accurate diagnosis is vital for parents who may have minimal or late onset symptoms but can have children who are more severely affected.

The advent of pre-implantation genetics allows for the detection of embryos that are homozygous for inheritable recessive disorders. Through detection of a mutation of the dystrophin gene in a fertilized cell, a couple can avoid bringing to term a child with Duchenne's muscular dystrophy. Approximately 30 other diseases like Huntington's disease, TaySachs, cystic fibrosis and familial dysautonomia can be diagnosed on an embryo prior to implantation in the womb.

20.5.11 Multiple Endocrine Neoplasia

Multiple endocrine neoplasia, type 2 (MEN 2) presents as two syndromes, 2A and 2B. In MEN 2A, medullary thyroid carcinoma (MTC) involving the thyroid interstitial C-cells is ultimately found in some 90–95% of affected individuals, pheochromocytoma in 25–50% and hyperparathyroidism in 15–20%. MEN 2B is characterized by MTC, pheochromocytoma, mucosal neuromata and marfanoid habitus.

Classical linkage studies initiated in the early 1980s led to mapping of the MEN 2 gene to a centromeric chromosome 10 locus in 1987. Mutations of the RET proto-oncogene causative for MEN 2 were identified in 1993. Clinical laboratory testing became available a few years later. (Hoff et al., 2001).

The discovery of molecular abnormalities of the RET proto-oncogene in MEN 2 and FMTC has significantly improved clinical management of these disorders. The identified mutations are responsible for 90–95% of all hereditary MTC. Six to seven percent of patients with apparent sporadic MTC have been found to have germline mutations of RET indicative of hereditary MTC. Application of these tests in the management of MTC has improved diagnostic accuracy, lessened the likelihood that hereditary disease will be missed in the context of apparent sporadic MTC and improved the clinical management of this disease.

Many of the common disorders that plague humans are multifactorial gene disorders, including cardiovascular disease, diabetes and most types of cancer. Some genes associated with these disorders may indicate increased predisposition, just as total cholesterol and LDL cholesterol serve as important but not exclusive predictors of cardiovascular disease. With additional studies, more genes are being identified with behaviors. These genes may become more important in the future as we develop a fuller understanding of the impact of genetic factors on behavioral disorders and characteristics.

Common disorders for which intensive research is underway include cardiovascular disease, diabetes, cancer, neurological diseases (including Alzheimer's), bipolar disease, osteoporosis and behavioral disorders. The diagnostic tests that will be developed will complement current and future tests, including those involving proteomics and metabolic analysis. Many of these diagnostic tests may not live up to initial expectations and others may exceed them. Academic centers and large reference laboratories will be on the forefront of this frontier. The prospectors and pioneers will be followed by a growing group of laboratories as molecular diagnostic tests are more widely accepted and adopted. This is the same pattern that was observed with testing for HIV.

20.5.12 Factor V (Leiden) Mutation

The factor V (Leiden) mutation (1691G > A) occurs primarily in the Caucasian population and is a major risk factor for venous thrombosis (a lifetime risk of 12–30% in affected individuals) and a lesser risk factor for arterial thrombosis (cardiovascular disease). Additionally, factor V (Leiden) mutation is associated with arterial thrombosis (especially in smokers), complications of pregnancy (including fetal loss) and increased levels of factor VIII.

The factor V (Leiden) mutation leads to the laboratory finding of activated protein C resistance (APCR) and a sevenfold increase in venous thromboembolic events in heterozygous individuals and an eightyfold increase in homozygous subjects. Due to a synergistic increase in venous thrombosis risk, individuals heterozygous for the factor V mutation are at greater risk when taking oral contraceptives. When a heterozygous mutation is coupled with oral contraceptive use, risk increases synergistically to thirtyfold (Grody et al., 2001).

Since laboratory tests for APCR are highly sensitive, specific and simpler to perform, APCR is usually the test of choice; however, factor V mutation analysis is recommended to confirm positive APCR tests. It is also recommended in place of APCR for patients with lupus anticoagulant, since such patients often have a false-positive APCR test. Although this test is highly specific, identification of a mutation may occur in the absence of APCR in rare cases. Sensitivity of this test for APCR is 94%; thus, a negative result does not rule out APCR or an increased risk of venous thrombosis.

More than half of thromboembolic events associated with factor V (Leiden) mutation occur in the presence of additional risk factors, such as surgery and use of oral contraceptives. Thus, factor V (Leiden) mutation is a risk factor and not an indication of thromboembolic disease.

20.5.13 Hemochromatosis Gene, HFE

Hemochomatosis is an excess accumulation of iron that causes damage to organs, leading to such diseases as cirrhosis, cardiomyopathy, diabetes and arthritis. Two mutations in the HFE gene – C282Y and H63D – are associated with hemochromatosis (Hanson et al., 2001). Disease develops in less than 1% of individuals with these genotypes. These mutations have low penetrance. Other factors, such as diet, hepatotoxins and likely other genes, are important factors that lead to hemochromatosis. The role of testing may be limited to family members of individuals with hemochromatosis. Even for these individuals, biochemical testing remains the cornerstone for diagnosis.

Many molecular diagnostic tests may be similar to the HFE gene tests that have a limited role and must be interpreted in the context of family, clinical history, and other risk factors and laboratory tests.

20.5.14 Colon Cancer Gene, APC

Dr. Bert Vogelstein generated enormous excitement 15 years ago when he and his colleagues at Johns Hopkins described a series of genetic alterations leading to colorectal cancer (Vogelstein et al., 1998), that occurs in the different phases of cancer development, starting from normal epithelium and moving through adenomatous polyps to cancer. This suggested a genetic pathway in tumor development. Unfortunately, the number and nature of the genetic alterations may vary in a population. There may be a spectrum of routes that describes the genetic alterations leading to cancer. The interaction with environmental factors, including diet and other genetic factors, is open for exploration. Today, genetic testing for colorectal cancer is quite limited.

20.6 Microarray Analysis for Detection of Complex Pattern of Genes

Comparative genome hybridization technique used in microarrays hold enormous hope in the field of oncology. Many solid and hematologic malignancies have gene duplications and/or gene deletions associated with them. Amplification of the protooncogene Her2/neu has been demonstrated to correlate with a higher grade of malignancy and with tumor response to the chemotherapeutic agent Herceptin. Comparative genome hybridization microarrays have the ability to scan the entire genome for such insertions and deletions and may revolutionize drug development and prognostic testing. Multiple SNP analysis is currently necessary for cystic fibrosis carrier detection, which requires 25 mutations and six polymorphisms to be analyzed simultaneously. A low-density microarray has been developed for this purpose. Suggestions have been made to screen patients for multiple drug sensitivity pharmacogenetic SNPs using a single chip. This approach suffers from HIPAA and compliance issues, as genetic testing would be performed without a definite indication.

Microarrays will likely be used in one of four ways: expression arrays, resequencing microarrays, multiple genotyping arrays and comparative genome hybridization arrays. Expression arrays consist of DNA probes immobilized on chips that are used to detect mRNA, primarily in solid tumors. Promising data suggests these expression arrays can be used to predict malignant potential in stage 1 breast cancer and other malignancies. Resequencing microarrays are being developed to replace expensive DNA sequencing assays for large genes, such as BRCA1 and BRCA2. These chips may reduce the cost of BRCA1 sequencing. If this can be accomplished, the indications for BRCA1 and BRCA2 testing may be broadened.

Genetic testing, as all clinical laboratory testing, must always be interpreted in the light of clinical findings. Molecular testing for hemochromatosis, alone, is not capable of making the diagnosis of hemochromatosis. Less than 5% of patients homozygous for C282Y in the HFe gene will ever develop symptoms of hemochromatosis. Molecular genotyping tests are extremely sensitive and specific for the mutations they are designed to detect, but may be less so for disorders associated with those diseases. These complex tests may need to be interpreted with additional clinical and routine test results. Regulatory oversight and ethical debate will strive to keep pace with the rapid advances and applications.

20.7 Conclusion

Molecular diagnosis of inherited diseases will expand rapidly into clinical labs in the coming years. Laboratorians must develop internal capabilities to perform and interpret selective tests that are appropriate for their setting, understand the clinical application of these tests, examine the technical issues with performing these tests and maintain the expertise to interpret the test results in the full context of the patient and the family being tested. The future will certainly include more demand for information that will improve the lives of individuals, including testing of embryo and foetus. Technology will make it easier for laboratories to perform molecular diagnostics of common disorders. The application of microarray will aid in the discovery of complex patterns of genes, whose functions may not be understood, that define prognostic patterns and lead to therapeutic recommendations.

From Mendel's rudimentary study of peapods a century and a half ago, to the sophisticated clinical laboratories of today, the rapid growth of molecular genetics provides unsurpassed diagnostic insights into complicated hereditary diseases that affect the lives of human beings worldwide. From these insights, medical science can contribute solutions to the families who struggle to resolve hereditary issues. After all, life should be as simple as "she has your smile and my eyes."

References

Bouchara, J.P., Hsieh, H.Y., Croquefer, S., Barton, R., Marchais, V., Pihet, M., Chang, T.C., 2009, Development of an oligonucleotide array for direct detection of fungi in sputum samples from patients with cystic fibrosis. *J Clin Microbiol* 47(1):142–152.

Grody, W.W., 1999, Cystic fibrosis: molecular diagnosis, population screening and public policy. *Arch Pathol Lab Med* 123:1041–1046.

Grody, W.W., Griffin, J.H., Taylor, A.K., Korf, B.R., Heit, J.A., ACMG Factor V. Leiden Working Group, 2001, American College of Medical Genetics consensus statement on factor V Leiden mutation testing. *Genet Med* 3(2):139–148.

Hanson, E.H., Imparatore, G., Burke, W., 2001, HFE gene and hereditary hemochromatosis: a HuGE review. *Am J Epidemiol* 154:193–206.

Harper, P.S., Lim, C., Craufurd, D., 2008, Ten years of presymptomatic testing for Huntington's disease: the experience of the UK Huntington's Disease Prediction Consortium. *J Med Genet* 37:567–571.

Hoff, A.O., Cote, G.J., Gagal, R.F., 2001, Multiple endocrine neoplasias. *Annu Rev Physiol* 62:377–411.

Holtzman, N.A., Watson, M.S., 2002, Promoting safe and effective genetic testing in the United States: Final Report of the Task Force on Genetic Testing. Baltimore Burke, W. Genetic Testing. *N Engl J Med* 347:1867–1875.

Vogelstein, B., Fearon, E.R., Hamilton, S.R., Kern, S.E., Preisinger, A.C., Leppert, M., Nakamura, Y., White, R., Smits, A.M., Bos, J.L., 1998, Genetic alterations daring colorectal-tumor development. *N Engl J Med* 319:525–532.

Chapter 21
Diagnosis of Complex Diseases

Abstract Although many types of diagnostic and carrier testing for complex disorders have been available for decades, the use of molecular methods is a relatively recent phenomenon. Such testing has expanded the range of disorders that can be diagnosed and has enhanced the ability of clinicians to provide accurate prognostic information and formulate appropriate health supervision measures. However, the proper application of these tests may be difficult because of their scientific complexity and the potential for negative, sometimes unexpected, consequences for many patients.

Keywords Complex diseases · SNP · Alzheimers · Cancer · Haemophilia · Multiple sclerosis · Obesity · Diabetes · Diagnosis · Complex diseases · X-linked inheritance · Multifactorial disorders · Monogenic disorders · Marfan syndrome · Quantitative trait locus · QTL · Tumor markers

21.1 Prologue

Molecular Diagnostics is constantly translating new discoveries and novel technologies into useful clinical tests that provide a molecular fingerprint of tumours and that are predictive of the response to specific therapies. We have entered a new era in medical research. Today's technologies permit us to survey a sufficient number of letters throughout the human genome to provide a clearer picture of how life works and ultimately allow better clinical management of diseases. The most common diseases are the toughest to crack. Heart disease, cancer, diabetes, psychiatric illness, all of these are "complex" or "multifactorial" diseases, meaning that they cannot be ascribed to mutations in a single gene or to a single environmental factor. Rather they arise from the combined action of many genes, environmental factors, and risk-conferring behaviors. One of the greatest challenges facing biomedical researchers today is to sort out how these contributing factors interact in a way that translates into effective strategies for disease diagnosis, prevention, and therapy. Advances in the Human Genome Project have profoundly influenced not only the search for new genes, but also the molecular diagnosis of genetic disorders (Vilain, 1998).

M. Debnath et al., *Molecular Diagnostics: Promises and Possibilities*,
DOI 10.1007/978-90-481-3261-4_21,

21.2 Concept

Beginning in 1986, map-based gene discovery (positional cloning) became the leading method for elucidating the molecular basis of genetic diseases. Almost all medical specialties have used this approach to identify the genetic causes of some of the most puzzling human disorders. Positional cloning has also been used reasonably successfully in the study of common diseases with multiple causes (so-called complex disorders), such as type I diabetes mellitus and asthma. With the availability of the human genome sequence and those of an increasing number of other species, sequence-based gene discovery is complementing and will eventually replace map-based gene discovery.

Complex diseases have complex phenotypes, and proper diagnosis requires that the analysis take into account the patient's history and exposure to environmental factors, as well as genetic information. Signaling information is one aspect of a grandeur "biomedical informatics" approach advocated for a better understanding of a patient's medically relevant disease phenotype. Complex diseases are genetic diseases that do not obey the single-gene dominant or single-gene recessive Mendelian law. The term complex trait is also used for phenotypes that may not be considered as diseases. The etiology of most common diseases is multifactorial, but some Mendelian conditions have a high prevalence (e.g., hereditary hemochromatosis and the genetic predisposition toward breast or colorectal cancer). The same holds true for chromosomal anomalies.

The diagnosis of the complex diseases is a very difficult process and it is a very tough job in the development of the diagnostic techniques for them. The major diagnostics applied today deal mostly with the physical appearances of these diseases, but slowly the application of the modern techniques like the use of DNA analysis, cytogenetics and other molecular diagnostics are picking up the pace. These techniques have got immense potential to boost the early diagnosis of the complex diseases and in the designing of the treatment for them. The complex diseases comprise of diabetes, cancer, obesity, Alzheimer, schizophrenia, hemophilia, multiple sclerosis etc. All these diseases have been shown to be linked to one or more than one genes and have been found to be passed from one generation to the other (inheritance). The example of the X-linked inheritance is hemophilia which is passed from mother to her children, in this the son would be the patient and the daughter would be carrier, both having the probability of 50% each.

The exponential increase in the discovery of genes over the past few years has transformed the DNA diagnosis of genetic disorders from a minor research-based activity to a major professional operation. For any genetic disease, once the defective gene is identified, knowledge of the pathogenic mutations is indispensable to offer DNA diagnosis. DNA diagnosis is offered at pre- and postnatal levels either by direct or indirect approaches. Direct mutational analysis and linkage studies with highly polymorphic intragenic markers are carried out depending on the feasibility of their detection. Apart from the genes that cause monogenic disorders, the discovery of functional sequence variants that confer genetic susceptibility to common multifactorial disorders, such as cardiovascular disease, psychiatric disorders, autoimmune disorders and cancer, also creates demand for high throughput testing for clinically relevant polymorphisms.

An SNP involves any nucleotide of the human genome in which two different bases can be found in the population. SNPs are the most common type of human genetic variation. Because they have only two alleles, they are less informative than microsatellites. This drawback is overcome by their high density throughout the genome and the simplicity and possible automation of their analysis. They may become one of the most powerful tools in human genetics for identifying disease genes and mapping complex traits by linkage.

As larger portions of the human genome are sequenced, one might speculate that mapping tools would become obsolete for diagnostic studies. Direct mutation detection is considered the gold standard of molecular diagnosis because all of the methods based on linkage of polymorphic markers may be impeded by a lack of informative markers or by misinterpretations resulting from recombination events, new mutations, or misattribution of paternity. SNPs will most likely be used widely for the localization of loci for complex traits. It remains to be seen whether SNPs will also be major tools for molecular diagnosis. One of their unique characteristics is the possibility of a relatively simple automation of their analysis, which may lead in the future to molecular screening of large populations. Because of their potential for high throughput, SNPs will have increasing application for predisposition traits when linkage disequilibrium is demonstrated with disease genes.

21.3 Background Information

It is estimated that anywhere from 50,000 to 100,000 genes are contained in the 46 chromosomes present in each human cell. A genetic locus is the place on homologous chromosome pairs where genes are located. Each gene is composed of 2 alternative copies known as alleles, one originating from the maternally derived chromosome and the other originating from the paternally derived chromosome of each chromosome pair. Genes are composed of DNA, and the products of genes are most often proteins that may be used for a variety of purposes, including structural development, regulation of cellular function, enzyme activity, and control of metabolic pathways.

Although most changes in the DNA base-pair composition of genes do not result in disease and are known as *polymorphisms,* some gene changes alter gene function to such a degree that clinical disease is manifested, and these are known as *mutations.* Mostly genetic disease is caused by single base-pair deletions, additions, or substitutions. However, some disorders are caused by large-scale gene abnormalities, such as deletions of the entire gene that can be detected by newer methods of diagnosis, such as molecular cytogenetic analysis. As in all diagnostic testing, it is most important that the clinician have a reasonable index of suspicion based on clinical signs and symptoms that suggest a specific diagnosis.

Advances in genetics are likely to uncover new genes involved in complex diseases such as diabetes, hypertension, cancer, Alzheimer's disease, and others (Mouulsky, 2006). This may increase the demand for genetic testing to predict individual predisposition to future illness in individuals with low-to-average risk and provide opportunities for preventive measures and early intervention. The advances in genetics and the constant changes in gene testing technologies present special challenges to health professionals. There is a need for better education of health professionals about the genetic testing process to improve patient care.

21.4 Detecting New Metabolic Disease Pathways

The majority of publications reporting genetic studies of complex diseases investigate candidate genes and known metabolic pathways. The major problem with any strategy for analyzing a candidate gene or metabolic pathway is that we look wherever we can (that is, among the candidates that we already know) and we most probably overlook other essential genes or pathways because of our ignorance of human biology. Proteins, the products of genes, perform their functions by interacting with each other in coordinated networks. But, only a fraction of these networks have been identified and characterized through classical biochemistry, structural analyses, and assays of activity. A much more comprehensive view of how proteins interact with each other inside cells will become possible with sequence information from multiple species, including humans. Knowing the full complement of our genes, we should be able to identify all of the metabolic pathways in the human body, no matter how short the half-life of the participating proteins or how small the developmental window during which the pathway is turned on.

21.5 Phenotypic Variations in Simplex Diseases

There are major problems associated with dissecting the molecular basis of even simple monogenic diseases caused by mutations in a single gene. Principal among these are the modifying effects of other genes. No gene operates in a vacuum; rather, each gene busily interacts either directly or through its protein product with many other genes and gene products. This results in marked variations in the symptoms of patients with the same disease. Of the 1500 or so monogenic diseases for which the mutated gene has been identified, there are only a few where the effects of other genes on disease pathogenesis have been studied. Existing information about monogenic diseases, such as cystic fibrosis and Hirschsprung disease, demonstrates that certain modifier genes cause variations in the clinical phenotype of these disorders. It is probable that many diseases that are considered monogenic will turn out to be "complex" disorders. Such complexity may be attributable to the typically unpredictable effects of gene mutations on the encoded protein and on the metabolic pathway in which the protein acts. As exemplified by Marfan syndrome, there could be an expression threshold below which the mutant protein does not cause a disease phenotype. In the hemoglobinopathies, for example, modifier genes play an important part in the appearance of clinical symptoms. Finally, the recognition that metabolic pathways are only rarely controlled by single rate-limiting steps greatly complicates the prediction of which symptoms a patient will develop based on the gene that is mutated. These possibilities all require further analysis in human patients as well as in model organisms before we can understand why the severity of monogenic diseases varies not only with different mutations in the same gene, but also among affected individuals within the same family.

21.6 The Genetic Background of Complex Disorders

If there is a challenge to identify the genes involved in so-called monogenic diseases; clearly this challenge will be far greater for oligo- and polygenic disorders, which have multiple causes. Many of the diagnostic features of these complex diseases – called quantitative trait locus (QTL) disorders – are probably regulated by at least several genes.

In monogenic diseases, mutations in a single gene are both necessary and sufficient to produce the clinical phenotype and to cause the disease (Fig. 21.1). The impact of the gene on genetic risk for the disease is the same in all families. In complex disorders with multiple causes, variations in a number of genes encoding different proteins result in a genetic predisposition to a clinical phenotype. Pedigrees reveal no Mendelian inheritance pattern, and gene mutations are often neither sufficient nor necessary to explain the disease phenotype. Environment and life-style are major contributors to the pathogenesis of complex diseases. In a given population, epidemiological studies expose the relative impact of individual genes on the disease phenotype. However, between families the impact of these same genes might be totally different. In one family, a rare gene C (Family 3) might have a large impact on genetic predisposition to a disease. However, because of its rarity in the general population, the overall population effect of this gene would be small. Some genes that predispose individuals to disease might have minuscule effects in some families.

21.7 Dissecting Interactions Between Genes and Environment

Most common human diseases represent the culmination of lifelong interactions between our genome and the environment. Predicting the contribution of genes to complex disorders is still a challenge, and determining the interactions between genes and the environment during any disease process is a daunting task (Wu et al., 2008). Many human diseases, such as hypertension, coronary artery disease, and even some psychiatric disorders, represent quantitative traits that are caused by interactions among genes and between genes and the environment. The functional importance of identified DNA variations can be established by taking advantage of experimental opportunities in other species – for example, by introducing specific

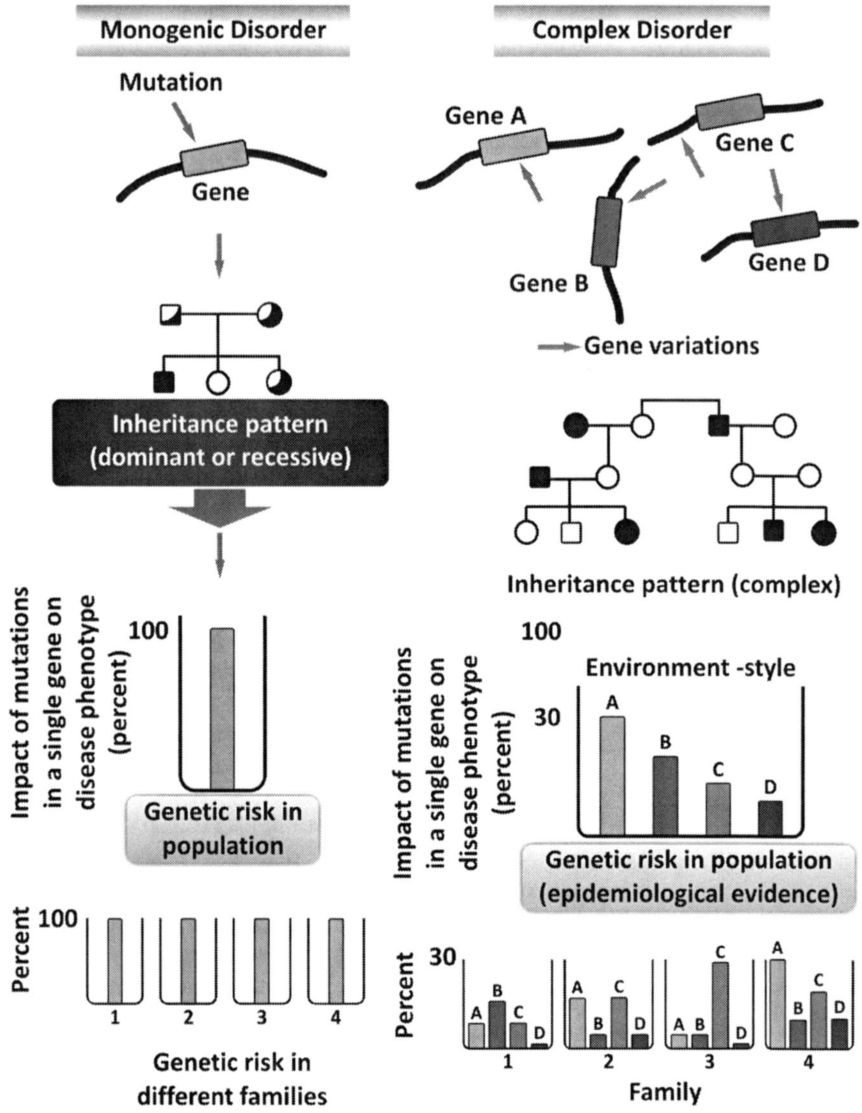

Fig 21.1 Trends of inheritance pattern and genetic risk in monogenic and complex disorders

DNA variations into animal strains that have a well-defined genetic background or that live in an environment that can be precisely controlled.

The genes that contribute to complex disease are difficult to identify, because they typically exert small effects on disease risk; in addition, the magnitude of their effects is likely to be modified by other unrelated genes as well as environmental factors. Perhaps reflecting these difficulties, susceptibility loci for complex diseases identified in one study population often cannot be replicated in other populations. There is no unified story in genetic analysis of complex diseases.

We will now consider seven diseases that offer contrasting examples of genetic determination. The selection is more or less arbitrary. We could equally well have chosen other diseases, but our examples illustrate some typical situations. We have not concentrated on presenting success stories. In some causes there is little progress to report.

21.7.1 Alzheimer Disease

Alzheimer disease is the most common form of progressive dementia in the elderly. It is a neurodegenerative disorder characterized by the neuropathologic findings of intracellular neurofibrillary tangles (NFT) and extracellular amyloid plaques that accumulate in vulnerable brain regions (Sennvik et al., 2000). Within the higher-functioning portions of their brains (the areas responsible for thought and memory), twisting tangles of threads made up of chains of tiny "tau" proteins are being assembled inside billions of nerve cells (neurons). Outside the neurons, other amyloid-beta (AB) proteins are fusing together into sticky clumps (plaque) akin to the substance that clogs heart arteries. Together, these tangles and plaques disrupt the normal functioning of the nerve cells, destroying the pathways along which packets of chemical "information" move. Memories cannot be stored or retrieved, and, eventually, the brain cannot control the body. The axonal defects studied in the mouse model were found to be similar to the axonal defects in the early stages of the Alzheimer disease in humans (Stokin et al., 2004). Bian et al. (2005) found no association of 6 A2M gene polymorphisms with Alzheimer disease in a study of 216 late-onset Alzheimer's disease (AD) patients and 200 control subjects from the Han Chinese population. An estimated 26.6 million people worldwide had Alzheimer's in 2006; this number may quadruple by 2050. The diagnosis of the disease is very difficult leading to its progress for years before getting fully diagnosed. Diagnosis is with the use of the diagnostic aids like medical imaging with computed tomography, magnetic resonance imaging, single photon emission computed tomography and positron emission tomography.

The search for other disease loci gathered pace following the recognition of the genetic heterogeneity of AD. One of the early genetic investigations of Familial Alzheimer's disease (FAD) that failed to show linkage to Chromosome 21 used a mixture of families, but the majority were of late onset (mean age: >60 years). Recently, several independent lines of evidence have provided support for the hypothesis that the disease locus on Chromosome 19 is the apolipoprotein E (ApoE) gene. ApoE is one of the many different proteins found to associate with A4 amyloid fibrils. Daw et al. (2000) estimated the number of additional quantitative trait loci (QTLs) and their contribution to the variance in age at onset of AD, as well as the contribution of apoE and sex. They also estimated that the apoE genotype can make a difference of as many as 17 years in age on the onset of the Alzheimer disease. The study in the Arab population in the Wadi Ara region couldnot explain the prevalence of the Alzheimer disease in relation with the data on the ApoE4 allele (Bowirrat et al., 2002).

Olson et al. (2001) reported convincing evidence of a major role for the APP locus in late-onset AD. They used a covariate-based affected-sib-pair linkage method to analyze the chromosome 21 clinical and genetic data obtained on affected sibships by the Alzheimer Disease Genetics Initiative of the National Institute of Mental Health. In a comparison of 59 unrelated patients with AD and over 1000 controls, it was found that a combination of low head circumference and presence of the ApoE4 allele strongly predicted earlier onset of Alzheimer disease (Graves et al., 1990). The study of the two different populations (Dominican Republic and Puerto Rico) suggested that ApoE4 allele was associated with a nearly two fold increased risk of Alzheimer disease (Romas et al., 2002). In further use of a covariate-based linkage method to reanalyze genome scan data, it was determined that a region on chromosome 20p, showed the same linkage pattern to very-late-onset AD as APP. Two-locus analysis provided evidence of strong epistasis between 20p and the APP region, limited to the oldest age group and to those lacking E4 alleles at the ApoE locus (Olson et al., 2002). Among 563 AD patients and 118 controls, it was found that presence of the ApoE4 allele was strongly associated with reduced cerebrospinal fluid (CSF) levels of beta-amyloid-42 in both patients and controls (Prince et al., 2004). Elkins et al. (2004) concluded that the ACE I allele is associated with an increased risk of late-onset AD, but noted that the risk is very small compared to the effects of other alleles, especially ApoE4. Holmans et al. (2005) performed linkage analyses on 28 sib pairs with late-onset AD. Linkage was observed with chromosome 21 for age-at-onset effects. He also in the same year observed suggestive evidence of increased identical by descent in APOE4 homozygotes on chromosome. Although the APOE epsilon 4 allele is the only proven genetic risk factor for the late form of the disease, genetic epidemiological studies suggest that other loci are also involved (Hoenicka, 2006). The cerebral amyloid angiopathy(CAA) associated with the Alzheimer was also found in the Down syndrome and it was reasoned that due to the amyloid precursor protein gene (APP) locus located on the chromosome 21q21 might be affected by the gene dosage alterations in the subset of the demented individuals (Rovelet et al., 2006).

Alzheimer's disease has been identified as a protein misfolding disease, or proteopathy, due to the accumulation of abnormally folded A-beta and tau proteins in the brains of AD patients. Itoh et al. (2001) noted that the tau-199 test exceeds both sensitivity and specificity over 85% as a sole biomarker of AD; however, they also noted that many of the non-AD tauopathy and degenerative dementias also showed increased tau-199 levels.

Sunderland et al. (2003) suggested that CSF beta-amyloid and tau are biologic markers of AD pathophysiology and that the measures may have potential clinical utility in the future diagnosis of AD. Scott et al. (2003) considered age of onset as a covariant in the analysis of data from 336 markers in 437 multiplex white AD families, they concluded that the linkage to the regions 2q34 and 15q22 were linked to early onset alzheimer disease and very late onset alzheimer disease respectively. In a study it was that the CSF markers are particularly useful to discriminate early or incipient AD from age-associated memory impairment, depression, and some secondary dementias (Blennow et al., 2001). In a study it was found that CSF markers may be useful as diagnostic aids (Sjogren et al., 2003), especially to discriminate early or incipient AD from age-associated memory impairment, depression, and some secondary dementias (Blennow and Vanmechelen, 2003). The work on the CSF for the diagnosis of Alzheimer lead to the conclusion that changes in CSF biomarkers occur early in the course of AD in most patients (Herukka et al., 2005).

Mace et al. (2005) found a significant association between a C-T SNP in exon 14 of the ABCA2 gene and Alzheimer disease in a large case-control study involving 440 AD patients. Li et al. (2005) found the SNP in theAPBB2 gene and reported using this marker the association of late-onset Alzheimer disease was most pronounced in subjects with disease onset before 75 years of age. Sillen et al. (2006) conducted a genome wide linkage study on 188 individuals with AD from 71 Swedish families, using 365 markers (average intermarker distance 8.97 cm). The results suggested that the disorder in these families was tightly linked to the ApoE4 region (19q13). Reiman et al. (2007) used a genome wide SNP survey to examine 1411 individuals with late-onset AD and controls, including 644 carriers of the ApoE4 allele and 767 noncarriers. There was a significant association between AD and 6 SNPs in the GAB2 gene that are part of a common haplotype block.

An in vivo correlate of pathologic involvement, structural imaging measures are potential surrogate markers for disease progression in patients with established AD and in patients with prodromal AD, who will benefit most from disease-modifying therapies underway (Kantarci and Jack, 2003). It has been observed that simple MRI biomarkers, in addition to their diagnostic use, have a prognostic value with respect to mortality in a memory clinic population. A number of neuroimaging candidate markers are promising, such as hippocampus and entorhinal cortex volumes, basal forebrain nuclei, cortical thickness, deformation-based and voxel-based morphometry, structural and effective connectivity by using diffusion tensor imaging, tractography, and functional magnetic resonance imaging. CSF Abeta42, BACE1, total tau, and p-tau are substantially altered in MCI and clinical AD. Other interesting novel marker candidates derived from blood are being currently proposed (phase I). Biomarker discovery through proteomic approaches requires further research. Large-scale international controlled multicenter trials (such as the U.S., European, Australian, and Japanese Alzheimer's Disease Neuroimaging Initiative and the German Dementia Network) are engaged in phase III development of the core feasible imaging and CSF biomarker candidates in AD. Biomarkers are in the process of implementation as primary outcome variables into regulatory guideline documents regarding study design and approval for compounds claiming disease modification (Hampel et al., 2008; Henneman et al., 2008).

Finckh et al. (2000) investigated the proportion of early- onset dementia attributable to known genes. They screened for mutations in 4 genes, PSEN1, PSEN2, APP, and the prion protein gene PRNP, in patients with early-onset dementia before age 60 years. In 12 patients, they found 5 novel mutations and 5 previously reported mutations that were all considered to be disease - causing. In a study it has been reported that a deletion polymorphism at intron 16 of the ACE gene are also associated with AD (Wakutani et al., 2002). Rossi et al. (2004) reported a family in which at least 6 members spanning 3 generations had Alzheimer disease and strokes associated with a heterozygous mutation in the APP gene. Edwards-Lee et al. (2005) reported an African American family in which multiple members spanning 3 generations had early-onset Alzheimer disease. In this two siblings who were tested were found to be heterozygous for the mutation in the APP gene. Rovelet Lecrux (2006) estimated that in their whole cohort of 65 Autosomal dominant early onset Alzheimer's disease (ADEOAD) families, the frequency of the APP locus duplication was roughly 8% (5 of 65), which corresponds to half of the contribution of APP missense mutations to ADEOAD.

Alzheimer's disease is a progressive complex disorder, where genetic predisposition interacts with the environmental factors. With the current knowledge of the pathogenesis of the Alzheimer's disease the pharmaceutical industry is focussing on the development of the novel strategies with an etiopathogenic orientation (Gupta et al., 2008).

21.7.2 Cancer

Cancer is a group of many related diseases that begin in cells, the body's basic unit of life. Normally, cells grow and divide to produce more cells only when the body needs them. Sometimes, however, cells become abnormal and keep dividing to form more cells without control or order, creating a mass of excess tissue called a tumor. Tumors can be malignant (cancerous) or benign (not cancerous).

Cancer can begin in any organ or tissue of the body. The primary, or original, tumor is usually named for the part of the body or the type of tissue in which the cancer begins. The disease can spread (metastasize) from the primary tumor and form metastatic tumors in other parts of the body. For example, breast cancer cells can metastasize to the lungs and cause the growth of a new tumor. When this happens, the tumor in the lung is called metastatic breast cancer because it is composed of breast cancer cells, not lung cancer cells (Anieta, 2009).

Cancer is a diverse class of diseases which differ widely in their causes and biology. The common thread in all known cancers is the acquisition of abnormalities in the genetic material of the cancer cell and its progeny. Research into the pathogenesis of cancer can be divided into three broad areas of focus. The first area of research focuses on the agents and events which cause or facilitate genetic changes in cells destined to become cancer. Second, it is important to uncover the precise nature of the genetic damage, and the genes which are affected by it. The third focus is on the consequences of those genetic changes on the biology of the cell, both in generating the defining properties of a cancer cell, and in facilitating additional genetic events, leading to further progression of the cancer. Some cancers can be caused by infection with pathogens. Many cancers originate from a viral infection; this is especially true in animals such as birds, but also in humans, as viruses are responsible for 15% of human cancers worldwide. The main viruses associated with human cancers are human papillomavirus, hepatitis B and hepatitis C virus, Epstein-Barr virus, and human T-lymphotropic virus. Most forms of cancer are "sporadic", and have no basis in heredity. There are, however, a number of recognized syndromes of cancer with a hereditary component, often a defective tumor suppressor allele. Famous examples are:

- certain inherited mutations in the genes *BRCA1* and *BRCA2* associated with an elevated risk of breast cancer and ovarian cancer
- tumors of various endocrine organs in multiple endocrine neoplasia (MEN types 1, 2a, 2b)
- Li-Fraumeni syndrome (various tumors such as osteosarcoma, breast cancer, soft tissue sarcoma, brain tumors) due to mutations of p53
- Turcot syndrome (brain tumors and colonic polyposis)
- Familial adenomatous polyposis, an inherited mutation of the *APC* gene that leads to early onset of colon carcinoma.
- Hereditary nonpolyposis colorectal cancer (HNPCC, also known as Lynch syndrome) can include familial cases of colon cancer, uterine cancer, gastric cancer, and ovarian cancer, without a preponderance of colon polyps.
- Retinoblastoma, when occurring in young children, is due to a hereditary mutation in the retinoblastoma gene.
- Down syndrome patients, who have an extra chromosome 21, are known to develop malignancies such as leukemia and testicular cancer, though the reasons for this difference are not well understood.

At present number of tumor markers have been identified and they are playing an important role in the diagnosis of the various cancers (Table 21.1).

1. Genetic Testing Genetic testing for high-risk individuals is already available for certain cancer-related genetic mutations. Carriers of genetic mutations that increase risk for cancer incidence can undergo enhanced surveillance, chemoprevention, or risk-reducing surgery. Early identification of inherited genetic risk for cancer, along with cancer-preventing interventions such as surgery or enhanced surveillance, can be lifesaving for high-risk individuals.

2. The Other Diagnostic Techniques

1. Imaging techniques- X-ray, Ultrasonography, computed tomography, magnetic resonance imaging.
2. Histopathology- studying of the tissue section of the affected tissue or cell.
3. Blood test- This is especially done for the diagnosis of the leukemia.
4. Biopsy- This involves the surgical removal of the affected organ or the tissue.

3. Molecular Marker Analysis Moul (1999) has shown that even though p53 and Bcl-2 are predictive biomarkers when sampling the radical prostatectomy specimen, they are not useful to predict postoperative recurrence when sampling the pretreatment needle biopsy. They also concluded that more research is needed to assess new biomarkers and, most importantly, to standardize the methodology for sampling and assaying biomarkers in heterogeneous and multifocal prostate cancer.

Table 21.1 Selected tumor markers in diagnosis of disease

Tumor marker	Description	Comment about texting
Alpha-fetoprotein (AFP)	Levels may be raised in blood of people with cancer of the colon. Blood levels may also be elevated in patients with other cancers or noncancerous conditions	Testing can be useful in diagonosing these cancers and in monitoring treatement
Beta-human chorionic gonadortropin (β-HCG)	This hormone is produced during pregnancy but also occurs in women who have a cancer originating in the placenta and in men with various types of testicular cancer.	Testing can be useful in diagnosing such cancers and in monitoring treatment.
Beta2 (β2)-microglobulin	Levels may be raised in people with multiple mycloma or other cancers of blood cells.	This test cannot be recommended for cancer screening.
Calcitonin	Produced by certain cells in the thyroid gland (C cells). Blood levels elevated in medullary thyroid cancer.	May be used to monitor response to treatment of medullary thyroid cancer.
Carbohydrate antigen 125 (CA-125)	Levels may be increased in women with a variety of gynecological diseased, including ovarian cancer.	This is not recommended for routine cancer screening.
Carbohydrate antigen 19-9 (CA 19-9)	Levels may be increased raised in people with cancers of the digestive tract, particularly pancreatic cancer.	This text cannot be recommended for cancer screening
Carbohydrate antigen 27.29 (CA27.29)	Levels may be increased in people with breast cancer.	This text cannot be recommended for cancer screening.
Carcinoembryonic antigen (CEA)	Levels may be raised in the blood of people with cancer of the colon. Blood levels may also be elevated in patients with other cancers or noncancerous conditions.	After surgery for colon cancer, texting can be useful in monitoring treatment and detecting recurrence.
Lactate dehydrogenase	Levels can be raised for a variety of reasons.	This text cannot be recommended for cancer screening. However, it is useful in assessing prognosis and monitoring treatment, particularly for people with testicular cancer, melanomas, and lymphomas.
Prostate-specific antigen (PSA)	Levels are raised in men with noncancerous (benign) enlargement of the prostate and ofter are considerably higher in men with prostate cancer. What constitutes a meaningfully abnormal level is somewhat uncertain, but men with an elvated PSA level should be evaluated further by a doctor.	Testing can be useful in screening for cancer and in monitoring its treatment
Thyroglobulin	Elevated blood levels may occur in patients with thyroid cancer or benigh thyroid conditions.	This text cannot be recommended for routine screening but may be helpful for monitoring response to treatment of thyroid cancer.

Bast et al. (2005) used two proteomic approaches: one examines the pattern of peaks on mass spectroscopy and the other uses proteomic analysis to identify a limited number of critical markers that can be assayed by more conventional methods. More than 30 serum markers have been evaluated alone and in combination with CA125 by different investigators. Some of the most promising include: HE4, mesothelin, M-CSF, osteopontin, kallikrein(s), and soluble EGF receptor.

Badgwell and Bast (2007) worked on the early detection of the ovarian cancer and measured multiple serum markers simultaneously. Peracaula et al. (2008) focused on the glycan changes of two serum glycoproteins, prostate specific antigen – currently used as a tumour marker of prostate cancer – and human pancreatic ribonuclease in pancreatic adenocarcinoma. The detection of glycan changes, associated with subsets of glycoforms in serum glycoproteins that are specific to the tumour situation, could be the basis for developing more specific biomarkers. Researchers at the University of Southern California (USC) and USC/Norris Comprehensive Cancer Center have identified genetic markers in cancer cells that predicted the benefit of a novel cancer drug prior to chemotherapy (Wilson et al., 2008). About 111 prostate cancer patients with bone metastasis at the diagnosis were enrolled. Thirteen genetic polymorphisms were genotyped using polymerase chain reaction-restriction fragment length polymorphism or an automated sequencer with a genotyping software. There prognosis suggested the metastasis to be influenced by intrinsic genetic factors (Hansel, et al., 2009). The *IGF-1* (CA) repeat and *CYP19* (TTTA) repeat polymorphisms may be novel markers in prostate cancer patients (Norihiko et al., 2007) with bone metastasis at the diagnosis (Tsuchiya et al., 2006).

HPV testing has been suggested for primary screening, triage of equivocal Pap smears or low-grade lesions and follow-up after treatment for cervical intraepithelial neoplasia. Determination of HPV genotype, viral load, integration status and RNA expression could further improve the effectiveness of HPV-based screening and triage strategies (Boulet et al., 2008). Wang et al. (2009) studied on the systematic investigation of the differentially expressed genes during carcinogenesis in hepatocellular carcinoma (HCC) using cDNA microarray technology. This study describes a gene expression profiling of HCC, which provides an extensive list of potential molecular markers for early diagnosis and molecular targets for the development of drugs to treat patients with primary HCC.

A significant correlation has been found between EGFR overexpression and PCNA labeling index in Grade III and Grade II astrocytomas and glioblastoma. These suggest that the tumor proliferation, at least in higher grades of astrocytomas is dependent in some measure on EGF and EGFR-related signaling pathways (Maiti et al., 2008). PMS2 protein levels have been shown to be a predictor of time-to-recurrence after surgery. It is reported that the elevation of a mismatch repair protein negatively correlates with prognosis and has implications in patient diagnosis and molecular profiling (Norris et al., 2009). MRI has a role in well defined indications, but its role is still controversial even in some of these indications. MRI is not indicated in diagnosis in which image-guided percutaneous biopsy is the method of choice. Further studies must be performed (Van Goethem et al., 2009).

In a study the effect of the TSP2 gene introduction into the Human malignant melanoma cell lines A375 was done and it was concluded that the TSP2 suppresses hematogenous metastasis through microenvironment-modification including PAI up-regulation and anti-vascularization in human malignant melanoma. Thrombospondin (TSP) 2 interacts with matrix metalloproteinases (MMPs) and matrix serine proteases such as plasminogen activator (PA) (Chijiwa et al., 2009). Cancer Researchers at the Max Delbrück Center for Molecular Medicine (MDC) Berlin-Buch and the Charité – Universitäts Medizin Berlin (Germany) have identified a gene which enables them to predict for the first time with high probability if colon cancer is going to metastasize. They demonstrated that the gene MACC1 (Metastasis-Associated in Colon Cancer 1) not only promotes tumor growth but also the development of metastasis (Ulrike et al., 2009). When MACC1 gene activity is low, the life expectancy of patients with colon cancer is longer in comparison to patients with high MACC1 levels (Stein et al., 2009). There is no convincing evidence that changing dietary pattern following breast cancer diagnosis will improve prognosis for most women with early stage breast cancer. However, it would appear to be important for some subgroups. Further investigation of mechanisms for such selective action is needed (Pierce, 2009). The developments in the methods of the diagnosis and even more the in depth knowledge of the genetic level of the various types of the cancer has made it possible for the designing of the novel drugs, which aim at the precise cause of the disease.

21.7.3 Multiple Sclerosis

Multiple sclerosis (abbreviated MS, also known as *disseminated sclerosis* or *encephalomyelitis disseminata*) is an autoimmune condition in which the immune system attacks the central nervous system, leading to demyelination. Disease onset usually occurs in young adults, and it is more common in women. MS affects the ability of nerve cells in the brain and spinal cord to communicate with each other. Nerve cells communicate by sending electrical signals called action potentials down long fibers called axons, which are wrapped in an insulating substance called myelin. In MS, the body's own immune system attacks and damages the myelin. When myelin is lost, the axons can no longer effectively conduct signals. The name *multiple sclerosis* refers to scars (scleroses – better known as plaques or lesions) in the white matter of the brain and spinal cord, which is mainly composed of myelin. Although much is known about the mechanisms involved in the disease process, the cause remains unknown. Theories include genetics or infections. Different environmental risk factors have also been found.

Almost any neurological symptom can appear with the disease, and often progresses to physical and cognitive disability. MS takes several forms, with new symptoms occurring either in discrete attacks (relapsing forms) or slowly accumulating over time (progressive forms). Between attacks, symptoms may go away completely, but permanent neurological problems often occur, especially as the disease advances. There is no known cure for MS. Treatments attempt to return function after an attack, prevent new attacks, and prevent disability. MS medications can have adverse effects or be poorly tolerated, and many patients pursue alternative treatments, despite the lack of supporting scientific study. The prognosis is difficult to predict; it depends on the subtype of the disease, the individual patient's disease characteristics, the initial symptoms and the degree of disability the person experiences as time advances. Life expectancy of patients is nearly the same as that of the unaffected population. MS affects over 400,000 people in the United States and may affect 2.5 million people worldwide. Multiple sclerosis affects 2–3 times as many women as men.

Further indications that more than one gene is involved in MS susceptibility comes from studies of families in which more than one member has MS. Several research teams found that people with MS inherit certain regions on individual genes more frequently than people without MS. Of particular interest is the *human leukocyte antigen (HLA)* or *major histocompatibility complex* region on chromosome 6. HLAs are genetically determined proteins that influence the immune system. The HLA patterns of MS patients tend to be different from those of people without the disease. Investigations in northern Europe and America have detected three HLAs that are more prevalent in people with MS than in the general population. Studies of American MS patients have shown that people with MS also tend to exhibit these HLAs in combination-that is, they have more than one of the three HLAs-more frequently than the rest of the population. Furthermore, there is evidence that different combinations of the HLAs may correspond to variations in disease severity and progression.

Studies of families with multiple cases of MS and research comparing genetic regions of humans to those of mice with EAE suggest that another area related to MS susceptibility may be located on chromosome 5. Other regions on chromosomes 2, 3, 7, 11, 17, 19, and X have also been identified as possibly containing genes involved in the development of MS.

These studies strengthen the theory that MS is the result of a number of factors rather than a single gene or other agent. Development of MS is likely to be influenced by the interactions of a number of genes, each of which (individually) has only a modest effect. Additional studies are needed to specifically pinpoint which genes are involved, determine their function, and learn how each gene's interactions with other genes and with the environment make an individual susceptible to MS. In addition to leading to better ways to diagnose MS, such studies should yield clues to the underlying causes of MS and, eventually, to better treatments or a way to prevent the disease.

Rosati (2001) did systematic study of multiple sclerosis in populations. Because the environmental and genetic determinants of geographic gradients are by no means mutually exclusive, the race versus place controversy is, to some extent, a useless and sterile debate. There is a small but significant north-to-south gradient of MS prevalence rates in Japan, suggesting that environmental factors varying with latitude also play a role in developing MS even in the low prevalence areas. Conventional MS in Japanese is, like MS in Caucasians, associated with HLA-DRB1*1501 whereas opticospinal MS is associated with HLA-DPB1∗0501 (Kira,

2003). Dharmasaroja (2003) reviewed and focused on the specificity of autoantibodies to the epitopes of myelin proteins and correlate this to the structures of proteins. Factors that influence the expression of myelin-protein epitopes such as the alpha-helical or beta-sheet structure of the protein, the tri-proline site, and the post-translational modifications as well as physicochemical properties of amino acid changed are included.

Despite the number of issues that still need to be solved, the measurement of brain atrophy by MRI is sufficiently precise and accurate. It represents one of most promising in vivo measures of neuroaxonal degeneration in MS, and it should be used extensively in the future to assess and monitor pathological evolution and treatment efficacy in this disease (Stefano et al., 2007). With increased sophistication and improved analysis of these techniques of NMR and MRI scanning, understanding of the pathology underlying MS may increase, and objective quantification of the natural history of MS is possible (Rashid and Miller, 2008). MR imaging has a major impact on understanding the dynamic neuropathologic findings of multiple sclerosis (MS), early diagnosis of the disease, and clinical trial conduct. With new ways of monitoring the disease, new treatment targets should become practical, helping to translate advances in the understanding of immunology and regenerative medicine into novel therapies (Matthews, 2009).

Bielekova et al. (2004) reviewed the general concepts of biomarkers and their potential use as surrogate endpoints and tailor these concepts to specific applications in multiple sclerosis research. Recent studies have pointed out new interesting candidates, driving new energies and hopes in the MS research community using the immunological markers (Rinaldi and Gallo, 2005). Reindl et al. (2006) concluded that antibodies serving as biomarkers will help to establish a differential therapeutic concept in MS, which will allow to treat individuals selectively according to their pathogenetic subtype and disease status. Berger and Reindl (2007) reviewed the current status and potential applicability of antibodies as biological markers for the diagnosis, classification, disease activity and prediction of clinical courses in MS. Lutterotti et al. (2007) worked on the identification of biomarkers for MS in cerebrospinal fluid and/or blood. They focussed on antibodies to myelin and non-myelin antigens, cells and soluble molecules of the immune system and the brain as biomarkers for (1) the diagnosis and prediction of clinical courses, (2) disease activity and (3) treatment response in MS. Wu et al. (2008) systematically evaluated a series of stilbene derivatives as myelin imaging agents. Spectrophotometry-based and radioligand-based binding assays showed that these stilbene derivatives exhibited relatively high myelin-binding affinities. These studies suggested that these stilbene derivatives can be used as myelin-imaging probes to monitor myelin pathology in vivo. Satoh (2008) identified the principal molecular network involved in development of MS and induction of acute relapse. Thus, DNA microarray technology is highly valuable to identify molecular mechanism-based biomarkers (Bibiana and Roland, 2004) for classification of MS subgroups and prediction of MS relapse. Autoantibody profiles against epitopes derived from MS brain tissue could serve as diagnostic markers or form the basis for the identification of a subgroup of MS patients (Somers et al., 2008). Anti-MBP autoantibody-mediated, epitope-specific binding and cleavage may be regarded as a specific characteristic of MS compared with OND and healthy donors and may serve as an additional biomarker of disease progression (Belogurov et al., 2008).

21.7.4 Diabetes

Diabetes and its associated complications have affected about 200 million people worldwide representing 6% of the population. Diabetes is a disorder of metabolism. Diabetes mellitus (DM) is one of the most common chronic diseases in children and adolescents, and type II DM accounts for more than 95% of cases. Nevertheless, over the last years it has become apparent that not all cases of DM presenting in children have an autoimmune basis. In addition to type 2 DM, which continues to be an infrequent diagnosis among pediatric patients in most countries worldwide, several forms of monogenic DM may present during childhood and are responsible for the disease in 1–3% of patients. Molecular diagnosis, increasingly available, improves both clinical management and quality of life, and is also important for genetic counselling (Rubio-Cabezas and Argente, 2008).

Genetic alterations in expression profiles of pancreatic cells result in phenotypic changes. Therefore, mapping the chromosomal location of the genes that encode differentially expressed transcripts found by microarrays, may lead to the identification of underlying genetic differences accounting for diabetes. The generation of transcriptome maps for tissues from diabetic and non -diabetic individuals could provide tools to identify candidate genes that are overexpressed or silenced in this disease. Cell-based therapies offer the possibilities of a permanent cure for diabetes. Recently, success in the transplantation of pancreatic islets in the pancreas of type 1 diabetics has afforded the opportunity for a potential cure.

Now, the genes underlying certain forms of diabetes are being discovered, and are having a direct impact on treatment. Type 2 diabetes is a classic "polygenic" disease, caused by the interaction of multiple genes and the environment. But this is not always the case: in 1–2% of cases, a mutation in a single gene can cause diabetes. The commonest type of single gene diabetes was originally classified clinically as "maturity onset diabetes of the young" (MODY). This is a form of young-onset diabetes (typically diagnosed under 25) that is not insulin-dependent and is inherited. Since 1992, mutations in six genes have been found to cause MODY, most cases involving the enzyme glucokinase or one of three gene-regulatory proteins: hepatic nuclear factor-1a (HNF-1a), HNF-4a and HNF-1b.

Mutations in these genes affect insulin production by the beta cells, but the outcomes are markedly different. People with glucokinase mutations are born with mildly raised blood glucose levels, and their condition deteriorates little throughout their life. With a careful diet, there is usually no need for insulin injections or drug treatment. People with mutations in one of the MODY transcription factors are usually born with normal blood glucose but usually develop diabetes between the ages of ten and 25. In fact, they are often misdiagnosed as having type 1diabetes, as they are in the right age range, are slim and can have high blood glucose levels.

Busch and Hegele (2001) studied many common variants in functional and positional candidate genes, including ADRB3, PPARG, ENPP1, and CAPN10, and also evaluated for their possible role as determinants of type 2 diabetes. Ksiazek et al. (2003) suggested that the ecNOS gene polymorphism can serve as a useful genetic marker of increased susceptibility to type 2 diabetes and its renal complications. Wasserfall and Atkinson (2006), worked on the four autoantibodies: islet cell cytoplasmic (ICA), insulin (IAA), glutamic acid decarboxylase (GADA), and IA2/ICA512 autoantigen (IA2A), for the detection of diabetes 1 and found that these markers combined with other metabolic and genetic markers, are extremely effective for predicting eventual development of type 1 diabetes in otherwise healthy individuals. In a study it was demonstrated that adiponectin was an independent predictor for gestational diabetes mellitus (GDM). As for GDM screening, adiponectin was not as strong a marker as glucose challenge test (GCT). However, with advantage of being less cumbersome, adiponectin could be used to rule out pregnant women at low risk of GDM (Weerakiet et al., 2006). Eyzaguirre and Mericq (2009) worked on the various currently known biochemical markers to evaluate insulin sensitivity and asses their utility as markers.

McCarthy and Hattersley (2001) surveyed present knowledge about the genetic basis of Type 2 diabetes and discussed the current and future role of genetic diagnostics, with an emphasis on lessons learned from study of monogenic forms of the disease. Hansen (2009) studied and supported the candidate gene approach as a feasible method for directly either identifying or excluding any gene as a diabetes-susceptibility gene ("diabetogene"). Broedl et al. (2006) studied the genetic aspect of the diagnosis of diabetes and concluded that genetic tests are crucial for the correct classification of the type of diabetes, genetic counseling, and initiation of the appropriate therapy regimen. Recent study demonstrated that M55V variant in SUMO4 at IDDM5 was associated with susceptibility to type 1 diabetes in the Asian population (Noso et al., 2006). Molecular analysis of chromosome 6 anomalies, the KCNJ11 and ABCC8 genes encoding Kir6.2 and SUR1 provide a tool to identify transient from permanent neonatal diabetes mellitus in the neonatal period (Flechtner et al., 2007). ABCC8 mutations cause PNDM, TNDM or permanent diabetes diagnosed outside the neonatal period. There is some evidence that the location of the mutation is correlated with the clinical phenotype (Patch et al., 2007).

Among the world's most rapidly spreading complex diseases is the diabetes. The exact genetic cause of the diabetes is not yet established and the latest research has lead to the multiple gene linkage of the disease and the better understanding of the disease. The biotechnology holds the future for the possible cure for the disease as stem cell therapy and the recombinant insulin with even more efficiency are in line to be approved by the governing bodies.

21.7.5 Hemophilia

Hemophilia is a bleeding disorder that slows the blood clotting process. People with this condition may experience prolonged bleeding or oozing following an injury, surgery, or having a tooth pulled or spontaneous bleeding in joints and muscles in severe cases of hemophilia. Serious complications can result from bleeding into the brain, abdomen or other internal organs. The major types of this condition are hemophilia A (also known as classic hemophilia) and hemophilia B (also known as Christmas disease). Although the two types have very similar signs and symptoms, they are caused by mutations in different genes. Both are X linked disorders and hence, manifest predominantly in males. Female carriers of the mutation are usually asymptomatic. The two major forms of hemophilia occur much more commonly in males than in females. Hemophilia A is characterized by deficiency in factor VIII clotting activity. Hemophilia B is inherited in an X-linked recessive pattern.

Hemophilia B is characterized by deficiency in factor IX clotting activity. The age of diagnosis and frequency of bleeding episodes are related to the level of factor VIII and IX clotting activity. The diagnosis of hemophilia B is established in individuals with low factor VIII and IX clotting activity. Molecular genetic testing of *F8* and *F9*, the gene encoding factor VIII and IX, identifies disease-causing mutations in more than 99% of individuals with hemophilia A and B. Detection of mutation is useful for the carrier detection and prenatal diagnosis (Ljunj 1996). As the number of the mutations is very high, mutation detection is difficult for the clinical purposes. Hence, for carrier detection of hemophilia in females at risk, Restriction fragment length polymorphism (RFLP) linkage analysis is most widely used. Linked markers are also used for the prenatal diagnosis.

The two types of polymorphism, namely; single nucleotide polymorphisms (SNPs) and length polymorphisms, also known as variable number tandem repeat sequences (VNTRs) or microsatellites are commonly used for linkage analysis. Polymorphisms have a scientific interest of their own and are useful in areas as diverse as; gene mapping, forensic science and the study of human evolution. They have clinical relevance in the context of hereditary disorders in that they can be used to track a defective (or normal) gene through an affected family. Such linkage studies have permitted carrier status investigation and prenatal diagnosis in hemophilia A and hemophilia B. DNA analysis using intragenic or closely linked restriction fragment length polymorphisms (RFLPs) is rapidly gaining acceptance as a reliable and Practical approach to carrier detection in both hemophilia B and hemophilia A. It is easier and convenient than mutation detection by gene sequencing (Berber et al., 2006).

High throughput mutation scanning based on Southern blot analysis and direct sequencing of all PCR amplified coding exons and the exon-intron boundaries of the factor VIII gene was done in 89 hemophilia A patients and it was found that the correlation of phenotype with genotype as observed in this study was not absolute (Citron et al., 2002). Direct carrier testing was done in 54 at-risk female relatives of haemophilic patients by initially analysing 2·46 kb of the factor IX gene in 1 haemophiliac per family by genomic amplification with transcript sequencing. A presumptive mutation was found in all 14 haemophiliacs examined (Bottema et al., 1989). The technique has clinical application (Sorenson et al., 2003). Identification of a causative mutation leads to more precise carrier detection than does conventional polymorphism-based linkage analysis. This can effectively be used to establish genotype/ phenotype relationships (Mahajan et al., 2004). Jayandharan et al. (2005), analysed DNA from 109 unrelated Indian patients with HA for their

FVIII gene defects, and were screened for point mutations by a multiplex PCR and conformation sensitive gel electrophoresis strategy.

Pandey and Mittal (2001) demonstrated that the combine use of the RFLP markers and short tandem repeats can be used to identify carriers and provide prenatal diagnosis in hemophilia A upto 95% accuracy and help prevent the extent in the hemophilia A affected family. Using the polymorphic markers like intron 18 Bcl I, intron 19 Hind III, intron 22 xbaI and DXS52/St14 of the factor VIII gene and intron 1 DdeI, intron 4 TaqI, 3 HhaI and the residue 148 codon Mnl I of the factor nine gene, diagnosis of hemophilia can be made (Shetty et al., 2001). For first-trimester diagnosis in the chorionic villus samples, we have used both direct and indirect methods, that is, intron 22 and 1 inversions in the factor VIII gene, a multiplex PCR for the detection of gross deletions in the factor IX gene and RFLP analysis using a battery of markers within and outside the factor VIII/IX gene. This method yielded positive result and hold a lot of promise for the prenatal diagnosis of haemophilia (Shetty et al., 2006). De Carvalho et al. (2007) used three microsattelite repeats and one HindIII RFLP markers for the carrier detection of the hemophilia in the brazilian population. Molecular analysis of 76 unrelated Iranian haemophilia B patients was performed by PCR, single strand conformational polymorphism (SSCP) on important functional regions of the F9 gene followed by sequencing on samples with different migration pattern. Using this approach we found mutation in 52 out of 76 patients (Karimipoor et al., 2008). Saha et al. (2007) examined variations of single nucleotide polymorphism (SNP) in F8 in the Indian population and established the utility of a combination of SNP and microsatellite markers for the successful identification of carriers in the affected families. Polymerase chain reaction (PCR) and restriction enzyme analysis were used to study the polymorphism in bcl1, and long-distance PCR for detection of VNTR (ST14) alleles. Carrier detection and prenatal diagnosis is possible in haemophilia A families using both DNA markers. They suggest screening haemophilic families first for bcl1 polymorphism followed by analysis of St14 locus (Hussein et al., 2008). Study was done on the 33 unrelated hemophilia B patients with PCR amplification of the exon sequence and multiplex ligation-dependent probe amplification (MLPA) was performed for the exon dosage test. It was concluded that MLPA can be a feasible platform at clinical laboratories to detect large deletion mutations in the suspected cases (Kwon et al., 2008).

Kim et al. (2005) demonstrated the utility of two intragenic markers in intron 1 and 24 and the other markers intron 13 and 22, using fluorescent PCR in factor 8 gene for carrier detection and prenatal diagnosis of hemophilic families. Diagnostic tests were validated using previously studied samples. IS-PCR evaluated carrier mosaicisms and performed robustly over wide ranges of DNA qualities and procedural conditions. IS-PCR improved the molecular diagnosis of Hemophilia A. This genotyping strategy may potentially be adapted to virtually all known rearrangements in the human genome (Rossetti et al., 2008).

High-resolution melting analysis is an appealing technology for F8 gene screening. It is rapid and quickly identifies mutations in the majority of HA patients; samples in which no mutation is detected require further testing by DNA sequencing. The LC480 and Light Scanner platforms performed similarly (Laurie et al., 2007).

Haemophilia is a genetic disorder of the blood. The recent advancement in the field of diagnosis has lead to the early diagnosis of the disease at the foetal level and helps the parents with the family history of the disease to decide on their future with the help of the genetic counselling along with the diagnosis. The advancement (Rossetti et al., 2008) in the genetic level of the disease has helped in the development of the more targeted form of gene therapy.

21.7.6 Obesity

Obesity is a condition of increased adipose tissue mass. Obesity can also be defined as an increase in body weight beyond the limits of physical requirement, as the result of an excessive accumulation of fat. Accumulation of fat, or triacylglycerol, is essentially the only way that body weight can become excessive, as other energy storage (e.g. carbohydrate glycogen or protein in liver and muscle) does not have the potential of adipose tissue to exceed the limits of requirement. Obesity is associated with a significant increase in morbidity and mortality and is a major public health problem. For reasons that are not fully known, obesity is associated with an increased risk of hypertension, heart disease, diabetes and cancer. Even modest weight loss ameliorates these associated conditions. In addition to the prospect of diminished health, obese people are often stigmatized both socially and in the workplace. Although the premium on leanness has become especially prominent in late twentieth-century Western societies (at least among the affluent), this view is very dependent on the cultural context. In many cultures obesity is considered to be a sign of affluence and prestige, particularly among those cultures where food is less available. In modern times, however, intense pressure to be thin is felt by most individuals, lean and obese. Despite this, obesity affects a significant and increasing number of individuals. The heritability of obesity is roughly equivalent to that of height and exceeds that of many disorders that are generally considered to have a genetic basis. The identity of several of these genes is now known and in these instances the evidence that obesity is not simply a personal failing is overwhelming (James and Trayhum, 1976).

In search for the genetic etiology of the obesity-related metabolic adversities, phenotyping and genotyping of individuals from 507 Caucasian families were conducted. An obese proband was identified by quantitative trait loci (QTLs) on chromosomes 3 (3q27) and 17 (17p12), which were strongly linked to variation in total body adiposity, abdominal fat partitioning, and insulin resistance. The strong evidence for a genetic component to human obesity has been unequivocally established over the past years with the identification of the genetic defects responsible for different monogenic forms, being involved in 4% of cases of human obesity (Clement et al. 2002). However, the role of genetic factors in common obesity is complex, being determined by the interaction of several genes (polygenic), each of which may have relatively small effects (i.e., they are susceptibility genes) and which may work in combination with each other as well as with environmental factors (e.g., nutrient intake and physical activity). The examination of candidate genes for involvement in the susceptibility to common obesity has not yet yielded convincing results (Chagnon et al. 2003). Another approach used for identifying genes underlying common polygenic obesity utilizes genome-wide scans in order to detect chromosomal regions showing linkage with obesity in large collections of nuclear families. The power of this approach was proven in other complex traits (Horikawa et al. 2000; Gretarsdottir et al. 2003).

In another study, a genome-wide scan performed in 158 multiplex French obese Caucasian families (514 individuals) having at least one subject with a body mass index (BMI) of greater than 40 kg/m^2 and at least one further sibling with a BMI of greater than 27 kg/m^2 reported significant evidence for linkage of obesity to a Chromosome 10p locus (Hager et al. 1998), with a maximal logarithm of odds (LOD) score (MLS) near the D10S197 marker was done. Replication studies in both European–American and African–American cohorts confirmed the maximum nonparametric linkage peak at D10S197 for the combined sample set (Price et al. 2001) as well as for a German young obese sib cohort (Saar et al. 2003). All together, these independent linkage studies strengthen the hypothesis of a susceptibility gene for obesity in the Chromosome 10p11–12 locus. Marker D10S197 is located in intron 7 of the *GAD2* gene encoding the glutamic acid decarboxylase enzyme (GAD65). GAD65 catalyzes the formation of γ-aminobutyric acid (GABA) from L-glutamic acid and is expressed in both pancreatic islets and brain (Erdo and Wolff, 1990). GABA is colocalized in neuropeptide Y (NPY) neurons and is involved in the leptin pathway through the arcuate nucleus in the hypothalamus (Ovesjo et al. 2001). GABA interacts with NPY in the paraventricular nucleus to stimulate food intake (Pu et al. 1999). The bilateral injection of *GAD2* antisense oligodeoxynucleotide into rat ventromedial hypothalamus decreased the content of GABA by 50% 24 h after the injection, decreasing food intake, while also enhancing locomotor activity (Bannai et al. 1998). The gene *GAD2* encoding the glutamic acid decarboxylase enzyme (GAD65) is a positional candidate gene for obesity on Chromosome 10p11–12, a susceptibility locus for morbid obesity in four independent ethnic populations. GAD65 catalyzes the formation of γ-aminobutyric acid (GABA), which interacts with neuropeptide Y in the paraventricular nucleus to contribute to stimulate food intake (Boutin et al., 2003).

It can be concluded that genetic and functional arguments favor the *GAD2* gene as a positional candidate gene for obesity on the Chromosome 10p locus. QTL on human chromosome 7q35–q36 that affects both plasma TG and LDL-C variation in Caucasian families of predominantly northern European ancestry have also been localised. Several suggestive QTLs were also localized, including a region on chromosome 12p12.3, which affects plasma HDL-C variation. The large cohort of families studied, the selection of families including both obese and never-obese individuals, and the magnitude of the LOD scores identified emphasize the important contributions of these loci to the genetic architecture associated with the obesity-related lipid disorder.

The additive genetic heritability estimates for the lipid-lipoprotein phenotypes analyzed in our cohort range from 0.30 to 0.47 and strongly suggest the existence of genetic factors, which influence expression of this adverse lipid profile. Similar degrees of heritability influencing plasma levels of the same lipid components have previously been reported in twin, adoption, and family studies in nonobese normal subjects and individuals with established coronary artery disease. Our confidence in the stability of the QTL on chromosome 7 (7q35–q36) is also heightened by its replication in other studies. In a cohort of Mexican-Americans, Duggirala et al. identified a locus harboring susceptibility genes for hypertriglyceridemia in this same region, while analysis of results from Caucasian families in the Framingham Heart Study provided evidence for a gene near this locus that affects the ratio of plasma TG and HDL-C. Moreover, a susceptibility locus for hereditary pancreatitis, a disorder frequently resulting in hypertriglyceridemia, has been mapped to this location.

Li and Loos (2008) studied the genetic progress of obesity and concluded that Genome-wide association promises to enhance greatly our understanding of the genetic basis of common obesity, although candidate gene studies will remain a valuable approach because they allow more detailed analyses of biologically relevant candidates. A key factor contributing to continued success lies in large-scale data integration through international collaboration, which will provide the sample sizes required to identify genetic association with conclusive evidence. Pantsulaia et al. (2009) studied and suggested that genetic factors play a significant role in regulation of variation of the examined traits. The variation of fat distribution measures (OB) traits is almost fully due to genes influencing variation of adipokines (AC), whereas the correlation between blood pressure (BP) and AC is only marginally significant and caused only by shared environment.

Studies in humans and rodents have suggested a central role for FTO through regulation of food intake, whereas others have proposed a peripheral role through an effect on lipolytic activity in adipose tissue (Loos and Bouchard, 2008). Recent advancement in scanning the loci, mapping to the fat mass and obesity associated gene (FTO), influences diabetes risk through a primary effect on fat mass, making this the first common variant known to influence weight and individual risk of obesity (Lindgren and McCarthy, 2008).

Percy et al. (2008) suggested that obesity and hypertension have differing oxidant handling and signalling pathways that act in the pathogenesis of age-related Chronic Kidney disease (CKD). Gaillard et al. (2008) gave an overview of postoral regulation of food intake by lipids and then highlights recent data suggesting the existence of a "fatty taste" which might contribute to lipid overeating and hence to the risk of obesity. Armutcu et al. (2008) suggested that oxidative stress parameters and components of metabolic syndrome are closely related; therefore, significant alterations may occur in the antioxidant and inflammatory status. However, further studies are required to evaluate the possible molecular mechanisms of heat shock protein 70 levels in metabolic syndrome. Phipps et al. (2008) suggested that body size may also be related to the risk of postmenopausal triple-negative breast cancer among nonusers of hormone therapy. Given the expanding obesity epidemic, the widespread cessation of hormone therapy use, and the poor prognosis of triple-negative tumors, this novel finding merits confirmation.

Scheja et al. (2008) identified high fat (HFD)-fed mice as a suitable model to study A-SAA as a biomarker and a novel possible mediator of insulin resistance. A study suggested that Peripheral blood mononuclear cells (PBMC) are a suitable RNA source and model system to perform nutrigenomics studies related to obesity and development of personalized dietary treatments. IL8 gene expression warrant further research as a putative novel biomarker of changes in body fat percentage in response to an low-Calorie diet (LCD) (Crujeiras et al., 2008).

Obesity is a multiple symptomatic disease, which is characterized mainly by the overweight of the patient. Other symptoms arise due to malfunction in the metabolic processes. Its genetic study can help identify the genes responsible and their possible treatment, by treating the error in the metabolic pathway.

21.7.7 Schizophrenia

Schizophrenia is a chronic, severe, and disabling brain disease. It is found all over the world. The severity of the symptoms and the long-lasting, chronic pattern of schizophrenia often results in disability, and many patients need ongoing assistance to manage the most basic functions of independent living. People with schizophrenia may have perceptions of reality that are strikingly different from the reality seen and shared by others around them. Their behavior may seem odd, unusual or even bizarre at times. They sometimes hear voices, talk to themselves, or respond to imaginary fears. The recent study has evaluated various risks factors causing schizophrenia.

At times, normal individuals may feel, think, or act in ways that resemble schizophrenia. Normal people may sometimes be unable to "think straight." They may become extremely anxious, for example, when speaking in front of groups and may feel confused, be unable to pull their thoughts together, and forget what they had intended to say. This is not schizophrenia. At the same time, people with schizophrenia do not always act abnormally. Indeed, some people with the illness can appear completely normal and be perfectly responsible, even while they experience hallucinations or delusions. An individual's behavior may change over time, becoming bizarre if medication is stopped and returning closer to normal when receiving appropriate treatment. There is a common misconception that schizophrenia is the same as a "split personality". Another related misconception is that schizophrenia results in several different personalities, and the individual switches between these different persons. These perceptions are not correct. Although schizophrenia affects men and women with equal frequency, the disorder often appears earlier in men, usually in the late teens or early twenties, than in women, who are generally affected in the twenties to early thirties. People with schizophrenia often suffer terrifying symptoms such as hearing internal voices not heard by others, or believing that other people are reading their minds, controlling their thoughts, or plotting to harm them. These symptoms may leave them fearful and withdrawn. Their speech and behavior can be so disorganized that they may be incomprehensible or frightening to others. Available treatments can relieve many symptoms, but most people with schizophrenia continue to suffer some symptoms throughout their lives; it has been estimated that no more than one in five individuals recovers completely. This is a time of hope for people with schizophrenia and their families. Research is gradually leading to new and safer medications and unraveling the complex causes of the disease. Scientists are using many approaches from the study of molecular genetics to the study of populations to learn about schizophrenia. Methods of imaging the brain's structure and function hold the promise of new insights into the disorder.

In follow-up from evidence obtained in linkage studies, systematic linkage disequilibrium mapping within chromosomal region 13q33 has led to the identification of a schizophrenia susceptibility locus which harbors the genes *G72* and *G30*. These association findings have been replicated in several independent schizophrenia samples. Association has also been found between genetic variants at the *G72/G30* locus and bipolar affective disorder (BPAD), with replication in independent studies. Results from studies of more detailed psychiatric phenotypes show that association exists with symptom clusters that are common to several disorders as well as with specific psychiatric diagnoses. These findings may indicate that the association lies not with the diagnostic categories per se but with more specific aspects of the phenotype, such as affective symptoms and cognitive effects, which cross traditional psychiatric diagnostic boundaries. At the molecular level, the picture remains far from clear. No putative functional variants have been identified in the coding regions of *G72* or *G30*, and it is therefore likely that disease susceptibility is caused by as yet unidentified variants which alter gene expression or splicing. A further complication is the fact that inconsistencies are evident in the risk alleles and haplotypes observed to be associated across different samples and studies, which may suggest the presence of multiple susceptibility variants at this locus. Functional analyses indicate that the *G72* gene product plays a role in the activation of N-methyl-D-aspartate receptors, a molecular pathway implicated in both schizophrenia and BPAD, making it the most plausible candidate gene at this locus.

Three differentially expressed proteins in schizophrenia have been identified viz. manganese superoxide dismutase (MnSOD) was under-expressed; and collapsing response mediator protein 2 (CRMP-2) and t-complex protein 1 (TCP-1) were over-expressed. MnSOD catalyzes the dismutation of superoxide anion (O_2^-) into water (H_2O) and hydrogen peroxide (H_2O_2). In nervous tissue, it may protect the survival of cell membranes. CRMP-2 regulated axonal growth and polarity. Its over-expression may explain the vast neural interconnections in the schizophrenic brain.

Oxidative stress is particularly implicated in neurodegenerative diseases like Alzheimer's and Parkinson's diseases. Gene expression studies with schizophrenia have consistently identified oxidation related transcripts. Theories have postulated that when the production of harmful oxidants exceeds the rate of anti-oxidant compounds, macromolecules such as DNA and proteins become the targets of oxidative attack, which signals subsequent death. Moreover, Japanese scientists were the first to identify proteins targets for nitration in the brains of rats, which included TCP-1. The original research methodology did not allow for post-translational modification testing; the state of TCP-1 in schizophrenic hippocampal tissue therefore remains unknown.

Chromosomal location is another feature of the study that sheds light on schizophrenia. Three of the four characterized proteins were mapped to chromosomal arm 6q. Their vicinity reveals a region important in schizophrenia and confirms the susceptibility of loci found by the linkage study. Despite the informative studies using 2D-GE with schizophrenia, this modality proved very limited. Three important issues were raised that have instigated investigations into other resolution measures. First, 2D-GE analysis has limited reproducibility. Second, weakly soluble proteins cannot be easily resolved. Third, only a tiny portion of the proteome can be effectively stained. It is especially difficult for low-level expressed proteins and those masked by greater expression within a similar molecular weight or isoelectric point, or both. Heritability estimates for schizophrenia range from 70 to 80%. However, efforts to map susceptibility genes using traditional linkage studies have been inconsistent, probably because of complex patterns of inheritance and the relatively small risk conferred by individual genes. An alternative to linkage analyses is to test candidate genes for linkage and association using the transmission disequilibrium Test (TDT). Identification of such candidates has come more rapidly with the use of modern molecular methods, such as gene microarrays.

The expression of *RGS4*, but not other RGS family members, is decreased across the cerebral cortex in subjects with schizophrenia, but not among subjects with major depressive disorder. RGS proteins function primarily as GTPase-activating proteins (GAPs) for heterotrimeric G-protein $\alpha(G_\alpha)$ subunits, accelerating the hydrolysis of G_α-bound GTP. Thus, they shorten the duration of intracellular signaling of many G-protein-coupled receptors (GPCRs) belonging to dopamine, GABA, glutamate and other neurotransmitter systems. *RGS4* is an interesting candidate gene for additional reasons. It is highly expressed in brain regions implicated in the pathophysiology of schizophrenia, modulates function of multiple G-protein-coupled neurotransmitter receptors and exhibits robust transcriptional changes to stress. Furthermore, *RGS4* maps to chromosome 1q21–q22, a region implicated in prior schizophrenia linkage studies.

Emilien et al. (1999) studied the relationship between dopamine receptors and schizophrenia and lent some useful insights as to the potential advantages of the newer antipsychotics which appear to have a sparing or beneficial effect on various components of cognitive function. Golimbet (2008) studied the association between gene polymorphisms and cognitive dysfunctions in schizophrenic patients. The main emphasis was to the association between gene polymorphisms and cognitive dysfunctions in schizophrenic patients. The association studies on the genes for dopamine and serotonin receptors, brain-derived neurotrophic factor, dysbindin, DISC1, D-amino acid oxidase and D-amino acid oxidase activator are reviewed as well.

Post-mortem studies have revealed a lower density of glutamatergic receptors in patients with schizophrenia. Other studies of cerebrospinal fluid reported lower levels of glutamate in patients with schizophrenia in healthy comparison subjects. The most compelling evidence is provided by the psychomimetic effects of the NMDA antagonists' phencyclidine and ketamine. Pharmacological modulation of the effects of NMDA receptor antagonists, such as ketamine, may lead to development of novel therapeutic agents for psychiatric illnesses such as schizophrenia (Mechri et al., 2001). Bota et al. (2008) reviewed the development of the prodromal concept and its implications for early identification.

Morrens et al. (2008), gave considerable evidence to support the existence of a cluster of psychomotor symptoms, in addition to positive, negative and cognitive symptoms. The highly sensitized computerized neuropsychological testing was used for monitoring of the cognitive effects of antipsychotics. Computerized assessment assists in the identification of state- and trait-related cognitive impairments (Kertzman et al., 2008).

Nuechterlein et al. (2004) studied some schizophrenia patients and concluded that seven separable cognitive factors were replicable across studies and represent fundamental dimensions of cognitive deficit in schizophrenia: Speed of Processing, Attention/Vigilance, Working Memory, Verbal Learning and Memory, Visual Learning and Memory, Reasoning and Problem Solving, and Verbal Comprehension. In 2004 a meeting was held, based on discussions at this meeting, five criteria were considered essential for test selection of schizophrenia: good test-retest reliability, high utility as a repeated measure, relationship to functional outcome, potential response to pharmacologic agents, and practicality/tolerability (Green et al., 2004). Stip (2006) reviewed and concluded that from now on cognitive deficit should be recognized as a major element in social and professional integration of schizophrenia patients, and should become a standardized assessment approach in clinical practice.

Schwarz and Bahn (2008) reviewed molecular alterations of proteins and metabolites that have been identified in schizophrenia CSF and discuss their potential applicability as diagnostic markers (Sheja et al., 2008). Chopra et al. (2008) studied various patents that describe the use of nanoparticles to deliver various neurotherapeutics and neurodiagnostics to brain.

Advances in DNA technology have provided important new tools in the effort to discover the genetic basis of schizophrenia. One of the most valuable techniques is linkage analysis, which is based on the fact that, when chromosomes cross over and exchange segments of DNA (i.e. recombine) during meiosis, genetic loci that are close to each other are more likely to be co-inherited than are loci that are more distant. This is important to the identification of genes that cause disease because, if an unknown disease gene is "linked" (i.e. co-segregates) with a DNA marker having a known chromosomal location, the marker can be used to find the gene. This marker must be polymorphic (since variants must be specific to a disease). Linkage analysis became a powerful tool when molecular geneticists developed methods to identify many DNA markers throughout the genome.

Statistical methods of linkage analysis involve computing the probability that the co-segregation of genetic markers and disease within pedigrees exceeds what would be expected by chance. Many linkage studies have been performed under SML conditions, in which the odds for linkage are analysed for a specified degree of linkage. If the odds against a random finding (expressed as the logarithm of odds ratio, or LOD score) exceed 1000 to 1, a LOD of 3 is assigned, and evidence of linkage between a gene and a trait is provided. On the other hand, a LOD score of –2 is considered to be a cut-off point used to exclude the possibility of linkage. These statistical decision rules have proven to be reliable for single gene diseases; however, for complex disorders such as schizophrenia, other factors such as the presumed mode of transmission, the definition of the phenotype, the degree of penetrance, the sample size and the number of affected family members, must also be considered in determining linkage. Unfortunately, some of those factors, such as the mode of transmission and the degree of penetrance, are not yet known for schizophrenia. To compensate, linkage analyses can be performed repeatedly, with different values set for each parameter. However, a disadvantage of this approach is that positive results must be viewed conservatively, because the risk of false positive findings increases with the number of tests performed.

In general, linkage analysis is most useful in uncovering variance in important genes that follow Mendelian patterns of inheritance; it is less helpful in identifying genes that exert a small or moderate effect upon complex psychiatric disorders. This problem is exacerbated when definitions of phenotypes are not accurate. Despite these reservations, evidence has been aggregated that link schizophrenia to several chromosomal sites (Chromosome 1, 5, 6, 8, 10, 11, 13, 15, 18, and 22). Schizophrenia is a complex psychiatric disorder, who's diagnosis for long has been a challenge for the doctors and the scientists. The modern diagnostic methods have opened the new realm for the genetic study of the disease and the development of the psychotropic drugs, which can be even used in treating other psychological diseases.

21.8 Conclusion

We are rapidly advancing upon the postgenomic era in which genetic information will have to be examined in multiple health care situations throughout the lives of individuals. Currently, newborn babies can be screened for treatable genetic diseases such as phenylketonuria. Perhaps in the not-so-distant future, children at high risk for coronary artery disease will be identified and treated to prevent changes in their vascular walls during adulthood. Parents will have the option to be told their carrier status for many recessive diseases before they decide to start a family. For middle-aged and older populations, we will be able to determine risk profiles for numerous late-onset diseases, preferably before the appearance of symptoms, which at least could be partly prevented through dietary or pharmaceutical interventions. In the near future, the monitoring of individual drug response profiles with DNA tests throughout life will be standard practice. Soon, genetic testing will comprise a wide spectrum of different analyses with a host of consequences for individuals and their families – an issue worth emphasizing when explaining genetic testing to the public.

The challenge for health care professionals will be to correctly interpret the outcome of genetic testing for their patients, their patients′ families, and for society in general. Genetic counselors, who explain the purpose and results of genetic tests, will be crucial for helping individuals to make informed decisions, particularly when test results indicate the possibility of disease. Current training programs, including those in medical schools, do not adequately teach students how to deal with these challenges.

The tremendous potential for efficient information transfer via the Internet can and should be used to inform the public of the possibilities provided by the genomics era. However, when it comes to sensitive and very personal aspects of genetic information, traditional contact with health care professionals is still the most appropriate route. Reaping the fruits of the human genome sequencing project through alleviating the suffering of patients will only be possible if available genetic information is combined with the skilled professionalism of health care workers and ethically solid standards.

In the "post-genomic" era, the natural progression is towards interrogating the main effectors of physiological functions – proteins. The major genomic projects of the last decade have shaped proteome-wide sequencing, mapping, and analysis. For example, the creation of the Human Proteome Organization's Human Brain Proteome Project to foster the effective international communication of brain related proteomic data. Complex diseases are now rapidly investigated by novel high-throughput biochemical inquires to uncover disease dynamics, clinical markers, and drug targets.

Contrary to the genome, the proteome is composed of an active array of molecules constantly being modified and with special localization. Proteomic approaches are able to characterize post-translational modifications, a method by which the cell dynamically and quickly modifies protein function and regulates both creation and degradation in response to cellular perturbations (e.g., disease provocation). Protein profiling and identification techniques using mass spectrometry (MS) and bioinformatics can lead to the discovery, identification, and characterization of protein biomarkers differentially expressed in the diseased states versus the control.

References

Anieta, S., 2009, Prognostic test for breast cancer may not detect all tumor types. *Science Daily*, Jan. 8.

Armutcu, F., Ataymen, M., Atmaca, H., Gurel, A., 2008, Oxidative stress markers, C-reactive protein and heat shock protein 70 levels in subjects with metabolic syndrome. *Clin Chem Lab Med* 46:785–790.

Badgwell, D., Bast, R.C. Jr, 2007, Early detection of ovarian cancer. *Dis Markers* 23:397–410.

Bast, R.C. Jr, Badgwell, D., Lu, Z., Marquez, R., Rosen, D., Liu, J., Baggerly, K.A., Atkinson, E.N., Skates, S., Zhang, Z., Lokshin, A., Menon, U., Jacobs, I., Lu, K., 2005, New tumor markers: CA125 and beyond. *Int J Gynecol Cancer* 3:274–281.

Bannai, M., Ichikawa, M., Nishihara, M., Takahashi, M., 1998, Effect of injection of antisense oligodeoxynucleotides of GAD isozymes into rat ventromedial hypothalamus on food intake and locomotor activity. *Brain Res* 784:305–315.

Belogurov, A.A. Jr, Kurkova, I.N., Friboulet, A., Thomas, D., Misikov, V.K., Zakharova, M.Y., Suchkov, S.V., Kotov, S.V., Alehin, A.I., Avalle, B., Souslova, E.A., Morse, H.C. 3rd, Gabibov, A.G., Ponomarenko, N.A., 2008, Recognition and degradation of myelin basic protein peptides by serum autoantibodies: novel biomarker for multiple sclerosis. *J Immunol* 180:1258–1267.

Berber, E., Leggo, J., Brown, C., Berber, E., Gallo, N., Feilotter, H., Lillicrap, D., 2006, DNA microarray analysis for the detection of mutations in hemophilia A. *J Thromb Haemost* 4(8):1756–1762.

Berger, T., Reindl, M., 2007, Multiple sclerosis: disease biomarkers as indicated by pathophysiology. *J Neurol Sci* 259:21–26.

Bian, L., Yang, J.D., Guo, T.W., Sun, Y., Duan, S.W., Chen, W.Y., Pan, Y.X., Feng, G.Y., He, L., 2004, Insulin-degrading enzyme and Alzheimer disease: A genetic association study in the Han Chinese, *Neurology* 63(2):241–245.

Bibiana, B., Roland, M., 2004, Development of biomarkers in multiple sclerosis. *Brain* 127: 1463–1478.

Blennow, K., Vanmechelen, E., 2003, CSF markers for pathogenic processes in Alzheimer's disease: diagnostic implications and use in clinical neurochemistry. *Brain Res Bull* 61:235–242.

Blennow, K., Vanmechelen, E., Hampel, H., 2001, CSF total tau, Abeta42 and phosphorylated tau protein as biomarkers for Alzheimer's disease. *Mol Neurobiol* 24:87–97.

Bota, R.G., Sagduyu, K., Filin, E.E., Bota, D.A., Munro, S., 2008, Toward a better identification and treatment of schizophrenia prodrome. *Bull Menninger Clin Summer* 72:210–227.

Bottema, C.D., Koeberl, D.D., Sommer, S.S., 1989, Direct carrier testing in 14 families with haemophilia B. *Lancet* 2:526–529.

Boulet, G.A., Horvath, C.A., Berghmans, S., Bogers, J., 2008, Human papillomavirus in cervical cancer screening: important role as biomarker. *Cancer Epidemiol Biomarkers Prev* 17: 810–817.

Boutin, P., Dina, C., Vasseur, F., Dubois, S., Corset, L., et al., 2003, GAD2 on chromosome 10p12 is a candidate gene for human obesity. *PLoS Biol* 1:e68. doi: DOI: 10.1371/journal.pbio.0000068.

Bowirrat, A., Friedland, R.P., Korczyn, A.D., 2002, Vascular dementia among elderly Arabs in Wadi Ara. *J Neurol Sci* 203:73–76.

Broedl, U.C., Maugeais, C., Marchadier, D., Glick, J.M., Rader, D.J., 2006, Molecular diagnosis of diabetes mellitus. *Internist (Berl)* 47:47–54.

Busch, C.P., Hegele, R.A., 2001, Genetic determinants of type 2 diabetes mellitus. *Clin Genet* 60:243–254.

Chagnon, Y.C., Rankinen, T., Snyder, E.E., Weisnagel, S.J., Perusse, L., et al., 2003, The human obesity gene map: The 2002 update. *Obes Res* 11:313–367.

Chijiwal, T., Abe, Y., Ikoma, N., Yamazaki, H., Tsukamoto, H., Suemizu, H., Kawai, H., Wakui, M., Nishime, C., Matsumoto, H., Matsuyama, M., Mukai, M., Ueyama, Y., Nakamura, M., 2009, Thrombospondin 2 inhibits metastasis of human malignant melanoma through microenvironment-modification in NOD/SCID/ÁC null (NOG) mice. *Int J Oncol* 34:5–13.

Chopra, D., Gulati, M., Saluja, V., Pathak, P., Bansal, P., 2008, Brain permeable nanoparticles, recent patents CNS. *Drug Discov* 3:216–225.

Citron, M., Godmilow, L., Ganguly, T., Ganguly, A., 2002, High throughput mutation screening of the factor VIII gene (F8C) in hemophilia A: 37 novel mutations and genotype-phenotype correlation. *Hum Mutat* 20:267–274.

Clement, K., Boutin, P., Froguel, P., 2002, Genetics of obesity. *Am J Pharmacogenomics* 2: 177–187.

Crujeiras, A.B., Parra, D., Milagro, F.I., Goyenechea, E., Larrarte, E., Margareto, J., Martínez, J.A., 2008, Differential expression of oxidative stress and inflammation related genes in peripheral blood mononuclear cells in response to a low-calorie diet: a nutrigenomics study. *OMICS* 12:251–261.

Daw, E.W., Payami, H., Nemens, E.J., Nochlin, D., Bird, T.D., Schellenberg, G.D., Wijsman, E.M., 2000, The number of trait loci in late-onset Alzheimer disease. *Am J Hum Genet* 66:196–204.
De Carvalho, F.M., de Vargas Wolfgramm, E., Paneto, G.G., de Paula Careta, F., Spagnol Perrone, A.M., de Paula, F., Louro, I.D., 2007, Analysis of Factor VIII polymorphic markers as a means for carrier detection in Brazilian families with haemophilia A. *Haemophilia* 13:409–412.
Dharmasaroja, P., 2003, Specificity of autoantibodies to epitopes of myelin proteins in multiple sclerosis. *J Neurol Sci* 206:7–16.
Edwards-Lee, T., Ringman, J.M., Chung, J., Werner, J., Morgan, A., St. George Hyslop, P., Thompson, P., Dutton, R., Mlikotic, A., Rogaeva, E., Hardy, J., 2005, An African American family with early-onset Alzheimer disease and an APP (T714I) mutation. *Neurology* 64: 377–379.
Erdo, S.L., Wolff, J.R., 1990, Aminobutyric acid outside the mammalian brain. *J Neurochem* 54:363–372.
Elkins, J.S., Douglas, V.C., Johnston, S.C., 2004, Alzheimer's disease risk and genetic variation in ACE. *Neurology* 62:363–368.
Emilien, G., Maloteaux, J.M., Geurts, M., Owen, M.J., 1999, Dopamine receptors and schizophrenia: contribution of molecular genetics and clinical neuropsychology. *Int J Neuropsychopharmacol* 2:197–227.
Eyzaguirre, F., Mericq, V., 2009, Insulin resistance markers in children. *Horm Res* 71:65–74.
Finckh, U., Muller-Thomsen, T., Mann, U., Eggers, C., Marksteiner, J., Meins, W., Binetti, G., Alberici, A., Hock, C., Nitsch, R.M., Gal, A., 2000, High prevalence of pathogenic mutations in patients with early-onset dementia detected by sequence analyses of four different genes. *Am J Hum Genet* 66:110–117.
Flechtner, I., Vaxillaire, M., Cavé, H., Froguel, P., Polak, M., 2007, Neonatal diabetes: a disease linked to multiple mechanisms. *Arch Pediatr* 14:1356–1365.
Frazier, L., Meininger, J., Halsey Lea, D., Boerwinkle, E., 2001, Genetic discoveries and nursing implications for complex disease prevention and management. *J Prof Nurs* 20:222–229.
Gaillard, D., Passilly-Degrace, P., Besnard, P., 2008, Molecular mechanisms of fat preference and overeating. *Ann N Y Acad Sci* 1141:163–175.
Golimbet, V.E., 2008, Molecular genetics of cognitive deficit in schizophrenia. *Mol Biol (Mosk)* 42:830–839.
Graves, A.B., White, E., Koepsell, T.D., Reifler, B.V., van Belle, G., Larson, E.B., Raskind, M., 1990, A case-control study of Alzheimer's disease. *Ann Neurol* 28:766–774.
Green, M.F., Nuechterlein, K.H., Gold, J.M., Barch, D.M., Cohen, J., Essock, S., Fenton, W.S., Frese, F., Goldberg, T.E., Heaton, R.K., Keefe, R.S., Kern, R.S., Kraemer, H., Stover, E., Weinberger, D.R., Zalcman, S., Marder, S.R., 2004, Approaching a consensus cognitive battery for clinical trials in schizophrenia: the NIMH-MATRICS conference to select cognitive domains and test criteria. *Biol Psychiatry* 56:301–307.
Gretarsdottir, S., Thorleifsson, G., Reynisdottir, S.T., Manolescu, A., Jonsdottir, S., et al., 2003, The gene encoding phosphodiesterase 4D confers risk of ischemic stroke. *Nat Genet* 35: 131–138.
Gupta, M., Kaur, H., Grover, S., Kukreti, R., 2008, Pharmacogenomics and treatment for the dementia induced by Alzheimer's disease. *Pharmacogenomics* 9(7):895–903.
Hager, J., Dina, C., Francke, S., Dubois, S., Houari, M., et al., 1998, A genome-wide scan for human obesity genes reveals a major susceptibility locus on chromosome 10. *Nat Genet* 20:304–308.
Hampel, H., Bürger, K., Teipel, S.J., Bokde, A.L., Zetterberg, H., Blennow, K., 2008, Core candidate neurochemical and imaging biomarkers of Alzheimer's disease. *Alzheimers Dement* 4:38–48.
Herukka, S.-K., Hallikainen, M., Soininen, H., Pirttilä, T., 2005, CSF Aß42 and tau or phosphorylated tau and prediction of progressive mild cognitive impairment. *Neurology* 64: 1294–1297.
Hansen, L., 2009, Candidate genes and late-onset type 2 diabetes mellitus, susceptibility genes or common polymorphisms? *Dan Med Bull* 50:320–346.

Hansel, D.E., Nakayama, M., Luo, J., Abukhdeir, A.M., Park, B.H., Bieberich, C.J., Hicks, J.L., Eisenberger, M., Nelson, W.G., Mostwin, J.L., De Marzo, A.M., 2009, Shared TP53 gene mutation in morphologically and phenotypically distinct concurrent primary small cell neuroendocrine carcinoma and adenocarcinoma of the prostate. *Prostate* 69:603–609.

Henneman, W.J., Sluimer, J.D., Cordonnier, C., Baak, M.M., Scheltens, P., Barkhof, F., van der Flier, W.M., 2008, MRI biomarkers of vascular damage and atrophy predicting mortality in a memory clinic population. *Stroke* 40:492–498.

Hoenicka, J., 2006, Genes in Alzheimer's disease. *J Rev Neurol* 42:302–305.

Holmans, P., Hamshere, M., Hollingworth, P., Rice, F., Tunstall, N., Jones, S., Moore, P., Wavrant, DeVrieze, F., Myers, A., Crook, R., Compton, D., Marshall, H., Meyer, D., Shears, S., Booth, J., Ramic, D., Williams, N., Norton, N., Abraham, R., Kehoe, P., Williams, H., Rudrasingham, V., O'Donovan, M., Jones, L., Hardy, J., Goate, A., Lovestone, S., Owen, M., Williams, J., 2005, Genome screen for loci influencing age at onset and rate of decline in late onset Alzheimer's disease. *Am J Med Genet Part B* 135B:24–32.

Horikawa, Y., Oda, N., Cox, N.J., Li, X., Orho-Melander, M., et al., 2000, Genetic variation in the gene encoding calpain-10 is associated with type 2 diabetes mellitus. *Nat Genet* 26: 163–175.

Hussein, I.R., El-Beshlawy, A., Salem, A., Mosaad, R., Zaghloul, N., Ragab, L., Fayek, H., Gaber, K., El-Ekiabi, M., 2008, The use of DNA markers for carrier detection and prenatal diagnosis of haemophilia A in Egyptian families. *Haemophilia* 14:1082–1087.

Ito, T., Chiba, T., Ozawa, R., Yoshida, M., Hattori, M., Sakaki, Y., 2001, A comprehensive two-hybrid analysis to explore the yeast protein interactome. *Proc Natl Acad Sci USA* 98(8): 4569–4574.

James, W.P., Trayhurn, P., 1976, An integrated view of the metabolic and genetic basis for obesity. *Lancet* 2:770–773.

Jayandharan, G., Shaji, R.V., Baidya, S., Nair, S.C., Chandy, M., Srivastava, A., 2005, Identification of factor VIII gene mutations in 101 patients with haemophilia A: mutation analysis by inversion screening and multiplex PCR and CSGE and molecular modelling of 10 novel missense substitutions. *Haemophilia* 11:481–491.

Kantarci, K., Jack, C.R. Jr., 2003, Neuroimaging in Alzheimer disease: an evidence-based review. *Neuroimaging Clin N Am* 13:197–209.

Karimipoor, M., Kokabee, L., Kamali, E., Karizi, S.Z., Zeinali, S., 2008, Molecular analysis of factor IX gene in an Iranian female with severe Hemophilia B. *Acta Haematol* 119:151–153 (DOI: 10.1159/000128043).

Kertzman, S., Reznik, I., Grinspan, H., Weizman, A., Kotler, M., 2008, Antipsychotic treatment in schizophrenia: the role of computerized neuropsychological assessment. *Isr J Psychiatry Relat Sci* 45:114–120.

Kim, J.W., Park, S.Y., Kim, Y.M., Kim, J.M., Kim, D.J., Ryu, H.M., 2005, Identification of new dinucleotide-repeat polymorphisms in factor VIII gene using fluorescent PCR. *Haemophilia* 11:38–42.

Kira, J., 2003, Epidemiology of multiple sclerosis: environmental factors versus genetic factors. *Nippon Rinsho* 61:1300–1310.

Ksiazek, P., Wojewoda, P., Muc, K., Buraczynska, M., 2003, Endothelial nitric oxide synthase gene intron 4 polymorphism in type 2 diabetes mellitus. *Mol Diagn* 7:119–123.

Kwon, M.J., Yoo, K.Y., Kim, H.J., Kim, S.H., 2008, Identification of mutations in the F9 gene including exon deletion by multiplex ligation-dependent probe amplification in 33 unrelated Korean patients with haemophilia B. *Haemophilia* 14:1069–1075.

Li, M., Boehnke, M., Abecasis, G.R., 2005, Joint modeling of linkage and association: Identifying SNPs responsible for a linkage signal. *Am J Hum Genet* 76:934–949.

Lindgren, C.M., McCarthy, M.I., 2008, Mechanisms of disease: Genetic insights into the etiology of type 2 diabetes and obesity. *Nat Clin Pract Endocrinol Metab* 4(3):156–163.

Li, S., Loos, R.J., 2008, Progress in the genetics of common obesity:size matters. *Curre Opin Lipidol* 19:113–121.

Ljung, R.C., 1996, Prenatal diagnosis of haemophilia, Baillieres. *Clin Haematol* 9(2): 243–257.
Loos, R.J., Bouchard, C., 2008, MRC FTO: the first gene contributing to common forms of human obesity. *Obes Rev* 9:246–250.
Lutterotti, A., Berger, T., Reindl, M., 2007, Biological markers for multiple sclerosis. *Curr Med Chem* 14:1956–1965.
Mace, S., Cousin, E., Ricard, S., Genin, E., Spanakis, E., Lafargue-Soubigou, C., Genin, B., Fournel, R., Roche, S., Haussy, G., Massey, F., Soubigou, S., Brefort, G., Benoit, P., Brice, A., Campion, D., Hollis, M., Pradier, L., Benavides, J., Deleuze, J.-F., 2005, ABCA2 is a strong genetic risk factor for early-onset Alzheimer's disease. *Neurobiol Dis* 18:119–125.
Mahajan, A., Chavali, S., Kabra, M., Chowdhury, M.R., Bharadwaj, D., 2004, Molecular characterization of hemophilia B in North Indian families: identification of novel and recurrent molecular events in the factor IX gene. *Haematologica* 89:1498–1503.
Maiti, A.K., Ghosh, K., Chatterjee, U., Chakrobarti, S., Chatterjee, S., Basu, S., 2008, Epidermal growth factor receptor and proliferating cell nuclear antigen in astrocytomas. *Neurol India* 56:456–562.
Matthews, P.M., 2009, Brain imaging of multiple sclerosis: the next 10 years. *Neuroimaging Clin N Am* 19:101–112.
McCarthy, M.I., Hattersley, A.T., 2001, Molecular diagnostics in monogenic and multifactorial forms of Type 2 diabetes. *Exp Rev Mol Diagn* 1:403–412.
Mechri, A., Saoud, M., Khiari, G., d'Amato, T., Dalery, J., Gaha, L., 2001, Glutaminergic hypothesis of schizophrenia: clinical research studies with ketamine. *Encephale* 27: 53–59.
Morrens, M., Hulstijn, W., Sabbe, B.G., 2008, Psychomotor symptoms in schizophrenia: the importance of a forgotten syndrome. *Tijdschr Psychiatr* 50:713–724.
Moul, J.W., 1999, Angiogenesis, p 53, bcl-2 and Ki-67 in the progression of prostate cancer after radical prostatectomy. *Eur Urol* 35:399–407.
Mouulsky, A.G., 2006, Genetics of complex diseases. *J Zhejiang Univ Sci* B 7:167–168.
Norihiko, T., Lizhong, W., Hiroyoshi, S., Takehiko, S., Hisami, F., Shintaro, N., Masaki, S., Toshiyuki, K., Kenji, M., Tomohiko, I., Osamu, O., Akira, N., Tomonori, H., 2007, Impact of IGF-I and CYP19 gene polymorphisms on the survival of patients with metastatic prostate cancer, medline. *J Clin Oncol* 24:1982–1989.
Norris, A.M., Gentry, M., Peehl, D.M., D'Agostino, R. Jr, Scarpinato, K.D., 2009, The elevated expression of a mismatch repair protein is a predictor for biochemical recurrence after radical prostatectomy. *Cancer Epidemiol Biomarkers Prev* 18:57–64.
Noso, S., Ikegami, H., Fujisawa, T., Kawabata, Y., Asano, K., Hiromine, Y., Sugihara, S., Lee, I., Kawasaki, E., Awata, T., Ogihara, T., 2006, Association of SUMO4, as a candidate gene for IDDM5, with susceptibility to type 1 diabetes in Asian populations. *Ann N Y Acad Sci* 1079:41–46.
Nuechterlein, K.H., Barch, D.M., Gold, J.M., Goldberg, T.E., Green, M.F., Heaton, R.K., 2004, Identification of separable cognitive factors in schizophrenia *Schizophr Res* Dec 15, 72(1): 29–39.
Olson, J.M., Goddard, K.A.B., Dudek, D.M., 2001, The amyloid precursor protein locus and very-late-onset Alzheimer disease. *Am J Hum Genet* 69:895–899.
Olson, J.M., Goddard, K.A.B., and Dudek, D.M., 2002, A second locus for very-late-onset Alzheimer disease: A genome scan reveals linkage to 20p and epistasis between 20p and the amyloid precursor protein region. *Am J Hum Genet* 71(1):154–161.
Ovesjo, M.L., Gamstedt, M., Collin, M., Meister, B., 2001, GABAergic nature of hypothalamic leptin target neurons in the ventromedial arcuate nucleus. *J Neuroendocrinol* 13:505–516.
Pandey, G.S., Mittal, B., 2001, Molecular diagnosis in haemophilia A. *J Postgrad Med* 47:274–280.
Pantsulaia, Ia., Trofimova, S., Kobyliansky, E., Livshits, G., 2009, Relationship between obesity, adipocytokines, and blood pressure: possible common genetic and environmental factors. *Am J Hum Biol* 21:84–90.

Patch, A.M., Flanagan, S.E., Boustred, C., Hattersley, A.T., Ellard, S., 2007, Mutations in the ABCC8 gene encoding the SUR1 subunit of the KATP channel cause. *Diabetes Obes Metab* 9:28–39.

Peracaula, R., Barrabés, S., Sarrats, A., Rudd, P.M., de Llorens, R., 2008, Altered glycosylation in tumours focused to cancer diagnosis. *Dis Markers* 25:207–218.

Percy, C.J., Brown, L., Power, D.A., Johnson, D.W., Gobe, G.C., 2008, Obesity and hypertension have differing oxidant handling molecular pathways in age-related chronic kidney disease. *Mech Ageing* 130:129–138.

Phipps, A.I., Malone, K.E., Porter, P.L., Daling, J.R., Li, C.I., 2008, Body size and risk of luminal, HER2-overexpressing, and triple-negative breast cancer in postmenopausal women. *Cancer Epidemiol Biomarkers Prev* 17:2078–2086.

Pierce, J.P., 2009, Diet and breast cancer prognosis: making sense of the Women's Healthy Eating and Living and Women's Intervention Nutrition Study trials. *Curr Opin Obstet Gynecol* 21: 86–91.

Price, R.A., Li, W.D., Bernstein, A., Crystal, A., Golding, E.M., et al., 2001, A locus affecting obesity in human chromosome region 10p12. *Diabetologia* 44:363–366.

Prince, J.A., Zetterberg, H., Andreasen, N., Marcusson, J., Blennow, K., 2004, APOE ε4 allele is associated with reduced cerebrospinal fluid levels of Aß42. *Neurology* 62:2116–2118.

Pu, S., Jain, M.R., Horvath, T.L., Diano, S., Kalra, P.S., et al., 1999, Interactions between neuropeptide Y and gamma-aminobutyric acid in stimulation of feeding: A morphological and pharmacological analysis. *Endocrinology* 140:933–940.

Rashid, W., Miller, D.H., 2008, Recent advances in neuroimaging of multiple sclerosis. *Semin Neurol* 28:46–55.

Ray, L., 2007, Evaluation of gnetic markers linked to hemophilia A locus: an Indian experience. *Haematologica* 92(12):1725–1726.

Reiman, E.M., Webster, J.A., Myers, A.J., Hardy, J., Dunckley, T., Zismann, V.L., et al., 2007, GAB2 alleles modify Alzheimer's risk in APOE epsilon4 carriers. *Neuron* 54(5):713–720.

Reindl, M., Khalil, M., Berger, T., 2006, Antibodies as biological markers for pathophysiological processes in MS. *J Neuroimmunol* 180:50–62.

Rinaldi, L., Gallo, P., 2005, Immunological markers in multiple sclerosis: tackling the missing elements. *Neurol Sci* 26(Suppl 4):S215–S217.

Romas, S.N., Vincent Santana, V., Williamson, J., Ciappa, A., Lee, J.H., Rondon, H.Z., Estevez, P., Lantigua, R., Medrano, M., Torres, M., Stern, Y., Tycko, B. Mayeux, R., 2002, Familial Alzheimer disease among caribbean hispanics. *Arch Neurol* 59:87–91.

Rosati, G., 2001, The prevalence of multiple sclerosis in the world: an update. *Neurol Sci* 22: 117–139.

Rossetti, L.C., Radic, C.P., Larripa, I.B., De Brasi, C.D., 2008, Developing a new generation of tests for genotyping hemophilia-causative rearrangements involving int22h and int1h hotspots in the factor VIII gene. *J Thromb Haemost* 6:830–836.

Rossi, G., Giaccone, G., Maletta, R., Morbin, M., Capobianco, R., Mangieri, M., Giovagnoli, A.R., Bizzi, A., Tomaino, C., Perri, M., Di Natale, M., Tagliavini, F. Bugiani, O., Bruni, A.C., 2004, A family with Alzheimer disease and strokes associated with A713T mutation of the APP gene. *Neurology* 63:910–912.

Rovelet Lecrux, A., Hannequin, D., Raux, G., Le Meur, N., Laquerriere, A., Vital, A., et al. 2006, APP locus duplication causes autosomal dominant early-onset Alzheimer disease with cerebral amyloid angiopathy. *Nat Genet* 38:24–26.

Rubio-Cabezas, O., Argente, J., 2008, Current insights into the genetic basis of diabetes mellitus in children and adolescents. *J Pediatr Endocrinol Metab* 21:917–940.

Saar, K., Geller, F., Ruschendorf, F., Reis, A., Friedel, S., et al., 2003, Genome scan for childhood and adolescent obesity in German families. *Pediatrics* 111:321–327.

Saha, A., Mukherjee, S., Maulik, M., Chandak, G.R., Indian Genome Variation Consortium.

Saha, M.M., Ullah, S.M.J., Mondal, D.K., Bakar, M.A., Bhuiyan, J.H., 2007, Haemophilia: An update. *J Bangladesh Coll Phys Surg* 25:29–37.

Satoh, J., 2008, Molecular biomarkers for prediction of multiple sclerosis relapse. *Nippon Rinsho* 66:1103–1111.

Scott, W.K., Hauser, E.R., Schmechel, D.E., 2003, Ordered-subsets linkage analysis detects novel Alzheimer disease Loci on chromosomes 2q34 and 15q22. *Am J Hum Genet* 73:1041–1051.

Scheja, L., Heese, B., Zitzer, H., Michael, M.D., Siesky, A.M., Pospisil, H., Beisiegel, U., Seedorf, K., 2008, Acute-phase serum amyloid A as a marker of insulin resistance in mice. *Exp Diabetes Res* 230837.

Schwarz, E., Bahn, S., 2008, Cerebrospinal fluid: identification of diagnostic markers for schizophrenia. *Exp Rev Mol Diagn* 8(2):209–216.

Sennvik, K., Fastbom, J., Blomberg, M., Wahlund, L.O., Winblad, B., Benedikz, E., 2000, Levels of alpha- and beta-secretase cleaved amyloid precursor protein in the cerebrospinal fluid of Alzheimer's disease patients. *Neurosci Lett* 278:169–172.

Sieuwerts, A.M., Look, M.P., Gelder, M.M., Timmermans, M., Trapman, A.M.A.C., Garcia, R.R., Anneke, M.A., Goedheer, J.W. Weerd, V.De., Pottengen, H., Klijn, J.G.M., Foekens, J.A., 2009, Prognostic test for breast cancer may not detect all tumor types, *Science Daily*, Jan. 8.

Sillén, A., Forsell, C., Lilius, L., Axelman, K., Bjork, B.F., Onkamo, P., Kere, J., Winblad, B., Graff, C., 2006, Genome scan on Swedish Alzheimer's disease families. *Mol Psychiatr* 11: 182–186.

Shetty, S., Ghosh, K., Bhide, A., Mohanty, D., 2001, Carrier detection and prenatal diagnosis in families with haemophilia. *Natl Med J India* 14:81–83.

Shetty, S., Ghosh, K., Jijina, F., 2006, First-trimester prenatal diagnosis in haemophilia A and B families-10 years experience from a centre in India. *Prenat Diagn* 26:1015–1017.

Sjögren, M., Andreasen, N., Blennow, K., 2003, Advances in the detection of Alzheimer's disease-use of cerebrospinal fluid biomarkers. *Clin Chim Acta* 332:1–10.

Somers, V., Govarts, C., Somers, K., Hupperts, R., Medaer, R., Stinissen, P., 2008, Autoantibody profiling in multiple sclerosis reveals novel antigenic candidates. *J Immunol* 180:3957–3963.

Sorenson, J.R., Jennings-Grant, T., Newman J., 2003, Communication about carrier testing within hemophilia A families. *Am J Med Genet C: Semin Med Genet* 119C(1):3–10.

Stefano, N.D., Battaglini, M., Smith, S.M., 2007, Measuring brain atrophy in multiple sclerosis. *J Neuroimaging* 17(s1):10S–15S.

Stein, U., Walther, W., Arlt, F., Schwabe, H., Smith, J., Fichtner, I., Birchmeier, W., Schlag, P.M., 2009, MACC1, a newly identified key regulator of HGF-MET signaling, predicts colon cancer metastasis. *Nat Med* 15:59–67.

Stip, E., 2006, Cognition, schizophrenia and the effect of antipsychotics. *Encephale* 32:341–350.

Stokin, G.B., Lillo, C., Falzone, T.L., Brusch, R.G., Rockenstein, E., Mount, S.L., Raman, R., Davies, P., Masliah, E., Williams, D.S., Goldstein, L.S., 2005, Axonopathy and transport deficits early in the pathogenesis of Alzheimer's disease. *Science* 307(5713): 1282–1288.

Sunderland, T., Gur, R.E., Arnold, S.E., 2005, The use of biomarkers in the elderly: current and future challenges. *Biol Psychiatr* 58:272–276.

Tsuchiya, N., Wang, L., Suzuki, H., Segawa, H., Fukuda, H., Narita, S., Shimbo, M., Kamoto, T., Mitsumori, K., Ichikawa, K., Ogawa, O., Nakamura, A., Habuchi, A., 2006, Impact of IGF-I and CYP19 gene polymorphisms on the survival of patients with metastatic prostate cancer. *J Clin Oncol* 24(13):1982–1989.

Tsuji, S., 1998, Recent advances in the studies on Alzheimer's diseases. *Rinsho Shinkeigaku* 38:983–987.

Ulrike, S., Peter, M.S., Walter, B., 2009, Newly identified gene powerful predictor of colon cancer metastasis. *Science Daily* 5.

Van Goethem, M., Verslegers, I., Biltjes, I., Hufkens, G., Parizel, P.M., 2009, Role of MRI of the breast in the evaluation of the symptomatic patient. *Curr Opin Obstet Gynecol* 21: 74–79.

Vilain, E., 1998, CYPs, SNPs, and molecular diagnosis in the postgenomic era. *Clin Chem* 44:2403–2404.

Wakutani, Y., Kowa, H., Kusumi, M., Yamagata, K., Wada-Isoe, K., Adachi, Y., Takeshima, T., Urakami, K., Nakashima, K., 2002, Genetic analysis of vascular factors in Alzheimer's disease. *Ann N Y Acad Sci* 977:232–238.

Wang, W., Peng, J.X., Yang, J.Q., Yang, L.Y., 2009, Identification of gene expression profiling in hepatocellular carcinoma using cDNA microarrays. *Dig Dis Sci* [Epub ahead of print] PMID: 19117127.

Wasserfall, C.H., Atkinson, M.A., 2006, Autoantibody markers for the diagnosis and prediction of type 1 diabetes. *Autoimmun Rev* 5:424–428.

Weerakiet, S., Lertnarkorn, K., Panburana, P., Pitakitronakorn, S., Vesathada, K., Wansumrith, S., 2006, Can adiponectin predict gestational diabetes? *Gynecol Endocrinol* 22:362–368.

Wilson, P.M., Yang, D., Shi, M.M., 2008, Researchers identify genetic markers that predict efficacy of novel cancer drug. *ScienceDaily* May 30.

Wray, N.R., Goddard, M.E., Vissche, P.M., 2008, Prediction of individual genetic risk of complex disease, science direct. *National Intelligence Council* USA, 2000–2002 "global trends 2015".

Wu, C., Wei, J., Tian, D., Feng, Y., Miller, R.H., Wang, Y., 2008, Molecular probes for imaging myelinated white matter in CNS. *J Med Chem* 51:6682–6688.

Chapter 22
Biochips

Abstract Biochip technology is highly effective method that allows monitoring of thousands of genes/alleles at a time in computerized automatic operations with minimal volumes of necessary reagents. Biochips promise an important shift in molecular biology, DNA diagnostics, and pharmacology, research in carcinogenesis and other diseases and also the possibility of a better understanding of the world of biology in its globality.

Keywords Biochips · Microarray · Hybridization · Gene expression · DNA diagnostics · Peptide nucleic acid · Oligonucleotide · Protein arrays

22.1 Prologue

Technological advances in miniaturization have found a niche in biology and signal the beginning of a new revolution. The biological and physical sciences share a common interest in small structures (the definition of "small" depends on the application, but can range from 1 nm to 1 mm). A vigorous trade across the borders of these areas of science is developing around new materials and tools (largely from the physical sciences) and new phenomena (largely from the biological sciences). The physical sciences offer tools for synthesis and fabrication of devices for measuring the characteristics of cells and sub-cellular components, materials useful in cell and molecular biology; biology offers a window into the most sophisticated collection of functional nanostructures that exists (Whitesides, 2003). Most of the attention and advances have been made with DNA chips yet a lot of progress is being made in the use of other biomolecules and cells. Development of high-throughput "biochip" technologies has dramatically enhanced our ability to study biology and explore the molecular basis of disease. Biochips enable massively parallel molecular analyses to be carried out in a miniaturized format with a very high throughput.

22.2 Concept

A new revolution technology that may become promising for research, diagnostics and therapy enters into biology and medicine. Biochips containing microarrays

M. Debnath et al., *Molecular Diagnostics: Promises and Possibilities*,
DOI 10.1007/978-90-481-3261-4_22,

of genetic information promise to be important research tools in the post genomic era (Jain, 2001). The basic idea of the biochip technology is to convert the chemistry of life into a static form programmed to monitor genes, proteins and relations between them. Biochip programmed by known sequences of DNA/RNA or proteins can recognize the real genes, mutations and levels of expression.

Biochips are collections of miniaturized test sites (microarrays) arranged on a solid substrate onto which a large number of biomolecules are attached with high density. The word "biochip" derives from the computer term "chip". Although silicon surfaces bearing printed circuits can be used for DNA binding, the term biochip is now broadly used to describe all surfaces bearing microscopic spots, each one being formed by specific capture probes. The capture probes are chosen to complement the target sequence to be detected. Each capture probe will bind to its corresponding target sequence. Like a computer chip performing millions of mathematical operations in a few split seconds, a biochip allows for simultaneous analyses of thousands of biological reactions, such as decoding genes, in a few seconds. Biochip technologies can be applied to numerous fields including genomic, proteomic, and glycomic research, as well as pharmacology and toxicology. However, one of the most common applications is in the determination of gene expression in human cells and tissues. Global gene expression analysis has helped to identify important genes and signalling pathways in human malignant tumours.

Biochips are formed by in situ (on chip) synthesis of oligonucleotides (Braun et al., 2005) or peptide nucleic acids (PNAs) or spotting of DNA fragments. Hybridisation of RNA- or DNA-derived samples on chips allows the monitoring of expression of mRNAs or the occurrence of polymorphisms in genomic DNA. Basic types of DNA chips are the sequencing chip, the expression chip and chips for comparative genomic hybridisation. Like a computer chip performing millions of mathematical operations in a few split seconds, a biochip allows simultaneous analyses of thousands of biological reactions, such as decoding genes, in a few seconds. Advanced technologies used in automated microarray production are photolithography, mechanical microspotting and ink jets. Bioelectronic microchips contain numerous electronically active microelectrodes with specific DNA capture probes linked to the electrodes through molecular wires. Several biosensors have been used in combination with biochips. The purpose of the chips is to detect many genes present in a sample in one assay rather than performing individual gene assays as is the practice e.g. in so-called multiwells, plates with 96 wells, where the reactions take place. The huge amount of information coming from the genome sequence and other research genome programs cannot be utilised to the full without the availability of methods such as biochips which enable these genes or specific DNA sequences to be detected in biological samples. DNA chip technology is an example of the enormous efforts undertaken in the genomic field in the last few years.

22.3 Design of Biochip

Use of protein array technology over conventional assay methods has advantages that include simultaneous detection of multiple analytes, reduction in sample and reagent volumes, and high output of test results. The susceptibility of ligands to denaturation, however, has impeded production of a stable, reproducible biochip platform, limiting most array assays to manual or, at most, semi automated processing techniques. Such limitations may be overcome by novel biochip fabrication procedures (FitzGerald et al., 2005).

A design for a biochip memory device based on known materials and existing principles is presented by Robinson and Seeman, (1987). The fabrication of this memory system relies on the self-assembly of the nucleic acid junction system, which acts as the scaffolding for a molecular wire consisting of polyacetylene-like units. A molecular switch to control current is described which is based on the formation of a charge – transfer complex. A molecular-scale bit is presented which is based on oxidation – reduction potentials of metal atoms or clusters. The readable "bit" which can be made of these components has a volume of 3×10^7 Å^3 and should operate at electronic speeds over short distances. After selection of a suitable biochip substrate, biochip surfaces are chemically modified and assessed to enable optimization of biochip fabrication procedures for different test panels. Each biochip has hundreds to thousands of gel drops on a glass, plastic or membrane support, each about 100 microns in diameter. A segment of a DNA strand, protein, peptide or antibody is inserted into each drop, tailoring it to recognize a specific biological agent or biochemical signature. These drops are in known positions so when a sample reacts, the reaction position can be detected, identifying the sample. The biochip system can identify infectious disease strains in less than 15 minutes when testing protein arrays and in less than two hours when testing nucleic acid arrays.

22.4 Applications of Biochips

The complete sequence of the human genome and subsequent intensive searches for polymorphic variations are providing the prerequisite markers necessary to facilitate elucidation of the genetic variability in drug responses. Improvements in the sensitivity and precision of DNA microarrays permit a detailed and accurate scrutiny of the human genome. These advances have the potential to significantly improve health care management by improving disease diagnosis and targeting molecular therapy. Pharmacogenetic approaches, in limited use today, will become an integral part of therapeutic monitoring and health management, permitting patient stratification in advance of treatments, with the potential to eliminate adverse drug reactions (Hardiman, 2008).

Biochip technologies can be applied to numerous fields including genomic, proteomic, and glycomic research, as well as pharmacology and toxicology (Fig. 22.1).

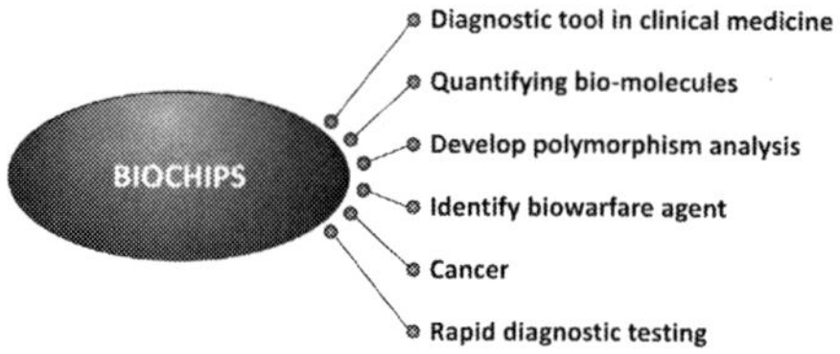

Fig. 22.1 Biochips is used as a molecular diagnostic tool in point-of care diagnosis

However, one of the most common applications is in the determination of gene expression in human cells and tissues. Global gene expression analysis has helped to identify important genes and signaling pathways in human malignant tumors. And there is hope that microarrays will make the step from "the (laboratory) bench to the bedside (of the patient)".

Biochips (e.g., GeneChip, CYP450, electrochemical biochips, protein biochips, microfluidic biochips and nanotechnology-based biochips) are assuming an important role in molecular diagnostics, and their application in point-of-care diagnosis is expected to facilitate the development of personalized medicine. Gene expression profiling by microarrays should advance the progress of personalized cancer treatment based on the molecular classification of subtypes. Refinements in biochip miniaturization with the advent of nanotechnology will further contribute to molecular diagnostics and the development of personalized medicine.

22.4.1 Biochip and Clinical Medicine

In the next years biochips will enter into clinical medicine. Development of high-throughput "biochip" technologies has dramatically enhanced our ability to study biology and explore the molecular basis of disease. Biochips enable massively parallel molecular analyses to be carried out in a miniaturized format with a very high throughput (Kallioniemi, 2001). Each larger diagnostics laboratory will offer genetic tests with this method.

The technology provides a point-of-care diagnostic system that would save time and money compared to current systems, which require sending samples to a centralized lab for confirmatory diagnosis Jain (2004). Various biochip technologies in cancer research, including analysis of disease predisposition by using single-nucleotide polymorphism (SNP), global gene expression patterns by cDNA microarrays, concentrations, functional activities or interactions of proteins with proteomic biochips, and cell types or tissues as well as clinical endpoints associated with molecular targets by using tissue microarrays. One can predict that individual cancer risks can, in the future, be estimated accurately by a microarray profile of multiple SNPs in critical genes. Diagnostics of cancer will be facilitated by biochip readout of activity levels of thousands of genes and proteins. Biochip diagnostics coupled with informatics solutions will form the basis of individualized treatment decisions for cancer patients. The 12 tumor markers' (TMs) biochip diagnostic (C12) system has been proven useful in some previous studies but its value for colorectal cancer (CRC) only was not systematically investigated (Chen et al., 2008). Biochip sensors have been used for the rapid and sensitive detection of viral disease (Livingston et al., 2005). The biochip could be a feasible tool for rapidly diagnosing mastitis-causing pathogens in milk and providing information for a more effective treatment to cure mastitis. The biochip can detect 7 common species of mastitis-causing pathogens, including *Corynebacterium bovis, Mycoplasma bovis, Staphylococcus aureus,* and the *Streptococcus spp. S. agalactiae, S. bovis, S. dysgalactiae,* and *S. uberis,* within 6 h. This technique is based on DNA amplification of genes specific to the target pathogens and consists of 4 basic steps: DNA extraction of bacteria, polymerase chain reaction, DNA hybridization, and colorimetric reaction.

22.4.2 Biochips and Biosensors

In the past two decades, the biological and medical fields have seen great advances in the development of biosensors and biochips capable of characterizing and quantifying biomolecules. Biochips (eg. Gene Chip, CYP450, electrochemical biochips, protein biochips, microfluidic biochips and nanotechnology-based biochips) are assuming an important role in molecular diagnostics. Their application in point-of-care diagnostics and enable parallel molecular analyses in a miniaturized format with a very high throughput. Refinements in biochip miniaturization with the advent of nanotechnology will further contribute to molecular diagnostics and the development of personalized medicine. Biochip platforms on gold can be manufactured by either nanoscale biotinylated self-assembled architectures to streptavidin surface or proteins containing free NH_2 groups to N-hydroxysuccinimide (NHS) – activated surfaces. The potential application of tumor necrosis factor – (TNF_α) and serodiagnosis of hemophagocytic lymphohistiocytosis (HLH) has been investigated. Interactions of TNF_α antigen and TNF_α antibody on the biochips have been optimized using an indirect immunofluorescence method (Weidong et al., 2008).

22.4.3 Biochip and Pharmacogenetics

Large-scale population researches, diagnostics of genetic predisposition to multifactorial diseases, screening of the polymorphic loci associated with individual sensitivity to pharmaceutical preparations require the development of effective, exact and rapid methods of analysis for detection of many mutations simultaneously. One of the most perspective methods to solve these problems is a method of allele-specific hybridization with biochips. Taking the analysis of mutations in genes CYP1A1, CYP2D6, GSTT1, NAT2, CYP2C9, CYP2C19 and MTHFR as an example, efficiency of using the approach for identification of individual genetic polymorphism can be studied. Biochips can also be a convenient tool in pharmacogenetics researchs.

22.4.4 Biochip and Biowarfare

The biochip can also be used to identify biological warfare agents. Easy sample preparation, standard operating protocols and a portable biochip reader that is smaller than a lunchbox make the system suitable for use in the field by first responders, military personnel and medical technicians. The system makes use of the polymerase chain reaction (PCR), a universal method for replicating billions of copies from one piece of genetic material. PCR allows trace quantities of DNA to be replicated to a level where they can be detected in the biochip system. A sample to be tested is applied to a biochip, which is then put in a reader and scanned using patented side illumination laser technology to detect reaction sites. Automated algorithms determine the agents present in the sample.

22.4.5 Biochip and Diagnostics

Biochip can be used for rapid diagnostic testing since the biochip allows technicians to test for so many different agents at once. The technology holds great promise for rapid diagnostic testing since the biochip allows technicians to test for so many different agents at once. Under current development is a respiratory syndrome chip that tests for strep throat, influenza A, and influenza B (Lee et al., 2008). This chip will allow doctors to make a faster and more precise diagnosis when patients display symptoms common to several different ailments. The TB-Biochip oligonucleotide microarray system is a rapid system to detect mutations associated with rifampin (RIF) resistance in mycobacteria (Caoili, et al., 2006). The lack of a sensitive immunoassay for quantitating serum prostate-specific membrane antigen (PSMA) hinders its clinical utility as a diagnostic/prognostic biomarker. An innovative protein biochip immunoassay was used to quantitate and compare serum PSMA levels in healthy men and patients with either benign or malignant prostate disease. These initial results suggest that serum PSMA may be a more effective biomarker than prostate-specific antigen for differentiating benign from malignant prostate disease and warrants additional evaluation of the surface-enhanced laser desorption/ionization PSMA immunoassay to determine its diagnostic utility (Xiao et al., 2001). An antibody sandwich assay with different capture antibodies on one chip surface and with the detection antibodies linked to a congruent surface via the DNA zippers was used to capture and quantify a recombinant hepatitis C antigen from solution. In this case, the DNA zippers enable not only discrimination between specific and nonspecific binding, but also allow for the local application of detection antibodies, thereby eliminating false-positive results caused by cross-reactive antibodies and nonspecific binding (Blank et al., 2003).

Technological advances in miniaturization have found a niche in biology and signal the beginning of a new revolution. Most of the attention and advances have been made with DNA chips yet a lot of progress is being made in the use of other biomolecules and cells. A variety of reviews have covered only different aspects and technologies but leading to the shared terminology of "biochips." (Ng and Ilag, 2003).

22.4.6 Biochip and Cancer

The various biochip technologies in cancer research, include analysis of (1) disease predisposition by using single-nucleotide polymorphism (SNP) microarrays, (2) global gene expression patterns by cDNA microarrays, (3) concentrations, functional activities or interactions of proteins with proteomic biochips, and (4) cell types or tissues as well as clinical endpoints associated with molecular targets by using tissue microarrays. One can predict that individual cancer risks can, in the future, be estimated accurately by a microarray profile of multiple SNPs in critical genes.

The biochip system also has great potential as a discovery tool. Current research aimed at the development of proteome chips has shown great promise. Proteome chips, which are biochips displaying all the proteins expressed by an organism at a specific time, provide the ability to screen for new cancer biomarkers, vaccine targets and therapeutic targets, as well as provide a means of characterizing disease states.

Comparative genomic hybridization (CGH) to metaphase chromosomes has been widely used for the genome-wide screening of genomic imbalances in tumour cells. Substitution of the chromosome targets by a matrix consisting of an ordered set of defined nucleic acid target sequences would greatly enhance the resolution and simplify the analysis procedure, both of which are prerequisites for a broad application of CGH as a diagnostic tool. However, hybridization of whole genomic human DNA to immobilized single-copy DNA fragments with complexities below the megabase pair level has been hampered by the low probability of specific binding because of the high probe complexity. High-copy-number amplifications contained in tumour cells are rapidly scored by use of target DNAs as small as a cosmid. Low-copy-number gains and losses are identified reliably by their ratios by use of chromosome-specific DNA libraries or genomic fragments as small as 75 kb cloned in PI or PAC vectors as targets, thus greatly improving the resolution achievable by chromosomal CGH. The ratios obtained for the same chromosomal imbalance by matrix CGH and by chromosomal CGH corresponded very well. The new matrix CGH protocol provides a basis for the development of automated diagnostic procedures with biochips designed to meet clinical needs (Solinas-Toldo et al., 1997).

22.4.7 Biochip Department of Defense

Development of an implantable biochip that could relay vital health information if a soldier is wounded in battle or a civilian is hurt in an accident has been done. The Department of Defense has awarded $1.6 million to the Center for Bioelectronics,

Biosensors and Biochips (C3B) at Clemson University for the development of an implantable biochip that could relay vital health information if a soldier is wounded in battle or a civilian is hurt in an accident. The biochip, about the size of a grain of rice, could measure and relay such information as lactate and glucose levels in the event of a major hemorrhage, whether on the battlefield, at home or on the highway. Biosensors are also widely used for biowarfare detection, including anthrax, for the Department of Defense. Biochip has other long-term potential applications, such as monitoring astronauts' vital signs during long-duration space flights and reading blood-sugar levels for diabetics.

22.5 Biochip Market

During the past ten years, major pharmaceutical ("pharma") companies have seen their R&D costs explode as their actual productivity has declined. Drug companies have looked outside of their labs to find drug products faster. As a result, pharma companies have increasingly become motivated research and financial partners with many biotechnology companies. At the same time, pharma companies also want to become customers and partners of biochip companies because these companies have technologies that might help them become more productive faster. The products from the biochip market help to accelerate the research processes and capabilities of bio-pharmaceutical drug discovery and basic academic bioresearch. New business models are being created. The ideas of pharmacogenomics and targeted therapies have received positive reception from government regulators. Roche and Affymetrix were the first companies to receive FDA approval for a biochip-based, molecular diagnostic test and instrument system. Basic academic research is helping as well. Accelerating the drug making process requires updated industrialization of R&D, done through the use of genomics and proteomics technologies such as biochips that the biotechnology companies provide.

22.6 Future of Biochips

Differentiation and prognosis in breast cancer and individualized drug therapy (pharmacogenetics) are the two potential applications of biochips (Gary, 2008). In cancer the gene expression profile permits an accuracy of differentiation and prognosis that was impossible so far. Soon oncologists will make their therapeutic decisions on the basis of biochip-based gene expression profiles. For an individualized drug therapy extensive genetic tests must be accomplished, above all of the cytochrome P450 system and an already available biochip seems to make this possible reliably and economically. Moreover, in science biochips will play a substantial role in the analysis of the genetic basis of common diseases. If these diseases are understood once, the clinical use of biochips will open an enormous potential for a predictive medicine. DNA chips will facilitate the integration of diagnosis and therapeutics, as well as the introduction of personalized medicines.

In recent years (Zhang et al., 2008) the application of gold nanoparticles instead of fluorescence dyes and enzyme-conjugation in biochips is very common. For example, Au nanoparticles labelling method was applied in many DNA-detection methods, and a novel readout scheme for gold nanoparticle-based DNA microarrays was studied relying on "Laser-Induced Scattering around a nanoAbsorber" and nanogold electrode, and the colorimetric detection using gold label plus silver stain was also developed. The quality analysis of probe spots can be obtained by using gold nanoparticles with positive charges to label DNA through electrostatic attraction. The probe spots can also be detected by a simple personal computer scanner. Gold nanoparticles deposited on a glass surface can be dissolved in bromine-bromide solution. The same microarray treated with gold particles staining and destaining can still be used for hybridization with nearly the same efficiency (Hsiao and Chen, 2009). The technology is a good combination of gene technology and nanotechnology. At the same time, a number of scientists from different countries have paid more attention to the application of nanoparticles in biochips and have some new patents recieved. This approach makes quality control of a microarray chip feasible and should be a valuable tool for biomarker discovery in the future.

Protein arrays are emerging to follow DNA chips as possible screening tools. Array-based protein technologies are emerging for basic biological research, molecular diagnostics and therapeutic development with the potential of providing parallel functional analysis of hundreds or perhaps hundreds of thousands of proteins simultaneously. Array-based methods are becoming prevalent within proteomics research due to the desire to analyze proteins in an analogous format to that of the DNA microarray. Novel protein biochips are under development in academic laboratories and emerging biotechnology companies to advance the pace and scope of scientific discovery. Biochips are a relevant topic for insurers partially already today, much more however in the future.

References

Blank, K., Mai, T., Gilbert, I., Schiffmann, S., Rankl, J., Zivin, R., Tackney, C., Nicolaus, T., Spinnler, K., Oesterhelt, F., Benoit, M., Clausen-Schaumann, H., Gaub, H.E., 2003, A force-based protein biochip. *Proc Natl Acad Sci USA* 100:11356–11360.

Braun, Y., Wai, D.H., Biemel, K., Schäfer, K.L., Gabbert, H.E., Poremba, C., 2005, Biochip analysis: status quo. *Verh Dtsch Ges Pathol* 89:116–124.

Caoili, J.C., Mayorova, A., Sikes, D., Hickman, L., Plikaytis, B.B., Shinnick, T.M., 2006, Evaluation of the TB-biochip oligonucleotide microarray system for rapid detection of rifampin resistance in *Mycobacterium tuberculosis*. *J Clin Microbiol* 44:2378–2381.

Chen, C., Chen, L.Q., Yang, G.L., Li, Y., 2008, The application of C12 biochip in the diagnosis and monitoring of colorectal cancer: systematic evaluation and suggestion for improvement. *Clin Rep* 54:186–190.

Fitzgerald, S.P., Lamont, J.V., McConnell, R.I., Benchikh, el, O., 2005, Development of a high-throughput automated analyzer using biochip array technology. *Clin Chem* 51: 1165–1176.

Gary, H., 2008, Applications of microarrays and biochips in pharmacogenomics pharmacogenomics in drug discovery and development: from bench to bedside. *Methods Mol Biol* 448:21–30.

Hardiman, G., 2008, Applications of microarrays and biochips in pharmacogenomics. *Methods Mol Biol* 448:21–30.
Hsiao, C.R., Chen, C.H., 2009, Characterization of DNA chips by nanogold staining. *Anal Biochem* 389:118–123.
Jain, K.K., 2001, Biochips for gene spotting. *Science* 294:621–623.
Jain, K.K., 2004, Applications of biochips: from diagnostics to personalized medicine. *Curr Opin Drug Discov Devel* 7:285–289.
Kallioniemi, O.P., 2001, Biochip technologies in cancer research. *Ann Med* 33:142–147.
Lee, K.H., Lee, J.W., Wang, S.W., Liu, L.Y., Lee, M.F., Chuang, S.T., Shy, Y.M., Chang, C.L., Wu, M.C., Chi, C.H., 2008, Development of a novel biochip for rapid multiplex detection of seven mastitis-causing pathogens in bovine milk samples. *J Vet Diagn Invest* 20:463–471.
Livingston, A.D., Campbell, C.J., Wagner, E.K., Ghazal, P., 2005, Biochip sensors for the rapid and sensitive detection of viral disease. *Genome Biol* 6:112.
Ng, J.H., Ilag, L.L., 2003, Biochips beyond DNA: technologies and applications. *Biotechnol Annu Rev* 9:1.
Robinson, B.H., Seeman, N.C., 1987, The design of a biochip: a self-assembling molecular-scale memory device. *Protein Eng* 1:295–300.
Solinas-Toldo, S., Lampel, S., Stilgenbauer, S., Nickolenko, J., Benner, A., Döhner, H., Cremer, T., Lichter, P., 1997, Matrix-based comparative genomic hybridization: biochips to screen for genomic imbalances. *Genes Chrom Cancer* 20:399–407.
Weidong, Du., Xueling, Ma., Schneider, E.M., 2008, A direct immunoassay assessment of streptavidin- and n-hydroxysuccinimide-modified biochips in validation of serological TNF responses in hemophagocytic lymphohistiocytosis. *J Biomol Screening* 13:515–526.
Whitesides, G.M., 2003, The 'right' size in nanobiotechnology. *Nat Biotechnol* 21:1161–1165.
Xiao, Z., Adam, B.-L., Cazares, L.H., Clements, M.A., Davis, J.W., Schellhammer, P.F., Dalmasso, E.A., Wright, G.L. Jr., 2001, Quantitation of serum prostate-specific membrane antigen by a novel protein biochip immunoassay discriminates benign from malignant prostate disease. *Cancer Res* 61:6029–6033.
Zhang, Y.Q., Wang, Y.F., Jiang, X.D., 2008, The application of nanoparticles in biochips. *Recent Pat Biotechnol* 2:55–59.

Chapter 23
Personalised Medicine

Abstract Towards the post genomic sequencing era, conventional drug discovery is drastically improving genomic technologies and computational advances. One of the potential benefits is the accelerated introduction of new diagnostics and treatments to the general public. Personalized medicine is typically described as the use of molecular or genetic characteristics to customize therapy. The right medication for the right patient is the goal of personalized medicine, which directly benefits from many of biotechnology's biggest and most recent advances. Personalized medicine approach is increasingly incorporated into health care, understanding patients' needs and their readiness to adopt these novel technologies will become progressively more important for the development of appropriate health policies.

Keywords Pharmacogenomics · Drug · Predisposition · Personalised medicine · Drug discovery · Pharmaceutical industry · Genetic markers · Cytochrome P450 · Patent · Ethical legal social issues

23.1 Prologue

With the beginning of the post-genome era, the promise of tailoring the practice of medicine to the individual has started. The completion of the entire genome sequence of many experimental organisms as well as the human allow us to compare several genomic sequences, to get valuable information for gene discovery and functional genomics. Driven by chemistry but increasingly guided by pharmacology and the clinical sciences, drug research has contributed more to the progress of medicine during the past century than any other scientific factor. The advent of molecular biology and, in particular, of genomic sciences is having a deep impact on drug discovery. Recombinant proteins and monoclonal antibodies have greatly enriched our therapeutic armamentarium. Genome sciences, combined with bioinformatics tools, allow us to dissect the genetic basis of multifactor diseases and to determine the most suitable points of attack for future medicines, thereby increasing the number of treatment options (Priori et al., 1999). The dramatic increase in the complexity of drug research is enforcing changes in the institutional basis of

M. Debnath et al., *Molecular Diagnostics: Promises and Possibilities*,
DOI 10.1007/978-90-481-3261-4_23, © Springer Science+Business Media B.V. 2010

this interdisciplinary endeavour. The biotechnology industry is establishing itself as the discovery arm of the pharmaceutical industry. In bridging the gap between academia and large pharmaceutical companies, the biotech firms have been effective instruments of technology transfer. Molecular, pharmacological and patient clinical data will be captured at various phases and integrated in a "knowledge management system" that will be used to facilitate rational drug design around molecular diseases.

23.2 Concept

In recent years, tools for the development of new drugs have been dramatically improved. These include genomic and proteomic research, numerous biophysical methods, combinatorial chemistry and screening technologies (Schwardt et al., 2003). As new therapeutic treatments emerge, molecular diagnostics will increasingly spin off products from drug development data to compliment the therapies allowing a more and more targeted approach to drug selection.

Initial applications of molecular diagnostics were mostly for infections but are now increasing in the areas of genetic disorders, preimplantation screening and cancer. Genetic screening tests, despite some restrictions are a promising area for future expansion of in vitro diagnostic market. Molecular diagnostics is being combined with therapeutics and forms an important component of integrated healthcare. Molecular diagnostic technologies are also involved in development of personalized medicine based on pharmacogenetics and pharmacogenomics. Currently, there has been a considerable interest in developing rapid diagnostic methods for point-of-care and biowarfare agents such as anthrax.

Personalized medicine simply means the prescription of specific therapeutics best suited for an individual based on pharmacogenetic and pharmacogenomic information (Jain, 2002). Several technologies are used including single nucleotide polymorphism genotyping, haplotyping, gene expression studies by biochip/microarrays and proteomics. Molecular diagnostics will play an important role in the development of personalized medicine, in which therapy and diagnosis will be integrated. There are several examples of the personalized medical approach, which include genotype-based selection of patients for effective cancer therapy, to spare those who would not respond or would suffer undesirable side effects. This newly emerging field uses the patient's genetic composition to tailor strategies for patient-specific disease detection, treatment, or prevention. The regulations governing personalized medicine can be complicated because they encompass in vitro diagnostic systems and laboratory tests as well as methods of disease treatment and patient care (Richmond, 2008).

Personalized therapy is financially feasible, as it will reduce the costs of drug development by shortening the drug development cycle. The introduction of pharmacogenomics into clinical trials is reducing the chances of failed clinical trials and increasing the prospects of safer and more effective therapies for specific groups of patients. Several advantages, as well as challenges to the development of personalized medicine are examined. Personalized medicine is anticipated to be an acceptable part of medical practice by the year 2010 (Jain, 2002). Personalization of medicine requires two characterizations: a well-grounded understanding of who the patient is and an equally robust understanding of the subpopulation that most resembles that patient in the context of the decisions at hand. These characterizations are readily represented and can be used to drive decision-making in a rational manner that maximizes the positive outcomes for the patient (Kohane, 2009).

Not only is personalized medicine tailoring the right drug, for the right person, at the right time but it also includes evaluating predisposition to disease sometimes decades in advance of its threatened onset (Ginsburg and McCarthy, 2001). The key task is to find genes and gene variations that play a role in a disease. The first step is to associate the occurrence of a particular gene variant with the incidence of a particular disease or disease predisposition – an association that can vary from one individual to another depending on many factors, including environmental circumstances. The outcome is the development of biomarkers which are stable and predictive. Today's biomarker is tomorrow's theranostic.

Personalized medicine will provide the link between an individual's molecular and clinical profiles, allowing physicians to make the right patient-care decisions and allowing patients the opportunity to make informed and directed lifestyle decisions for their future well-being. Molecular diagnostics, the use of DNA-, protein- or mRNA-based biological markers to predict the risk of developing disease or the molecular phenotype of an existing one, will change the way we currently define disease. Genomic analysis of diseases with homogeneous clinical phenotypes will unveil distinct molecular entities that require different treatment strategies for optimal outcomes. Clinical diseases as we know them will be replaced by molecular classification. Therapies directed at the root cause of disease will replace those that simply treat the symptoms of disease. Finally, a pharmacogenomic test that predicts therapy response based on a patient's genomic profile will accompany many drugs. Personalized medicine will involve radical changes in the pharmaceutical industry and medical practice and is likely to affect many aspects of society. Most importantly, the individual whose health is at stake will benefit enormously.

The promise of personalised medicines tailored to a person's genes has been "overhyped", particularly for developing countries. Personalized diagnostic tests are used to detect patient-to-patient variations in gene or protein expression levels, which act as indicators for drug treatments or disease prognosis. The aim of personalized medicine or individualized treatment is to match the right drug to the right patient and, in some cases, even to design the appropriate treatment for a patient according to his/her genotype. Basic technologies of molecular diagnostics play an important role, particularly those for single nucleotide polymorphism (SNP) genotyping. Diagnosis is integrated with therapy for selection of the treatment as well for monitoring the results. Biochip/microarray technologies are also important and finally bioinformatics is needed to analyze the immense amount of data generated by various technologies.

23.3 Practice of Medicine in the Twenty-First Century

In the next decade, medical care will undergo revolutionary changes. No longer will medical practice be limited to the empirical extrapolation of a patient's care from generalized clinical-trial results. Traditional medical practice, based on trial-and-error, results in both under-treatment and over-treatment, multiple office visits, the need for drug monitoring, and frequent regimen changes. More than 100,000 deaths per year are attributed to adverse drug (Lazarou et al., 1998). A personalized approach of tailored care for every individual based on their specific, molecular disease will become the standard of care. In the prototypical office visit of 2015, the physician will examine a patient's genetic profile (stored on CD ROMs or equivalent), lifestyle, and results from objective molecular screening and monitoring tests. Algorithms, derived from previous research efforts, will be used to compute the likelihood that a patient develops a host of chronic diseases. The focus of medicine at this juncture will be entirely preventive. Lifestyle modifications and the use of prophylactic therapy will be recommended based on what is best for that patient to avoid chronic disease to which they could be susceptible. The "office" of the future might be virtual; internet office visits might supplement some of the direct patient–physician contact Issa et al. (2009). Patients will be more knowledgeable of their own health and risk profiles and more active in directing their own healthcare (Triggle, 2007).

23.4 Role of Personalized Medicine in Drug Discovery

The personalized medicine strategy for drug discovery and development should yield a spectrum of product opportunities for the pharmaceutical industry. Diagnostic risk assessment and disease-monitoring tools that accurately quantify disease burden in patients will be a direct outcome of research during the early discovery process. Pharmacogenomic markers of efficacy and side effects will be used in conjunction with specific drugs to target drug therapy to those patients who will have an optimal response (Sadee, 2002). The business rationale for targeted therapies, which some argue will decrease market share, is that such products will eventually expand the market by recruiting patients from less effective therapies or by identifying less symptomatic individuals who might benefit from prophylactic therapy (Fig. 23.1).

The clinical phases of drug development (Fig. 23.2) afford the opportunity to capture patient clinical data, imaging and in vitro molecular response data simultaneously. Academic medical centres and clinical research organizations are now conducting clinical trials with future research in mind. Archiving biological specimens along with traditional clinical covariates is becoming routine. Some centres are also actively engaged in pharmacogenomic marker research. In the near future, clinical trials might be conducted in specialized units where detailed clinical, biological and genomic data are collected and integrated. Genome- and proteome-wide profiles together with biological pathway databases, imaging and clinical data

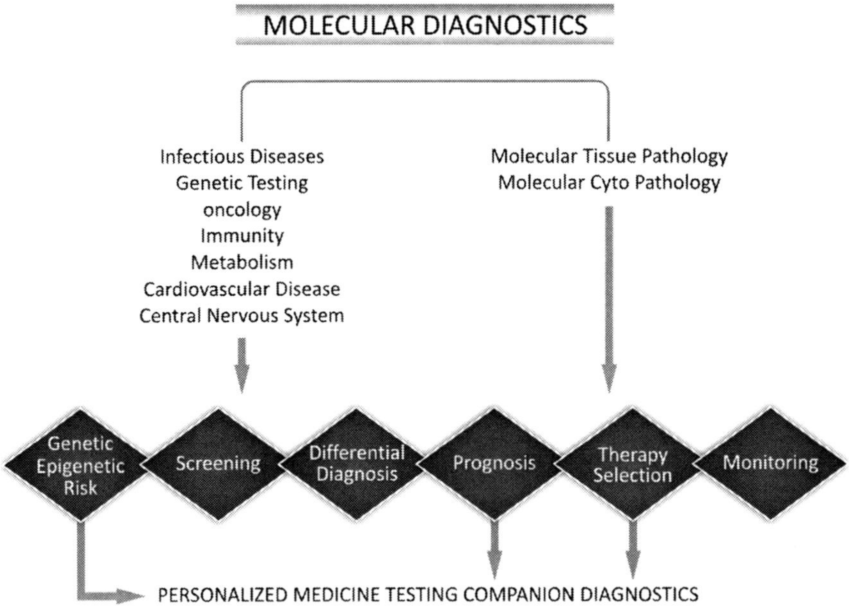

Fig. 23.1 The role of molecular diagnostics in the development of personalised medicine testing

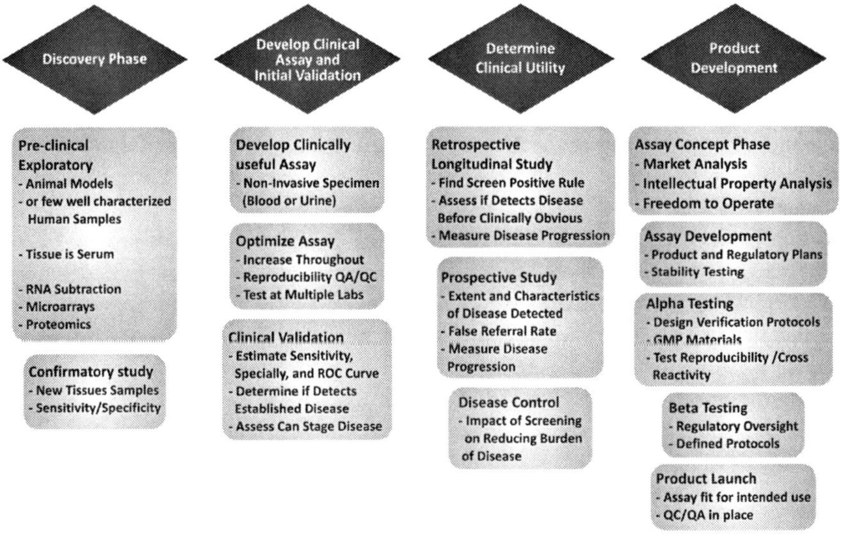

Fig. 23.2 Different phases of new drug discovery

on every patient will be used to analyze an individual's disease and drug response (Baba, 2001). The understanding of the biology of disease and drug action gleaned from these sophisticated new paradigms will dramatically accelerate the realization of truly personalized medicine.

Molecular pathology is in a state of rapid evolution featuring continuous technology developments and new clinical opportunities for drug selection, predicting efficacy and toxicity, and monitoring disease outcome. The approvals of trastuzumab (Herceptin, Genentech, South San Francisco, CA) for the treatment of HER-2/neu overexpressing breast cancer and imatinib (Gleevec, Novartis Pharmaceuticals, East Hanover, NJ) for the treatment of chronic myelogenous leukemia featuring a bcr/abl translocation and gastrointestinal stromal tumors with selective c-kit oncogene-activating mutations have brought to the diagnostic laboratory an expanding role for the testing of patients to determine their eligibility to receive these new therapies.The introduction of targeted therapeutics into clinical practice has created major opportunities for further development of the molecular pathology. A variety of new approaches to drug use have introduced genetically prescribed medications. Moreover, modern medicine now is driven by rapid communication and electronic information sharing, which has created more informed and demanding consumers.

The advantages of personalized medicine to the pharmaceutical industry are as follows:

- Increased efficiency and reduced costs of target and lead discovery
- Reduced timelines and costs of clinical trials
- Emergence of new gene targets for drug discovery
- Product differentiation in the market place.

Personalized medicine offers many advantages to patients and clinician like higher probability of desired outcome with a drug, low probability of untoward side effects, preventive strategies, focused therapies, reduced costs and better health and better healthcare.

Biological therapies such as those which use patient's own cells are considered to be personalized medicines. Vaccines are prepared from individual patient's tumor cells. Individualized therapeutic strategies using monoclonal bodies can be directed at specific genetic and immunologic targets. Ex vivo gene therapy involves the genetic modification of the patient's cells in vitro, prior to reimplantation of these cells in the patient's body.

Various technologies are integrated to develop personalized therapies for specific therapeutic areas described in the report. Examples of this are genotyping for drug resistance in HIV infection, personalized therapy of cancer (Veer and Bernand, 2008), antipsychotics for schizophrenia, antidepressant therapy, antihypertensive therapy and personalized approach to neurological disorders. Although genotyping is not yet a part of clinically accepted routine, it is expected to have this status by the year 2010. Several players are involved in the development of personalized therapy. Pharmaceutical and biotechnology companies have taken a leading role in this venture in keeping with their future role as healthcare enterprises rather than mere developers of technologies and manufacturers of medicines.

Drug discovery is a continuum of the processes, i.e. identification of disease-associated targets, screening of compound libraries, identification of leads and their optimization to drugs, preclinical and clinical drug development. Recent advances in genomic sciences will greatly reinforce processes such as early target identification and clinical researches. Particularly, exploitation of genomic information in clinical researches is expected to revolutionarily improve drug development and eventual clinical use. Ultimately, tailor-made medicine will provide us with great benefit in the near future therapy (Nohmi, 2002).

Genetic heterogeneity appears to be a significant source of variability observed in the response to drugs. This variability means that information pertaining to interethnic and interindividual genetic differences can be used to facilitate rational drug discovery and development and to avoid or minimize the incidence of adverse events in clinical trials. Thus, one could generate criteria for selecting patients most likely to benefit from a drug without incurring unnecessary risk. Early or preventive therapy guided by genotyping could significantly enhance clinical outcome. The need for a new, individualized approach to drug development and therapy is clear. Every year, approximately 3.1 billion prescriptions are issued in the United States, of which approximately 2.1 million results in an adverse reaction. One million prescriptions from this latter group may result in hospitalization, and of these more than 100,000 patients may die.

How can we reduce these severe adverse reactions by using pharmacogenomics? Over the near term, patient genotyping prior to therapy in a few but increasing number of instances will serve to avert or minimize severe drug toxicity. Alternatively, drugs may be designed a priori so that they are not subject to the differential metabolic patterns known to be caused by polymorphic variation. Looking farther ahead, and on a much broader scale, the efficacy of administered drugs may be improved, rather than avoiding toxicity as the main objective, by distinguishing good responders from poor responders prior to therapy. Often, effective drug response is limited to a portion of treated patients, whereas the majority benefits little or not at all. Predicting which patients are most likely to respond best to a particular drug, or which drug will yield optimal effects for a given patient, would represent a significant advance in therapy even with current drugs (Schwardt et al., 2003), let alone novel drugs developed with these criteria in mind. The success of this approach depends in large part on assembling an extensive, high-quality database of informative SNPs, a major focus for genomics companies. Ultimately the vision of pharmacogenomics encompasses a genetic profile for each individual, containing sufficient information to select which drugs are most likely to be safe and effective in that person. The same insight will serve to prevent disease to begin with, arguably the most desirable goal.

23.5 Molecular Diagnosis Will Determine Prognosis and Therapy

Personalized medicine is rooted in the hypothesis that diseases are heterogeneous, from their causes to rates of progression to their response to drugs. Each person's disease might be unique and therefore that person needs to be treated as an individual. With limited understanding of the molecular basis of disease, we have relied on non-specific clinical signs. As genomic tools are sharpened, so will be our ability to dissect disease into its component parts (Nohmi, 2002). Clinical phenotypes thought to be one disease will be subclassified by a new genomic taxonomy. Recent discoveries in the molecular pathology of cancer have highlighted important and clinically significant differences in the gene expression patterns of a variety of tumors, including leukemias and breast cancer (Golub et al., 1999). In cardiovascular disease, genetic heterogeneity has been identified in the Long QT syndrome, a disorder of ventricular depolarization where clinical manifestations range from no visible signs to sudden death (Zareba et al., 1998). The etiology of the Long QT syndrome is attributed to mutations in one of at least four different ion channels (HERG, KVLQT1, SCN5A or KCNE1). The clinical course of the disease, level of aggressive therapy and choice of therapy (Na^+-channel blocker versus K^+-channel blocker versus beta blocker) are now determined by the genetic etiology (Shimizu and Antzelevitch, 1997; Schwartz et al., 1995) of the syndrome.

Familial hypertrophic cardiomyopathy is another example of a genetically heterogeneous disease with a clinical phenotype of ventricular hypertrophy (Fananapazir, 1999). Familial cardiomyopathy results from >80 different mutations, each affecting the expression of a cardiac muscle sarcomeric protein. Mutation-specific prognoses have been established that mandate screening to determine who requires more frequent clinical monitoring, therapeutic intervention and family screening. As the underlying molecular architecture of other diseases is determined, medical practice will be tailored to properly diagnose and treat them.

23.6 Personalized Medicine and Genetic Markers

The past few years have seen major advances in technology and the growth of genomic information as a byproduct of the human genome project. As a result, new and innovative markers of disease are being uncovered at an unprecedented rate. At the DNA level, >350 genetic tests are currently available (http://www.genetests.org). Although most tests are for rare, monogenic disorders, some are becoming available for more common, complex diseases. Examples include ApoE testing among dementia patients for differential diagnosis of Alzheimer disease (Eldik et al., 2002), and Factor V Leiden testing for predisposition to venous thrombosis. Furthermore, for most common diseases, a large number of genetic markers suggesting association with disease are reported in the literature. Advances in SNP-discovery technologies are providing opportunities for large-scale candidate gene studies (Crouch, 2001; Cargill et al., 1999). Indeed, whole-genome association studies are being contemplated for finding genetic predisposition markers for common, complex diseases. The next three to five years will see an explosion of new information in this area and the development of new predictive tests for complex diseases.

Identification of RNA and protein markers for screening, diagnosis, prognosis and monitoring is also under way, facilitated through advances in transcript profiling and proteomics. The basic research methods used in the discovery of these markers require access to relevant disease tissues. Because tumor samples are routinely biopsied or removed, the first disease area likely to benefit from these technologies is cancer. Advances in the ability to classify disease are best illustrated in the work of Golub et al., 1999). Expression patterns (Hedenfalk et al., 2001; Waring et al., 2001) of 50 genes were determined to distinguish accurately between acute lymphoblastic leukemia (ALL) and acute myeloid leukemia (AML). Similar approaches have been taken to identify candidate prognostic markers for melanoma. Clark et al. (2000) and Bittner et al. (2000) used transcript profiling to compare metastatic with nonmetastatic human melanoma cell lines. As a result, they identified several genes that are selectively unregulated in the metastatic lines that could have use in patient management.

The acknowledgement that post-transcriptional modification of proteins might be an important determinant of disease is one factor driving the use of proteomic technologies for discovering molecular disease markers. These technologies include traditional 2D gel electrophoresis in addition to more advanced mass-spectrometry methods. Proteomic analysis might be used in medical microbiology in which the entire proteome of an organism can be studied or in diseases such as rheumatoid arthritis or central nervous system disorders in which protein-rich fluids at the site of injury, such as synovial joint and cerebral spinal fluid, respectively, are available for analysis. Cell line supernatants or explants from tumor tissues have already been used in large-scale expression profiling experiments to identify cancer markers. Page et al. (1999) compared the proteome of purified normal human luminal (from which most breast cancer is derived) with that of myoepithelial breast cells. They detected 170 proteins that differed between the two cell types. These experiments might shed light onto the process of cancer development and ultimately find use as cancer diagnosis or monitoring markers.

Pharmacogenomic tests are finding their way into practice in several disease areas. Genotype resistance testing of HIV isolates has demonstrated clinical utility and provides a way to assist therapeutic decision-making in patients whose levels of HIV RNA are rising (Durant et al., 1999). In addition, assays are available to detect the HER2 protein receptor or copies of the HER2 gene sequence to determine eligibility for herceptin treatment or Adriamycin treatment, respectively, in node-positive breast cancer patients (Tsongalis et al., 2000). The Food and Drug Administration (FDA) approval of the pharmacogenomic marker HER2 linked to herceptin represents an important precedent for regulatory approval of personalized medicine products.

23.7 Challenges of Realizing the Promise of Personalized Medicine

It is widely anticipated that during the next 5 years the molecular diagnostic industry will continue to grow at double-digit pace to meet increasing demand for personalized medicine. A wide variety of drugs in late preclinical and early clinical development are being targeted to disease-specific gene and protein defects that will require co-approval of diagnostic and therapeutic products by regulatory agencies. An increasingly educated public will demand more information about their predisposition for serious diseases and how these potential illnesses can be detected in an early stage when they can be arrested or cured with new therapies custom-designed for their individual clinical status. To respond to this demand, major pharmaceutical companies will partner with diagnostics companies or develop their own in-house capabilities that will permit efficient production of more effective and less toxic integrated personalized medicine drug and test products. For clinical laboratories and pathologists, this integration of diagnostics and therapeutics represents a major new opportunity to emerge as leaders of the new medicine, guiding the selection, dosage, route of administration, and multidrug combinations and producing increased efficacy and reduced toxicity of pharmaceutical products.

In spite of the achievement of a complete human genome sequence, there are numerous challenges in realizing the vision of personalized medicine. Identifying genetic variants that are markers of disease or drug response requires sifting through several million SNPs in the human genome to find those that contribute to the disease and then demonstrating that the SNPs are clinically valid markers and are useful for managing patients (Halushka et al., 1999). To uncover DNA variants that predict common, complex diseases that result from a combination of genes and environmental factors will require cost-effective, high-throughput genotyping; large, well-characterized patient populations; sophisticated computational methodologies; and a detailed understanding of the biological pathways of disease (Fig. 23.3). Uncovering mRNA and protein markers for use in screening, diagnosis,

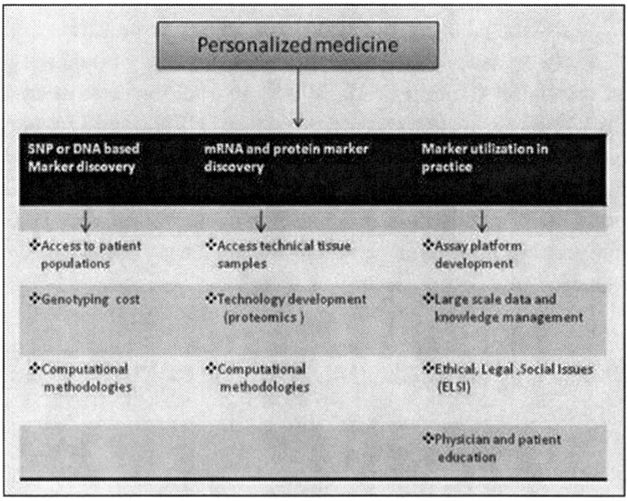

Fig. 23.3 The promise of personalized medicine

prognosis and monitoring of disease will have its own set of challenges. Access to optimal relevant tissues might not be possible for many diseases. Proteomic technologies require further development, as do computational approaches for analyzing massive amounts of gene and protein profile data. To realize the vision of personalized medicine, the agenda for medical and pharmaceutical research must include the assembly and integration data from many sources on large numbers of patients. Clinical investigations should incorporate genotyping and molecular profiling technologies along with traditional clinical data collection and should establish a repository of patient samples where possible.

New molecular markers might face many hurdles before they can be implemented in patient care. The issues range from FDA regulation and acceptance of these new markers, to developing tractable assay platforms, to resolving issues around the ethical, legal and social implications of obtaining highly sensitive genetic information. Foremost among these, in our opinion, is the education and engagement of physicians and patients in the paradigm shift to objective, quantitative marker-based clinical care. If appropriate patient management systems, integrated databases, educational tools and genetic counselling are not in place, then it will be difficult to realize the significant benefits forecast from this approach. Fortunately, we have already learned valuable lessons from past efforts to implement genetic screening for sickle cell anaemia, and from more recent efforts to screen for BRCA1 mutations in breast cancer families. The ethical, legal and social implications (ELSI) of human genetic research are the subject of a government-funded program (http://www.nhgri.nih.gov/ELSI/). The US government is playing an active role in addressing public concern over genetic information by, among other things, drafting legislation to protect patients from discrimination by employers and insurance companies. Furthermore, health professionals are rising to the challenge of educating both their members and the public. The American Medical Association has co-founded the National Coalition for Health Professional Education in Genetics (http://www.nchpeg.org) to promote health professional education and access to genetic information.

23.8 Personalised Medicine and Pharmacogenomics

Personalized medicine and pharmacogenomics are inextricably linked. Pharmacogenomics is the use of genetic variations (such as SNPs, gene expression variability, or other molecular signatures) to understand and correlate with differential response to pharmaceutical agents (drugs). Pharmacogenomics can be deployed clinically to stratify patients into responders and non-responders and this practice is termed personalized medicine. To frame the context of current approaches, pharmacogenomics seeks to identify and validate the signature(s) of molecular analytes and these are converted to assays using the tools of molecular diagnostics. The deployment of these molecular diagnostics assays on defining and targeting patient populations is the domain of personalized medicine. This study includes all of the generally accepted imaging activities that are currently used in personalized medicine, including pharmacogenomics, genomics and theranostics. It examines associated clinical market segments in which personalized medicine has taken a prominent role including cancer treatment, cardiology and neurology markets.

Medicine, as we move into the third millennium, still targets therapy to the broadest patient population that might possibly benefit from it, and it relies on statistical analysis of this population's response for predicting therapeutic outcome in individual patients. Therapists of necessity make decisions about the choice of drug and appropriate dosage based on information derived from population averages. This "one drug fits all" approach could, with the fruits of pharmacogenomic research, evolve into an individualized approach to therapy where optimally effective drugs are matched to a patient's unique genetic profile. This involves classifying patients with the same phenotypic disease profile into smaller subpopulations, defined by genetic variations associated with disease, drug response, or both. The assumption underlying this approach is that drug therapy in genetically defined subpopulations can be more efficacious and less toxic than in a broad population.

Individualizing drug therapy raises a number of issues with enormous practical consequences. Currently, the pharmaceutical industry is in a consolidation and merger phase, with ever larger corporations emerging at a steady pace. This consolidation is done in the expectation that many novel drugs can be brought to market with high efficacy against major diseases, driven by genomics-based drug discovery. Indeed, large corporations depend on generating "blockbuster" drugs-; drugs that raise in excess of a billion dollars in revenue each year by targeting large patient populations. However, it remains to be seen whether betting on a "one drug fits all" approach is realistic. Certainly, a few blockbuster drugs continue to emerge, for example, the Cox-2 selective inhibitors in the therapy of inflammatory joint diseases. Efficacy does not appear to exceed substantially that of traditional nonsteroidal anti-inflammatory drugs (NSAIDs), which inhibit both Cox-1 and Cox-2 to varying degrees; however, the incidence of gastrointestinal lesions is reduced. Yet, only a portion of patients receiving conventional NSAIDs develop these lesions, and the traditional drugs are much less expensive. Moreover, it remains to be seen what long-term sequelae arise from treatment with Cox-2 selective inhibitors. These sequelae might be beneficial (for example, the possible prevention of colon cancer or neurodegenerative disorders associated with inflammation in the CNS), but the physiological functions of Cox-2 remain poorly understood. Trials over longer time periods will be needed to address these questions fully. As three quarters of all health care costs are used for the treatment of chronic illness, mostly of the aged, long-term issues will be the battleground where optimal therapies will be decided.

Whether a single drug emerges superior to others in a broad patient population or whether best clinical response requires differential therapy of small subpopulations is the subject of fierce debates. Bringing a new drug to the market currently costs approximately $500 million, making it economically impossible to target small patient populations. If smaller patient populations are to be served, we need to change the entire process, up to final regulatory agency approval for clinical use. Conceivably, targeting well-defined patient populations will sharpen our analysis of risk/benefit ratios and permit clinical trials to be substantially reduced in size. Laws and FDA regulations may have to be changed to accommodate the need for targeting patients with rare diseases or with subtypes of otherwise common diseases. This approach will set the stage for testing whether targeting small patient populations with select drugs is superior to treating many patients with the best drug available for a given disease. The outcome may vary from one case to another.

Thus, individualizing drug therapy with the use of pharmacogenomics holds the potential to revolutionize medical therapeutics, by challenging the "one drug fits all" approach. Furthermore, pharmacogenomics could also enhance the value of currently approved drugs with limited market share because of significant toxicity or limited efficacy, enabling prescribers to identify patients for whom they will be both effective and safe.

The emergence of personalized medicine raises issues for those who pay for treatment. The cost of new diagnostic tests and individualized medications may be more expensive, but it is hoped that the predictive potential of personalized medicine could avert more costly treatments required after the onset of a disease. Currently, less than 5% of all US private companies reimburse for genetic tests, indicating that the current health care delivery system may not be able to deliver effective "personalized medicine".

Pharmacogenetics applies not only to traditional drugs but also to bioengineered proteins and gene therapy. Human genetic variability can be expected to affect all treatment modalities. For example, breast cancer treatment with trastuzumab (Herceptin), a humanized monoclonal antibody against the HER2 receptor developed by Genentech, Inc., is linked to HER2 overexpression. This reaction correlates with poor clinical prognosis and serves as a marker for responsiveness to trastuzumab therapy, either alone or in combination with chemotherapy.

The cytochrome P450 monooxygenase system of enzymes is responsible for a major portion of drug metabolism in humans. Although commonly serving to detoxify xenobiotics, these enzymes are also principally responsible for the activation of procarcinogens and promutagens in the human body. This scenario is particularly important for lipophilic drugs such as CNS-active drugs, which generally must be lipophilic to penetrate the blood-brain barrier. Because renal excretion is minimal for these compounds, P450 metabolism provides the primary means of drug elimination. This large family of genes has been intensely studied, and among the numerous P450 subtypes, CYP2D6, 3A4/3A5, 1A2, 2E1, 2C9, and 2C19 play particularly critical roles in genetically determined responses to a broad spectrum of drugs. Determination of a patient's CYP2D6 phenotype/genotype may prove useful in treatment with antipsychotic drugs, while comprehensive genotyping assays for all relevant P450 isotypes and their main sequence variants are being developed.

Cytochrome P450s inactivate or in some cases activate xenobiotics. Therefore, P450 polymorphisms affect an individual's susceptibility to environmental toxins. As a result, sequence variation of P450 isotypes attracts special attention in toxicogenetics. Recently the US National Institute of Environmental Health Sciences launched the Environmental Genome Project with the stated goal of understanding the genetic factors governing an individual's response to the environment on a genome-wide scale. This effort parallels the study of genetic variability in drug response.

Examples of personalized cancer management include:

- Testing for disease-causing mutations in the *BRCA1* and *BRCA2* genes, which are implicated in familial breast and ovarian cancer syndromes. Discovery of a disease-causing mutation in a family can inform "at-risk" individuals as to whether they are at higher risk for cancer and may prompt individualized prophylactic therapy including mastectomy and removal of the ovaries. This testing involves complicated personal decisions and is undertaken in the context of detailed genetic counseling.
- Minimal residual disease (MRD) tests are used to quantify residual cancer, enabling detection of tumor markers before physical signs and symptoms return. This assists physicians in making clinical decisions sooner than previously possible.
- Targeted therapy is the use of medications designed to target aberrant molecular pathways in a subset of patients with a given cancer type. For example, Herceptin is used in the treatment of women with breast cancer in which HER2 is overexpressed. Tyrosine kinase inhibitors such as Gleevec have been developed to treat chronic myeloid leukemia (CML), in which the BCR-ABL fusion gene (the product of a reciprocal translocation between chromosome 9 and chromosome 22) is present in >95% of cases. These medications specifically inhibit the Ableson tyrosine kinase (ABL) and are thus a prime example of "rational drug design" based on knowledge of disease pathophysiology.

23.9 Personalized Medicine and Diseases

Personalized medicine aims to identify individuals at risk for common diseases such as cancer, heart disease, and diabetes. The simple family history has long been used by physicians to identify individuals at increased risk and to advise preventive measures such as lifestyle modifications (changes in diet, cessation of toxic habits, increased exercise) earlier screening, or even prophylactic medications or surgery. Scientific advancements offer the potential to define an individual's risk based on their genetic make-up. Fields of biomedical research termed "-omics" (genomics, proteomics, and metabolomics) study the contribution of genes, proteins, and metabolic pathways to human physiology and variations of these pathways that can lead to disease susceptibility. It is hoped that these fields will enable new approaches to diagnosis, drug development, and individualized therapy. Nanotechnology will also play a major role in this area (Jain, 2005).

23.10 Personalized Medicine and Diagnostics Industry

Although pharmacogenomics-based diagnostics and therapeutics are increasingly being translated into personalized medicine applications, relatively little evidence exists about how novel pharmacogenomics-based technologies will be accepted and adopted by patients. It is important to understand the characteristics of genomic diagnostics and targeted therapeutics that might impact utilization or serve as barriers to adoption of these novel technologies in order to formulate appropriate policies and procedures.

The traditional diagnostics industry is mature and only achieving a high growth rate. The diagnostics industry has not been as successful as the pharmaceutical industry in attracting investment funding. However, the advent of molecular diagnostic tests, or theranostics, opens new opportunities in a small but believed to be rapidly growing niche market. New relationships are likely to develop between industry partners committed to personalized medicine embracing the approach of successful, specialised pharmaceutical firms. Major pharmaceutical firms have responded to the growing emphasis on individualized therapy to improve drug efficacy and safety with large investments in pharmacogenomics research (Posner, 2005). It is becoming apparent that genetic testing to identify patients in whom a particular drug can be given safely and effectively may provide those products with a competitive advantage. Several of the world's largest pharmaceutical firms, including AstraZeneca, Bayer, Pfizer, SmithKline Beecham, and Novartis have formed a consortium with 5 major academic centers with the goal of identifying 300,000 heritable SNPs within the next 2 years. The National Institutes of Health, in an independent effort, has made $30 million available over 3 years, starting in January of 1998, for the discovery and compilation of 100,000 SNPs. To top this all, scientists at Celera Genomics contend that they will have a collection of 6 to 10 million SNPs by mid 2000. With availability of high-resolution SNP maps and DNA microarray analytical capability, performing genome-wide association studies during clinical trials becomes feasible, enabling one to identify disease-susceptibility genes for prognosis, drug discovery, and selection of therapy. If risk for a given disease is predicted to be high, as judged by the SNP pattern of a patient, preventive therapy and lifestyle adjustments (diet, exercise, etc) may be implemented. A comprehensive SNP map will also contain genetic variants relevant to drug transport, metabolism, and receptor interaction and, therefore, needs to be considered in drug selection. Moreover, a comprehensive SNP map may also serve to alert the therapist when careful drug dosage monitoring is required. Stratifying patient populations using genome-wide SNP maps presents a major challenge to the pharmaceutical industry. The outcome from applying such an approach cannot be accurately gauged at present.

Today, many examples of variability in both drug response and toxicity associated with known genetic variability are documented. In a few cases, genetic tests are beginning to find their way into clinical practice, making a proactive approach to individualized therapy possible. In cancer chemotherapy of acute lymphocytic leukemia, administration of drugs such as 6-mercaptopurine, 6-thioguanine, and azathioprine can cause severe hematologic toxicity or even death in patients possessing nonfunctional ("null") variants of thiopurine methyltransferase (TPMT). Functional assays of TPMT in red blood cells, or alternatively genotyping, can identify those patients (approximately 1 in 300) who are homozygous for alleles encoding nonfunctional enzyme, and therefore unable to metabolize the drugs to their inactive methylated forms. These patients can be safely treated with doses 10–15 times less than commonly prescribed. Therefore, genotyping, or functional enzyme analysis, has become standard practice in major cancer treatment centers such as the Mayo Clinic (Rochester, MN) and St. Jude's Children Research Hospital (Memphis, TN). Nanobiotechnology is being used to refine discovery of biomarkers, molecular diagnostics, drug discovery and drug delivery, which are important basic components of personalized medicine and are applicable to management of cancer as well. Examples are given of the application of quantum dots, gold nanoparticles, and molecular imaging in diagnostics and combination with therapeutics another important feature of personalized medicine. Personalized medicine is beginning to be recognized and is expected to become a part of medical practice within the next decade. Personalized management of cancer, facilitated by nanobiotechnology, is expected to enable early detection of cancer, more effective and less toxic treatment increasing the chances of cure (Jain, 2005).

23.11 Impact of the US Patent System on the Promise of Personalized Medicine

The pharmaceutical industry faces a number of challenges that have cumulatively led to a decline in productivity, despite increasing levels of investment in research and development (Posner, 2005). The US patent system rewards innovation in medicine and other arts and sciences by granting innovators, for a period of time, the right to exclude others from using what was invented. One of the purposes of the patent system is to trade that right to exclude, and in its stead obtain the patent holder's obligation to fully and publicly disclose the essence of the innovations so that they can be improved, thus advancing the common welfare. A tension exists between personalized medicine's need for access to and use of scientific advances and the patent system's reward of exclusive use or nonuser to innovators. This tension may result in fewer diagnostic and therapeutic tools brought to the market and generally adopted. The risk seems particularly acute with respect to the diagnostic and therapeutic tools arising from genetic testing that hold specific value for a subset of the population. The judicial system has introduced ethical exceptions that overcome a patent holder's right to exclude; these judicial overrides relate to the provision of certain types of medical procedures and the development of certain types of new drugs, and not, apparently, to the use of diagnostic and therapeutic tools essential to the success of personalized medicine. A serious question exists as to whether legislative action is necessary to increase public access to genetic testing (Solomon and Sieczkiewicz, 2007).

Personalized medicine is a paradigm that exists more in conceptual terms than in reality, with only a few marketed drug-test companion products and not very many actual clinical practices set up to personalize medicine in the way that supporters have intended. Nevertheless, the reality of personalized medicine has become more imminent because of the increased awareness of the shortcomings in the delivery of drugs with adequate benefit/risk to patients, a better molecular understanding of how to optimize drug selection and dosing, and an increased demand for integrating more clinically relevant genetic information into the drug development process to improve both innovation and productivity,. "personalized medicine" is in many ways simply an extension of traditional clinical medicine taking advantage of the cutting edge of genetics research. However, obstacles to the implementation of this vision are formidable. The dynamic complexity of the human genome, multigenic disease origins, and involvement of numerous genes in drug response impede the effective application of genome-wide SNP scanning in the clinic. Drug responses will most likely be associated with patterns of multiple polymorphically expressed traits, rather than single causative polymorphisms. Such patterns of genetic variants differ among distinct ethnic groups. This factor could obscure prediction of disease susceptibility and drug response across patient populations, and it points to the need to genetically stratify patients for clinical pharmacogenomic studies.

23.12 Ethical Legal and Social Issues of Personalised Medicine

We are uncertain as to the overall direction of pharmacogenomics and personalized medicine over the next 10 years. Although new analytical systems introduced during the last decade have offered incremental improvements over previously available technology, they have not allowed scientists to maximize the benefit of multiple advancements in genomics, combinatorial chemistry, and assay technologies. The realization of an individualized approach to drug discovery and therapy will require new statistical methods and analytical systems providing an order-of-magnitude increase in throughput, along with corresponding decreases in operating costs, with enhanced accuracy and reduced complexity.

In addition to the daunting scientific challenges we have outlined, ethical issues need to be resolved. Information about an individual's genetic makeup raises privacy questions and ethical dilemmas about disease susceptibility, prognosis, and treatment options. Obviously, information of this type must be carefully safeguarded to ensure privacy. Many legal and economic issues will need to be resolved (Jeffords and Daschle, 2001). Whether or not these new genomic technologies find their way into everyday clinical use during the next 10 years, they will prove valuable tools in clinical research directed at optimizing drug therapy. The vision of pharmacogenomics is leading us to a more individualized approach to drug therapy, while revealing limits inherent to the treatment of disease in broad patient populations.

Ethical issues are involved in the development of personalized medicine mainly in the area of genetic testing. These along with social issues and consideration of race in the development of personalized medicine are being debated. Regulatory issues are discussed mainly with reference to the FDA guidelines on pharmacogenomics. Increase in efficacy and safety of treatment by individualizing it has benefits in financial terms. Information is presented to show that personalized medicine will be cost-effective in healthcare systems. For the pharmaceutical companies, segmentation of the market may not leave room for conventional blockbusters but smaller and exclusive markets for personalized medicines would be profitable. Marketing opportunities for such a system are described with market estimates from 2007 to 2017.

23.13 Conclusion

Personalized medicine promises to offer the right treatment for the right patient at the right time. Although that promise might seem far off, there is clear evidence that the traditional trial-and-error practice of medicine is eroding in favor of more precise marker-assisted diagnosis and treatment. For the patient, the benefits are clear: safer and more effective treatment of disease. For industry there appears an equally desirable outcome of this approach: increased efficiency, productivity and better product lines. Society as a whole will also realize a benefit: more focused application of precious healthcare resources to those in need of them most. The realization of personalized medicine is not without challenges, yet many of these challenges are being addressed. By encouraging public dialogue and debate, we expect that there will be continued progress forward. Lastly, as we take on more and more of the burden of our own health and well-being, educational forums must be developed for patients and physicians alike to understand the complex nature of the genomic information that is being used for decision making. Then we will have truly fulfilled the promise of the future.

References

Baba, Y., 2001, Development of novel biomedicine based on genome science. *Eur J Pharm Sci* 13:3–4.

Bittner, M., Meltzer, P., Chen, Y., Jiang, Y., Seftor, E., Hendrix, M., Radmacher, M., Simon, R., Yakhini, Z., Ben-Dor, A., Sampas, N., Dougherty, E., Wang, E., Marincola, F., Gooden, C., Lueders, J., Glatfelter, A., Pollock, P., Carpten, J., Gillanders, E., Leja, D., Dietrich, K., Beaudry, C., Berens, M., Alberts, D., Sondak, V., 2000, Molecular classification of cutaneous malignant melanoma by gene expression profiling. *Nature* 406:536–540.

Cargill, M., Altshuler, D., Ireland, J., Sklar, P., Ardlie, K., Patil, N., Shaw, N., Lane, C.R., Lim, E.P., Kalyanaraman, N., Nemesh, J., Ziaugra, L., Friedland, L., Rolfe, A., Warrington, J., Lipshutz, R., Daley, G.Q., Lander, E.S., 1999, Characterization of single-nucleotide polymorphisms in coding regions of human genes. *Nat Genet* 22:231–238.

Clark, T., Golub, R., Lander, E.S., Hynes, R.O., 2000, Genomic analysis of *metastasis* reveals an essential role for RhoC. *Nature* 406:532–535.

Crouch, M.A., 2001, Effective use of statins to prevent coronary heart disease. *Am Fam Physician* 63:309–320.

Durant, J., Clevenbergh, P., Halfon, P., Delgiudice, P., Porsin, S., Simonet, P., Montagne, N., Boucher, C.A., Schapiro, J.M., Dellamonica, P., 1999, Drug-resistance genotyping in HIV-1 therapy: the VIRADAPT randomised controlled trial. *Lancet* 353:2195–2199.

Fananapazir, L., 1999, Advances in molecular genetics and management of hypertrophic cardiomyopathy. *J Am Med Assoc* 281:1746–1752.

Ginsburg, G.S., McCarthy, J.J., 2001, Personalized medicine: revolutionizing drug discovery and patient care. *Trends Biotechnol* 19:491–496.

Golub, T.R., Slonim, D.K., Tamayo, P., Huard, C., Gaasenbeek, M., Mesirov, J.P., Coller, H., Loh, M.L., Downing, J.R., Caligiuri, M.A., Bloomfield, C.D., Lander, E.S., 1999, Molecular

classification of cancer: class discovery and class prediction by gene expression monitoring. *Science* 286:531–537.
Halushka, M.K., Fan, J.B., Bentley, K., Hsie, L., Shen, N., Weder, A., Cooper, R., Lipshutz, R., Chakravarti, A., 1999, Patterns of single-nucleotide polymorphisms in candidate genes for blood-pressure homeostasis. *Nat Genet* 22:239–247.
Hedenfalk, I., Duggan, D., Chen, Y., Radmacher, M., Bittner, M., Simon, R., Meltzer, P., Gusterson, B., Esteller, M., Kallioniemi, O.P., Wilfond, B., Borg, A., Trent, J., Raffeld, M., Yakhini, Z., Ben-Dor, A., Dougherty, E., Kononen, J., Bubendorf, L., Fehrle, W., Pittaluga, S., Gruvberger, S., Loman, N., Johannsson, O., Olsson, H., Sauter, G., 2001, Gene-expression profiles in hereditary breast cancer. *New Engl J Med* 344:539–544.
Issa, A.M., Tufail, W., Hutchinson, J., Tenorio, J., Baliga, M.P., 2009, Assessing patient readiness for the clinical adoption of personalized medicine. *Public Health Genomics* 12: 163–169.
Jain, K.K., 2002, Personalized medicine. *Curr Opin Mol Ther* 4:548–558.
Jain, K.K., 2005, Role of nanobiotechnology in developing personalized medicine for cancer. *Technol Cancer Res Treat* 4:645–650.
Jeffords, J.M., Daschle, T., 2001, Political issues in the genome era. *Science* 291:1249–1251.
Kohane, I.S., 2009, The twin questions of personalized medicine: who are you and whom do you most resemble?. *Genome Med* 1:4.
Lazarou, J., Pomeranz, B.H., Corey, P.N., 1998, Incidence of adverse drug reactions in hospitalized patients: a meta-analysis of prospective studies. *J Am Med Assoc* 279:1200–1205.
Nohmi, T., 2002, Drug discovery based on genomic information. *Rinsho Byori* 50:694–697.
Page, M.J., Amess, B., Townsend, R.R., Parekh, R., Herath, A., Brusten, L., Zvelebil, M.J., Stein, R.C., Waterfield, M.D., Davies, S.C., O'Hare, M.J., 1999, Proteomic definition of normal human luminal and myoepithelial breast cells purified from reduction mammoplasties. *Proc Natl Acad Sci USA* 96:12589–12594.
Posner, B.A., 2005, High-throughput screening-driven lead discovery: meeting the challenges of finding new therapeutics. *Curr Opin Drug Discov Devel* 8:487–494.
Priori, S.G., Barhanin, J., Hauer, R.N., Haverkamp, W., Jongsma, H.J., Kleber, A.G., McKenna, W.J., Roden, D.M., Rudy, Y., Schwartz, K., Schwartz, P.J., Towbin, J.A., Wilde, A.M., 1999, Genetic and molecular basis of cardiac arrhythmias: Impact on clinical management. Part III. *Circulation* 99:674–681.
Richmond, T.D., 2008, The current status and future potential of personalized diagnostics: Streamlining a customized process. *Biotechnol Annu Rev* 14:411–422.
Sadee, W., 2002, Pharmacogenomics: the implementation phase. *AAPS PharmSci* 4:E5.
Schwardt, O., Kolb, H., Ernst, B., 2003, Drug discovery today. *Curr Top Med Chem* 3:1–9.
Schwartz, P.J., Priori, S.G., Locati, E.H., Napolitano, C., Cantù, F., Towbin, J.A., Keating, M.T., Hammoude, H., Brown, A.M., Chen, L.S., 1995, Long QT syndrome patients with mutations of the SCN5A and HERG genes have differential responses to Na? Channel blockade and to increases in heart rate: implications for gene-specific therapy. *Circulation* 92:3381–3386.
Shimizu, W., Antzelevitch, C., 1997, Sodium channel block with mexiletine is effective in reducing dispersion of repolarization and preventing torsade de pointes in LQT2 as well as LQT3 models of the long-QT syndrome. *Circulation* 96:2038–2047.
Solomon, L.M., Sieczkiewicz, G.J., 2007, Impact of the US Patent System on the promise of personalized medicine. *Gend Med* 4:187–192.
Triggle, D.J., 2007, Drug discovery and delivery in the 21st century. *Med Princ Pract* 16:1–14.
Tsongalis, G.J., Cartun, R.W., Ricci, A. Jr., 2000, Gene amplification as means for determining therapeutic strategies in human cancers. *Clin Chem Lab Med* 38:837–839.
Van Eldik, L.J., Koppal, T., Watterson, D.M., 2002, Barriers to Alzheimer disease drug discovery and development in academia. *Alzheimer Dis Assoc Disord* 16(Suppl 1):S18–S28.
Veer, L.J., Bernards, R., 2008, Enabling personalized cancer medicine through analylsis of gene-expression patterns. *Nature* 452(7187): 564–570.

Waring, J.F., Ciurlionis, R., Jolly, R.A., Heindel, M., Ulrich, R.G., 2001, Microarray analysis of hepatotoxins in vitro reveals a correlation between gene expression profiles and mechanisms of toxicity. *Toxicol Lett* 120:359–368.

Zareba, W., Moss, A.J., Schwartz, P.J., Vincent, G.M., Robinson, J.L., Priori, S.G., Benhorin, J., Locati, E.H., Towbin, J.A., Keating, M.T., Lehmann, M.H., Hall, W.J., 1998, Influence of genotype on the clinical course of the long-QT syndrome. International Long-QT Syndrome Registry Research Group. Influence of genotype on clinical course of the long-QT syndrome. *New Engl J Med* 339:960–965.

Chapter 24
Biopharmaceutical Industry and Health Care

Abstract Global manufacturing of biopharmaceuticals has increased significantly over the last decade due to a number of reasons. Biopharmaceuticals offer several advantages such as highly effective and potent action, fewer side effects and the potential to actually cure diseases rather than merely treat the symptoms. These advantages, combined with the increasing number of new diseases that can be treated with biopharmaceuticals, are driving enhanced production of these drugs worldwide.

Keywords Biopharmaceuticals · Health care · Recombinant protein · Industry · Molecular farming · Pharmaceutical industry · Mimetics and peptidomimetics · Glycotherapeutics · Human monoclonal antibodies

24.1 Prologue

Today, the pharmaceutical industry is experiencing a profound transition. Globalization and technological advancement represent the principal pressures for change in the market, where it is increasingly more difficult for this type of industry to efficiently recoup the growing cost of innovation (Gonzalez Pier, 2008). The pharmaceutical industry maintain its dominant position in small molecules through the use of biotechnology in drug discovery, but the development and manufacture of protein based drugs requires a completely different set of core competencies. Broadly, the history of pharmaceutical biotechnology includes Alexander Fleming's discovery of penicillin in a common mold, in 1928, and the subsequent development, prompted by World War II injuries of large-scale manufacturing methods to grow the organism in tanks of broth. Pharmaceutical biotechnology has since changed enormously. Two breakthroughs of the late 1970s became the basis of the modern biotech industry: the interspecies transplantation of genetic material, and the fusion of tumor cells and certain leukocytes. The cells resulting from such fusion, hybridomas replicate endlessly and can be geared to produce specific antibodies in bulk.

Modern pharmaceutical biotechnology encompasses gene cloning and recombinant DNA technology. Gene cloning comprises isolating a segment of DNA

M. Debnath et al., *Molecular Diagnostics: Promises and Possibilities*,
DOI 10.1007/978-90-481-3261-4_24, © Springer Science+Business Media B.V. 2010

molecule that corresponds to a single gene and synthesizing ("copying") the segment (Steinberg and Raso, 1998).The first drug produced via genetic engineering was human insulin which appeared on the market in 1982. By mid-2000, 84 biopharmaceuticals had been approved for marketing with almost half launched during the past three years. Worldwide sales have grown more than seven-fold over the past decade to reach US$15 billion by 1998. The US represents 46% of the market, compared to 36% for conventional drugs, due to a combination of earlier regulatory approval, easier market acceptance, and greater pricing flexibility than other countries. Although biopharmaceuticals comprise only 5% of world prescription drug sales, they account for six of the top 50 selling drugs, 13% of new medicines approved by the FDA in the 1990s and about 18% of all drugs in development. At the end of 1999, there were 369 biotechnology drugs in US clinical development against 438 disease indications with 25% in Phase III. Pharmaceutical policy offers a rare example for a complementary approach between a sound health policy and an efficient economic policy; that is, a "healthy pharmaceutical policy."

24.2 Concept

The discovery of recombinant DNA and monoclonal antibody technologies in the 1970s marked the birth of the biopharmaceutical industry. Biopharmaceuticals are complex macromolecules derived from recombinant DNA technology, cell fusion, or processes involving genetic manipulation. They include recombinant proteins, genetically engineered vaccines; therapeutic monoclonal antibodies; and nucleic acid based therapeutics (i.e. DNA based drugs), including gene therapy vectors. Unlike orally delivered small molecule drugs that underpin the traditional pharmaceutical industry, biopharmaceuticals are usually administered by subcutaneous, intravenous, or intramuscular injection (Hu et al., 2006).

Biopharmaceutical is a subset of pharmaceutical. Biopharmaceuticals are pharmaceuticals inherently biological in nature and manufactured using biotechnology (involving use of live organisms), while drugs comprise the other major subset of pharmaceuticals, with their source and manufacture being chemical (non-biological) in nature. These molecules manufactured by biotechnology methods, with the products obviously having biological sources, usually involving live organisms or their active components (bioprocessing; also usually very obvious; or directly involving surrogates, e.g., protein/gene sequences). This broad view has been adopted by *Biopharmaceuticals in the U.S. and European Markets*, which includes all recombinant proteins, (monoclonal) antibodies, vaccines, blood/plasma-derived products, nonrecombinant culture-derived proteins, and cultured cells and tissues. Diagnostics, vaccines and recombinant therapeutic proteins are the three main segments in bio pharma industry.

Since the last two decades emphasis in new drug development has shifted from small-molecule chemicals to large-molecule proteins and other biopharmaceuticals such as human insulin, gene therapies and specialized antibiotic treatments. Unlike chemically synthesized small molecule drugs that have long underpinned the traditional pharmaceutical industry, biopharmaceuticals are complex macromolecules created through the genetic manipulation of living organisms using gene cloning, recombinant DNA (gene splicing), or cell fusion technologies. In terms of product type, these include: recombinant proteins; recombinant antigen vaccines and vaccines crafted from genetic material such as DNA; therapeutic monoclonal antibodies; and oligonucleotides (short sequences of DNA or RNA) such as antisense molecules which interrupt the production of disease causing proteins by inhibiting gene function and gene therapy which can enhance the production of a missing protein through the addition of a synthetic gene.

The first recombinant protein, human insulin was launched in 1982, the first recombinant vaccine against hepatitis B in 1986, the first therapeutic monoclonal antibody against kidney transplant rejection also in 1986, and the first and only oligonucleotide in 1998 (against cytomegalovirus retinitis in AIDS patients). No gene therapy product has yet been approved. Recombinant proteins dominate the biopharmaceutical market accounting for the bulk of sales to date.

The biopharmaceutical industry cannot be identified using the Standard Industrial Classification which distinguishes firms on the basis of their output rather than their technology or production process, and international data are difficult to compare. Many biotechnology firms, for example, are identified as biopharmaceutical companies to differentiate them from the mainstream pharmaceutical industry, but their prime focus are small molecule drugs targeted against proteins thought to be important in the disease pathway (proteins can be used both as drugs i.e. biopharmaceuticals or drug targets). Biologics, an area that consists of blood derived polyclonal antibodies and clotting factors, antibiotics, and classical vaccines based on live or killed viruses are frequently classified as biopharmaceuticals, but these long predate the emergence of recombinant DNA and monoclonal antibodies. Insulin, for example, was originally obtained from porcine or bovine pancreas while human growth hormone was extracted from the pituitary glands of cadavers. The biopharmaceutical category also often includes drugs derived from plants, fungi or marine organisms, but these are more in the realm of traditional medicinal chemistry research based on the random screening of natural compounds.

24.3 Biopharmaceutical Research

Biopharmaceutical research encompasses a wide range of scientific disciplines: genetics, molecular biology, biochemistry, microbiology, physics, pharmacology, and information technologies. Modern drug discovery is built on four core technologies: genomics (source of novel targets), combinatorial chemistry (source of molecules that interact against those targets), high-throughput screening (testing one against the other); and bioinformatics which is crucial to the analysis of the vast amounts of data generated. While current pharmaceuticals are active against some 500 biological targets, genomics is expected to lead to the identification of up to 10,000 new targets for development (Dutta and Garner, 2003). Whether these will be amenable to small molecule intervention or require protein and nucleic acid based therapies remains to be seen. Despite the heavy investment in genomics, only a handful of genomics-based drugs (defined as those based on the identification of an unknown gene sequence followed by elucidation of its function and therapeutic potential) have reached the clinical stage. All are protein drugs.

24.4 Opportunities in Healthcare

The biopharmaceuticals industry is a rapidly growing sector within the pharmaceuticals industry with immense opportunities and implications for healthcare. The success achieved by companies with a few revolutionary new drugs has brought this industry into the limelight in recent years. However, the demand and supply equation for manufacturing capacities has recently become a cause for much speculation. The capacity gap, the difference between the demand and supply of manufacturing capacities represents a fundamental challenge to the industry (Serajuddin and Serajuddin, 2003). Deficit of capacity is expected to significantly impede industry progression.

Quantifying and understanding this capacity gap is essential for biopharmaceuticals developers. Developers also need to examine and evaluate their own financial positions. This will enable them to decide whether to outsource the manufacturing to contract manufacturing organisations chiefly on long-term basis or invest in setting up manufacturing plants. Manufacturing capacity supply is expected to be greater than the worldwide demand throughout 2005–2011. The global manufacturing capacity of biopharmaceuticals was around 2.27 million litres in 2004. This includes the capacity held by both captive use and contract manufacturers. It is expected to increase to 3.69 million litres in 2011 at a compound annual growth rate (CAGR) of 7.2%. The annual biopharmaceutical technology infrastructure expenditures are shown in Fig. 24.1.

Today indeed, the discovery and marketing of innovative medicines is linked: not only to Companies abilities to solve relatively basic and traditional difficulties which have deeply changed over the past few years, for instance: the regulatory constraints, the dramatic increase in development costs, the critical mass required for investments, the development time periods and the protection of innovation, human resources and organization; but also to the acceptance of recent factors which the Research Pharmaceutical Industry must imperatively overcome, such as: the new Research channels and namely biotechnologies, the building of a pharmaceutical Europe and the internationalization of molecules, the needs of the third world, the innovation financing and the control of Health expenditures, the public opinion's shift in expectations regarding ethical and environmental issues. The Pharmaceutical Industry is currently in a shake-up phase during which it will undergo deep changes, with regard both to its activities and in relations with its environment (Chalchat, 1993). The Research Pharmaceutical Industry is currently facing a number of increasingly complex challenges.

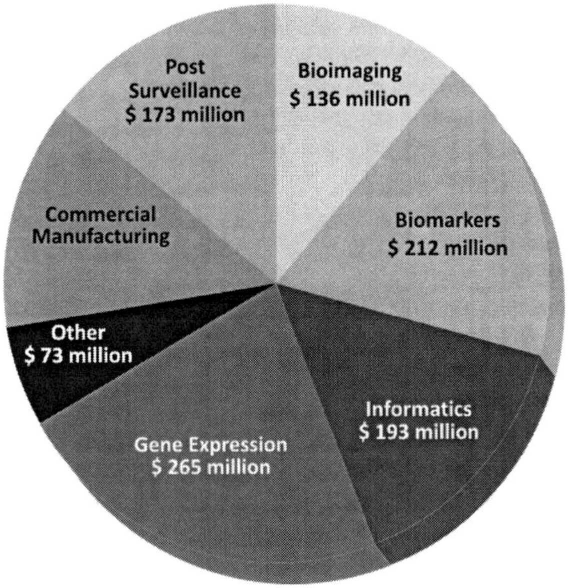

Fig. 24.1 Annual biopharmaceutical technology infrastructure expenditures

The high cost of biopharmaceuticals and the low reimbursement levels from insurance companies present a major challenge. Companies need to explore various strategies to counter this challenge such as developing technologies that bring down the cost of drugs as well as manufacturing pharmaceuticals with substantial market size.

24.5 Molecular Diagnostics and Health Care

Since the last century pharmaceuticals has been an industry of intensive innovation. (Marini, 2006). The rate and direction of technological innovation are affected by a complex of supply- and demand side factors, including biomedical research, education, patent law, regulation, health care payment, law, and more (Goodman and Gelijns, 1996). Molecular biology has revolutionized the understanding of many aspects of human disease. Ongoing developments in DNA diagnostics, the analysis of disease at the nucleic acid level will soon provide automated, rapid, and inexpensive analyses for DNA or RNA sequences associated with genetic, malignant, and infectious diseases. DNA diagnostics will also facilitate the identification of disease-associated genes at birth, thus creating new opportunities for preventive medicine. Pharmaceutical products can be categorized according to their intended purpose that is, diagnostic (to aid in the detection of a disease), therapeutic (treatment), and vaccines and other biological products (prevention). Products included in the diagnostics category are instruments and reagents used for the screening, diagnosis, and monitoring of diseases. They consist, for the most part, of immunoassays (tests for hormones, allergies, HIV, etc.), clinical chemistry (enzymatic, electrochemical, and chromatographic techniques), hematology testing (blood counts), diabetes (omnipresent on the home-testing market), and microbiology tests (supplies of bacterial cultures and various probe tests for specific microorganisms). From the industry's standpoint, therapeutic drugs are typically labelled according to their patent status. For instance, patented drugs (also known as innovative or brand-name drugs) provide a manufacturer with the exclusive right to make and sell a drug for a certain period of time. Nonpatented drugs, in contrast, refer to generic copies of existing patented drugs and other specialty products that were previously subject to patent protection (Grabowski, 2004).

But therapeutics also includes *natural health* remedies, the functional foods and nutraceuticals technology. Their emergence into the sphere of pharmaceuticals can be attributed, among other things, to the increasing use of traditional medicines derived from plants, herbs, and other natural sources as alternatives to modern medicine and pharmaceuticals. Finally, the emergence of biotechnologies has also injected new life into vaccines and other biological products. Combined with the fact that vaccination is no longer limited to infants, a growing number of companies are focusing their resources on the development of new and improved vaccines. Advances in vaccines against sexually transmitted diseases, flu vaccines, adjuvant (i.e., substances pooled with antigens to enhance the immune response), and high-tech vaccines such as naked DNA vaccines (which involve genes instead of proteins) are expanding the boundaries of this field. In addition to vaccines, this group of products contains other biological materials such as plasma and blood products, insulin (a hormone produced by the beta cells of the pancreas that helps to regulate the amount of glucose in the blood), as well as scores of other hormones, serums, and enzymes.

24.6 Global Context

Already, bioengineered drugs against hairy cell leukaemia as well as medicines for heart attacks (thrombolytic agents), Crohn's disease, and rheumatoid arthritis (TNF α antibody) and numerous vaccines against infectious agents (hepatitis B virus, HIV, etc.) have been approved. So far, approximately sixty-three biotechnology products have been approved and are available on the market. More than ever, the influx of biotechnologies is breathing new life into the pharmaceutical industry, increasing the efficiency of drug development, and suggesting innovative avenues of research. This is a very welcome sign for companies that are increasingly feeling the pressures of bottom-line financial performances and rising cost containment measures. The World's Top Fifteen Pharmaceutical Companies includes United States Merck & Co., Bristol-Myers Squibb, Pfizer, American Home Products, Johnson & Johnson, Lilly, Abbott, Schering-Plough, United Kingdom Glaxo Wellcome, AstraZeneca, SmithKline Beecham, Switzerland Novartis, Roche, Germany Aventis, France Sanofi-Synthélabo.

Products and markets: By mid-2000, 84 biopharmaceuticals had been approved for marketing with almost half launched during the past three years. Worldwide sales in last 10 years ago totalled US$15 billion now it is much more. Although representing only 5% of world prescription drug sales of $300billion, biopharmaceuticals comprise six of the top 50 selling drugs, 13% of new medicines approved by the FDA in the 1990s, and 11% of all drugs in development. The US is the largest and most rapidly growing market, accounting for 46% of sales, compared to 36% for conventional prescription drugs, reflecting a combination of earlier regulatory approval, easier market acceptance and greater pricing flexibility. Europe and Japan are the next most important markets, accounting respectively for 30 and 17% (Federsel, 2006).

Industry size: There are approximately 3000 dedicated biotechnology companies worldwide employing 230,000 people with the US accounting for 40% of firms and 70% of employment. Publicly traded firms – including an estimated 100 that floated IPOs during 2000 – number approximately 460 of which 74% are based in the US and 18% in the EU. The segment involved in biopharmaceuticals numbers about 150 firms. Only a handful of biotech firms – primarily those involved in biopharmaceuticals- have a product on the market. The industry as a whole continues to experience major losses because of high R&D costs. The largest and most successful (e.g. Amgen, Chiron, Biogen and Genentech) are mainly US based. With the exceptions of Serono (Switzerland), Celltech (UK), and Bio-Technology General (Israel), few non-US firms have brought a biopharmaceutical to market, reflecting their later startup dates. The total market capitalization of all public European biotechnology companies, for example, is only a little larger than Amgen. Japan has few dedicated biotechnology companies, with the industry primarily consisting of large food and pharma firms with historic strengths in fermentation attempting to diversify into biopharmaceuticals.

There has been a steady migration of European R&D to the US because the US is the most profitable pharmaceutical market in the world, has a more rapid uptake of new drugs, and pricing is largely unregulated by the federal government. As the major European pharmaceutical companies have research operations in the US and are therefore closely tuned to US technical developments, they have looked first to the US biotech industry for collaborative venture, making it difficult to attract investment to Canada. However, factors such as Canadian research excellence, lower wages and infrastructure costs, and access to capital for Québec based start-ups proved instrumental in attracting companies such as Intellivax, Methylgene, DSM Biologics, and Amgen. A number of them have also executed major alliance deals attesting to the commercial of their technology: SemBioSys/Dow agro Sciences (US$12 million) for molecular farming; GlycoDesign/Seikagaku (US$56 million) for anti-inflammatory agents; Inflazyme/Aventis Pharma (US$91 million) for respiratory diseases; and Xenon Genetics/Warner Lambert (US$58 million) for elevating high-density lipoprotein to clear cholesterol from blood. 369 biotechnology drugs were in US clinical development against 438 indications at the end of 1999, with one quarter in Phase III. The three largest categories were vaccines (27%), monoclonal antibodies (16%), and gene therapy (7%). Cancer, mainly breast, melanoma, and prostate, accounted for 40% of indications. Canadian products in the survey included human parathyroid hormone for osteoporosis (NPS Allelix); GMCSF – granulocyte-macrophage colony-stimulating factor (Cangene); stress protein against papillomaviris-related dysplasia (StressGen); group A Strep vaccine (ID Biomedical); breast cancer vaccine (Biomira); and anticancer oligonucleotides (Inex).

Most biotechnology firms do not have the interest or resources in bringing a product to market, preferring instead to focus on research and early clinical development where they feel they can add the most value; commercialization for these companies means developing the technology to a point where it can be licensed out using royalties to fund the next generation of products.

Manufacturing: Only a handful of companies (e.g. Amgen, Genentech, Biogen) have in-house manufacturing capability. The majority contract out their production needs because it is faster and less risky than building a facility and ramping up expertise in-house. Proteins are more difficult to manufacture than conventional drugs, and the time and cost of building and validating a facility certified to current Good Manufacturing Practices (cGMP) standards (up to three years and US$100 million) is a costly and risky commitment, particularly if there is only a single product in advanced stages of clinical development. An added challenge for Canadian companies is the scarcity of bio-chemical engineers with expertise in taking products from the bench through to pilot and commercial scale. The contract biomanufacturing industry is facing capacity constraints due to the large number of protein-based drugs in development combined with the market success of monoclonal antibodies and their great demands for mammalian cell culture. Biotech firms are beginning to queue up for production time and long lead times are expected over the next few years, threatening to increase the time and costs of product development. A number of factors will affect future capacity: the success rates of the products presently in the clinic, actual dose levels of these drugs, the possibility of improving existing manufacturing yields, facility expansions, and the success of molecular farming technologies.

24.7 Emerging Biopharmaceuticals

The success of the pharmaceutical industry will continue to depend on its ability to satisfy the clinical needs of established market economies. The number and quality of new drugs emerging from development pipelines seems likely to rise due to increased research and development budgets of the merged pharmaceutical companies, efficiencies across all facets of the development process, increasing use of new technologies and availability of new targets from the ongoing work on the role of human genes in disease pathways. In addition to the traditional small-molecule drugs, the market for protein products, including monoclonal antibodies and therapeutic vaccines, is likely to expand as advances in recombinant and formulation technologies are made. Current work on relatively newer fields of pharmaceutical research, such as novel G-protein-coupled receptors, chemokines/cytokines, integrins and control of cell cycle regulation and signal transduction pathways (kinases, phosphatases and transcription factors) will lead to new drugs over the next decade. Some of the emerging biopharmaceuticals are

1. *Mimetics* and *peptidomimetics* (small molecule compounds that mimic proteins and peptides) have advantages over proteins in terms of ease of delivery and cost, but there are considerable technical challenges in mimicking the large and intricate protein-protein interacting surfaces and overcoming the weak binding forces of small molecules, a major cause of side effects. Examples include mimetics of insulin and granulocyte colony stimulating factor (in early stage development), peptide based mimetics of erythropoietin (late stage), and Montreal based Neurochem's mimetics of glycosaminoglycans (GAGs – complex carbohydrates that promote amyloid fibril formation characteristic of Alzheimer's) which compete with naturally occurring GAGs to inhibit deposition of amyloids (early stage).
2. *Glycotherapeutics* (complex sugars/carbohydrates) have significant potential both as drugs and drug targets because of their diverse biological roles. Cell surface carbohydrates act as binding sites for other molecules, playing structural roles in cancer transformation, immune system regulation, tissue repair, and anti-infection responses, as well as genetic disorders such as Gaucher's and Fabry's diseases and their enormous structural diversity (more than 10 million polysaccharides can be formed from nine common monosaccharides compared to 16,000 peptides from 20 amino acids). However, their application as biopharmaceuticals has lagged far behind protein drugs because they are difficult molecules to analyse and synthesize.
3. *Fully human monoclonal antibodies* First generation mouse-based antibodies provoked a strong immune reaction in humans limiting their clinical utility. Strategies to generate more human compatible antibodies include mouse-human chimeras (60–70% human protein) first introduced in 1995, followed by humanized antibodies (90–95% human protein) in 1998 in which certain mouse antibody amino acid fragments are grafted onto a human antibody. The next generation now in clinical development are fully human antibodies produced via methods such as transgenic mice containing human antibody genes, phage display, or collection of human B cells from infected individuals which are then fused with human tumor cells to create antibody producing hybridomas. Because of their high costs and large dose requirements, advances in manufacturing capability are needed to meet market demand.
4. *Vaccines* Conventional vaccines based on inactivated (killed) viruses; live, attenuated (weakened) viruses; microbial toxoids; or antigenic fragments (subunit vaccines), have a numberof disadvantages.

   New vaccine strategies include:

   - *recombinant antigen vaccines* where the antigen is produced in bacteria or yeast rather than extracting it from chronic human or animal carriers (e.g. hepatitis B)

- *recombinant vector vaccines* using weakened viruses such as poliovirus, vaccinia virus or salmonella bacterium as carriers of the disease causing organism to overcome the inability of inactivated whole antigens or recombinant antigen vaccines to enter cells and express the desired gene product (hepatitis B, HIV, herpes simplex, in development)
- *DNA vaccines*, produced from a core gene of a virus, are effective against those pathogens that change their exterior protein coat by attacking the interior proteins of a virus (HIV, malaria, and a number of anticancer applications are the closest to commercialization); and
- *RNA vaccines* (particularly applicable to yellow fever and polio caused by RNA viruses).

Genomics has the potential to improve the effectiveness of existing vaccines and design vaccines for diseases for which no protection currently exists: by sequencing invading bacterial and viral organisms, those critical genes that encode the proteins that cause disease can be identified, leading to the development of novel and more effective antigens. Complete DNA sequences of most human pathogenic viruses and more than 20 bacterial pathogens have already been determined. Vaccines are also being developed as anti-cancer agents, involving attempts to activate immune responses against antigens in the tumour to which the immune system has already been exposed. Research is also being undertaken on edible vaccines e.g. transgenic bananas in order to offer an improved mode of delivery.

5. *Gene therapy:* Whereas recombinant proteins and antibodies are produced outside the body and administered to the patient, *gene therapy* has the potential for the body to produce the desired protein on its own if a synthetic gene with the correct DNA sequence for that protein can be inserted into targeted defective cells. A number of major challenges have to be solved: design of a safe and effective gene delivery system, the short duration of gene expression, control over the loci where the transferred gene is expressed, and possible immune reaction caused by the vectors, genes or new proteins. Many efforts now concentrate on conditions that require only transient gene expression, e.g. cancer and coronary artery disease, because sustained gene expression is not necessary once the targeted tumor cells die or new blood vessels grow to bypass blocked arteries.

Biopharmaceuticals are typically produced by microbial fermentation (e.g. insulin and human growth hormone in *E. coli*, recombinant hepatitis B vaccine in yeast) or mammalian cell culture (e.g. tPA, Factor VIII, in Chinese hamster ovary cell lines). Production is not the primary cost driver except for those proteins that must be administered in large amounts e.g. antibodies, or over a period of many years. Transgenic plants, the mammary gland of transgenic animals, and eggs from transgenic chickens may offer more attractive alternatives because of their low cost (factor of 10 to 100 over mammalian cell culture), ease of scale-up, and ability to express complex, large, correctly folded proteins (not possible with microbial fermentation). No products derived via molecular farming have yet been approved. The most advanced is antithrombin for managing coagulation during open heart surgery expressed in transgenic goats (Phase III completed). Others in clinical development include alpha-1-antitrypsin for cystic fibrosis expressed in sheep (Phase II); alpha glucosidase against Pompe's disease expressed in transgenic rabbits (Phase II); and the anti-coagulant hirudin in canola, monoclonal antibodies in corn, and interleukin-10 in tobacco leaves (all in Phase I).

24.8 Challenges of the Biopharmaceutical Industry

The pharmaceutical industry is investing heavily in new biotechnologies, building this capability internally so it can move from traditional screening and medicinal chemistry capabilities to the front end of drug discovery research, in direct competition with biotechnology firms (Vilian, 1998).

Access to capital is critical due to the high costs of drug development and length of time it takes to get a product approved. Most firms do not have a product on the market that can be used to generate cash flow and instead are dependent on funds from venture capital, public equity markets, and strategic alliances used to share development costs and spread risk. The industry predominantly consists of research-oriented, royalty-based companies. Product rights are usually licensed at an early stage with multinational pharmaceutical partner's gaining most of the commercial benefits. With the exception of recombinant insulin; biopharmaceuticals firms are much more expensive than traditional drugs. There is difficulty developing the data needed on long-term outcomes and evaluating benefits, some of which take the form of quality of life improvements. At the same time, there is a need for governments to recognize the overall health care and social economic benefits in evaluating formulary inclusion. The biotechnology industry has been characterized by significant litigation involving intellectual property rights. Genomic companies have filed for patent protection on thousands of genes including the proteins associated with them which is a major area of concern. Although the US PTO has recently tightened gene patents, it will take court precedents to ultimately clarify the issues.

Despite decreases in regulatory approval times in the US and the EU, there has been a substantial increase in the time biopharmaceuticals spend in clinical development, from 39.4 months for products approved in the 1980s to 61.5 months in the late 1990s. A large number have also been discontinued in Phase III. Many of the earlier products were recombinant versions of natural hormones with relatively well understood properties compared to newer biopharmaceuticals with more difficult therapeutic targets and undefined disease mechanisms. Also, the hurdle for discontinuing development is higher for a biotechnology company because it only has a handful of products in clinical development and its stock price is largely determined by its lead product.

In short, the pharmaceutical industry is made up of a wide variety of players, each of which engages in one or more functions in the production process (Kola 2008). To understand the industry's dynamic evolution, however, we need to look at some of the leading trends in pharmaceuticals over the last few years.

References

Chalchat, B., 1993, Current challenges for the pharmaceutical research industry. *Ann Pharm Fr* 51:16–25.

Dutta, A.S., Garner, A., 2003, The pharmaceutical industry and research in 2002 and beyond. *Drug News Perspect* 16:637–648.

Federsel, H.J., 2006, In search of sustainability: process R&D in light of current pharmaceutical industry challenges. *Drug Discov Today* 11:966–974.

Gonzalez Pier, E., 2008, Healthy pharmaceutical policy. *Salud Publica Mex* 50(Suppl 4): S488–S495.

Goodman, C.S., Gelijns, A.C., 1996, The changing environment for technological innovation in health care. *Baxter Health Policy Rev* 2:267–315.

Grabowski, H., 2004, Are the economics of pharmaceutical research and development changing?: productivity, patents and political pressures. *Pharmacoeconomics* 22:15–24.

Hu, X., Ma, Q., Zhang, S., 2006, Biopharmaceuticals in China. *Biotechnol J* 1:1215–1224.

Kola, I., 2008, The state of innovation in drug development. *Clin Pharmacol Ther* 83:227–230.

Marini, G., 2006, Biopharmaceutical industry's commitment to innovation and to the future of health care. *Recenti Prog Med* 97:626–633.

Serajuddin, H.K., Serajuddin, A.T., 2003, Value of pharmaceuticals: ensuring the future of research and development. *J Am Pharm Assoc* 46:511–516.

Steinberg, F.M., Raso, J., 1998, Biotech pharmaceuticals and biotherapy: an overview. *J Pharm Pharm Sci* 1:48–59.

Vilain, E., 1998, CYPs, SNPs, and molecular diagnosis in the postgenomic era. *Clin Chem* 44:2403–2404.

Chapter 25
Forensic Medicine

Abstract The knowledge and technique of medical science applied to assist in the resolution of crimes, legal disputes, etc., constitute forensic medicine. Establishing the identity of victims (in cases of murder, accidents, etc.), criminals (in case of rape, murder, etc.) and the father (in case of paternity disputes), etc., is critical in solving the problem of crimes and disputes. This field was completely revolutionized by the technique of DNA fingerprinting. Using this molecular diagnostic technique, the identity of a person can be traced with the help of bloodstain, semen stain, hair roots, tears, saliva and even perspiration.

Keywords Forensic medicine · DNA fingerprinting · Pathology · Autopsy · Anthropometry · Pathology · RFLPs · mtDNA · Sex determination · Single locus probe · SLP · STR · Microsatellite · Antemortem · DNA methylation · Clinical forensic medicine · Post mortem

25.1 Prologue

In the large majority of homicide trials, there is nothing controversial in the pathologist's testimony. This is not to say that the testimony is unimportant. It is just that, the medical evidence, usually does not establish the guilt or innocence of the defendant nor does it independently establish a degree of the guilt. What is really important, in the majority of instances, is to provide expert opinion based upon objective, indisputable facts which help to evaluate the reliability and credibility of other witnesses. The pathologist is a middle man between the facts of the crime and the testimony of persons whose version of the episode is at issue.

Trust and confidence in scientific evidence have undoubtedly improved over the past few years and forensic sciences themselves have progressed in reliability and effectiveness. Law enforcement agencies rely on scientific investigation to a constantly growing extent. Objectivity, reliability and completeness are the hallmarks of scientific evidence for it to be acceptable in courts. A scientific investigation should have pertinence to the problem under question. Irrelevant scientific investigation will only add to the confusion and may even lead to miscarriage of justice. The quality and reliability of scientific investigations depends on many factors.

M. Debnath et al., *Molecular Diagnostics: Promises and Possibilities*,
DOI 10.1007/978-90-481-3261-4_25,

Failure to take notice of these factors may result in erroneous conclusions and wrong interpretation of the scientific observations. The investigator, therefore, must take all necessary precautions to see that extraneous factors do not interfere with the scientific experiment and vitiate the findings.

25.2 Concept

Forensic medicine and pathology have traditionally been regarded under the auspices of the medical examiner and/or coroner, with medico-legal investigation of unexpected, unexplained, mysterious and/or violent deaths being its principal application. The field of forensic medicine embraces not only traumatic and/or drug-related pathology, however, but also addresses issues of child and elder abuse, sudden and unexpected deaths of a broad constellation, public health and many issues within the overarching fields of the law, medicine and the judiciary. Applications of forensic medicine has impact in almost all fields of clinical medicine to an at least limited degree, with assessment of patient mortality and proper completion of the death certificate merely one of a number of examples which illustrate this relationship.

The knowledge and technique of forensic science help in solving cases of crimes, legal disputes, etc. This also helps in solving the problem of crimes and disputes for the identification of victims in cases of murder, accidents, etc., criminals in case of rape, murder, etc. and the father in case of paternity disputes. The scope of forensic medicine is boundless (Honda et al., 1990). The duties of the investigators include finding out the cause of death in suspected cases of homicide, infanticide, suicide and accidental deaths, to provide expert evidence on injuries regarding their manner of causation, their age and their ability to cause death or disability, to help the police in establishing identity of criminals and victims of crime with the help of medical knowledge to establish the time of death in medico legal deaths, to establish proof of foul play or rules it out.

The study of the above problems and scientific assembly of evidence for presentation to the legal authorities form the basis of this speciality. All branches of medicine, anatomy, pathology therapeutics and obstetrics provide the basic knowledge and applications of which shapes to conform to the needs of the law to form the body of the subject. The study of the above problems and scientific assembly of evidence for presentation to the legal authorities form the basis of this speciality. The ultimate and sole aim is to arrive at the truth or the nearest approach to it so that justice may prevail on the cases that come up for trial. To attain this objective, an expert in forensic medicine should aim at thoroughness in his investigations, impartial reporting to the concerned authorities and he should depend on scientific truth in the formulation of his findings and conclusions.

Through these principles he seeks the constant improvement of the relationship between science and justice in the interest of personal rights, responsibilities and freedoms.

Proper collection and preservation of the materials used in the scientific experiment will ensure reliability in the results obtained. The next pitfall is in the transmission of material for scientific experiments by specialists in their chosen fields. To make sure that the material is not lost, misplaced or tempered with it is necessary that an unbreakable chain of custody is established. Each person in the chain should take responsibility for the proper transmission of the material to the next person and should maintain a record of the date and time of the recovery of the specimen, what was done with it, when it left possession and to whom given. Needless to mention that the chain should be as short as possible. Scientific observations, often allow more than one interpretation depending upon the circumstances of the situation. It is necessary, therefore, that the investigator is, provided with adequate information of the particular incident, regarding the background and circumstances surrounding the incident, so that, he may draw the correct conclusions from his scientific observations keeping in view the correct perspective of the situation as a whole.

Modern scientific investigation is not a one man's job. It is a team work involving several experts working individually or together with a common aim, namely to arrive at the truth. This concept of team work may be clearly understood if it is agreed that a single incident may present a number of different clues which require the employment of different specialists for their interpretation. Conversely, a single item of evidence may require the combined examination by several specialists. There is now growing recognition of the importance of team work of investigations, particularly in the investigation of murders, mass catastrophes, vehicular accidents, arsons etc. A well trained and experienced forensic pathologist at the scene of crime can be of great assistance to the investigating team to assure that the dead body is properly preserved and handled. Necessary photographs and sketches are made, all potential valuable evidence recognised and collected and marked by appropriate officials. This information will be of great value to the forensic pathologist in his procedures with respect to autopsy of the deceased in determining the cause of death and in forming other opinions concerning how death occurred (accident, suicide and homicide).

An autopsy report gives vital clues about the nature of crime and helps in solving the case. Though all doctors are qualified to perform autopsies, experts say vital clues can be missed by them. A forensic medicine expert can pick up fine details during autopsy, which can be missed otherwise. The time, cause and manner of death can be identified during autopsy. Though the aim and object of the system of justice is to discover the facts and to ensure that truth has its full sway on the judgements passed in the courts of law, it is not able to adhere to this aim and object because of some inherent shortcomings in the system. Though the aim of the court is to place responsibility upon the contending parties, in a litigation to present relevant facts and hopefully determine the truth, in reality, the adversary system often undermines the pursuit of truth with opposing sides seeking to win, at all costs, without obligation to reveal information which may be detrimental to them.

The lawyer aims at winning in the fight, not at aiding the court to discover the facts. He does not want the trial court to reach a sound and educated guess if it is likely to be contrary to his scientific investigation and this will only add to the confusion and may even lead to miscarriage of justice. The quality and reliability of scientific investigations depends on many factors. Failure to take notice of these factors may result in client's interest. This is a special privilege enjoyed by lawyers. An expert witness has no such privilege and he has no axe to grind. Nor does he have a personal interest to protect a client. His evidence, therefore, is objective, unbiased and impartial as he is not concerned with the outcome of the judgement. A medical witness is not expected to take more responsibility upon himself that he can cope with.

Contrary to general belief, forensic medicine is not a static science. It is vital and is continuously growing, taking into its fold any new scientific discovery and effectively turn it to its own use. Many a scientific principle such as photography, neutron activation analysis, Coombs test, etc. has, from time to time, found a permanent and accepted place as significant armaments in the hands of the forensic pathologist in his relentless war against social injustice, misdemeanour and felony.

25.3 Forensic Medicine and DNA Fingerprinting

DNA fingerprinting without doubt represents one of the most significant advances in forensic medicine in this century. Some of the commonly used presumptive test reagents for identification of blood and semen could potentially affect the recovery of intact high-molecular-weight deoxyribonucleic acid (DNA) from evidentiary samples (Semba et al., 1994). DNA can be used not only for convicting the guilty but also for exonerating the innocent. It was demonstrated for the first time that a DNA fingerprint could be used to find a perpetrator from within a population. In 1985, a year after the development of DNA fingerprinting, the polymerase chain reaction (PCR) was discovered (Saiki et al., 1985). This discovery has revolutionize the field of molecular biology, though the method would not come into routine use in forensic cases until the early 1990s, since new platforms and biochemical tools were needed in order to take full advantage of the potential of PCR. In particular, new automation technology is the key, and the advent of the automated fluorescent DNA sequencer in the early 1990s is a major step forward. More generally, forensic DNA analysis has benefited substantially from the Human Genome Project, for the genome could be sequenced only with automated equipment that permitted high-throughput processing. Because forensic science could use the same equipment and biochemical tools that gene sequencing used, new methods were rapidly developed in the early 1990s that would have been considered impossible just a few years earlier.

Perhaps the best example of this adjunct benefit of genomics is the development of national DNA databases. Since its inception in 1995, the National DNA Database for England and Wales has expanded to include more than 2.75 million reference DNA profiles, against which all specimens obtained from the scene of a crime ("crime stains") are routinely compared (Werrett et al., 1982). The likelihood that a match will be found is approximately 30%. Many other countries have since followed suit, and the benefits of such databases are considerable, since persons who commit serious crimes such as murder usually have a previous criminal record. The United Kingdom's policy permits the collection of DNA profiles from all convicted criminals, as well as from anyone suspected of committing a crime that could lead to a prison sentence and the law allows authorities to retain the DNA profile even if the suspect is found innocent. Consequently, persons who later commit more crimes can be identified and apprehended quickly (Gill et al., 2000; Marchi, 2004).

The analysis of DNA isolated from forensic biological evidence provides valuable information relating to the identification of the source of the sample. Sex determination from blood and bloodstains by polymerase chain reaction (PCR) was also performed (Semba et al., 1994) from the viewpoint of forensic medicine (Fig. 25.1). In a typical case where PCR method was used, Y chromosome specific sequence (DYZ3) and X chromosome specific sequence (DXZ1) belonging to alphoid (alpha) centromeric repeat family were specifically amplified. The limit of detection of the specific sequences by PCR corresponds to 0.00001 μl of the whole blood. In case of diluted blood, it was possible to detect X and Y specific sequences in the specimen. diluted up to 100,000 times. X and Y specific sequences

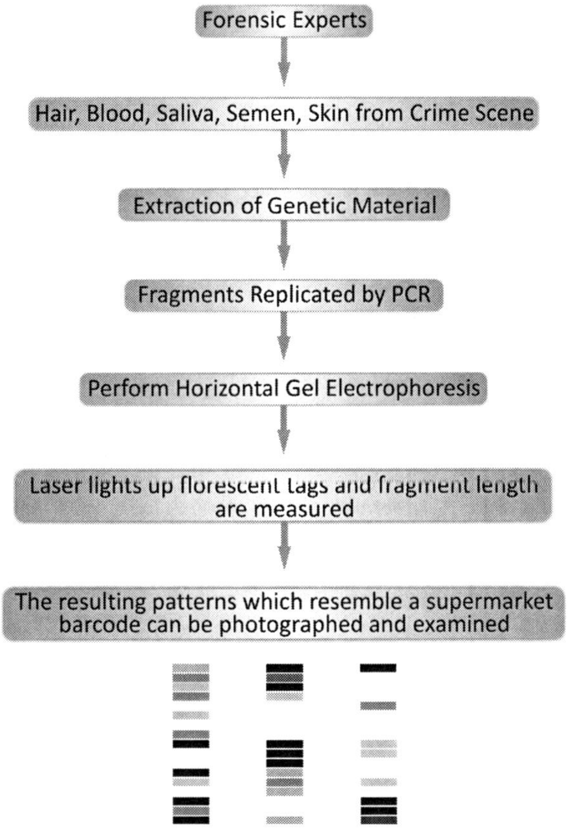

Fig 25.1 Method for analysis of DNA from forensic samples

could be detected from a cotton cloth in size of 1 mm^2, on which blood diluted by 10,000 times had been attached. In case of bloodstains, X and Y specific sequences could be detected even from 1 mm of a single cotton fiber and sex could be determined. X and Y specific sequences could be detected even from blood specimen left at room temperature for 7 months, and from bloodstains left at room temperature for two years. Further, sex determination could be achieved from aged bloodstains preserved at room temperature for 22 years. X and Y specific sequences could be distinctly detected after blood specimen was heated in a water bath at 100 degrees C for 9 h. After bloodstains were heated in an electric furnace at 150 degrees C for 30 min, both specific sequences could be detected. When male and female blood specimens were mixed together, X and Y specific sequences were amplified satisfactorily when male-female mixing ratio was from 100:1 to 1:1. From the mixing ratio of 1:10, amplified band of Y specific sequence began to gradually weaken, and only weak band was detected when the mixing ratio was 1:1000. These results reveal that the present sex determination method by PCR can be performed in simple and quick manner and has high detecting sensitivity, and sex can be determined even from putrefied, heated or aged specimens. Thus, it is expected to be one of useful examination methods in forensic practices.

RFLPs and 9-bp deletion type of mtDNA may suggest the characteristics of the human races (Misawa, 1994). In a study sex determination using polymerase chain reaction (PCR) on tooth material was evaluated from the viewpoint of forensic medicine by Murakami et al. (2000). Using the method of DNA analysis, Restriction fragment length polymorphism testing, using a combination of single locus probe (SLP) that vary highly among individuals, produces a DNA fingerprint or profile. The PCR method using STR (microsatellite) and mitochondrial DNA analysis (mtDNA) is suitable for examination of the forensic biological samples (bloodstains, hairs, seminal stains, bones, tooth). For sex identification of bloodstains, bleached skeletons and teeth, Southern blot hybridization with Y-chromosome specific probe (pHY10; 3.4 kb) and PCR amplifying with sex chromosome specific fragments, can be used. Mitochondrial DNA from bones and teeth can also help in sex determination (Harihara et al., 1990). The antemortem dental information including dental records and radiological records provided by families of the victims, play a major role in speedy identification (Yamamoto, 1996).

Central to this technology, which is based on the analysis of the genetic component of cells, is the use of DNA probes to regions of the human genome that exhibit great variability between individuals. These probes fall into two main categories. The first group comprises those that can detect a large number of these "hyper variable" loci simultaneously, namely multilocus probes (MLPs). On autoradiography, these give rise to a band pattern that is reminiscent of the bar code on supermarket goods, the main advantage of which is that a single such test provides a lot of information very rapidly. MLPs are, therefore, the probes of choice when the amount of material for testing is not limiting, e.g., a blood sample for paternity testing. In many forensic cases, however, the material evidence available for testing is minute, such as a few hair roots or a tiny semen stain, and the situation is often complicated by the presence of tissue from more than one person.

25.4 Applications of DNA Methylation Markers in Forensic Medicine

The methylation pattern of human genome is space-time specific, sex-specific, parent-of-origin specific and disease specific, providing us an alternative way to solve forensic problems (Zhao and Yang, 2005). DNA methylation profiles represent a more chemically and biologically stable source of molecular diagnostic information than RNA or most proteins. Recent advances attest to the great promise of DNA methylation markers as powerful future tools in the clinic. DNA methylation is a post-replication modification that is predominantly found in cytosines of the dinucleotide sequence CpG. Epigenetic information is stored in the distribution of the modified base 5-methylcytosine. In the past decade, DNA methylation analysis has been revolutionized by two technological advances viz. bisulphite modification of DNA and methylation-specific polymerase chain reaction (MSP).

25.5 Forensic Medicine and Anthropometry

Forensic medicine is an interdisciplinary science which in everyday practice applies all the knowledge that medical sciences, have accepted as reliable and scientifically solid facts or processes, and qualitative and quantitative definitions with the help of which accurate and reliable statements can be made. The use of anthropometry in the field of forensic science and medicine dates back to 1882 when Alphonse Bertillon, a French police expert invented a system of criminal identification based on anthropometric measurements. His system was based on three fundamental ideas- the fixed condition of the bone system from the age of twenty till death; the extreme diversity of dimensions present in the skeleton of one individual compared to those in another; the ease and relative precision with which certain dimensions of the bone structure of a living person can be measured using simply constructed calipers. This system of identification spread rapidly through much of the world but the system was not accepted much in view of some major drawbacks and discovery of other identification systems e.g. dactylography.

As anthropometry is an important part of biological/physical anthropology, hence the persons specializing in anthropometry are familiar with range of biological variability present in the human populations and its causes, and are well trained in comparative osteology, human osteology, craniometry, osteometry, racial morphology, skeletal anatomy and function. They are well aware of the knowledge of archaeological field techniques and methods which serve well in crime scene recoveries involving buried and surface remains. The term "forensic anthropometry" can be coined for this branch of applied physical anthropology, involving the use of methods/techniques of anthropometry in forensic/legal context. In other words, "forensic anthropometry is a scientific specialization emerged from the discipline of forensic anthropology dealing with identification of human remains with the help of metric techniques".

Anthropometric characteristics have direct relationship with sex, shape and form of an individual and these factors are intimately linked with each other and are manifestation of the internal structure and tissue components which in turn, are influenced by environmental and genetic factors. Anthropometric data are believed to be objective and they allow the forensic examiner to go beyond subjective assessments such as "similar" or "different". With measurement data, the examiner is able to quantify the degree of difference or similarity and state how much confidence can be placed in this interpretation.

The main aim of an anthropometrist employed in the forensic medicine/medico-legal department, working with unknown variables, is to describe the remains in such terms so that one can achieve the goal of estimating age at the time of death, sex, stock/race/ancestry/ethnicity, stature, body weight/body build, details of individualizing characteristics i.e. amputations, fractures, ankyloses, deformities and bone pathologies and to some extent the cause of death if reflected in the remains/bones. The objective is to enable the law enforcement agencies to achieve the ultimate goal of personal identification.

Forensic anthropometry incorporates most of the techniques originating with the analysis of human skeletal material from archaeological sites; the two disciplines have been closely linked. A good forensic anthropologist must, by definition, be a good skeletal biologist. He helps a forensic pathologist to reconstruct the biological nature of the individual at the time of postmortem examination, and sometimes giving clues and reconstructing the circumstances surrounding death. He is prepared for this by his training in describing the prehistoric skeletons from archaeological sites and usually by special experience in identifying unknown modern skeletons.

25.6 Clinical Forensic Medicine

Clinical forensic medicine (CFM) is "the application of appropriate forensic practices and principles, reserved for use by the pathologist at autopsy, to living patients in a clinical setting." "Living forensic" patients include survivors of trauma and potentially catastrophic experiences resulting in injury. CFM arose from "clinically" affirming that not all abuse or assault victims sustain fatal injuries. Appropriate medical documentation and interpretation of physical findings may aid law enforcement and/or social services in the legal evaluation of a case or situation. Additionally, timely collection of pertinent evidence may be performed as the case necessitates. (Recktenwald et al., 2005)

Criminal violence and its associated trauma comprise a critical health problem throughout the world. Clinical forensic medicine represents a new discipline of medical practice that is evolving in direct response to the sequelae of criminal and interpersonal violence. The application of the principles and standards of the forensic specialist has been increasingly recognized as playing a crucial role in trauma care; the results of the extremes of human behaviour-abused children, individuals suffering from blatant neglect and maltreatment, or self-inflicted injury, and victims of road-traffic accidents, firearm injuries and other assaults. These cases must be reported to a legal agency for investigation and follow-up. As trends in crime and violence change, new antiviolence legislation is likely to be implemented; consequently, new personnel resources are required to ensure that these legislative mandates effectively meet the needs of society (Sharma, 2006). Recently it was reported that post-mortem histopathological investigations of the bone marrow in forensic medicine is an important issue for both the forensic and clinical pathologist (Roll et al., 2009).

25.7 Conclusion

The technological advancement in forensic medicine will be imposing challenges, including those concerning genetic manipulation and the keeping of confidentiality regarding electronic files. On the other hand, old problems will recur, the most important being the vital question of research with prisoners and the use of torture by agents of the state. There is a need for international legislative mechanisms that would establish a set of explicit rules of behavior, thus diminishing the possibility of ethical dilemmas. The subject of forensic medicine has grown from the days when it was taught as part of pathology, to being a specialty in itself. However, forensic medicine has to advance to keep pace with new developments and we need the committed efforts of all doctors in this profession to meet the expectations of society.

References

Gill, P., Whitaker, J., Flaxman, C., Brown, N., Buckleton, J., 2000, An investigation of the rigor of interpretation rules for STRs derived from less than 100 pg of DNA. *Forensic Sci Int* 112:17–40.

Harihara, S., Nakanura T., Hirai, M., et al., 1990, Sex indentification by analysis of DNA extracted from hard tissues. *Nihon Hoigaku Zasshi* 44(4):293–301.

Honda, K., Harihara, S., Nakamura, T., Hirai, M., Misawa, S., 1990, Sex identification by analysis of DNA extracted from hard tissues. *Nihon Hoigaku Zasshi* 44:293–301.

Marchi, E., 2004, Methods developed to identify victims of the World Trade Center disaster. *Am Lab* 36:30–36.

Misawa, S., 1994, Application of DNA polymorphism to forensic medicine. *Rinsho Byori* 42: 636–642.

Murakami, H., Yamamoto, Y., Yoshitome, K., Ono, T., Okamoto, O., Shigeta, Y., Doi, Y., Miyaishi, S., Ishizu, H., 2000, Forensic study of sex determination using PCR on teeth samples. *Acta Med Okayama* 54:21–32.

Recktenwald, K., Hunsaker, D.M., Corey, T.S., Weakley-Jones, B., 2005, Clinical forensic medicine introduction for healthcare providers. *J Ky Med Assoc* 103:433–435.

Roll, P., Beham, A., Beham-Schmid, Ch., 2009, Post-mortem histopathological investigations of the bone marrow in forensic medicine: an important issue for both the forensic and clinical pathologist. *Forensic Sci Int* 186:e17–e20.

Saiki, R.K., Scharf, S., Faloona, F., Mullis, K.B., Horn, G.T., Erlich, H.A., Arnheim, N., 1985, Enzymatic amplification of beta-globin genomic sequences and restriction site analysis for diagnosis of sickle cell anemia. *Science* 230:1350–1354.

Semba, S., Yamamoto, Y., Ishizu, H., 1994, Sex determination from blood and bloodstains by polymerase chain reaction (PCR). *Nihon Hoigaku Zasshi* 48(1):7–18.
Sharma, B.R., 2006, Clinical forensic medicine in the present day trauma-care system-an overview. *Injury* 37:595–601.
Werrett, D., Pinchin, R., Hale, R., et al., 1982, Problem solving: DNA data acquisition and analysis. *Prof DNA* 2:1–6.
Werrett, D., Pinchin, R., Hale, R., 1982, Problem solving: DNA data acquisition and analysis. *Prof DNA* 2:1–6.
Yamamoto, K., 1996, Molecular biological studies on teeth, and inquests. *Forensic Sci Int* 80: 79–87.
Zhao, G.S., Yang, Q.E., 2005, Applications of DNA methylation markers in forensic medicine. *Fa Yi Xue Za Zhi* 21:61–64.

Chapter 26
Pharmacogenomics

Abstract Pharmacogenomics explores the contribution of genetics to drug safety and specifically examines the single gene interactions with drugs. It also makes use of genetic tests to distinguish among patients whose genetic characteristics predispose them to respond in certain ways to certain medicines. An understanding of the genetic variables that influence drug response could also help pharmaceutical companies design new, more effective therapies.

Keywords Drugs · Pharmacogenomics · Drug response · Gene variation · Personalised medicine · SNP · Drug discovery · Pharmacogenomic tests

26.1 Prologue

Revolutionary new technologies, capable of transforming the economics of sequencing, are providing an unparalleled opportunity to analyze human genetic variation comprehensively at the whole-genome level within a realistic timeframe and at affordable costs. Current estimates suggest that it would cost somewhere in the region of US$30 million to sequence an entire human genome using Sanger-based sequencing, and on one machine it would take about 60 years. When applied over a large enough genomic region, these new approaches to resequencing will enable the simultaneous detection and typing of known, as well as unknown, polymorphisms, and will also offer information about patterns of linkage disequilibrium in the population being studied (Shi and Chen, 2002). Technological progress, leading to the advent of single-molecule-based approaches, is beginning to dramatically drive down costs and increase throughput to unprecedented levels, each being several orders of magnitude better than that which is currently available (Lee, 2009). A new sequencing paradigm based on single molecules will be faster, cheaper and more sensitive, and will permit routine analysis at the whole-genome level (Johnson and Evans, 2002).

M. Debnath et al., *Molecular Diagnostics: Promises and Possibilities*,
DOI 10.1007/978-90-481-3261-4_26,

26.2 Concept

Despite advances in modern medicine, in practice it remains difficult for physicians to determine a patient's response to a treatment. A drug that is safe and effective for one patient may prove to be dangerous or ineffective in another. In some cases, the lack of effectiveness of a drug or the occurrence of side-effects can be linked to factors such as medication errors. However, it is now known that a significant proportion of differential drug response is genetic in origin (López-López et al., 2004). This genetic angle has often not been fully appreciated, but new developments are driving the use of this information in medicine.

Pharmacogenomics is the study of how an individual's genetic inheritance affects the body's response to drugs (Evans and Relling, 1999). The term comes from the words pharmacology and genomics and is thus the intersection of pharmaceuticals and genetics. Pharmacogenomics refers to the general study of all of the many different genes that determine drug behaviour. The way a person responds to a drug (this includes both positive and negative reactions) is a complex trait that is influenced by many different genes. Without knowing all of the genes involved in drug response, scientists have found it difficult to develop genetic tests that could predict a person's response to a particular drug. Scientists discovered that genes show small variations (or changes) in their nucleotide (DNA base) content. Genetic testing for predicting drug response is now possible. The inherited variations in genes that dictate drug response and explores the ways these variations can be used to predict whether a patient will have a good response to a drug, a bad response to a drug, or no response at all .This study also allows to know the influence of genetic variation on drug response in patients by correlating gene expression or single-nucleotide polymorphisms with a drug's efficacy or toxicity (Johnson and Evans, 2001). By doing so, pharmacogenomics aims to develop rational means to optimise drug therapy, with respect to the patients' genotype, to ensure maximum efficacy with minimal adverse effects.

Pharmacogenomics holds the promise that drugs might one day be tailor-made for individuals and adapted to each person's own genetic makeup (Evans and Relling, 1999). Environment, diet, age, lifestyle, and state of health all can influence a person's response to medicines, but understanding an individual's genetic makeup is thought to be the key to creating personalized drugs with greater efficacy and safety (Mancinelli et al., 2000). Pharmacogenomics combines traditional pharmaceutical sciences (Fig. 26.1) such as biochemistry with annotated knowledge of genes, proteins, and single nucleotide polymorphisms (Blake and Sobel, 2008).

Pharmacogenomics seeks to uncover significant associations between genomic patterns and clinical outcomes. It also helps to produce useful predictive knowledge, allowing clinical treatment decision making to be based upon more rational criteria than today's probabilistic approach, which is largely based upon educated guesswork. Most of the attention on pharmacogenomics has focused on its role in bringing future drugs to market, but the technology could also be used to identify genetic risk factors for adverse drug reactions associated with drugs.

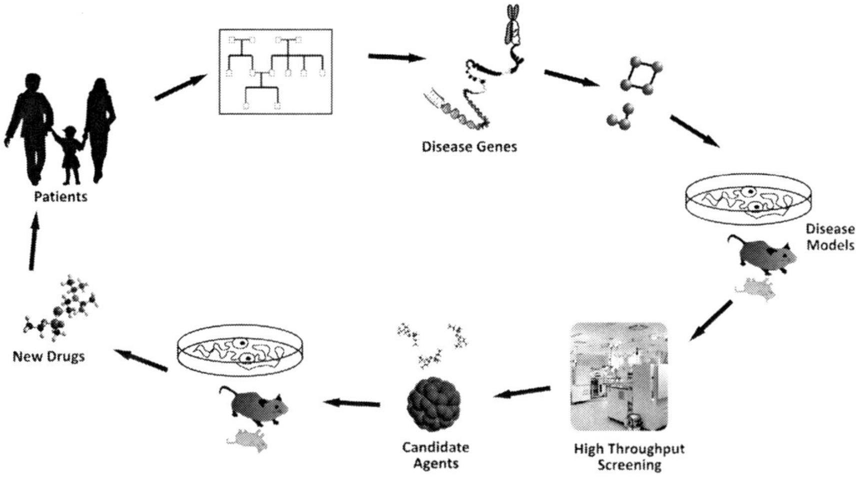

Fig 26.1 Pharmacogenomics, a key to creating personalized drugs

26.3 Predicting Drug Response on Gene Variation

DNA sequencing is the determination of the order of nucleotides (the base sequence) in a DNA molecule. As many of the genetic variations found within the human genome as possible. These variations or SNPs can be used as a diagnostic tool to predict a person's drug response. For SNPs to be used in this way, a person's DNA must be examined (sequenced) for the presence of specific SNPs. The problem is, however, that traditional gene sequencing technology is very slow and expensive and has therefore impeded the widespread use of SNPs as a diagnostic tool (Dan et al., 2006). DNA microarrays (or DNA chips) are an evolving technology that should make it possible for doctors to examine their patients for the presence of specific SNPs quickly and affordably. A single microarray can now be used to screen 100,000 SNPs found in a patient's genome in a matter of hours. As DNA microarray technology is developed further, SNP screening in the doctor's office to determine a patient's response to a drug, prior to drug prescription, will be commonplace.

26.4 Drug Development and Testing Benefit from Pharmacogenomics

Pre-screening should allow clinical trials to be smaller, faster, and less expensive; therefore, the consumer could benefit in reduced drug costs. SNP screenings will benefit drug development and testing because pharmaceutical companies could exclude from clinical trials those people whose pharmacogenomic screening would show that the drug being tested would be harmful or ineffective for them. Excluding these people will increase the chance that a drug will show it useful to a particular population group and will thus increase the chance that the same drug will make it into the marketplace. Pre-screening clinical trial subjects should also allow the clinical trials to be smaller, faster, and therefore less expensive; therefore, the consumer could benefit in reduced drug costs. Finally, the ability to assess an individual's reaction to a drug before it is prescribed will increase a physician's confidence in prescribing the drug and the patient's confidence in taking the drug, which in turn should encourage the development of new drugs tested in a like manner.

26.5 Applications and Benefits of Pharmacogenomics

Pharmaceutical companies will be able to create drugs based on the proteins, enzymes, and RNA molecules associated with genes and diseases. This will facilitate drug discovery and allow drug makers to produce a therapy more targeted to specific diseases. This accuracy not only will maximize therapeutic effects but also decrease damage to nearby healthy cells (Evans and Relling, 1999).

26.5.1 Better, Safer Drugs the First Time

Instead of the standard trial-and-error method of matching patients with the right drugs, doctors will be able to analyze a patient's genetic profile and prescribe the best available drug therapy from the beginning. Not only will this take the guesswork out of finding the right drug, it will speed recovery time and increase safety as the likelihood of adverse reactions is eliminated. Pharmacogenomics has the potential to dramatically reduce the estimated 100,000 deaths and 2 million hospitalizations that occur each year in the United States as the result of adverse drug response.

26.5.2 More Accurate Methods of Determining Appropriate Drug Dosages

Current methods of dosages on weight and age will be replaced with dosages based on a person's genetics, how well the body processes the medicine and the time it takes to metabolize it. This will maximize the therapy's value and decrease the likelihood of overdose.

- **Advanced Screening for Disease** Knowing one's genetic code will allow a person to make adequate lifestyle and environmental changes at an early age so as to avoid or lessen the severity of a genetic disease. Likewise, advance knowledge of particular disease susceptibility will allow careful monitoring, and treatments can be introduced at the most appropriate stage to maximize their therapy.

- **Better Vaccines** Vaccines made of genetic material, either DNA or RNA; promise all the benefits of existing vaccines without all the risks. They will activate the immune system but will be unable to cause infections. They will be inexpensive, stable, easy to store, and capable of being engineered to carry several strains of a pathogen at once.
- **Improvements in the Drug Discovery and Approval Process** Pharmaceutical companies will be able to discover potential therapies more easily using genome targets. Previously failed drug candidates may be revived as they are matched with the niche population they serve. The drug approval process should be facilitated as trials are targeted for specific genetic population groups providing greater degrees of success. The cost and risk of clinical trials will be reduced by targeting only those persons capable of responding to a drug.
- **Decrease in the Overall Cost of Health Care** Decreases in the number of adverse drug reactions, the number of failed drug trials, the time it takes to get a drug approved, the length of time patients are on medication, the number of medications patients must take to find an effective therapy, the effects of a disease on the body (through early detection), and an increase in the range of possible drug targets will promote a net decrease in the cost of health care.

26.6 Benefits of Pharmacogenomic Testing

As with other types of genetic testing, pharmacogenomic tests usually require only a small sample, such as blood or a scraping from inside the cheek. Results would let the physician predict whether someone will respond positively to drug therapy within hours rather than in the days or weeks it might take with the trial-and-error method and with substantially less risk to the patient. In the end, these tests may give doctors invaluable information about their patients not otherwise available to them (López-López et al., 2004).

26.6.1 Patient's Ability to Metabolize Drugs

Testing patients prior to initiating drug therapy to determine their ability to metabolize different classes of drugs is a key emerging area of investigation. Such metabolic information could prove useful to both the doctor and patient when choosing current and future drug therapies and drug doses.

There are a number of types of enzymes in the liver that metabolize medications. Genetic variations in these enzymes that affect metabolic rate are relatively common, but the prevalence of the variations differs significantly by ethnic background. Some of these enzymes include:

- The Cytochrome P450 family
- N-acetyltransferase
- Thiopurine methyltransferase (TPMT)
- UDP-glucuronosyltransferase

The Cytochrome P450 family: Some of the most studied enzymes are the members of the Cytochrome P450 (CYP) family of about 50 liver enzymes. These enzymes metabolize more than 30 classes of drugs, including antidepressants, antiepileptics and cardiovascular drugs. Patients can be separated into poor, normal and ultra-rapid metabolizers of drugs by the CYP enzymes. These classifications are due to variations in the associated CYP gene. When a poor metabolizer of a particular drug is given a standard dose of that drug, he will process the drug more slowly, resulting in increased levels of the drug in his bloodstream, the potential for side effects, and an increased risk of toxicity. For an ultra-rapid metabolizer, the same dose may be ineffective as the drug is processed too rapidly to have its full effect. Dosages of these drugs must be altered to accommodate the rate of metabolism. The CYP family is important as it affects the metabolism of a significant percentage of available drugs and because a significant proportion of the population are poor or ultra-rapid metabolizes (Kirchheiner and Seeringer, 2007; Kirchheiner et al., 2005).

N-acetyltransferase: This is a liver enzyme that activates some drugs and deactivates others. Some patients can acetylate (a type of metabolic change) drugs slowly while other patients acetylate drugs quickly. Those who are slow acetylators may experience toxicity when taking drugs such as procainamide, isoniazid, hydralazine, and sulfonamides. Those who are fast acetylators may not respond to isoniazid or hydralazine. About 40–70% of Caucasians and African-Americans are considered slow acetylators.

Thiopurine methyltransferase (TPMT): This enzyme metabolizes the immune suppressant azathioprine and other thiopurine medications such as 6-mercaptopurine and 6-thioguanine (used to treat children with acute lymphocytic leukemia and also used to treat autoimmune diseases). Each copy of the TPMT gene will produce some TPMT enzyme. This leads to three different groups of enzyme activity levels (low/low, low/high, and high/high or deficient, intermediate, and normal). About 1 in 300 Caucasians and African-Americans are TPMT- deficient. If these patients are given a standard drug dose, they may suffer severe hematopoietic (red blood cell producing) toxicity. Many are able to achieve the desired therapeutic effect from a dose that is one tenth of the "normal" dose.

UDP-glucuronosyltransferase: This enzyme is involved in the metabolism of the chemotherapy drug irinotecan, which is used in the treatment of metastatic colorectal cancer. Variation in the gene that code for this enzyme can influence the patient's ability to break down the major active metabolite. The inability to break the metabolite down can lead to increased levels of it in the blood and a higher risk of side effects, which include reduced white blood cell count and severe diarrhea.

26.6.2 Age-Related Genetic Variations

Some researchers are looking at changes in genetic variation over time to help evaluate how age may affect genetic response to drug therapy.

26.7 Recent Reports of Pharmacogenomics Use

For many drugs, pharmacogenetic polymorphisms are known affecting biotransformation and clinical outcome. The clinical importance of these variants depends on allele-frequency and the effect size of the clinical outcome parameters. Further, it depends on the therapeutic range of the drug which is affected, on predictability of drug response as well as on duration until onset of therapeutic efficacy. Consequences which arise from genotyping might be: adjustment of dose according to genotype, choice of therapeutic strategy or even choice of drug.

In antidepressant drug treatment, most drugs are metabolized via the polymorphic cytochrome P450 enzyme CYP2D6. Huge differences in pharmacokinetic parameters have been consistently shown for many tricyclics, some SSRIs, and other antidepressant drugs whereas the effects on therapeutic efficacy and adverse events have been described controversially. In cardiovascular disease, oral anticoagulants, nonsteroidal anti-inflammatory drugs, oral hypoglycemic drugs and other drugs are affected by genetic polymorphisms of the cytochrome P450 drug metabolizing enzyme CYP2C9. Studies in patients or healthy volunteers revealed up to 10-fold differences in pharmacokinetic parameters due to genetic polymorphisms of CYP2C9. Pharmacogenetics based dose adjustments are one tool to individualize drug treatment according to genetic factors. They can be derived from pharmacokinetic data with the aim to obtain equal drug concentrations in each individual. Prospective validation of dose adjustments based on pharmacogenetics should be performed before routine application of such strategies.

A controlled prospective clinical trial with one arm receiving genotype-based dose adjustments and the other arm receiving therapy as usual will elucidate the benefit of pharmacogenomics-based individualization of certain drug therapies. Pharmacogenetic variability plays an important role in the pharmacokinetics of oral antidiabetic drugs; however, to date, the impact of this variability on clinical outcomes in patients is mostly unknown and prospective studies on the medical benefit of CYP genotyping are required (Kirchheiner et al., 2006). Loss of function of thiopurine S-methyltransferase (TPMT) results in severe and life-threatening hematopoietic toxicity if patients receive standard doses of mercaptopurine and azathioprine. Gene duplication of cytochrome P4502D6 (CYP2D6), which metabolizes many antidepressants, has been identified as a mechanism of poor response in the treatment of depression. There is also a growing list of genetic polymorphisms in drug targets that have been shown to influence drug response. A major limitation that has moderated the use of pharmacogenetic testing in the clinical setting is the lack of prospective clinical trials demonstrating that such testing can improve the benefit/risk ratio of drug therapy.

As research has graduated from studying single candidate genes to whole-genome scans, pharmacogenomics is beginning to make its impact on the therapeutics of complex CNS disorders, such as schizophrenia, Parkinson's disease and Alzheimer's disease. Alzheimer's disease is a progressive complex disorder, where genetic predisposition interacts with environmental factors. With conventional therapeutics only providing symptomatic treatment, the current focus of the pharmaceutical industry is on novel strategies with an etiopathogenic orientation (Gupta et al., 2008). Today, clinical trials researchers use genetic tests for variations in cytochrome P450 genes to screen and monitor patients. In addition, many pharmaceutical companies screen their chemical compounds to see how well they are broken down by variant forms of CYP enzymes (Kirchheiner and Seeringer, 2007). Another enzyme called TPMT (thiopurine methyltransferase) plays an important role in the chemotherapy treatment of a common childhood leukemia by breaking down a class of therapeutic compounds called thiopurines. A small percentage of Caucasians have genetic variants that prevent them from producing an active form of this protein. As a result, thiopurines elevate to toxic levels in the patient because the inactive form of TMPT is unable to break down the drug. Today, doctors can use a genetic test to screen patients for this deficiency, and the TMPT activity is monitored to determine appropriate thiopurine dosage levels.

Pharmacogenomics can help in translating this knowledge of human genome variability into efficacious and safer therapeutics. In another research the potential contributions of pharmacogenomics to improved treatment in the 21st century in context to Brain cancer. Patients with oligodendrogliomas have benefited from phamacogenomics, as there is a clear relationship between response to chemotherapy and chromosomal profile. Drug efficacy, safety and response could be improved by using pharmacogenomics to identify genetic markers that differentiate responder from nonresponder patient groups, as well as identifying patients likely to develop adverse drug reactions (Shai et al., 2008). Thus, pharmacogenomics by microarray studies may lead to much more accurate tumor classification, drug and biomarker discovery, and drug efficacy testing.

26.8 Interpreting Pharmacogenomic Tests

Pharmacogenomic test results can be difficult to interpret. There are limitations to interpretation because enzymes involved in drug metabolism arise from multiple genes and that process is often complex. The test results are predictions based on information about the specific genetic variations and on information about the associated diseases, adverse drug reactions, and patient outcomes that have been gathered during studies and clinical trials. In many cases, the predictions will be very accurate, but they cannot say with 100% certainty what will happen with an individual patient and they do not incorporate or make allowances for the other factors in a patient's life related to the disease condition or to the individual that may also affect their response to treatment. This is one of the reasons why the results are intended to be used in conjunction with other relevant clinical findings.

Some currently available pharmacogenomic tests include:

- A DNA microarray that tests for 29 CYP2D6 genetic variants and 2 for the CYP2C19. It is meant to be used as an aid in individualizing treatment selection and dosing for drugs metabolized through these genes. It helps predict poor, intermediate, extensive, or ultra-rapid metabolizers.

- A test that detects variations in the UGT1A1 gene, which produces the enzyme UDP-glucuronosyltransferase. The enzyme is active in the metabolism of drugs such as irinotecan, a drug used in metastatic colorectal cancer treatment. The test is used to identify patients who may be at increased risk of adverse reaction to the drug. The gene that the test detects has been shown to be an effective genetic marker for predicting irinotecan-induced toxicity.
- Tests that detect genetic variants of the CYP2C9 and VKORC1 (vitamin K epoxide reductase) enzymes. These enzymes are involved in the efficacy of warfarin as an anticoagulant. Warfarin is used to prevent dangerous blood clots from forming in the blood vessels of certain patients, but it can significantly increase the risk of bleeding into the head or gastrointestinal tract. These tests identify patients who have genetic variations and so need a reduced dose of warfarin to avoid bleeding episodes. An FDA advisory committee voted in November 2005 in favor of changing Warfarin's label to reflect the fact that pharmacogenomic information can be useful in deciding a patient's individual dose.

Pharmacogenomic tests are promising but there are questions and concerns about their utility, use, interpretation, ability to predict actual drug response, their effect on patient outcome, and about the potential for the results of testing to be used to discriminate against the patient or to prevent them from access to treatment.

26.9 Barriers to Pharmacogenomics Progress

Pharmacogenomics is a developing research field that is still in its infancy. Several of the following barriers will have to be overcome before many pharmacogenomics benefits can be realized.

- **Complexity of finding gene variations that affect drug response** – Single nucleotide polymorphisms (SNPs) are DNA sequence variations that occur when a single nucleotide (A, T, C, or G) in the genome sequence is altered. SNPs occur every 100–300 bases along the 3-billion-base human genome, therefore millions of SNPs must be identified and analyzed to determine their involvement (if any) in drug response. Further complicating the process is our limited knowledge of which genes are involved with each drug response. Since many genes are likely to influence responses, obtaining the big picture on the impact of gene variations is highly time-consuming and complicated.
- **Limited drug alternatives** – Only one or two approved drugs may be available for treatment of a particular condition. If patients have gene variations that prevent them using these drugs, they may be left without any alternatives for treatment.
- **Disincentives for drug companies to make multiple pharmacogenomic products** – Most pharmaceutical companies have been successful with their "one size fits all" approach to drug development. Since it costs hundreds of millions of dollars to bring a drug to market, will these companies be willing to develop alternative drugs that serve only a small portion of the population?

- **Educating healthcare providers** – Introducing multiple pharmacogenomic products to treat the same condition for different population subsets undoubtedly will complicate the process of prescribing and dispensing drugs. Physicians must execute an extra diagnostic step to determine which drug is best suited to each patient. To interpret the diagnostic accurately and recommend the best course of treatment for each patient, all prescribing physicians, regardless of specialty, will need a better understanding of genetics.

26.10 The Promise of Pharmacogenomics

Right now, in doctors' offices all over the world, patients are given medications that either don't work or have bad side effects. Often, a patient must return to their doctor over and over again until the doctor can find a drug that is right for them. Pharmacogenomics offers a very appealing alternative. Imagine a day when you go into your doctor's office and, after a simple and rapid test of your DNA, your doctor changes her/his mind about a drug considered for you because your genetic test indicates that you could suffer a severe negative reaction to the medication. However, upon further examination of your test results, your doctor finds that you would benefit greatly from a new drug on the market, and that there would be little likelihood that you would react negatively to it. A day like this will be coming to your doctor's office soon, brought to you by pharmacogenomics (Philips et al., 2000).

Although pharmacogenomics has the potential to radically change the way health care is provided, it is only in its infancy. In the future, pharmacogenomics could find uses along the entire drug discovery and development timeline, all the way from target discovery and validation to late-stage clinical trials (Fig. 26.2) Beyond that, pharmacogenomic tests could find their way into the doctor's office as a means to get the right medicine to the right patient at the right time (Terra and Johnson, 2002).

While genetics and genomics are often used synonymously, pharmacogenetics is more focused in scope than and is viewed as a subset of pharmacogenomics, which encompasses factors beyond those that are inherited. Some people believe that pharmacogenomics will lead to the stratification of diseases into genetically defined categories. For pharmacogenetics to be effective, markers must be found that

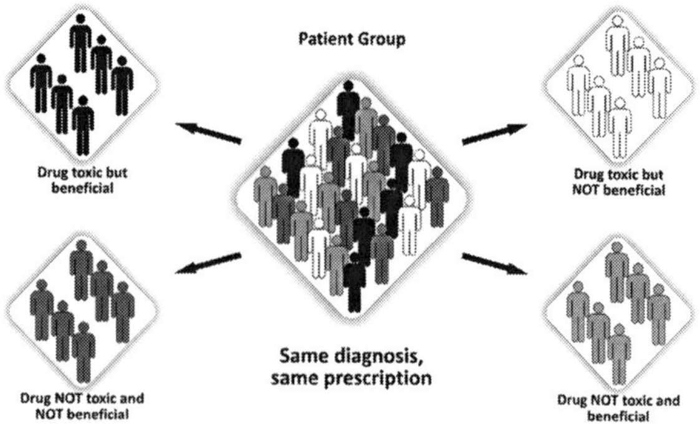

Fig. 26.2 Promises of pharmacogenomics: a new drug for a individualised therapy

indicate the connection between drug response and genetic makeup. The markers that companies are pursuing most diligently are known as single-nucleotide polymorphisms, or SNPs. SNPs, which are defined only in relation to a population, are variations in DNA at a single base that are found in at least 1% of the population (Tolle, 2001).

Thinking that breast cancer markers such as BRCA1 and BRCA2 only provide information about susceptibility. Obviously, breast cancer caused by BRCA1 is a different kind of cancer than cancers not caused by BRCA1 mutations. Pharmacogenomics can be used in many ways during drug discovery and development. For example, it can identify variations in drug targets and ensure that companies are screening against the most common variant. However, pharmacogenomics is most likely to be used during the clinical development process (Krynetski and Evans, 1998). Pharmaceutical companies will be able to use pharmacogenetics to help differentiate their drugs in a crowded marketplace.

Meanwhile, search is on for pharmacogenomic markers for osteoarthritis, which encompasses about half of all the cases of arthritis. In this disease, which affects primarily women older than 45, cartilage in the joints breaks down. One of the reasons drugs haven't been developed is that the progression time is long, so the clinical trials are long. The progression is monitored by taking X-rays of the joints and measuring the thinning of the distance between the bones. Clinical trials are expensive because they require many patients and a long time.

Pharmacogenomics will probably be most successful in areas such as oncology (Shai et al., 2008), where many therapies are available but each one works for only a small percentage of patients. Getting the treatment right the first time can be important, even life-saving. If a patient has cancer and is going to be treated with a drug for an extended period before you find out if the drug has worked, and if you know that a given drug is only going to work in 10% of patients, there's a very strong reason for a physician to do a test to try to get the drug right from day one. Toxicity is another important application of pharmacogenomics. One such test, developed by researchers at St. Jude Children's Research Hospital in Memphis, identifies patients who have mutations in the thiopurine S-methyltransferase gene, which is involved in the metabolism of mercaptopurine drugs. Children who metabolize these drugs too slowly can literally overdose on their treatment.

Technology must also advance if the field is to move forward. A platform that can do large amounts of genotyping inexpensively is required, Right now, the cheapest real way to do genotyping costs about a dollar per SNP. If you want to look at 100,000 SNPs, it's $100,000 a patient, which just isn't practical. You need to get that down to a penny [per SNP]. Then suddenly there are a lot of interesting experiments that you can do that you can't do today. Technology that is inexpensive could be available in as early as three, but no later than 10 years from now, he predicts. In addition to technological improvements, gaps in the knowledge of the human genome need to be filled.

Pharmacogenomics is the study of how drugs are metabolized in the body and the variations in the genes that produce the metabolizing enzymes. By studying the genes that produce the specific enzymes that metabolize a drug that is to be prescribed, a doctor may decide to raise or lower the dose, or even change to a different drug. The decision about which drug to prescribe may also be influenced by other drugs the patient is taking, to avoid drug-drug interactions. Currently, doctors typically prescribe one of several appropriate drugs for their patients. They prescribe a "standard" dose based on factors such as weight, sex, and age, and then adjust the dose over time, depending on whether the patient's condition is responding to the medication and whether the patient is experiencing unpleasant or dangerous side effects. The concentrations of ome drugs are monitored with blood tests and the dosages increased or decreased to maintain the drug level in an established "therapeutic" range. Follow-up of such processes is called "Therapeutic Drug Monitoring." If the drug is not effective in treating or controlling the patient's condition, then the patient is given a different drug and the process is started again.

Instead, pharmacogenomics offers physicians the opportunity to individualize drug therapy for patients based on their genetic make-up. For certain drugs, pharmacogenomics is already helping physicians predetermine dosages to have a better chance of achieving the desired therapeutic effect while reducing the likelihood of adverse effects (Johnson, 2001).

26.11 Impact of Drug Discovery, Development, and Marketing Affected by Pharmacogenomics

In the past, drug companies have focused primarily on blockbuster drugs, drugs that could be marketed to a large portion of the general population. Each of these drugs took hundreds of millions of dollars and years of research and clinical testing before it was cleared by the FDA and released to the market. For every blockbuster, there were numerous drugs that were orphaned (shelved) during the development process. Some of these drugs were only effective for a small percentage of those enrolled in clinical trials; others had too many associated side effects and complications (Blake and Sobel, 2008).

Each of the blockbusters released was also associated with rare but sometimes serious complications, and some complications that were not evident during clinical trials emerged with long term use. When a large number of people began taking these medications, even a tiny percentage of rare complications could become a significant number of affected patients. In some cases, these blockbusters had to be removed from the market and/or their benefits versus their risks weighed carefully by the patients taking them and the doctors prescribing them.

With pharmacogenomics, new drug development could be sped up in a couple of ways. First, the human genome project and advances in the mapping and collecting of specific types of genetic variations like SNPs, haplotypes, and microsatellites, have made it potentially easier for drug companies to identify genes of interest (those in which genetic variation is strongly associated with an identifiable disease) and targets of interest (the functioning of specific enzymes, proteins, or receptors that can be used as drug targets to block, inhibit, replace, or enhance the action of the target and have an effect on the condition).

And second, it can make it easier to identify the population most likely to benefit from the drug being developed. The results of phase I and phase II clinical trials (performed for safety and effectiveness) can be evaluated to determine common factors in patients who responded versus those who did not or who had significant side effects. Some of the orphan (shelved) drugs may be able to be reevaluated and retargeted at those most likely to respond. In the future, drugs and the tests used to determine who would be likely to benefit from them may be developed simultaneously.

Some patients and their advocates have concerns that pharmacogenomic test results might be used to discriminate against patients who have a genetic risk for health problems. Insurance carriers may mandate testing prior to approving drug therapy or require testing of family members. Patient privacy is an issue as with other genetic tests. These and other issues are currently being debated and addressed by members of the medical community, insurers, government agencies, and a variety of national organizations (Lazarou et al., 1998).

Pharmacogenomics is already having an impact in facilities where it is being used to guide drug therapy. As more data are reported, it will become more widely accepted until it is the standard of practice. This will happen on a drug-by-drug basis over the next few years.

Some classes of drugs that interest researchers at present and hold promise for the future are those used for post-transplant immunosuppression and targeted tumor therapy. Immunosuppressive medications given to patients after an organ transplant must remain within a narrow range in the blood to help prevent rejection of the transplant without causing significant side effects or toxicity. Testing patients for their likely response to these medications prior to transplant would help determine the right drug(s) and the appropriate dose.

With cancer, the patient's ability to tolerate a drug and the tumor's response to treatment are both important (Krynetski and Evans, 1998). Pharmacogenomic testing could be performed to determine which drug(s) the patient is likely to be able to metabolize efficiently and at what doses, and the tumor tissue could be tested to determine which treatment options it is likely to respond to. Pharmacogenomics has the potential to aid in the development of drugs that target cancer cells specifically without affecting normal tissues and cells. This could lead to safer and more effective treatment of cancers.

Pharmacogenomics will help physicians and their patients by enabling pharmaceutical companies to bring more drugs to market that are targeted at those who are most likely to benefit from them. Ultimately, the goal of pharmacogenomic testing is to help medicine and pharmacotherapy become less uncertain disciplines and more capable of improving the quality of an individual's life through a more personalized approach to drug therapy. In recognizing that patients respond to drugs differently and in working to individualize drug therapy, pharmacogenomics has become an integral part of personalized medicine (Eichelbaum et al., 2006).

The emergence of pharmacogenetics in the twentieth century followed a path forged by advances in molecular biology and genetics (Terra and Johnson, 2002). Before technology allowed the study of individual genetic variation, the field concentrated on identifying racial and ethnic variations in response to drugs. Most notable among several landmark studies, perhaps, was University of Toronto Professor emeritus Werner Kalow's investigation in the 1950s of the occurrence of prolonged paralysis and rare, unexplained deaths in surgical patients receiving succinylcholine, a neuromuscular blocker tolerated well by most patients. Kalow discovered that a genetically based deficiency in the metabolizing enzyme pseudo-cholinesterase was responsible, and proceeded to describe the population incidence of the various alleles responsible for the deficiency. Similar studies confirmed the genetic basis of the variability, seen in response to a wide variety of drugs.

As the wider fields of genetics and molecular biology progressed, so did pharmacogenetics. Now essentially folded into the burgeoning science of pharmacogenomics, the discoveries that have emerged from the progenitor field are today available and in use in the diagnostic arena, helping to screen patients who fall into broad populations that, due to their metabolic genotypes, should not receive specific drugs.

26.12 Pharmacogenomics Initiatives

The rapid development of pharmacogenomics has led to an encouraging amount of scientific cooperation and collaboration among government, industry, and academia. Large-scale collaborative initiatives are making significant contributions to the effort to eventually bring the benefits of pharmacogenomics to the bedside. One such effort, the Pharmacogenetics Research Network (PGRN), was established in 2000 and currently funds 13 academic research groups conducting basic research describing pharmacogenetic phenotypes and relating them to genetic and genomic information. The PGRN is spearheaded by the NIGMS, with the participation of the National Cancer Institute, the National Heart, Lung, and Blood Institute, the National Human Genome Research Institute, the National Library of Medicine, and the NIEHS, making it a true trans-NIH effort.

26.13 Pharmacogenomics and the Pharmaceutical Industry

The pharmaceutical industry has seen the future. In a 9 April 2003 presentation to a U.S. Food and Drug Administration (FDA) Science Board Advisory Committee meeting, Brian Spear, Director of pharmacogenomics at Abbott Laboratories in Abbott Park, Illinois, put it succinctly: "Pharmacogenomics is not something that a company here and a company there have taken a chance on. This is now a standard part of the drug discovery and development process in every one of the drug discovery and research companies."

The industry's avid pursuit of pharmacogenomics, as evidenced by a recent spate of acquisitions of pharmacogenomically oriented biotechnology firms by the major pharmaceutical companies, runs the gamut from drug discovery to enhancing the safety and efficacy of drugs that have been on the market for many years. Just as the days of the one-drug-fits-all treatment approach may be numbered, so too may be those of the present pharmaceutical industry business model, which relies heavily on the periodic introduction of blockbuster drugs (typically defined as products with annual revenues in excess of $1 billion) to generate profits and fund research and development. The hope is that the application of pharmacogenomics and other genomics technologies will enable a new paradigm to emerge, with lower development costs, fewer candidate drug failures, revitalized existing products, the possible resuscitation of withdrawn drugs, and a "portfolio" approach to the introduction of new agents, with drugs available in different formulations to maximize safety and efficacy in specific phenotypic populations.

According to a report, it currently takes on average $880 million and 15 years to bring a new drug to market. Failed candidate compounds represent a large proportion of that development cost. Right now, only about 10% of compounds that enter clinical development make it to the marketplace. It was estimated that the effective application of genomics technologies could reduce that staggering investment by as much as $300 million and two years.

Many are optimistic that pharmacogenomics can play a major role in increasing drug development productivity. The real cost savings will be that we have fewer failures going into development.

Some anticipate that pharmacogenomics will contribute to all pharmaceutical products, from new candidates to old warhorses. Many pharmaceutical companies are already using pharmacogenomics to screen participants in clinical trials. The FDA appears to be solidly on board the pharmacogenomics bandwagon as well. The agency has reportedly been working hard recently to acquire the necessary in-house expertise to facilitate the submission, interpretation, and implementation of pharmacogenomics data.

26.14 Applied Pharmacogenetics

Pharmacogenomics will undoubtedly be used to improve future health care and clinical research in different ways (Roden et al., 2006). Whereas treatment allocation has been based mainly on phenotype, genetic characterization will help researchers to identify suitable subjects for clinical trials, to facilitate interpretation of the results of clinical trials, and to identify novel targets for future drugs or new markets for current products. As interindividual variability in drug response is a substantial clinical problem, the second major objective of pharmacogenomic research is to decrease adverse responses to therapy through determination of adequate therapeutic targets and genetic polymorphisms that alter drug specificity and toxicity. Most of the promise of pharmacogenomics remains to be fulfilled. However, the concept of using known genetic associations to prevent patients from taking drugs that would likely be ineffective or harmful is already available and used in clinical practice in certain specific arenas (Phillips et al., 2001), thanks mainly to the steady progress made in pharmacogenetics over the past several decades (Cariou et al., 2002).

Cancer therapy today includes two shining examples of applied pharmacogenetics. First, there is now a commercially available diagnostic test measuring a patient's ability to produce the metabolic enzyme thiopurine S-methyltransferase (TPMT), which is essential for the metabolism of thiopurine medications used to treat acute lymphoblastic leukemia (ALL), the most common form of childhood cancer. Genetic testing gives clinicians the ability to classify ALL patients according to their TPMT genotype, which allows optimized dosing. Doses in patients with alleles rendering them deficient in TPMT (who are thus less tolerant of thiopurine medications) are reduced by as much as 95%. This means TPMT-deficient patients can tolerate the drug, yet enough is still metabolized to retain efficacy.

Second, the breast cancer drug trastuzumab (trade name Herceptin), which is marketed in tandem with a diagnostic test, is often cited as an early indicator of the value of the concept. Trastuzumab is effective only sin the 25–30% of breast cancer patients whose tumors overexpress the human epidermal growth factor receptor (HER2) protein. The drug was developed specifically to exploit that characteristic; it binds to HER2, which slows tumor growth. The diagnostic test measures HER2 expression in the tumor and is thus predictive of the potential efficacy of the drug; patients who do not overexpress HER2 are not given the drug, because it will not work.

This is a unique combination today. But such diagnostic-agent pairings will become more commonplace as pharmacogenomics progresses and strides are made in disease genetics, in which a variety of diseases (particularly cancer) are being genetically subclassified, often significantly redefining treatment strategies.

An intermediate step toward such pairings is illustrated by work being done with the cytochrome P450 (CYP450) family of enzymes, which is responsible for a large segment of human drug metabolism. It is the metabolic pathway of choice for about 60% of the drugs on the market today. It has also been the focus of a great deal of research attention through the years, and the numerous CYP450 subtypes are well characterized, as are the important phenotypes of variation in response. Several companies now offer CYP450 genotyping tests to the pharmaceutical industry for clinical trial subject inclusion/exclusion based upon metabolic profile, and now such tests are making their way into the clinical diagnostic marketplace. Gentris, for example, soon expects to market five kits to physicians for pharmacogenetic testing of their patients. Genelex Corporation of Seattle, Washington, has taken the concept one step further, marketing tests directly to the public for three of the major CYP450 pathways–CYP2D6, CYP2C9, and CYP2C19. Once a consumer has placed an order for the test, Genelex sends them a blood collection kit, and the consumer either sees their own doctor or Genelex will refer them to a phlebotomist in their area.

26.15 Ethical Concerns

On one level, the ethical issues involved with pharmacogenomics are similar to those raised by genomics.In general, broad concerns about research integrity, privacy, confidentiality, informed consent, the specter of genetic discrimination or stigmatization, and access to information or to specialized care (Moldrup, 2001) are some of the chief concerns.

Many observers, however, feel that the ethical issues specific to pharmacogenomics are actually somewhat less problematic than the ethical lightning rods attached to genetics or even to pharmacogenetics. One crucial difference lies in the nature of the information itself, which typically applies to individual patients as opposed to larger groups. Society, industry, groups and individuals appreciate the prospect of pharmacogenomics very differently. Secondly, there is a lack of research into the post-marketing implications of pharmacogenomics.

An extensive focus on the ethical, social and legal implications of pharmacogenomics, in terms of both pre- as well as post-marketing issues, is essential. Also, a multidisciplinary approach which includes individual and group opinions in an upfront manner in the research and development process is essential. Otherwise, there is a substantial risk that the positive prospects of pharmacogenomics will not survive due to fear and a lack of acceptance and understanding on the part of the general public. The ethical issues connected to pharmacogenomics are perhaps less urgent. The reason is that the core of pharmacogenomics is pharmaceutical intervention (Bellver, 2002). The fact that we're talking about genomics that a 'pharmaco-' attached to it means that the purpose is treatment, not primarily diagnosis of the disease.

Today, pharmacogenomics is still predominantly a research endeavor, and the nature of informed consent is presently the most prominent ethical consideration for investigators. Achieving the appropriate level of informed consent could prove to be a major challenge, as participants will need to comprehend that their DNA could be used for multiple experiments over a long period of time, and that in many cases, their samples will not be anonymous. In many pharmacogenomics studies, anonymization of samples defeats the purpose of drawing associations between drug response and populations.

On the other hand, participants' access to the information discovered about their genomes is also a thorny issue for researchers. Obligations need to be negotiated in advance during the informed-consent process, and must be in place in situations when an individual is found to have a particular abnormality in his or her response to certain drugs, or perhaps an increased risk of disease due to a genetic susceptibility.

26.16 Molecular Diagnostic Methods for Optimizing Drug Therapy

The potential is enormous for pharmacogenomics to yield a powerful set of molecular diagnostic methods that will become routine tools with which clinicians will select medications and drug doses for individual patients. A patient's genotype needs to be determined only once for any given gene, because except for rare somatic mutations, it does not change. Genotyping methods are improving so rapidly that it will soon be simple to test for thousands of single-nucleotide polymorphisms in one assay. It may be possible to collect a single blood sample from a patient, submit a small aliquot for analysis of a panel of genotypes (e.g., 20,000 single-nucleotide polymorphisms in 5000 genes), and test for those that are important determinants of drug disposition and effects. In our opinion, genotyping results will be of greatest clinical value if they are reported and interpreted according to the patient's diagnosis and recommended treatment options (Evans and Relling, 1999).

26.17 Challenges for the Future

There are a number of critical issues that must be considered as strategies are developed to elucidate the inherited determinants of drug effects. A formidable one is that the inherited component of the response to drugs is often polygenic. Approaches for elucidating polygenic determinants of drug response include the use of anonymous single-nucleotide polymorphism maps to perform genome-wide searches for polymorphisms associated with drug effects, and candidate-gene strategies based on existing knowledge of a medication's mechanisms of action and pathways of metabolism and disposition. Both these strategies have potential value and limitations. However, the candidate-gene strategy has the advantage of focusing resources on a manageable number of genes and polymorphisms that are likely to be important, and it has produced encouraging results in a number of studies. The limitations of this approach are the incompleteness of knowledge of a medication's pharmacokinetics and mechanisms of action. Gene-expression profiling and proteomic studies are evolving strategies for identifying genes that may influence drug response. Pharmacogenomics offers the hope of using the genetic profiling to personalized and thereby improve medical treatment of a variety of a diseases, including hypertension, depression, cancer, and HIV.

One of the most important challenges in defining pharmacogenetic traits is the need for well-characterized patients who have been uniformly treated and systematically evaluated to make it possible to quantitate drug response objectively (Sadee, 2002, 1999). To this end, the norm should be to obtain genomic DNA from all patients enrolled in clinical drug trials, along with appropriate consent to permit pharmacogenetic studies. Because of marked population heterogeneity, a specific genotype may be important in determining the effects of a medication for one population or disease but not for another; therefore, pharmacogenomic relations must be validated for each therapeutic indication and in different racial and ethnic groups. Remaining cognizant of these caveats will help ensure accurate elucidation of genetic determinants of drug response and facilitate the translation of pharmacogenomics into widespread clinical practice.

Much is needed before the potential benefits of pharmacogenomics can be fully realized. Identification and analysis of millions of SNPs that might influence the response to a specific drug or class of drugs is complex. Even after SNPs of interest have been catalogued, drug companies may limit drug development to agents targeting treatment of those comprising large groups. Development of drugs anticipated to be effective in only small cohorts may lie fallow. Improved DNA microarray and related technologies are needed to expedite SNP screening in a timely and cost effective manner. Extensive training of physicians will be required. Diverse ethical and legal issues including ownership of rights to genetic information, privacy, and insurance implications require resolution

The outcome of drug therapy is often unpredictable, ranging from beneficial effects to lack of efficacy to serious adverse effects (Dan et al., 2006). How quickly pharmacogenomics develops into a useful clinical toolkit will depend greatly on how policymakers respond to policy issues associated with both genetic testing and targeted drug development, including concerns about safety, access, cost, and potential ethical and social implications. Overcoming these challenges holds the promise of improving new drug development and ultimately individualizing the selection of appropriate drugs and dosages for individual patients. Finally, pharmacogenomics is likely to transform the way clinical trials are conducted by allowing for the selection of a more homogeneous study population, thereby reducing the size and cost of clinical investigation.

References

Bellver Capella, V., 2002, Ethics, politics, and rights in pharmacogenomics. *Law Hum Genome Rev* 17:31–55.

Blake, C.A., Sobel, B.E., 2008, Progress in pharmacogenomics and its promise for medicine. *Exp Biol Med* 233:1482–1483.

Cariou, A., Chiche, J.D., Charpentier, J., Dhainaut, J.F., Mira, J.P., 2002, The era of genomics: impact on sepsis clinical trial design. *Crit Care Med* 5:341–348.

Dan, M., Roden, M.D., Russ, B., et al., 2006, Pharmacogenomics: challenges and Opportunities. *Annals of Internal Medicine* 145(10):749–757.

Eichelbaum, M., Ingelman-Sundberg, M., Evans, W.E., 2006, Pharmacogenomics and individualized drug therapy. *Annu Rev Med* 57:119–137.

Evans, W.E., Relling, M.V., 1999, Pharmacogenomics: translating functional genomics into rational therapeutics. *Science* 286:487–491.

Gupta, M., Kaur, H., Grover, S., Kukreti, R., 2008, Pharmacogenomics and treatment for dementia induced by Alzheimer's disease. *Pharmacogenomics* 9:895–903.

Johnson, J.A., 2001, Drug target pharmacogenomics: an overview. *Am J Pharmacogenomics* 1:271–281.

Johnson, J.A., Evans, W.E., 2001, Pharmacogenomics: the inherited basis for inter-individual differences in drug response. *Annu Rev Genomics Hum Genet* 2:9–39.

Johnson, J.A., Evans, W.E., 2002, Molecular diagnostics as a predictive tool: genetics of drug efficacy and toxicity. *Trends Mol Med* 8:300–305.

Kirchheiner, J., Roots, I., Goldammer, M., Rosenkranz, B., Brockmöller, J., 2005, Effect of genetic polymorphisms in cytochrome p450 (CYP) 2C9 and CYP2C8 on the pharmacokinetics of oral antidiabetic drugs: clinical relevance. *Clin Pharmacokinet* 44:1209–1225.

Kirchheiner, J., Seeringer, A., 2007, Clinical implications of pharmacogenetics of cytochrome P450 drug metabolizing enzymes. *Biochim Biophys Acta* 1770:489–494.

Kirchheiner, J., Seeringer, A., Brockmöller, J., 2006, State of the art of pharmacogenetic diagnostics in drug therapy. *Bundesgesundheitsblatt Gesundheitsforschung Gesundheitsschutz* 49:995–1003.

Krynetski, E.Y., Evans, W.E., 1998, Pharmacogenetics of cancer therapy: getting personal. *Am J Hum Genet* 63:11–16.

Lazarou, J., Pomeranz, B.H., Corey, P.N., 1998, Incidence of adverse drug reactions in hospitalized patients. *JAMA* 279:1200–1205.

Lee, S.S., 2009, Pharmacogenomics and the challenge of health disparities. *Public Health Genomics* 2:170–179.

López-López, M., Guerrero-Camacho, J.L., Familiar-López, I.M., 2004, Pharmacogenomics: the quest for individualized therapy. *Rev Neurol* 39:1063–1071.
Mancinelli, L., Cronin, M., Sadee, W., 2000, Pharmacogenomics: The promise of personalized medicine. *AAPS Pharm Sci* 2:4.
Moldrup, C., 2001, Ethical, social and legal implications of pharmacogenomics: a critical review. *Community Genet* 4:204–214.
Phillips, K.A., Veenstra, D.L., Oren, E., Lee, J.K., Sadee, W., 2001, The potential role of pharmacogenomics in reducing adverse drug reactions: A systematic review. *JAMA* 286:2270–2279.
Phillips, K.A., Veenstra, D., Sadee, W., 2000, Implications of the genetic revolution for health services research: Pharmacogenomics and improvements in drug therapy. *Health Serv Res* 35:128–140.
Roden, D.M., Altman, R.B., Benowitz, N.L., Flockhart, D.A., Giacomini, K.M., Johnson, J.A., Krauss, R.M., McLeod, H.L., Ratain, M.J., Relling, M.V., Ring, H.Z., Shuldiner, A.R., Weinshilboum, R.M., Weiss, S.T. Pharmacogenetics Research Network, 2006, Pharmacogenomics: challenges and opportunities. *Ann Intern Med* 145:749–757.
Sadee, W., 1999, Pharmacogenomics. *Brit Med J* 319:1286.
Sadee, W., 2002, Pharmacogenomics: The Implementation Phase. *AAPS Pharm Sci* 4:611–614.
Shai, R.M., Reichardt, J.K., Chen, T.C., 2008, Pharmacogenomics of brain cancer and personalized medicine in malignant gliomas. *Future Oncol* 4:525–534.
Shi, J., Chen, S., 2002, Pharmacogenomics genomics approaches to optimizing drug therapy. *Zhonghua Yi Xue Yi Chuan Xue Za Zhi* 19:156–158.
Terra, S.G., Johnson, J.A., 2002, Pharmacogenetics, pharmacogenomics, and cardiovascular therapeutics: the way forward. *Am J Cardiovasc Drugs* 2:287–296.
Tolle, R., 2001, Information technology tools for efficient SNP studies. *Am J Pharmacogenomics* 1:303–314.

Chapter 27
Gene Technology in Forensic Sciences

Abstract Recombinant DNA technology can provide novel and powerful methods for forensic science application. Human genomic DNA can be analyzed directly for individual identification and paternity testing on the basis of polymorphism in its sequence. Restriction fragment length polymorphism (RFLP) testing, STR (microsatellite) and mitochondrial DNA analysis (mtDNA) is suitable for examination of the forensic biological samples (bloodstains, hairs, seminal stains, bones, tooth). Using a combination of single locus probe (SLP) that varies highly among individuals, a DNA fingerprint or profile can be made. Mitochondrial DNA RFLPs may also suggest the characteristics of the human races.

Keywords Forensic · Genetic markers · DNA typing · RFLP · STR · Gene technology · DNA profiling · RFLP · STR · Fingerprint · PCR based methods · Polymorphism · Amplification

27.1 Prologue

Forensic science has witnessed dramatic changes in the field of human identification. DNA can be used to identify the perpetrator of a murder (Linacre and Graham, 2002). The forensic potential of DNA profiling for human identification burst upon the scene in late 1985. Since then, forensic DNA testing has gone through several phases of technological and operational advance resulting in more rapid analysis, extension of testing to ever smaller biological samples, and a de facto international standardization based on use of commercial kits. The technology has altered by adopting novel methods developed originally for use in the field of medical genetics. There can be few areas of endeavour in molecular genetics that have captured the imagination and interest of the public as much as DNA typing. DNA typing is renowned as a technique that is used in forensic investigations to match criminals against samples obtained from crime scenes, to identify individuals from their remains and to determine paternity. It is also an excellent example of a new technology based on sound scientific principles and on the successful application of those principles to a commercial field.

M. Debnath et al., *Molecular Diagnostics: Promises and Possibilities*,
DOI 10.1007/978-90-481-3261-4_27,

27.2 Concept

Like many applications of molecular diagnostics, the field of forensic biology is undergoing a phase of expansion and diversification. Forensic science is the application of a broad spectrum of sciences to answer questions of interest to the legal system. Among the branches of Forensic sciences, criminalistics and Forensic DNA analysis rely on the gene technology using DNA profiling technique. Criminalistics is the application of various sciences to answer questions relating to examination and comparison of biological evidence, trace evidence, impression evidence (such as fingerprints, footwear impressions, and tyre tracks), controlled substances, ballistics, firearm and toolmark examination, and other evidence in criminal investigations. Typically, evidence is processed in a crime lab. Forensic DNA analysis takes advantage of the uniqueness of an individual's DNA to answer forensic questions such as determining paternity/maternity or placing a suspect at a crime scene.

In the last few years, DNA typing procedures have become increasingly important in the fields of forensic science and forensic medicine. The growth of forensic DNA databases and adoption of sophisticated analytical methods have catalyzed this increasing role. The range of molecular markers exploited in the fight against crime is beginning to increase too, and genes implying personal or physical characteristics are emerging in the research literature. DNA profiling has greatly changed the way in which human identification is performed for the purpose of a forensic investigation.

DNA typing is probably one of the most important advances in the forensic sciences in recent years. While most DNA is the same from one person to another, several sections can show marked variability in the sequence of the monomers. The DNA sequence of a person is therefore individual and can be shown to be theirs beyond reasonable doubt. The first application of DNA typing to forensic science was based on restriction fragment length polymorphisms (RFLP). More recent methods of DNA typing are based on the technique of polymerase chain reaction (PCR). This allows the amplification of minute amounts of DNA (as small as 1 ng of material) giving higher sensitivity for DNA typing. DNA typing is now possible from samples as small as a few cells on the end of a hair that has been pulled out.

Genetic characterization of individuals at the DNA level enables identity testing from a minimal amount of biological specimen in cases of sexual assault, homicide, and unknown human remains. Also paternity testing is changing from level of gene products to the genomic level (Sajantila and Budowle, 1991). Currently, millions of samples from blood, semen, hair and tissues are analyzed to determine the origin of the samples.

27.3 DNA Typing and Genetic Markers

Since the introduction of DNA polymorphism analysis techniques to forensic science, forensic identification research has made radical, astonishing progress at a rate that has already rendered the initial methodologies obsolete. The ability to extract and type DNA from forensic samples has revolutionized the field of forensic serology. DNA extraction now can be quickly and efficiently performed by various kinds of commercially available kits. DNA can also be purified by the potassium iodine method (Shiono, 1996) or phenol extraction method. The analysis of DNA isolated from forensic biological evidence provides valuable information relating to the identification of the source of the sample (Misawa, 1994). The advent of PCR has enabled the use of relatively crude and minute DNA as amplification templates while many kinds of new detection methods for analyzing the amplified products have also been developed. Organism can be identified by examination of DNA sequences unique to that species. In criminal cases, this generally involves obtaining samples from crime-scene evidence and a suspect, extracting the DNA, and analyzing it for the presence of a set of specific DNA regions called genetic markers. To identify individuals, forensic scientists scan 13 DNA regions that vary from person to person and use the data to create a DNA profile of that individual (sometimes called a DNA fingerprint). There is an extremely small chance that another person has the same DNA profile for a particular set of regions (Fig. 27.1)

Scientists find the markers in a DNA sample by designing small pieces of DNA or probes that will each seek out and bind to a complementary DNA sequence in the sample. A series of probes bound to a DNA sample creates a distinctive pattern for an individual. Forensic scientists compare these DNA profiles to determine whether the suspect's sample matches the evidence sample. A marker by itself usually is not unique to an individual; if, however, two DNA samples are alike at four or five regions, odds are great that the samples are from the same person.

If the sample profiles don't match, the person did not contribute the DNA at the crime scene. If the patterns match, the suspect may have contributed the evidence sample. While there is a chance that someone else has the same DNA profile for a particular probe set, the odds are exceedingly slim. Experts point out that using DNA forensic technology is far superior to eyewitness accounts, where the odds for correct identification are about 50:50.

The more probes used in DNA analysis, the greater the odds for a unique pattern and against a coincidental match, but each additional probe adds greatly to the time and expense of testing. Four to six probes are recommended to detect polymorphism. Since the introduction of DNA polymorphism analysis techniques to forensic science, forensic identification research has made radical, astonishing progress at a rate that has already rendered the initial methodologies introduced fifteen years ago obsolete (Fukushima, 1999). Although many minisatellites such as MCT118, YNZ22, COL2A1, and ApoB were highlighted at the beginning of 1980s, none of these loci, with the exception of MCT118, have proved useful for forensic DNA application due to their low amplification efficiency. On the other hand, STR loci containing four base pair repeat sequences have been used routinely for human identification since the mid-1990s. In the near future, the highly efficient STR should be selected as a consensus core marker in Japan. STR systems located on the Y chromosome are widely used in forensic science for the identification of male individuals. These systems have a special significance in forensic science cases where mixtures of male and female DNA are analyzed, as happens in cases of rape or

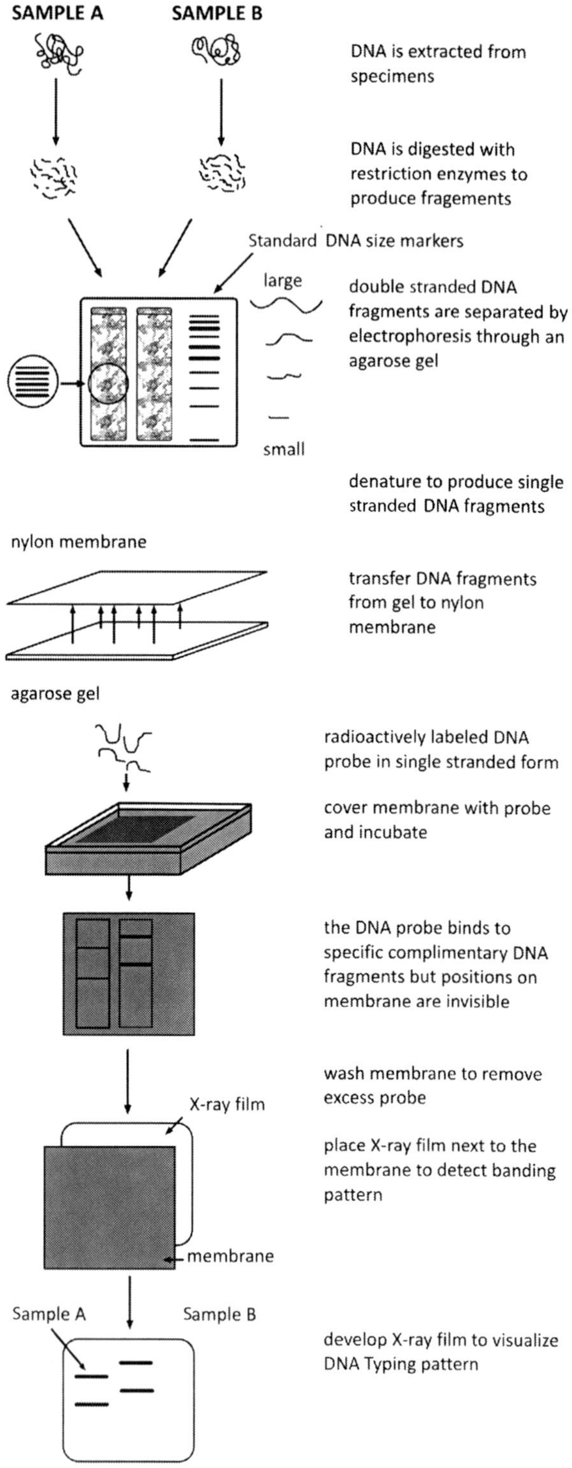

Fig 27.1 Method of DNA fingerprinting

other sexual crimes. The characteristics of high copy number, maternal inheritance, and high degree of sequence variability make mtDNA a powerful tool for forensic identification. Most of the variations in mtDNA among individuals are found within the displacement loop (D-loop). In all population groups, mtDNA sequences can be useful for discriminating among unrelated individuals. Now it is necessary to get as much as possible individual genetic information as quickly as possible in order to enable individual identification.

27.4 DNA Based Methods for Identification of Individuals

Previously, genetic marker typing was limited to the analysis of blood group markers (Sasaki and Shiono, 1996) and soluble polymorphic protein markers. Because the number of suitable markers expressed in particular fluids and tissues is relatively small, and because mixtures of fluids cannot be separated for conventional genetic marker typing, a suspect frequently cannot be included or excluded as a fluid donor in a case. However, the development of methods to extract DNA from virtually all biological specimens has greatly expanded the potential for individual identification. Of particular importance was the ability to extract mixtures of sperm cells and epithelial cells found in sexual assault cases such that the DNA from the sperm cells could be typed independently of the DNA from the victim's epithelial cells.

27.4.1 Non-PCR Based Methods

Restriction fragment length polymorphism (RFLP) analysis was the first DNA-based method applied to problems of individual identification (Reynolds et al., 1999). It can arise when an individual may have an additional recognition site for a particular restriction endonuclease in a given area of the genome or, an individual may be missing a recognition site for a particular restriction endonuclease in a given area of the genome. This method, while powerful in its ability to differentiate individuals, is limited by the quantity and quality of DNA required for an unambiguous result and by the amount of time it takes to obtain a result. Despite these limitations, several laboratories are using RFLP analysis successfully for the detection of polymorphisms in forensic DNA case samples.

Although either of the above situations can lead to polymorphic differences in the population, many of the specific loci that are used in criminal forensic cases and in paternity cases involve repetitions of a core sequence at a given genetic locus. The length of each repeat unit and the number of repeat units varies. Therefore, the length of the fragment generated by a restriction enzyme that cuts on either side of (but not within) the locus also varies. These types of loci are referred to as Variable Number of Tandem Repeats loci (VNTRs).

27.4.2 PCR Based Methods

While the field of forensic serology was being revolutionized by the prospect of DNA analysis, the field of molecular biology was being revolutionized by the invention of the polymerase chain reaction (PCR), which ultimately has an impact on every area of biological science (Lee et al., 1994). The PCR DNA amplification technology is ideally suited for the analysis of forensic DNA samples in that it is sensitive and rapid and not as limited by the quality of DNA as the RFLP method.

a. Ampli Type PM & DQA1 – Sequence Polymorphisms. The Commercial availability of the AmpliType PM & DQA1 system from Applied Biosystems, INC (ABI; formerly Perkin-Elmer) was the first PCR based DNA Typing systems. The loci involved in this system are all sequence polymorphisms that are detected/ delineated by hybridization to allele-specific oligonucleotide (ASO) probes. The discriminatory power of this system is much more limited than RFLP analysis (approximately 1:2000) because only a limited number of alleles (and therefore, a limited number of genotypes) exist in the population. Nevertheless, the system is useful when evidence samples yield limited amounts of DNA and/or DNA that is too degraded for RFLP analysis.

The AmpliType PM & DQA1 system tests for polymorphisms at 6 loci namely, low Density Lipoprotein Receptor (LDLR), Glycophorin A (GLYPA), Hemoglobin G Gamma Globulin (HBGG), D7S8, Group-Specific Component (GC), HLA-DQA1 (Comey et al., 1993).

The basic steps in the procedure are:

1. Simultaneous PCR amplification of the six loci using primers that have been conjugated to biotin.
2. Hybridization (allele-specific) of the PCR products using a reverse dot-blot method in which the ASO probes have been impregnated onto membrane filter strips in specific locations; two strips are used: one for the DQA1 locus and one for the remaining 5 loci of the Polymarker (PM).
3. Binding of a horseradish peroxidase/streptavidin conjugate to the hybridized biotinylated PCR product.
4. Colorimetric detection of the hybridized PCR product-biotin/HRP-streptavidin complex using a chromogenic substrate for the HRP.

Because of its low Power of Discrimination (Pd) this system is seldom used anymore. However, it is still commercially available and used occasionally.

b. D1S80 – Amplified Fragment Length Polymorphism (AFLP). VNTRs, such as described above for RFLP analysis, can also be detected by Polymerase Chain Reaction (PCR) Amplification. Primers are designed to anneal to sequences in the DNA that flank the VNTR. PCR Amplification of the material between the primer locations includes the VNTR region. Therefore, any allelic differences between individuals will be evidenced by different lengths for the amplification product among individuals tested. The D1S80 locus has been used this way. It has a core repeat unit of 16 bp, which is repeated 14–41 times.

The basic process involves:

1. Processing appropriate samples to obtain the DNA.
2. PCR Amplification of a region of the DNA containing a polymorphic VNTR (i.e. D1S80).
3. Separating the resulting DNA amplification products by electrophoresis.
4. Analysis of the resulting pattern by UV transillumination of Ethidium bromide stained gels or by Silver Staining to determine inclusion or exclusion.

c. Short Tandem Repeats (STRs; Microsatellites). Short Tandem Repeat sequences (STRs) are similar to VNTRs in that they involve tandem repeats of a core sequence in variable numbers among the population to produce a polymorphic distribution. The major difference is that the core sequence is usually only 3 or 4 nucleotides in length (VNTR core sequences can be 16 or more nucleotides). Usually a tighter range of allele's results. The small size of the STRs used in forensic DNA profiling (amplimers range from 100 to 500 bp) allows for more efficient amplification by PCR and also allows the use of DNA that has been degraded more significantly because even small pieces of DNA may contain intact STR sites. Primers are designed to anneal to sequences in the DNA that flank the STR. PCR Amplification of the material between the primer locations includes the STR region. Therefore, any allelic differences between individuals will be evidenced by different lengths for the amplification product among individuals tested (Divne and Allen, 2005).

d. Multiplex STRs. A number of STR loci have been identified with non-overlapping allelic distributions. This allows the use of multiplex PCR amplification and separation of multiple loci in a single lane of a polyacrylamide gel. The use of different fluorescent dyes allows the multiplex amplification of STR loci that have overlapping allele ranges as long as a separate dye is used for each overlapping STR. The dyes are coupled to the primers used for the amplification so that the different loci can be readily identified.

e. Mitochondrial D-Loop DNA (a sequence polymorphism). Recently, it has been recognized that information in the mitochondrial DNA (mtDNA) coding region can provide additional forensic discrimination with respect to the standard typing of the D-loop region, increasing the forensic power of mtDNA testing, which is sometimes rather limited(Wu et al., 2008). The characteristics of high copy number, maternal inheritance, and high degree of sequence variability make mtDNA a powerful tool for forensic identification (Fukushima, 1999). Two particular sequences within this region tend to mutate with extremely high frequency and thus allow for highly variable sequence polymorphisms among unrelated individuals. The PCR method using mitochondrial DNA analysis (mtDNA) is suitable for examination of the forensic biological samples (bloodstains, hairs, seminal stains, bones, tooth). It is also useful for tracking new variations arising by mutation. They are present in 100s–1000s of copies per cell. Due to their small size and circular form, it is resistant to degradation. So they can also be found in old samples (e.g. bones that are found years after the crime was committed), in dead cells such as hair shafts, bones and teeth. Mitochondrial DNA RFLPs and 9-bp deletion type of mtDNA may suggest the characteristics of the human races (Misawa, 1994)

f. Y-Chromosome STRs. The Y chromosome is inherited in a direct patrilinear manner. Since it is only found in males it can be useful in a variety of situations where a mixture of male and female donors could obscure the results. For sex identification of bloodstains, bleached skeletons and teeth, Southern blot hybridization with Y-chromosome specific probe (pHY10; 3.4 kb) can be an appropriate method (Misawa, 1994). This method is also used for tracking patrilineal descent, hemizygous individuals. It can also be used on mixed samples regardless of the relative proportions of male and female material (vaginal smears; fingernail scrapings; fingernail clippings, etc.).

g. SNP. The range of molecular markers exploited in the fight against crime is beginning to increase too, and genes implying personal or physical characteristics are emerging in the research literature. Of a number of DNA marker typing techniques for personal identification in the field of forensic medicine, polymorphic short tandem repeat (STR) typing is currently the most frequently used technique. However, the multiplex STR method is time consuming. In contrast, single nucleotide polymorphism (SNP) detection methods are relatively rapid and amenable to high throughput (Hu and Wu, 2001). Single nucleotide polymorphisms (SNPs) are emerging as new markers of interest to the forensic community because of their abundance in the human genome, their low mutation rate, the opportunity they present of analyzing smaller fragments of deoxyribonucleic acid (DNA) than with short tandem repeats. This method is also important in degraded DNA samples and the possibility of automating the analysis with high-throughput technologies (Sobrino and Carracedo, 2005).

The discrimination power of each SNP is inferior to that of an STR, but a combination of many SNPs could realize a high discriminating power (Hiratsuka et al., 2005). These have a particular advantage in the analysis of degraded or poor samples, which are often all that is available in forensics (Petkovski et al., 2005).

h. DNA Forensics Databases. Like many applications of molecular diagnostics, the field of forensic biology is undergoing a phase of expansion and diversification. The growth of forensic DNA databases and adoption of sophisticated analytical methods have catalyzed this increasing role (Walsh, 2004).

The Combined DNA Index System, CODIS, the national DNA databank, blends computer and DNA technologies into a tool for fighting violent crime. The current version of CODIS uses two indexes to generate investigative leads in crimes where biological evidence is recovered from the crime scene. The Convicted Offender Index contains DNA profiles of individuals convicted of felony sex offenses (and other violent crimes). The Forensic Index contains DNA profiles developed from crime scene evidence. All DNA profiles stored in CODIS are generated using STR (short tandem repeat) analysis.

CODIS utilize computer software to automatically search its two indexes for matching DNA profiles. Law enforcement agencies at federal, state, and local levels take DNA from biological evidence (e.g., blood and saliva) gathered in crimes that have no suspect and compare it to the DNA in the profiles stored in the CODIS systems. If a match is made between a sample and a stored profile, CODIS can identify the perpetrator. This technology is authorized by the DNA Identification Act of 1994.

As more offender DNA samples are collected and law enforcement officers become better trained and equipped to collect DNA samples at crime scenes, the backlog of samples awaiting testing throughout the criminal justice system is increasing dramatically.

27.5 Ethical, Legal, and Social Concerns About DNA Databanking

The primary concern is privacy. DNA profiles are different from fingerprints, which are useful only for identification. DNA can provide insights into many intimate aspects of people and their families including susceptibility to particular diseases, legitimacy of birth, and perhaps predispositions to certain behaviors and sexual orientation. This information increases the potential for genetic discrimination by government, insurers, employers, schools, banks, and others.

Collected samples are stored, and many state laws do not require the destruction of a DNA record or sample after a conviction has been overturned. So there is a chance that a person's entire genome may be available regardless of whether they were convicted or not. Although the DNA used is considered "junk DNA", single tandem repeated DNA bases (STRs), which are not known to code for proteins, in the future this information may be found to reveal personal information such as susceptibilities to disease and certain behaviors.

Practicality is a concern for DNA sampling and storage. An enormous backlog of over half a million DNA samples waits to be entered into the CODIS system. The statute of limitations has expired in many cases in which the evidence would have been useful for conviction.

27.6 Future of DNA Typing Systems

The major problem for the foreseeable future is not increasing the Power of Discrimination but rather increasing throughput so that more samples can be processed in a reasonable period of time to ensure speedy trials. New instruments and new methodologies are being developed with this goal in mind. Adoption of new methodologies in the forensic community can be slow because of the need for extensive validation and the development of QA/QC protocols.

- Capillary Array Electrophoresis – multiple capillary electrophoresis runs simultaneously (e.g. ABI PRISM 3700 runs 96 samples simultaneously)

- MALDI-TOF Mass Spectrometry (Matrix Assisted Laser Desorption-Ionization-Time of Flight) – analysis times on the order of seconds
- STR Determination by Hybridization Arrays – silicon microchip hybridization – size sensitive results without electrophoretic separation
- SNP Determination by Hybridization Arrays and other techniques for SNP detection
- Laboratory automation – liquid handling robots
- Computerized sample tracking programs
- Computerized STR Interpretation programs
- Portable systems that can be used at the crime scene – probably microchip systems.

However, the operational context of forensic biology is unlike many other fields of science. Harmonizing technological breakthroughs with the requirements of law enforcement agencies and the complexities of the legal system is an added challenge and one which evokes ongoing debate. A new era in which forensic identification can be performed using microarray technology can be visualized in future.

References

Comey, C.T., Budowle, B., Adams, D.E., Baumstark, A.L., Lindsey, J.A., Presley, L.A., 1993, PCR amplification and typing of the HLA DQ alpha gene in forensic samples. *J Forensic Sci* 38:239–249.

Divne, A.M., Allen, M., 2005, A DNA microarray system for forensic SNP analysis. *Forensic Sci Int* 154:111–121.

Fukushima, H., 1999, Forensic DNA analysis – past and future. *Nihon Hoigaku Zasshi* 53: 276–284.

Hiratsuka, M., Tsukamoto, N., Konno, Y., Nata, M., Hashiyada, M., Funayama, M., Mizugaki, M., 2005, Forensic assessment of 16 single nucleotide polymorphisms analyzed by hybridization probe assay. *Tohoku J Exp Med* 207:255–261.

Hu, R., Wu, X.Y., 2001, The single nucleotide polymorphisms and its application to forensic medicine. *Fa Yi Xue Za Zhi* 17:249–251.

Lee, H.C., Ladd, C., Bourke, M.T., Pagliaro, E., Tirnady, F., 1994, DNA typing in forensic science. I. Theory and background. *Am J Forensic Med Pathol* 15:269–282.

Linacre, A., Graham, D., 2002, Role of molecular diagnostics in forensic science. *Expert Rev Mol Diagn* 2:346–353.

Misawa, S., 1994, Application of DNA polymorphism to forensic medicine. *Rinsho Byori* 42: 636–642.

Petkovski, E., Keyser-Tracqui, C., Hienne, R., Ludes, B., 2005, SNPs and MALDI-TOF MS: tools for DNA typing in forensic paternity testing and anthropology. *J Forensic Sci* 50:535–541.

Reynolds, R., Sensabaugh, G., Blake, E., 1999, Analysis of genetic markers in forensic DNA samples using the polymerase chain reaction. *Anal Chem* 63:2–15.

Sajantila, A., Budowle, B., 1991, Identification of individuals with DNA testing. *Ann Med* 23: 637–642.

Sasaki, M., Shiono, H., 1996, ABO genotyping of suspects from sperm DNA isolated from postcoital samples in sex crimes. *J Forensic Sci* 41:275–278.

Shiono, H., 1996, Personal identification using DNA polymorphism–the identification of forensic biological materials. *Nihon Hoigaku Zasshi* 50:320–330.

Sorbrino, B., Carracedo, A., 2005, SNP typing in forensic genetics: a review. *Methods Mol Biol* 297:107–126.

Walsh, S.J., 2004, Recent advances in forensic genetics. *Expert Rev Mol Diagn* 4:31–40.

Wu, J., Ran, P., Zhang, B., Li, Y., Yan, J., Liao, M., Hou, Y., Zhang, L., 2008, Multiplex mutagenically separated polymerase chain reaction assay for rapid detection of human mitochondrial DNA variations in coding area. *Croat Med J* 49:32–38.

Chapter 28
Food Diagnostics

Abstract Food diagnostics is a relatively new and emerging area fuelled in large part by the ever-increasing demand for food safety. As a derivative of the larger medical diagnostic industry and other diagnostic areas (e.g., veterinary, environmental, and agricultural), the technology of food diagnostics is rapidly changing. Therefore, this field of application is used to outline principles, challenges, and current developments of DNA analysis in foods. In addition, DNA-based approaches are also essential in the areas of food authentication, detection of food-borne pathogenic micro organisms, and screening for food allergens.

Keywords Food diagnostics · Food allergens · PCR · ELISA · Food safety · Food borne disease · Microbiology · Antibody · Immunomagnetic separation · Amplification · Sensor technology · Microarray

28.1 Prologue

Today's era is driving the development and commercialization of biotechnological products and processes for agriculture and food system. These processes and products have a major impact on maintaining and improving food safety. The ability of commercial food industry to identify food pathogens and unsafe residues resulted in an almost fivefold increase in food as recalled by major manufacturers since 1988. New technology allows Government regulatory agencies to identify a bacterial pathogen and trace it back to its source more rapidly. The key to this new technology is the availability of rapid food-safety diagnostics. These rapid food-safety diagnostics provide a quick answer before the food product enters the distribution system. For example, rapid screening for *Salmonella typhimurium* requires using PCR method is approximately 24 h, in contrast to 4–5 days of analysis time for the traditional culture method. This methodology can contribute to meeting the increasing demand of quality assurance laboratories for standard diagnostic methods (Malorny et al., 2004).

M. Debnath et al., *Molecular Diagnostics: Promises and Possibilities*,
DOI 10.1007/978-90-481-3261-4_28,

28.2 Concept

Food Diagnostics is an emerging field which applies "modern" methods of detection of bacteria, viruses, parasites, chemicals, biotoxins, heavy metals and prions in all steps of the food chain from raw materials to end products. Using the molecular diagnostic techniques, the detection of a fragment of genetic material (nucleic acids, i.e., DNA or RNA) that is unique to the target pathogenic organism can be successfully accomplished.

One of the most practical and useful applications of molecular tools is their specificity as they target genetic regions unique to the organism, and depending on the gene target, they can also yield valuable information about virulence properties of the organism. They are also invaluable in detecting and identifying infectious agents. In the last decade the need for methods to detect and to quantify DNA from genetically modified organisms (GMOs) has been a major driver for the development and optimization of PCR-based techniques.

Throughout the world, food production, preparation and distribution have become increasingly complex, and raw materials are often sourced globally. Changes in food processing techniques, food distribution and the emergence of new food pathogens have changed the epidemiology of food-borne diseases. Food-borne microorganisms are continuously changing due to their inherent ability to evolve and their amazing capacity to adapt to different forms of stress. New primary production technologies and food manufacturing practices are introduced all the time; food consumption patterns and the demographic structure of many countries continue to change. The implementation of Food Safety should be seen as an ongoing process, which is influenced by environmental, socio-economical, political and cultural factors. Food safety issues need to be managed on a continuous basis, from a regional, national, and global point of view. New, flexible tools are required for evaluating and managing new food safety challenges. To guarantee the safety of foodstuffs producers have therefore shifted their focus towards the use of food safety management tools, most important HACCP (Hazard Analysis Critical Control Point), and the consequent application of hygienic measures, based on Good Manufacturing/Hygienic practice (GMP/GHP). Food safety management tools use input of scientific information to identify critical contamination points in the food chain and the production process, and design measures to control them. However, the lack of reliable data is often limiting the usefulness of this approach and therefore data collection is one of the priorities for future food safety strategies. In the absence of relevant data, other strategies might still have to be used to control food hazards. The Food and Drug Administration (FDA) has been involved in the regulation of in vitro diagnostic devices (IVDs or laboratory tests) since the introduction of the Medical Device Amendments of 1976. IVDs developed as kits or systems intended for use in multiple laboratories require review by the FDA before being marketed to ensure appropriate performance and labelling (Gutman, 1999).

28.3 Tools of Molecular Diagnostics to Assess Food Quality

Food safety concerns are best met with an eye toward all aspects of production. Our growing line of products for on-site testing assures comprehensive food safety throughout the processing chain. It must assure food safety, meet regulatory requirements and measure up to customer expectations. Certain company specializes in developing immunoassay tests for the food industry. This technology provides users with fast, accurate, easy-to-use tests that are cost effective, require little space and minimal capital investment. Whether it's the rapid analysis of GMOs in crops, mycotoxins in grains or food pathogens in meat, dairy (Chotár et al., 2006) or processed foods. Molecular diagnostics help to reduce risks, protect brands and gain confidence in the safety of operations (Lauri and Mariani, 2008).

The modern diagnostic techniques used to assess food diagnostics are as follows:

(1) The application of ultra sounds to monitor the quality of food products without affecting their integrity, i.e., non-destructive or non-invasive practices;
(2) The use of biosensors and electronic noses for evaluating the safety and quality of agricultural and horticultural products;
(3) The immunoassays for detection of food toxins, pesticides, food allergens ect.
(4) The nucleic acid based assays to characterize foodborne pathogens and to identify genetically-modified food or food components;
(5) Protein-based technology to detect and identify genetically-modified food or food components, and
(6) The detection of biochemical markers with rapid chromatographic techniques.

Recent advances in molecular cloning and recombinant DNA techniques have revolutionized the detection of pathogens in foods. Traditional methods of identification of food-borne pathogens, which cause disease in humans, are time-consuming and laborious, so there is a need for the development of innovative methods for the rapid identification of food-borne pathogens (Naravaneni and Jamil, 2005). Nontyphoidal Salmonella are highly prevalent food-borne pathogens. High-throughput sequencing of Salmonella genomes is expanding our knowledge of the evolution of serovars and epidemic isolates (Andrews-Polymenis et al., 2009). Genome sequences have also allowed the creation of complete microarrays. Microarrays have improved the throughput of in vivo expression technology (IVET) used to uncover promoters active during infection. In another method, signature tagged mutagenesis (STM); pools of mutants are subjected to selection. Changes in the population are monitored on a microarray, revealing genes under selection. Techniques such as in vivo expression technology (IVET) and signature-tagged mutagenesis (STM) provide alternatives for studying bacterial gene expression and growth requirements within these settings (Dudley, 2008).

28.4 Food Allergy

Over the past 2 decades, the prevalence of food allergy has been doubled. Major food allergens have been identified for many common foods (Nowack-Wegrzyn and Sampson, 2006). Clinically these immediate-type food-induced hypersensitivity reactions are characterised by a variety of signs and symptoms that occur within minutes or hours after consumption of the offending food. Reactions may be limited or more generalised with involvement of the skin, nose, eyes, and/or lungs. In more severe cases, cardiovascular symptoms including hypotension, shock, cardiac dysrhythmias and death can occur (Ballmer and Vieths, 2008). In food-allergic individuals, IgE is produced against naturally occurring food components, primarily glycoproteins that usually retain their allergenicity after heating and/or proteolysis. While adults tend to be allergic to fish, crustaceans, peanuts and tree nuts, children tend to be allergic to cow's milk, egg white, wheat and soya products more frequently. "Emerging" food allergens include tropical fruits, sesame seeds, psyllium, spices and condiments. These allergies frequently represent a cross-allergy to an allergen derived from another source, e.g. pollens or natural rubber latex. Food allergy encompasses a variety of immune-mediated adverse reactions to foods (Ebo and Stevens, 2001). IgE-mediated, cell-mediated, and mixed-mechanism food allergy disorders are recognized. The diagnostic work-up of suspected food allergy includes the skin prick test (SPT), the measurement of food specific immunoglobulin E (IgE) antibodies using serologic assays, and more recently the atopy patch test (Niggemann et al., 2005). Laboratory diagnosis of food allergy relies heavily on the detection of food-specific IgE antibodies, but novel approaches include tests for T-cell-mediated disorders and tests for prediction of tolerance. OFC remains the diagnostic standard for food allergy (Lidholm et al., 2006). Management of food allergy now-a-days focuses on avoidance of the offending foods, nutritional support, and prompt recognition and treatment of acute food allergic reactions. Anti-IgE monoclonal antibody is the first potential therapy for food allergy that is under-going testing in clinical trials.

28.5 Food-Borne Disease

Through various food items, many high-risk pathogens that cause disease in humans are transmitted (Nugen and Baeumner, 2008). Food-borne disease across the world costs billions of dollars annually. Approximately 30% of these diseases are caused by bacteria and their related toxic products (Mead et al., 1999). The bacteria species that are most frequently responsible for food contaminations are *Salmonella typhimurium* (meat, milk, and eggs), Shiga toxin-producing *Escherichia coli O157:H7* (meat), *Staphylococcus aureus* (milk, cream, and meat), *Clostridium perfringens* (sausages, preserved food), *Campylobacter jejuni* (poultry, eggs), *Vibrio parahaemolyticus* (shellfish), *Yersinia enterocolitica* (meat, milk), and *Listeria monocytogenes* (dairy products) (Magliulo et al., 2006; Lukinmaa et al., 2004). Infections due to *Yersinia, Shigella, Listeria, Campylobacter, Escherichia coli* O157:H7 and *Salmonella* have recently reported to decrease dramatically, while infections due to *Vibrio* have increased. A more recent report indicated similar findings, with a decrease in *Yersinia, Shigella, Listeria* and *Campylobacter* cases, and again a significant increase in *Vibrio* infections. The declining rates of infection due to *Listeria monocytogenes* and *E. coli* O157:H7 are likely a result of increased awareness. The FDA, USDA and EU have all implemented a zero-tolerance rule for *L. monocytogenes* in ready-to-eat (RTE) foods. Similarly, the USDA's Food Safety and Inspection Service (USDA-FSIS) has declared *E. coli* O157:H7 in raw ground beef to be an adulterant and therefore unfit for human consumption. Food safety practices have vastly improved in the processing environment as a result of these regulatory actions, as evidenced by the decreasing rates of infection by both *Listeria* and *E. coli* O157:H7. However, of the 121 foodborne outbreaks reported through FoodNet in 2005, almost half (49%) of the reported cases were attributed to viruses, and the number of *Vibrio* infections is on the rise (Nugen and Baeumner, 2008). Therefore, while the "hot" organism may change and while food safety practices are improving, there remains a growing need for enhanced means of food pathogen detection.

The majority of food-borne disease outbreaks result from unintentional contamination of a product as a result of inappropriate processing or handling. Intentional contamination of our food supply with biological or chemical agents also is a significant threat. Bioterrorist attacks on our food supply could be covert or announced and could be accomplished with selected bacterial pathogens or toxins.

A key step in establishing an effective food-safety program in a forward-deployed theater is to have adequate laboratory diagnostics that can identify and characterize rapidly any agent that could cause a food-borne disease outbreak. To enhance current food-safety and -quality programs, the commercial rapid-screening food laboratory diagnostics are used. Some capabilities of these diagnostics include:

- Microbiology testing of both liquid and solid foods.
- Food surface sanitation validation.
- Dairy product shelf-life prediction testing.
- Pesticide analysis of fresh fruits and vegetables.
- Continuous internal product temperature monitoring of perishable food while in transit.

Commercial rapid food-safety diagnostics have been procured and deployed throughout the whole world as an integral part of active food-safety surveillance program.

28.6 Commercial Rapid Food-Safety Diagnostics

Today's consumers demand safe food. The commercial food industry's ability to identify bacterial pathogens and unsafe residues has resulted in an almost fivefold increase in food recalls by major manufacturers since 1988. New technology allows Government regulatory agencies to identify a bacterial pathogen and trace it back to its source more rapidly. The key to this new technology is the availability of rapid food-safety diagnostics.

These rapid food-safety diagnostics provide a quick, "positive or negative" answer before the food product enters the distribution system. Tests take up to 30 h to complete because of the requirement for a bacterial growth enrichment period. This growth enrichment period is necessary to increase the total number of bacteria so they can be detected using current technology. A negative answer means that the product does not contain that particular bacteria or toxin and no further testing is required. A positive answer means that further testing is needed at a reference laboratory using standard laboratory methods to confirm the actual presence of the bacteria or toxin. For example, rapid screening for *Salmonella typhimurium* requires approximately 24 h to complete.

28.7 Methods for Analysis of Food

Food is a difficult test matrix since it is extremely varied in its chemical composition, contains many different ingredients, and often contains an intrinsic microbial flora with varying amounts of shelf-limiting bacteria and even pathogens. Furthermore, food is processed with varying technologies and stored under different conditions. During processing bacteria might be damaged sub lethally with subsequent problems of detection. As a result, different enrichment protocols might be necessary.

Microbiological methods are divided into *traditional* or *conventional* methods (Cancrini and Iori, 2004) and *rapid* methods (Bennett et al., 1998). Culture-based methods used for microbial detection and identification are simple to use, relatively inexpensive, and sensitive. Traditional methods of identification of food-borne pathogens, which cause disease in humans, are time-consuming and laborious, so there is a need for the development of innovative methods for the rapid identification of food-borne pathogens. Conventional methods are still considered the *gold standard* and are often required by national and international regulatory agencies as official control methods routine microbiology laboratories. Similar to conventional tests, the first-generation in deoxyribonucleic acid assays determine only a single analyte. Recent improvements in detection technologies have paved the way for the development of multiparameter assays using macroarrays or microarrays, while the introduction of closed-tube real-time polymerase chain reaction systems has resulted in the development of rapid microbial diagnostics with a reduced contamination risk. The use of these new molecular technologies is not restricted to detection and identification of microbial pathogens but also can be used for genotyping, allowing one to determine antibiotic resistance or to perform microbial fingerprinting. These *Rapid Methods* offer advantages in analysis time but also give the possibility to eliminate labour intensive steps and give the potential for automation. Due to the speed-up of the sampling process, sample treatment, detection/ enumeration procedure, the output of the laboratory can be improved(Fig. 28.1).

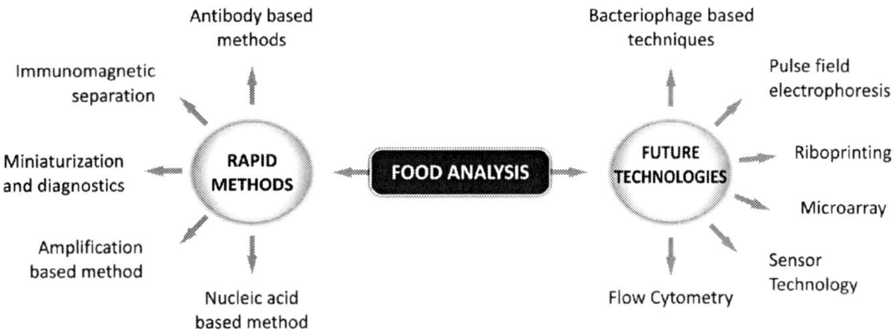

Fig 28.1 Diagnostic tools and strategies implemented for food safety

28.7.1 Rapid Methods

A new method must have high sensitivity, high specificity, high precision (repeatability), at the same time rapid, robust and cheap. There is currently no method that will fulfill all requirements. Antibody-based methods as represented by the ELISA assay and DNA based methods as the polymerase chain reaction are the most widely used technologies in food diagnostics today. Also immunomagnetic separation (IMS) is exploited in a number of commercially available kits and will become an even more important technology in the future. While some microarray based systems are commercially available, most of the biosensors and microarray based applications are still on the level of prototypes. Other promising methods with future potential include, Flow cytometry, Bacteriophage technology, biosensor based technology ect.

28.7.1.1 Rapid Biochemical Identification Techniques

The identification of microorganisms with conventional methods using different biochemical tests is labour intensive, with high usage of consumables and reagents. Miniaturized systems in the form of biochip based technology and diagnostic kits have been developed that are based on the use of dehydrated growth substrates, or on ready-to use media. These techniques usually termed modern biochemical identification techniques, are widely used and have a high importance for daily laboratory practice. Pure cultures are grown in a variety of liquid or solid media and detected by a colour change, gas formation or enzyme induced changes of the colour of the substrate. The results can be compared to a diagnostic chart, or analyzed with a database. A number of diagnostic kits are in the market for the identification of pathogens, spoilage organisms, starter cultures, etc. The systems offer visual analysis of the results or databases might be used to interpret and analyze the results in semi-automated and automated systems. In a comparison with conventional methods the use of kits was considered accurate, efficient, time-, space-, labour-saving and cheaper than conventional methods. They have a widespread use and conventional methods are often mainly relevant for reference testing (Ronald, 2001).

28.7.1.2 Antibody-Based Methods

Immunochemical analysis is a well-established technique with application to many different fields. Antibody-based methods have proven to be simple, rapid and sensitive for detecting and quantifying different types of food contaminants. Antibodies can be used to detect pathogens and spoilage organisms such as viruses, bacteria and moulds, but also low molecular weight food contaminants, such as mycotoxins, pesticides and veterinary drugs. However, the development has been constrained by the availability of high performance antibodies against the selected target analyte. The specificity of an immunoassay is largely determined by the intrinsic specificity of the antibody used for detection (Kerr et al., 2001).

There are three different types of antibodies available for assay development: polyclonal antibodies, monoclonal antibodies (Kerr et al., 2001) and recombinant antibodies. Recombinant antibodies are a very powerful new approach which offers many advantages. Also the emergence of molecular imprinting techniques opens possibilities for new applications. The selection of the appropriate antibodies allows the construction of tests with broad specificity narrow selectivity. Usually, an enrichment step is necessary. Tests can be single-use format like dip-sticks or done in miniplates, microplates or special formats. Only a few basic forms of antibody assay formats exist, but many different modifications are on the market. Both homogeneous and heterogeneous assays are widely used, but homogeneous assays are considered superior in their ease-of-use.

The different assays format include latex-agglutination test, immunodiffusion test, ELISA, and immunochromatography. Some recently developed immunoassays make use of the growth of the pathogens in the actual assay (Roda et al., 2002).

28.7.1.3 Immunomagnetic Separation

Immunomagnetic separation (IMS) techniques can be used to replace or supplement and speed up the enrichment step that is usually necessary before the detection of pathogens (Chapman et al., 1997). IMS differs from the assay format described above, as it is not a detection method in itself. It can be combined with different end-detection methods.

Often, PCR assays which are directly applied to food have a significantly decreased sensitivity which is based on the unspecific inhibition of the DNA polymerase by different food components. Through removal of these inhibiting food components from the sample, PCR ready DNA is easily obtained and the sensitivity of the PCR assay can be improved. Moreover, PCR assays use very small sample sizes. IMS allows the concentration of the sample by a factor 5–10 and thereby increases the sensitivity of the assay. A technique called magnetic capture-hybridization PCR (MCH-PCR) involves lysing the bacteria to release DNA and hybridization with pathogen- specific gene sequences using biotin labelled DNA probes. Following capture of the hybrids by streptavidin coated magnetic particles; the bound DNA is used for PCR amplification. IMS has proven to be a useful tool to enhance the speed and sensitivity of different detection methods. All of the steps have currently to be performed by hand, but an automation of the method can be expected in the next future.

28.7.1.4 Nucleic Acid-Based Assays

Nucleic acid-based assays can be either performed directly, or after amplification of the target sequence. Polynucleotides with a sequence complementary to the single-stranded DNA or RNA are used as probes. DNA sequences that are homologous to all members of the family of interest, or unique sequences that are particular to a species can be chosen for the construction of the probe. Probes that target different genera, species or different strains can thereby be constructed. Genes associated with virulence are often used for detection of pathogens. The ribosomal RNA represents an attractive target for probe design since it is present in multiple copies in most organisms, and databases with sequences for several thousand microorgansims are available. Ribosomal RNA contains stretches of conserved sequences interspersed with variable sequence regions, providing the scope to design single-stranded DNA probes for broad-range or specific target detection. The binding of the probe to the complementary cellular DNA/RNA is called hybridization. In order to detect hybridization, the probe needs to be labelled. Among others, Biotin, Digoxigenin and different fluorochromes have been used for labelling. Since there are many different assay formats available which are produced by many different manufacturers, only some will be introduced below.

28.7.1.5 Amplification-Based Methods

Molecular assays have become faster, less expensive and through the development of homogenous assays easier to use, making them more attractive for routine use in food laboratories. There are a number of methods available that allow the amplification of the target sequence or the detection signal, including branched DNA technology (bDNA), Nucleic Acid Sequence-based Amplification (NASBA), the Ligase Chain Reaction (LCR), Transcription- mediated amplification, strand displacement amplification (SDA) and rolling circle amplification (RCA).

PCR is now an accepted method to detect pathogens by amplification of the target DNA and detecting the target PCR products (Fenollar and Raoult, 2004). Detection of the amplification product has previously been done by gel electrophoresis, which is not a very convenient technique for routine use in a food laboratory. Recent advances have led to the development of homogenous assays for real-time detection which allow greater sensitivity and easy detection of PCR products that make PCR more suitable for routine-use in food laboratories (Poms et al., 2004). A major disadvantage of the PCR technology is that it will detect both viable and non-viable bacteria. The decreased sensitivity of the PCR reaction directly applied to a food sample is well known and can also lead to false-negative results. The inclusion of a short pre-enrichment step solves these issues, but also makes the analysis much slower. Several methods of sample preparation have been proposed to overcome this disadvantage including IMS, filtration, and centrifugation, the use of organic solvents and detergents and sample dilution. The use of intact RNA as target has been proposed to discriminate between viable and dead cells. The preparation and isolation becomes more difficult as RNA is much less stable. However, new specific isolation procedures for mRNA are currently being developed. The BAX system (Qualicon, Inc., Wilmington, DW. USA.) is the first commercial PCR system for the detection of food-pathogens (Bennett et al., 1998).

The FSIS (Food Safety and Inspection Sevice) USA has recently adopted this system for screening of ready-to-eat food, meat poultry and egg products for *Listeria monocytogenes.* The BAX system works directly from an overnight enrichment of the target organisms without a separate DNA isolation step. All PCR reagents required for the test are incorporated into a single tablet. The principle of the TDF-PCR detection is based on an intercalating dye, SYBR green I, binds to the double-stranded PCR product. The assay is an in situ process, simultaneously amplifying the target DNA and directly detecting the increasing fluorescence signal during the annealing/elongation phase of the PCR reaction. One additional thermal cycle is run consisting of a denaturation step and a product annealing step. Fluorescence is monitored directly in the PCR reaction tubes during this final cycle. The rate of increase in fluorescence over time during the product annealing step yields a characteristic pattern for positive samples that can be differentiated from negative samples.

28.8 Future Technologies

The most promising breakthroughs of the development of on-line or on-site, sensitive, low-cost, rapid methods for routine-use are expected to be made in the area of sensor technology. Many prototypes for food diagnostic application in the food and drink industry are currently being developed. They have high potential for automation and allow the construction of simple and portable equipment for fast analysis. These properties will open up many applications within quality and process control, control of fermentation processes, quality and safety control of raw materials, and for HACCP monitoring.

28.8.1 Microarrays

Microarray analysis is an emerging technology that has the potential to become a leading trend in bacterial identification in food and feed improvement (Rasooly and Herold, 2008). Microarrays often are also referred to as biochip, DNA chip, DNA microarray or gene array, and allow conducting many analyses in parallel. The technology uses fluorescent-labeled probes amplified from bacterial samples that are then hybridized to thousands of DNA sequences immobilized on chemically modified glass slides. The whole gene or open reading frame(s) is represented by a polymerase chain reaction fragment of double-strand DNA, approximately 1000 base pair (bp) or 20–70 bp single-strand oligonucleotides. The technology can be used to identity bacteria and to study gene expression in complex microbial populations, such as those found in food and gastrointestinal tracts. Data generated by microarray analysis can be potentially used to improve the safety of our food supply as well as ensure the efficiency of animal feed conversion to human food, e.g., in meat and milk production (Al-Khadi et al., 2002).

28.8.2 Sensor Technology

Sensor Technology covers a wide area of diverse techniques, including opto-chemical sensors and biosensors. Biosensors are a subgroup of chemical sensors where the analytical devices are composed of a biological recognition element such as enzymes, antibodies, receptors, proteins, oligonucleotides, or even a whole cell coupled to a chemical or physical transducer. A transducer measures the changes that occur when the sensor couples to its analyte. The sensitivity of the system is determined by the type of tranducers employed. Biosensors can be used for the detection of very different analytes such as pathogens, pesticides and toxins. Biosensors can be grouped according to their biological recognition element into immunosensors using antibodies and hybrid sensors using DNA or RNA probes. There have been many sensors developed for the detection of foodborne pathogens with the goal to overcome problems associated with traditional microbiological detection techniques such as time- and labor-intensive (Baeummer, 2003). In fact, biosensor advancements have greatly improved our ability to detect minute quantities of analytes as research into biosensors has mainly focused on detection platforms with very low detection limits (Rider et al., 2003). It has been estimated that 38% of reported pathogen biosensors in the past 20 years have been designed for the food industry. However, only a limited amount of methods are combined and currently exploited for their use in food diagnostics. As recognition elements, bioaffinity based receptors that use the selective interaction between ligand and receptor, antibody or nucleic acid are most widely used. As transducers, electrochemical and optical systems have gained practical importance. As nanobiotechnology progresses, sensors to detect pathogens or their constituents become smaller and more sensitive. Owing to the nature of these nanoscale sensors, the sample size from which the detection is being made is typically a microliter or smaller. Therefore, the challenge for scientists developing detection methods for pathogens in foods is in the sample preparation. Although the sample preparation requirements will vary from one food product to another, research into this step is required to bridge the emerging field of nanosensors with the food industry. Thus, while the organism with the largest number of diagnosed cases may fluctuate from year to year, the food industry will always be looking for detections system that will help identify all pathogens of concern in its food products (Rider et al., 2003).

Optical biosensors have been developed for rapid detection of contaminants in foods, including pathogens, and several have evolved into commercial prototype systems. The analyte in the food interacts with the bioactive molecule, usually an antibody. Antibodies can be immobilized directly on the fibre, either on the blunt end or along the sides of a fibre tip. The binding of antibody and analyte is detected as a change in an optical signal measured through the fibre-optic assembly. The light from a laser travels to the fibre tip and penetrates into the area outside the tip. A fluorescently labelled complex binds to the antibodies on the tip. The fluorescent signal then radiates in all directions, and some of it travels back up the fibre tip to the detector. Detection of molecules in solution can be made either by direct binding to the biosensor coating molecules, or by competition binding with soluble capture molecules added together with the sample. Future applications might include protein quality and the detection of allergens genetically modified (GM) proteins, Bovine Spongiform Encephalopathy prions, pathogens and biocide residues.

28.8.3 Flow Cytometry

Flow cytometry has mostly been used as a research tool for the study of microorgansims. However, the development of cheaper instrumentation should make it more useful for routine applications in the future. Flow cytometers can also be equipped with cell sorting devices allowing to sort out cells with a certain profile. By this method, *Staphylococcus aureus* could be separated form *E. coli* to a purity of 95%. The technique can provide rapid information on live and dead cells and information on their physiology and enables both qualitative and quantitative analysis of microbial cells in liquids. Due to the complexity and inherent cost of the equipment, the practical use of the method is still limited to research. However, with the appearance of cheaper systems on the market, it is considered to be a promising technology for the future.

28.8.4 Bacteriophage-Based Techniques

Bacteriophages allow discrimination between different bacterial strains and this has been exploited in "phage typing" of bacterial samples. Recently, a chemiluminiscence based method for detection of food-borne pathogens has been developed. The bacteriophages are engineered to carry the luxAB gene, expressing the enzyme luciferase. The infected cell will produce light that can be measured with a photodetector. Viability is essential for the light production, so that the assay can discriminate between viable and non-viable cells. The assay is more rapid due to the faster replication cycles of the virus compared to bacteria, generating a 20–100 fold increase of the virus particle within 30 min. At present, there are no commercial kits available using this technology.

28.8.5 Riboprinting and Pulse-Field Gel Electrophoresis

Riboprinting and Pulse-field gel electrophoresis are used for the genetical characterization of microbial cells. These methods are able to produce characteristic DNA finger printing patterns that can be used for surveillance programmes for food borne diseases. By comparing patterns collected in central databases the sources of contaminations can be traced even for geographically separated incidents. The use of these methods is currently limited to research laboratories. Many contract research organizations (CROs) offer the application of these methods for detection of contamination sources as a service to the food industry.

28.8.6 Applications and Limitations of Rapid Methods

The field of rapid methods is growing fast with numerous analytical options being available for different testing situation. These new rapid methods have the potential to offer considerable advantages regarding their sensitivity, rapidity and potential for automation, allowing increased sample through-put for analytical laboratories, and ease-of use (Magliulo et al., 2006; Lukinmaa et al., 2004). Most of the current rapid methods detect single target, but more and more methods with multiflexing properties are developed. Rapid methods are well suited for screening purposes for large numbers of food samples. Since most methods lack the required sensitivity, culture enrichment is still necessary before analysis. Benefits of enrichment include dilution of inhibitors, recovery of sub-lethally damaged cells and differentiation between viable and non- viable cells.

28.9 Business Outlook of Food Diagnostics

Food diagnostics as a business area is relatively new and can still be characterized as an emerging one, partly because the industry and the market is not well defined. The business of food analysis is expected to grow through the increasing demand for food safety; however, it is unclear whether this will lead to actual market growth of food diagnostics. Especially in Europe, the growing political interest into food safety issues has been driven by a need to re-establish consumer confidence into the public food safety system, which was badly disrupted as a consequence of recent food contamination incidents, e.g. Bovine Spongiform Encephalopathy, enterohemorragic *Escherichia coli*, multi-drug resistant strains, Dioxin, and GMOs. At this stage it is not clear what kinds of business models will prevail as profitable and viable, because it is dependent on whether we are concerned with food analysis, food diagnostics, microbiological testing, or microbiological food testing. In terms of trends and needs it is, however, clear that food safety has become a public issue. This need is most visible in the growing *public* concern, i.e. consumers and political actors. This is not to say that the corporate community is unresponsive to this need, but there is an evident conflict between what can be done with respect to what is economically viable and what is publicly desirable. It is expected by the public that more testing, especially for food pathogens, will result in an increase of food safety of the product. However, the concept of food testing is limited in guaranteeing food safety and therefore the HACCP concept has been applied to guarantee *process safety* rather than *product safety*. There is a clear corporate and public communication challenge that needs to be addressed. The question of whether implementation of rapid methods will lead to a sufficient degree of improvement in food safety or will have more impact on cost savings has to be further analyzed (Glynn et al., 2008).

It is not clear whether the consumers in general are willing to pay a premium price for food safety tests, on something which they already assume to be safe. Therefore, despite the growing public concerns over food safety, the economic incentives to develop new rapid methods are not exactly compelling. However, it is obvious that food safety has a very high impact on the image and branding of food processing companies and thus has a clear long-term corporate profit impact. In light of that it might be better to consider the issue of food safety and food diagnostics as an element of corporate communications. The business area of food diagnostics is currently a derivative of the much larger medical diagnostic industry and other diagnostic areas (veterinary, environmental, agricultural), because new technologies are developed there, which has implications for both technology and market management in terms of economies of scale and scope. There has been an increased and renewed interest in the US after September 11, 2001 to develop new methods for pathogen testing resulting in an increase of R&D funding in that area.

Food diagnostics in terms of market attractiveness is competing with other potentially more attractive markets such as pharmaceutical technologies and services measured in terms of market potential. Food diagnostics is a fairly small market in absolute terms, but also in comparison to the whole market for food analysis. There are approximately 50 established companies with some market leaders but no clear market dominator. The small market size has implications for market attractiveness, which may not provide incentives enough to develop new technologies or to adopt new technologies developed in, e.g. in vitro diagnostics to the special requirements for the food industry.

28.10 Conclusion

Diagnostic technologies can be applied in a variety of different areas either directly or after modification thus seemingly widening the commercial potential. Although there are apparent synergies between the different diagnostic sectors at the R&D level, the over all benefits might be meager as the markets are dramatically different, requiring for example modified product development processes, an entirely different sales force, technology transfer, and are subject to different regulatory requirements. The food industry itself is not expected to increase their R&D expenditures, which in most cases is currently keeping pace with inflation. R&D activities listed by importance are new product development, product safety research, clinical trial or support, quality control, quality assurance, ingredient technology, process development, market research and nutritional research analysis. Areas of food science in order of importance are healthy food, functional food, food safety from a process perspective, development of natural food, food safety from an ingredient point of view, organic food, reduced fat food, and methods development for quality control. Food diagnostics is set on an ongoing evolution. All the detection technology will be a part of the food safety surveillance programme using the latest molecular diagnostic tools.

References

Al-Khaldi, S.F., Martin, S.A., Rasooly, A., Evans, J.D., 2002, DNA microarray technology used for studying foodborne pathogens and microbial habitats: minireview. *J AOAC Int* 85:906–910.

Andrews-Polymenis, H.L., Santiviago, C.A., McClelland, M., 2009, Novel genetic tools for studying food-borne Salmonella. *Curr Opin Biotechnol* 20:149–157.

Baeummer, A.J., 2003, Biosensors for environmental pollutants and food contaminants. *Anal Bioandl Chem* 377:434–445.

Ballmer-Weber, B.K., Vieths, S., 2008, Potential pitfalls and prospects of food allergy diagnostics with recombinant allergens or synthetic sequential epitopes. *J Allergy Clin Immunol* 121: 1323–1330.

Bennett, A.R., Greenwood, D., Tennant, C., Banks, J.G., Betts, R.P., 1998, Rapid and definitive detection of Salmonella in foods by PCR. *Lett Appl Microbiol* 26:437–441.

Cancrini, G., Iori, A., 2004, Traditional and innovative diagnostic tools: when and why they should be applied. *Parassitologia* 46(1-2):173–176.

Chapman, P.A., Malo, A.T., Siddons, C.A., Harkin, M., 1997, Use of commercial enzyme immunoassays and immunomagnetic separation systems for detecting *Escherichia coli* O157 in bovine fecal samples. *Appl Environ Microbiol* 63:2549–2553.

Chotár, M., Vidová, B., Godány, A., 2006, Development of specific and rapid detection of bacterial pathogens in dairy products by PCR. *Folia Microbiol* 51:639–646.

Dudley, E.G., 2008, *In vivo* expression technology and signature-tagged mutagenesis screens for identifying mechanisms of survival of zoonotic foodborne pathogens. *Foodborne Pathog Dis* 5:473–485.

Ebo, D.G., Stevens, W.J., 2001, IgE-mediated food allergy – extensive review of the literature. *Acta Clin Belg* 56:234–247.

Fenollar, F., Raoult, D., 2004, Molecular genetic methods for the diagnosis of fastidious microorganisms. *APMIS* 112:785–807.

Glynn, B., Lacey, K., Reilly, J., Barry, T., Smith, T.J., Maher, M., 2008, Current and emerging molecular diagnostic technologies applicable to bacterial food safety. *Int J Dairy Technol* 59:126–139.

Gutman, S., 1999, The Role of Food and Drug Administration Regulation of *in vitro* diagnostic devices—applications to genetics testing. *Clin Chem* 45:746–749.

Kerr, P., Chart, H., Finlay, D., Pollock, D.A., MacKie, D.P., Ball, H.J., 2001, Development of a monoclonal sandwich ELISA for the detection of animal and human *Escherichia coli* O157 strains. *J Appl Microbiol* 90:543–549.

Lauri, A., Mariani, P.O., 2008, Potentials and limitations of molecular diagnostic methods in food safety. *Genes Nutr* 4:1–12.

Lidholm, J., Ballmer-Weber, B.K., Mari, A., Vieths, S., 2006, Component-resolved diagnostics in food allergy. *Curr Opin Allergy Clin Immunol* 6:234–240.

Lukinmaa, S., Nakari, U.M., Eklund, M., Siitonen, A., 2004, Application of molecular genetic methods in diagnostics and epidemiology of food-borne bacterial pathogens. *APMIS* 112: 908–929.

Magliulo, M., Roda, B., Zattoni, A., Michelini, E., Luciani, M., Lelli, R., Reschiglian, P., Roda, A., 2006, An innovative, flow-assisted, noncompetitive chemiluminescent immunoassay for the detection of pathogenic bacteria. *Clin Chem* 52:2151–2155.

Malorny, B., Paccassoni, E., Fach, P., Bunge, C., Martin, A., Helmuth, R., 2004, Diagnostic real-time pcr for detection of *Salmonella* in food. *Appl Environ Microbiol* 70:7046–7052.

Mead, P.S., Slutsker, L., Dietz, V., McCaig, L.F., Bresee, J.S., Shapiro, C., 1999, Food-related illness and death in the United States. *Emerg Infect Dis* 5:607–625.

Naravaneni, R., Jamil, K., 2005, Rapid detection of food-borne pathogens by using molecular techniques. *Med Microbiol* 54:51–54.

Niggemann, B., Rolinck-Werninghaus, C., Mehl, A., Binder, C., Ziegert, M., Beyer, K., 2005, Controlled oral food challenges in children – when indicated, when superfluous? *Allergy* 60:865–870.

Nowak-Wegrzyn, A., Sampson, H.A., 2006, Adverse reactions to foods. *Med Clin North Am* 90:97–127.

Nugen, S.R., Baeumner, A.J., 2008, Trends and opportunities in food pathogen detection. *Anal Bioanal Chem* 391:451–454.

Poms, R.E., Anklam, E., Kuhn, M., 2004, Polymerase chain reaction techniques for food allergen detection. *J AOAC Inter* 87:1391–1397.

Rasooly, A., Herold, K.E., 2008, Food microbial pathogen detection and analysis using DNA microarray technologies. *Food borne Pathog Dis* 5:531–550.

Rider, T.H., Petrovick, M.S., Nargi, F.E., Harper, J.D., Schwoebel, E.D., Mathews, R.H., Blanchard, D.J., Bortolin, L.T., Young, A.M., Chen, J., Hollis, M.A., 2003, A B cell-based sensor for rapid identification of pathogens. *Science* 301:213–215.

Roda, A., Mirasoli, M., Venturoli, S., Cricca, M., Bonvicini, F., Baraldini, M., Pasini, P., Zerbini, M., Musiani, M., 2002, Microtiter format for simultaneous multianalyte detection and development of a PCR-chemiluminescent enzyme immunoassay for typing human papillomavirus DNAs. *Clin Chem* 48:1654–1660.

Ronald, C., 2001, McGlennen miniaturization technologies for molecular diagnostics. *Clin Chem* 47:393–402.

Chapter 29
Rapid Diagnostic Methods for Biowarfare

Abstract There is a great concern within the scientific and security communities on the ongoing revolution in Genomics research. Genomics revolution may be used to enhance detection, protection and treatment so that biological warfare agents are never used. Biological weapons have the greatest potential for lethality of any weapon. Biological weapons are accessible to all countries; there are few barriers to developing such weapons with a modest level of effort. The current level of sophistication for many biological agents is low, but there is enormous potential based on advances in modern molecular biology, fermentation, and drug delivery technology for making more sophisticated weapons.

Keywords Bioterrorism · Bioweapons · Biodefence · Genomics · Anthrax · Diagnostics · Mycotoxins · Vaccines

29.1 Prologue

The advent of biotechnology, namely in the fields of microbiology, molecular chemistry and genetic engineering have opened new ways for human race to propel itself to a better future. However, like all things, there is a darker, more sinister side to biotechnology. One "dark side" is the manufacture and proliferation of biological and chemical weapons. Biological warfare agents have gained attention in recent years. Only a few years ago bioterrorism was considered a remote concern but today it has reached the forefront of the public imagination following recent terrorist attacks around the world. Advances in biotechnology and genetic engineering may facilitate the development of potentially new and more deadly biological warfare agents. The ability to modify microbial agents at a molecular level has existed since the 1960s, when revolutionary new genetic engineering techniques were introduced, but the enterprise tended to be slow and unpredictable. With today's more powerful techniques, infectious organisms can be modified to bring about disease in different ways. Many bioengineering companies (both U.S. and foreign) now sell all-in-one kits to enable even high school-level students to perform recombinant

M. Debnath et al., *Molecular Diagnostics: Promises and Possibilities*,
DOI 10.1007/978-90-481-3261-4_29,

DNA experiments. The availability of free on-line gene sequence databases and analytic software over the Internet further simplifies and disseminates this capability. It is now possible to transform relatively benign organisms to cause harmful effects. Genetic engineering gives biological warfare developers powerful tools with which to pursue agents that defeat the protective and treatment protocols of the prospective adversary. Genetically engineered micro-organisms also raise the technological hurdle that must be overcome to provide for effective detection, identification, and early warning of biological warfare attacks.

29.2 Concept

During the twentieth century, major military powers have established secret biological warfare programs driven by visions of an extraordinary weapon that could penetrate behind enemy lines and kill civilians (Guillemin, 2005). Some weapons not only inflict casualties but manage to strike considerable fear in the enemy (Arora, 2005). "Biological Warfare" (BW) is use of living pathogens to produce casualties in man or animals or damage to plants. A BW weapon may be intended to kill, incapacitate or seriously impede on an individual as well as entire cities or places. Civilians may encounter these agents during a terrorist or military attack, a riot control operation, or due to accidental exposure during industrial or agricultural use. Military personnel may be exposed during their active use in the battlefield or exposure to areas where the disease may be chronically endemic. A number of biological agents (Török et al., 1997) could be used in a terrorist attack, including anthrax, botulinum, salmonellosis, plague, smallpox, staphylococcal and streptococcal toxins, and the list of emerging pathogens is evolving rapidly.

In spite of major successes against infectious diseases in the twentieth century, new infectious diseases have emerged and old ones re-emerged in recent decades in different parts of the world (John, 1996).The serious diseases that their causal agents produce could cause considerable morbidity and mortality if used in a terrorist attack (Henderson et al., 1999). Of the weapons of mass destruction, the biological ones are the most feared and bioterrorism has become one of the most vicious threats to civilized society in recent times. Biological weapons have been sporadically used for centuries. Before the twentieth century, the use of biological agents took three major forms:

- Deliberate poisoning of food and water with infectious material (Török et al., 1997)
- Use of microorganisms, toxins or animals, living or dead, in a weapon system (Polyak et al., 2002)
- Use of biologically inoculated fabrics.

Despite international regulations, there has been a global re-emergence of the threat of biological warfare. As many as 17 countries are suspected of either including or developing biological agents in their weapons programmes. In the past decade, a number of terrorist organizations with access to bioweapons technology have emerged. Current surveillance systems may be inadequate to detect biological attacks. The onset of illness is often delayed, thus the timing and location of such an event may be extremely difficult to identify. We are unfamiliar with most of the agents of biological warfare and are ill-equipped to handle the consequences of such an attack. In addition, there is no apparent coherent policy to handle a biological terrorist incident. Given the enormity of what is possible in the event of a biological attack, one must be prepared to detect, diagnose, epidemiologically characterize and respond appropriately to biological weapons. Of the potential biological weapons, smallpox and anthrax pose the greatest threats (Rotz et al., 2002). The question of what disease-causing organisms might supplant classic biological warfare agents is critical to understanding future biological warfare threats. Biological warfare agents may emerge in two likely categories: man-made manipulations of classic biological warfare agents and newly discovered or emerging infectious agents that result from natural occurrences. The future likelihood of infectious agents being created for biological warfare purposes will be influenced by several technological trends, of which four of the most significant are:

- Genetically engineered vectors in the form of modified infectious organisms will be increasingly employed as tools in medicine and the techniques will become more widely available.
- Strides will be made in the understanding of infectious disease mechanisms and in microbial genetics that are responsible for disease processes.
- An increased understanding of the human immune system function and disease mechanisms will shed light on the circumstances that cause individual susceptibility to infectious disease.
- Vaccines and antidotes will be improved over the long term, perhaps to the point where classic biological warfare agents will offer less utility as a means of causing casualties.

The goal of biodefense is to integrate the sustained efforts of the national and homeland security, medical, public health, intelligence, diplomatic, and law enforcement communities. Health care providers and public health officers are among the first lines of defense. In some countries private, local, and State (province) capabilities are being augmented by and coordinated with federal assets, to provide layered defenses against biological weapons attacks. The traditional approach toward protecting agriculture, food, and water: focusing on the natural or unintentional introduction of a disease being strengthened by focused efforts to address current and anticipated future biological weapons threats that may be deliberate, multiple, and repetitive.

29.3 Characteristics of Biological Weapons

The revolution in biotechnology facilitates an evolution in the biological warfare threat. The revolution in biotechnology began in 1977 with the successful cloning of a protein using a synthetic, recombinant gene. Scientific and technological advances have facilitated the development of genetically engineered agents. The extreme lethality of biological warfare agents has long been known. The most lethal biological toxins are hundreds to thousands of times more lethal per unit than the most lethal chemical warfare agents. However, lethality is only one of many characteristics necessary to consider in the development, production, and employment of a biological warfare agent. Numerous characteristics need to be controlled for a highly effective biological warfare agent. Historically, the accentuation of one characteristic often resulted in the attenuation of one or more other characteristics, possibly even rendering the modified agent ineffective as a weapon. Advances in biotechnology, genetic engineering, and related scientific fields provide increasing potential to control more of these factors, possibly leading to the ability to use biological warfare agents as tactical battlefield weapons.

- The potential types of novel biological agents (microorganisms) that could be produced through genetic engineering methodologies:
- Benign microorganisms are genetically altered to produce a toxin, venom, or bioregulator.
- Microorganisms resistant to antibiotics, standard vaccines, and therapeutics.
- Microorganisms with enhanced aerosol and environmental stability.
- Immunologically-altered microorganisms able to defeat standard identification, detection, and diagnostic methods.

It is noteworthy that each of these techniques seeks to capitalize on the extreme lethality, virulence, or infectivity of biological warfare agents and exploit this potential by developing methods to deliver more efficiently and to control these agents on the battlefield.

Ideal characteristics of biological weapons targeting humans are of high infectivity, high potency, non-availability of vaccines, and delivery as an aerosol (Atlas 1998). Diseases most likely to be considered for use as biological weapons are contenders because of their lethality (if delivered efficiently), and robustness. The biological agents used in biological weapons can often be manufactured quickly and easily. The primary difficulty is not the production of the biological agent but delivery in an effective form to a vulnerable target. Diseases considered for weaponization, or known to be weaponized are discussed below. Ongoing scientific research into the functioning of disease organisms also should provide insights for the development of advanced medical defenses against new and emerging biological warfare threats. Current examples of infectious organisms that are attracting particular attention are hantaviruses; other hemorrhagic fever-causing agents, such as Ebola; and the bacteria invasive Group A streptococcus (commonly known as flesh-eating bacteria). The streptococcus example is illustrative. Once it is well established, the infection is very difficult to control with antibiotics. Although the natural form of this organism may not have significant potential as an aerosol threat agent, those seeking new infectious agents for military use could investigate its mechanisms of action.

29.3.1 Anthrax

Anthrax is primarily a zoonotic disease caused by *Bacillus anthracis* that produces a toxin consisting of a protective antigen, edema factor and a lethal factor (Witkowski and Parish, 2002). Anthrax is considered an effective agent for several reasons. First, it forms hardy spores, perfect for dispersal aerosols. Second, pneumonic (lung) infections of anthrax usually do not cause secondary infections in other people. Thus, the effect of the agent is usually confined to the target. A pneumonic anthrax infection starts with ordinary "cold" symptoms and quickly becomes lethal, with a fatality rate that is 90% or higher. Finally, friendly personnel can be protected with suitable antibiotics. Diagnosis is based on gram stain, cultures or direct immunofluorescence or ELISA or PCR. Anthrax toxin may be detected in the blood by immunoassays (Polyak et al., 2002). A simple mAb based dot-ELISA is available for detection of toxin-producing strains of *Bacillus anthracis* (Shastry et al., 2003).

29.3.2 Plague

Infection to *Yersinia pestis* has resulted in high mortality and thus become a potent BW agent. On inoculation 4–10% individuals may develop a pustule at the inoculation site. Terminal pneumonic and septicaemia plague patients would develop livid cyanosis and large ecchymoses on the back. Septicemia could cause petechiae, purpura, ecchymoses, and acral cyanosis and necrosis leading to the infamous description, "the Black Death." (Inglesby et al., 2000). Investigations include bubo aspiration for direct microscopy, and culture on blood agar of bubo aspirate, cerebrospinal fluid, blood and sputum. PCR- and ELISA-based systems are also available for detection or strain typing of *Y. pestis* alone or in conjunction with virulence markers such as F1 (fraction 1) and Pla (plasminogen activator) (Khushiramani et al., 2009).

29.3.3 Tularemia

Tularemia is a zoonosis caused by *Francisella tularensis*. The commoner forms include the ulceroglandular form, which involves the skin and lymph nodes, and the typhoidal form. A cutaneous ulcer occurs in approximately 60% of patients and is the most common sign of tularemia. Ulcers are generally single lesions of 0.4–3.0 cm in diameter, with heaped-up edges and accompanied with systemic upset. The organism is difficult to culture. Hence the diagnosis is usually established by serology. A live, attenuated vaccine is effective against aerosol infection (Friedlander and Evans, 1997).

29.3.4 Melioidosis

Burkholderia (formerly *Pseudomonas*) *pseudomallei* is a gram-negative bacillus. Cutaneous manifestations include severe urticaria flushing and cyanosis. Inhalational melioidosis could lead to pustules or cutaneous abscesses associated with lymphangitis, cellulitis, or regional lymphadenitis, but only after metastatic abscesses to the skin are formed, and this would be likely to take months (Wuthiekanun et al., 2004).

29.3.5 Viral Hemorrhagic Fevers

Hemorrhagic fever viruses, with the exception of dengue, are potentially transmitted via aerosol. Cutaneous manifestations are non-specific and include flushing, petechiae, purpura, ecchymoses, and edema. ELISA to detect specific IgM antibody responses must be done. One should not wait for viral cultures, which consume time, to initiate management. Management includes supportive care for prevention of hypotension and shock. Medical personnel need to be protected. Isolation of the patient during active disease and strict barrier nursing must be adopted. Specific treatment may include ribavirin. Only yellow fever has a licensed vaccine (Barrett, 2004).

29.3.6 Other Viral Fevers

The viral species important from the warfare context include variola (smallpox), monkeypox, and vaccinia. Smallpox has an incubation period of 7–17 days and a prodrome of 2–4 days. An enanthem on the buccal and pharyngeal mucosa starts on about the second day. An orderly progress of macules to papules, vesicles, pustules (sometimes umbilicated), and crusts distributed in a centrifugal pattern (in distinction to varicella) over a 1–2 week period is typical. Crusts are infective and detach in about 3 weeks leaving depressed, hypopigmented scars. Monkeypox greatly resembles variola, the only distinguishing feature being cervical and inguinal lymphadenopathy.

29.3.7 Trichothecene Mycotoxins

Trichothecene mycotoxins ("Yellow rain") are the only potential BW toxins with cutaneous activity and manifestations. Mycotoxins are a diverse group of small molecular weight compounds produced by the fungi of five fungal genera: *Aspergillus, Penicillium, Fusarium, Alternaria,* and *Claviceps.* "Yellow rain" was used in Southeast Asia between 1974 and 1981. Most attacks used yellow pigment, but some used red, green, white, or brown smoke or vapor. Vesication often occurred with "yellow rain" attacks. At low doses (nanograms), severe skin irritation with erythema, edema, and necrosis is observed (Seely et al., 1985). A mask and full-body clothing are protective. The clothing gear and contaminated areas of skin should be washed with soap and water followed by a water rinse. Washing within 4–6 h of exposure removes 80–98% of the toxin and prevents death and dermal lesions in experimental animals. High doses of systemic steroids decrease toxin injury.

29.3.8 Aflatoxin

This is a class of toxin, produced by *Aspergillus flavus* that induce liver cancer. Since there is difficulty in differentiating cancer origins between accidental and intentional exposure, even recognizing that a target population had been "attacked" would be laborious; this would be a case of a "stealth BW attack".

29.3.8.1 Ricin

Ricin is a protein toxin extracted from the castor bean plant and exhibits toxic effects on cells of different visceral organs (Kumar et al., 2007). Ricin is already being investigated for its "magic bullet" properties as an agent that might selectively destroy cancer cells (Rao et al., 2005). That same principle could be used to specifically target an enemy; in theory one could be specific enough to use this procedure to target a single individual for assassination.

29.3.8.2 *Fusarium oxysporum*

The potential use of genetic engineering in the production of BW is illustrated by the on-going studies on the possible use of the mould *Fusarium oxysporum* as a candidate for drug plant eradication. This fungus, which has devastated commercial crops (e.g. bananas and muskmelon), is being investigated for its potential to destroy coca and cannabis plants. Preliminary studies indicate that host specificity is narrow and species "jumping" is rare; i.e., targets can be carefully selected without posing danger to other commercial crops. Obviously, the same technology could be applied by terrorists to assail the commercial crops of perceived enemy states.

29.4 Biodefense

It is important to note that all of the classical and modern biological weapons organisms are animal diseases, the only exception being smallpox (Henderson et al., 1999). Thus, in any use of biological weapons, it is highly likely that animals will become ill either simultaneously with, or perhaps earlier than humans. Indeed, in the largest biological weapons accident known the anthrax outbreak, sheep became ill with anthrax as far as 200 km from the release point of the organism from a military facility in the southeastern portion of the city.Thus, a robust surveillance system involving human clinicians and veterinarians may identify a bioweapons attack early in the course of an epidemic, permitting the prophylaxis of disease in the vast majority of people (and/or animals) exposed but not yet ill.

29.4.1 Viral Agents – Poxviridae

Poxviruses, the largest of all viruses, differ from other DNA viruses by replicating in the cytoplasm where they produce eosinophilic inclusion bodies. They are relatively resistant to drying and many disinfectants. The Orthopox genus includes at least nine species.Three viruses interest us in a BW context: variola, monkeypox, and vaccinia.

Variola is an orthopox virus very similar to vaccinia but with different host predilections. There was no animal reservoir (although monkeys are susceptible to infection); this factor enabled global eradication of this disease. Variola retains its transmissibility for one year in dust and cloth. Person-to-person transmission requires close contact. Patients were most infective 4–6 days after the illness started, and respiratory spread was probably the most common route of transmission. Only 30% of susceptible contacts became infected.

Monkeypox was first identified in 1958 as a pathogen of cynomolgus monkeys; in 1971 it was linked to human disease. The virus exists in an enzootic state in arboreal squirrels of tropical rain forests of western and central Africa. Person-to-person transmission by respiratory droplet occurs.

29.4.2 Botulinal Toxins

Botulinal Toxins (botox) are produced by a microscopic organism *Clostridium botulinum.* Clostridial neurotoxins are among the most toxic substances known and cause severe illnesses in both humans and animals (Dixi et al., 2006). A deadly poison even in small doses. It can be absorbed through the skin, lungs, eyes and mucous membranes. The poison occurs in the soil and naturally in low acid foods. It can be distributed through aerosol methods, water and food. The symptoms are delayed (2–14 days), and the irreversible damage is done (walking dead) before victims realize what has occurred (Arora et al., 2002).

A vaccine for the disease is available but to obtain complete protection from the toxins, the amount needed would cause death so it is not usually recommended, except to certain military personnel. If administered within 48 h from exposure, an antitoxin can only stop the progress of the disease. Oxygen should also be provided for assisted breathing. Formaldehyde kills botulinum spores. Boiling in water for at least 10 min will kill the poison, not the spores. To kill the spores, it takes more than 6 h of boiling in water.

29.4.3 Mycotoxins

Trichothecene Mycotoxins are produced by fungi of the genera *Fusarium, Myrotecium, Trichoderma, Stachybotrys* and others. They inhibit protein synthesis, impair DNA synthesis, alter cell membrane structure and function, and inhibit mitochondrial respiration. The toxins, protein in nature, have a low molecular weight. They also contain food refusal and emetic factors.

Trichothecene mycotoxins are highly persistent and stable for long periods of time. 0.5 mg of the poison is enough to kill half the exposed humans. The skin of the victims can be irritated if the skin is exposed to the toxins. They can also cause radiomimetic injury of intestines, bone marrow, lymph nodes, spleen and thymus, leading to leukopenia and bone marrow atrophy.

Effects are also found on central nervous, circulatory and reproductive systems. After about 8 weeks from exposure to Normocyclic Trichothecenes, one will suffer from Alimentary Toxic Aleukia; burning sensation in the alimentary tract, vomiting, tachycardia, leukopenia, petechial hemorrhages with necrosis in skin and internal hemorrhages. After about 8 weeks from exposure to Macrocyclic Trichothecenes, one will suffer from Stachybotryotoxicosis, conjunctivitis, rhinitis, leukopenia, dematis and pulmonary fibrosis (Table 29.1).

29.5 Combating Detection of Biowarfare Agents

A growing demand for methods to detect the presence of biological warfare (BW) agents and other pathogens in samples from the environment, the battlefield, and food is currently driving a need for new detection technologies. Especially for biodefense and risk evaluation, an early and definite technology is needed. A protein chip has been developed that allows the simultaneous detection of a multitude of different biowarfare agents.

29.5.1 Prophylatic & Therapeutic Approaches

Traditionally, vaccines consisted of a preparation of the infectious agent itself – either living, weakened or killed. Introducing the vaccine into the body activates the immune system, resulting in the production of antibodies against that particular agent. If a vaccinated person is later exposed to the infectious agent, he or she will already have built up immunity against it. More recently, researchers have started using fragments of the pathogen's DNA genome as a vaccine, rather than the entire organism. This approach helps eliminate the risk of infection that comes with using traditional vaccines. Certain natural products and anti-oxidants are found highly promising against chemical as well as biological warfare agents (Vijayaraghavan et al., 2006).

Table 29.1 Selected biowarfare agent characteristics

Disease	Symptoms	Person-to-person transmission	Infective dose (Aerosol)	Incubation period	Duration of illness	Lethality	Persistence of organism	Treatment
Inhalation anthrax	Fever, malaise, cough, respiratory distress	No	8,000–50,000 spores	1–6 days	3–5 days (usually fatal if untreated)	High	spores remain viable in soil for > 40 years	Ciprofloxacin Doxycycline
Pneumonic plague	High fever, chills, headache, productive cough – watery then bloody	High	<100 organisms	2–3 days	1–6 days (usually fatal)	High unless treated within 12–24 hours	For up to 1 year in soil; 270 days in live tissue	Streptomycin Gentamycin or Chloramphenicol
Botulism	Dry throat, blurred vision, slurred speech, difficulty swallowing, progressive descending symmetrical paralysis	No	0.001 ng/kg is LD50 for type A	12–36 hours (range up to several days)	Death in 24–72 hours; lasts month if not lethal	High without respiratory support	For weeks in non-moving water and food	Antitoxin supportive care
Smallpox	Non-specific flu-like prodrome (malaise, fever, headache) then synchronously evolving maculopapular rash progressing to vesicles then pustules	High	Assumed low (10–100 organisms)	12–14 days (range 7–17 days)	4 weeks	High to moderate	Very stable	Cidofovir
Brucellosis	Irregular fever, chills, headache, malaise, cough and chest pain in 20%, osteoarticular disease	No	10–100 organisms	5–60 days (average 1–2 months)	Weeks to months	5% untreated	6 weeks in dust and 10 weeks in soil or water	Doxycycline + Rifampin

Table 29.1 (continued)

Disease	Symptoms	Person-to-person transmission	Infective dose (Aerosol)	Incubation period	Duration of illness	Lethality	Persistence of organism	Treatment
Tularemia	Fever, headache, malaise, weight loss, nonproductive cough	No	10–50 organisms	3–6 days (range 1–21 days)	2 weeks	Moderate if untreated	For months in moist soil or other media	Streptomycin Gentamycin
Q Fever	Fever, chills, headache, diaphoresis, malaise, fatigue, anorexia, and weight loss	Rare	1–10 organisms	7 days (range 2–14 days)	Weeks	Very low	Able to withstand heat and drying; persists in environment for weeks to months	Tetracycline Doxycycline
Viral Encephalitides	Fever, rigors, severe headache, photophobia, malaise, nausea, vomiting, diarrhea may follow	Low	10–100 organisms	1–5 days	Days to weeks	Variable	Relatively unstable in the environment	Supportive care
Viral Hemorrhagic Fevers	Fever, malaise, myalgia, prostration; vascular permeability may present as conjunctival injection and petechial hemorrhage and progress to mucous membrane hemorrhage and shock	Moderate	1–10 organisms	4–21 days	Days to weeks	5–90% case fatality rate depending on virus	Relatively unstable in the environment	Ribavirin Supportive care

Table 29.1 (continued)

Disease	Symptoms	Person-to-person transmission	Infective dose (Aerosol)	Incubation period	Duration of illness	Lethality	Persistence of organism	Treatment
Staph Enterotoxin B	Sudden onset fever, chills, headache, myalgias, non-productive cough	No	30 ng/person incapacitation	3–12 hours after inhalation	Days	<1%	Resistant to freezing	Supportive care
Ricin	Depends on route of exposure. Aerosol route: fever, chest tightness, cough, hypothermia. Oral route: gastrointestinal hemorrhage	No	3–5 g/kg is LD50	18–24 hours	Days. Death within 10–12 days for ingestion	High	Stable	Inhalation: supportive GI: lavage, charcoal, cathartics
T-2 Mycotoxins	Skin pain, redness, necrosis, sloughing of epidermis, wheezing, chest pain, hemoptysis	No	Moderate	Minutes to hours	Variable. Death may occur in minutes, hours or days	Moderate	For years at room temperature	Supportive care

29.5.2 Detection Methods

While vaccination helps protect a population from known infectious agents, rapid detection of a suspected act of biowarfare allows fast action to be taken to control the spread of disease. Current detection methods take advantage of the fact that each biological agent maintains its own unique DNA signature. Rapid detection methods use Polymerase Chain Reaction to make a billion copies of a single DNA strand within minutes. This method allows researchers to positively identify an infectious agent, by means of its DNA signature, using even the tiniest samples.

Following the progress of DNA chip technology, protein microarrays have emerged for these applications since they are not restricted to the detection of DNA or RNA carrying microbes, spores and viruses. Antibody arrays can also be applied to detect toxins. Furthermore, antibodies can detect molecules on microbial surfaces, so that in comparison to nucleic acid detection technologies and devices no additional time is needed to break open the target cells. However, the protein chip technology and its applications are still in its infancy. Protein chips are used in research applications but not for routine microorganism identification and routine diagnostics. The chip was developed for the ArrayTube platform providing a cheap and easy to handle technology solution that combines a microtube-integrated protein chip with the classical procedure of a sandwich-enzyme-linked immunosorbent assay and signal amplification by streptavidin-poly-horseradish peroxidase. Specific immunoassays for *Staphylococcus* enterotoxin B, ricin, Venezuelan equine encephalitis virus, St. Louis encephalitis virus, West Nile virus, Yellow fever virus, Orthopox virus species, *Francisella tularensis, Yersinia pestis, Brucella melitensis, Burkholderia mallei* and *Escherichia coli* EHEC O157:H7 were developed and optimized. All assays could be completed within 2 h and detection levels are demonstrated to be as low as in well established ELISAs. Most interesting, as a result of careful antibody screening and testing, it is currently possible to analyse at least five of the "dirty dozen" agents on one single protein chip in parallel. Application of nanotechnological approaches for detection of toxins like ricin decontamination of sulfur mustard are being pursued actively (Pradhan et al., 2009., Prasad et al., 2009).

29.6 Impact of Biological Weapons

Probably the biggest advantage is the fact that biological weapons are extremely efficient. It is hypothesized that one gram of purified botulinum toxin could kill 10 million people. This is approximately 3 million times more deadly than Sarin, a popular nerve agent. Yet another advantage is that gram for deadly gram, biological weapons are the cheapest of the lot. To "affect 1 km^2, it would cost about $2000 using conventional weapons, $800 using nuclear weapons, $600 using chemical weapons and a grand sum of $1 using biological weapons.

Disadvantages to using BW agents as weapons include their dependence on optimal weather conditions to result in effective dispersal, and their possible inactivation by solar irradiation and other climatic conditions. BW attacks would most likely occur late at night or early in the morning when agents would be less likely to undergo inactivation by ultraviolet radiation. At these times, atmospheric temperature inversions would allow an agent cloud to travel at low altitude to cover its target. The biggest disadvantage of using biological weapons is that they are really quite unpredictable. Who's to say that you won't end up infecting your own troops? Another disadvantage is that these agents last for quite some time. Anthrax, for example, can live for up to 50 years in soil. Therefore, it would be impractical to send in troops to occupy the area.

29.7 Tools for Self Defence Against Bioweapons

The tools for specific defense against bioweapons consist of vaccines against both viruses and bacteria, and of antibiotics and drugs against bacteria. Vaccines and antimicrobials are of limited usefulness because of the large number of possible microbes that can be used for weapons, because of antimicrobial resistance to drugs and antibiotics, and because of limitations in technical feasibility for developing vaccines and antibacterials against certain of the agents. Induction of non-specific innate immunity by immunostimulatory vaccines (at one time licensed) needs to be explored for possible immunoprophylactic-therapeutic activity when administered immediately following exposure to bioweapon pathogens.

The ideal solution to the bioweapons problem lies in measures to end their development and application throughout the world. Emphasis was made at the recent World Economic Forum for the need to end poverty and hunger in the world as a means to reduce the incentive to engage in warfare. Added to this is betterment of health, focused mainly on preventable diseases. A further solution to the problem may lie in the development of modern robotic systems for rapid forensic detection of development and production of bioweapons by "rogue" nations and even by individuals.

Research into the offensive use of biological weapons has been carried out all over the world. The investigation on the use of *Bacillus anthracis* (anthrax), botulinum toxin (botulism), *Yersinia pestis* (plague), *Francisella tularensis* (tularemia), *Coxiella burnetii* (Q fever), Venezuelan equine encephalitis virus, *Brucella suis* (brucellosis), and Staphylococcal enterotoxin B have already been made. In 1972, the United States and many other countries signed the Convention on the Prohibition of the Development, Production and Stockpiling of Bacteriological (Biological) and Toxin Weapons and on Their Destruction, commonly called the Biological Weapons Convention. This treaty prohibits the stockpiling of biological agents for offensive military purposes, and also forbids research into offensive use of biological agents. From 1975 to 1983, Soviet-backed forces in Laos, Cambodia, and Afghanistan allegedly used tricothecene mycotoxins (T-2 toxins) in what was called "Yellow Rain"(Seely et al., 1985) After being exposed, people and animals became disoriented and ill, and a small percentage of those stricken died. The use of T-2 toxins has been denied and the presence of the yellow spots was reported as being caused by deficating bees. Various molecular methods of detection of the agents by PCR, hybridization and strain typing made on the basis of bacterial total cell protein profiles, RAPD, ribotyping as well as of plasmid and DNA microrestriction analyses were used for the identification of pathogenic Burkholderia – *Burkholderia mallei* and *Burkholderia pseudomallei*. They are causative agents of glanders and melioidosis to man and animals and are regarded as potential agents of bioterrorism. The existing bacteriological and immunological methods of identification of *B. mallei* and *B. pseudomallei* are not efficient enough for the rapid diagnosis and typing of strains (Antonov and Iliukhin, 2005). Recently Schmoock and his colleagues (2009) reported a DNA microarray-based detection and identification of *Burkholderia mallei, Burkholderia pseudomallei* and *Burkholderia* spp.

Iraq, also a signatory of the Bioweapons Convention of 1972, admitted in 1991, that they had conducted research into the offensive use of *Bacillus anthracis*, botulinum toxins, and *Clostridium perfringens* (presumably one of its toxins). It was further discovered that they had also worked on development of aflatoxins, wheat cover smut, and ricin. In December 1990, the Iraqis filled 100 bombs with botulinum toxin, 50 with anthrax, and 16 with aflatoxin. In addition, 13 SCUD warheads were filled with botulinum toxin, 10 with anthrax, and 2 with aflatoxin. These weapons were deployed in January 1991 to four locations. In all, Iraq produced 19,000 of concentrated botulinum toxin, 8500 of concentrated anthrax, and 2200 of aflatoxin.

29.8 Genetic Engineering and Biological Warfare

By using genetic engineering, biological researchers have already developed new weapons that are much more effective than their natural counterparts (Atlas, 1998). Indeed, many research projects in basic science show sometimes unwillingly and unwittingly how to overcome current scientific and technological limits in the military use of pathogenic agents. Furthermore, genetic engineering is not merely a theoretical possibility for future biowarfare: it has already been applied in past weapons programmes, particularly in the former Soviet Union. One example is the USSR's "invisible anthrax", resulting from the introduction of an alien gene into *Bacillus anthracis* that altered its immunological properties (Pomerantsev et al., 1997). Existing vaccines proved to be ineffective against this new genetically engineered strain.

The natural pathogens are sufficiently dangerous and deadly, and genetic engineering is not necessary to turn them into more effective biological weapons. This is indeed true in that biological weapons can be used without genetic engineering or, for that matter, without any scientific knowledge as has been shown by their effective use in past centuries. In fact, genetic engineering has no role in a biowarfare programme. The development of reliable, effective biological weapons requires an intense and resource-demanding research programme that must, step by step, solve increasingly complex problems: the procurement of virulent strains of suitable agents, the mass production of the agents without loss of pathogenicity, and the development of an effective means of delivery. In particular, the third step is very demanding, and has rarely been accomplished.

The polio virus itself is not an effective biological weapon, but the experiment shows the tremendous potential of genetic engineering and also highlights its problems, particularly when applied to smallpox. The current risk assessments with regard to this virus rate the likelihood of an attack as being rather low, because it is highly unlikely although not completely impossible that countries other than Russia and the USA have access to it. If it should become possible to rebuild *Variola major*, the smallpox virus, in the laboratory from scratch – and the virus's genome sequence is available from biological databases this risk could change greatly. Smallpox is an ideal biological weapon, particularly for terrorist groups, because it is highly infectious and lethal and there is no effective treatment available. The relative safety that can be assumed today will then be gone.

However, the genetic enhancement of classical pathogens is only a small part of the broad array of possibilities that new biomedical techniques have created. From the point of view of disarmament, another trend is much more alarming: new types of biological weapons are becoming possible that were entirely fictitious until a few years ago. This is especially true of so-called "non-lethal" weapons that are designed for use outside classical warfare. The danger is that these new possibilities generate desires even in countries that previously renounced the use and development of classical biological weapons.

The global norm against biological weapons, laid down in the 1925 Geneva Convention and the 1972 Biological and Toxin Weapons Convention, clearly contributed to the fact that few countries have been engaged in research into offensive biowarfare during recent decades. This moral barrier seems to be lower for "non-lethal" weapons that are targeted against materials or drug-producing plants. Indeed, today's technical possibilities are creating a new interest in this area that might be leading to a new biological arms race.

Molecular biology and genetic engineering are still in their infancy, and more technical possibilities will arise in the years to come for military abuse too (Fraser and Dando, 2001). More efficient classical biowarfare agents will probably have only a marginal role, even if the genetically engineered "superbug" is still routinely featured in newspaper reports. More likely and more alarming are weapons for new types of conflicts and warfare scenarios, namely low-intensity warfare or secret operations, for economic warfare or for sabotage activities. To prevent the hostile exploitation of biology now and forever, a bundle of measures must be taken, from strengthening the Biological and Toxin Weapons Convention to building awareness in the scientific community about the possibilities and dangers of abuse. Any kind of biotechnological or biomedical research, development or production must be performed in an internationally transparent and controlled manner. In cases in which military abuse seems to be imminent and likely, alternative ways to pursue the same research goal have to be developed.

Biological warfare agents were one of the most feared classes of weapons that coalition troops believed Iraq might use against them. These agents can either be sufficiently toxic that they kill their victims quickly, or they may induce a sufficient degree of acute disability that troops cannot fight effectively or defend themselves against an attack (Török et al., 1997).

29.9 Impact of Genomics and Genetic Technology

We are in the midst of a revolution in biology that began with recombinant DNA technology in the 1970s, and is continuing with the rapidly accelerating technologies of genomics, proteomics, bioinformatics, etc. This revolution is transforming the sciences that underlie the development of chemical and biological weapons, as well as protection against such weapons. With regard to anti-agricultural biological agents, some of the implications are as follows:

1. *Genomics will facilitate protection of plants and animals against biological attack*

   Genomic technologies will make the development of new vaccines, new pesticides, new diagnostic reagents, and new genetic varieties easier and more rapid. This should facilitate a coordinated defense of agriculture against biological attack, at least for countries willing to make this a priority and spend the necessary funds.
2. *Genomics makes genotype-specific weapons possible*

   The emerging sciences of genomics and proteomics are already beginning to transform biology and medicine. This will continue, at an accelerating pace. One outcome will be the possibility of constructing genotype-specific biological control or weapons agents. Although there has been much concern about the possibility of ethnic weapons targeting specific groups of humans, the likelihood is that this will prove impossible (due to the high and increasing amount of intragroup genetic heterogeneity). However, equivalent weapons targeting specific agricultural varieties is a very real possibility.
3. *Agriculture is highly vulnerable to genotype-specific weapons*

   Agriculture, particularly in many developed countries, has several properties that make it vulnerable to attack with genotype-specific weapons. Typically it employs monocropping of large acreages with genetically identical cultivars, and high-density husbandry of genetically inbred animal strains. These agronomic practices reduce the genetic variability that makes populations resistant to genotype-specific weapons, and it creates conditions (large, dense populations) that facilitate disease spread.
4. *High-tech agent design is an option available only to sophisticated player*

   At least for the near future, the construction of novel toxin weapons or genetically engineered microbes as genotype-specific weapons would require a substantial scientific infrastructure, ranging from molecular biology to genomics to agronomy. It is unlikely that most terrorist groups could assemble the necessary expertise and materials for this. Thus genotype-specific weapons are likely to remain accessible only to states (and possibly their sub-state clients), and to large agricultural corporations.

29.10 Conclusion

The revolution in biotechnology, namely the new tools for analysing and altering an organism's genetic material, has led to an increased risk of biowarfare. The expansion of modern biotechnology in medical and pharmaceutical research and production has led to a worldwide availability of knowledge and facilities. Many countries and regions, where 30 years ago biotechnology merely meant brewing beer and baking bread, have established high-tech facilities for vaccine or single-cell-protein production that could be subverted for the production of biological weapons. Today, nearly all countries have the technological potential to produce large amounts of pathogenic microorganisms safely. Classical biowarfare agents can be made much more efficiently than their natural counterparts, with even the simplest genetic techniques. With modern biotechnology it becomes possible to create completely new biological weapons. And for technical and/or moral reasons, they might be more likely to be used than classical biowarfare agents. These possibilities have generated new military desires around the world, including within those countries that have publicly renounced biological weapons in the past. Genetic engineering has the potential to revolutionize biological warfare. Through techniques of genetic analysis and modification, steadily increasing in sophistication since the creation of the Human Genome Project, scientists now have the ability to modify deadly disease-causing agents to enhance their resilience and ease of infection, by selecting genes that hasten the onset of particular symptoms when put in contact with the body or possess a resistance to common treatments.

References

Antonov, V.A., Iliukhin, V.I., 2005, Molecular-genetic approaches to diagnosis and intraspecific typing of causative agents of glanders and melioidosis. *Mol Gen Microbiol Virusol* 2:3–9.

Arora, S., 2005, Cutaneous reactions in nuclear, biological and chemical warfare. *Indian J Dermatol Venereol Leprol* 71:80–86.

Arora, D.R., Gautam, V., Arora, B., 2002, Biological warfare: bioterrorism. *Indian J Med Microbiol* 20:6–11.

Atlas, R.M., 1998, Biological weapons pose challenge for microbiological community. *ASM* 64:383–388.

Barrett, A.D., 2004, Current status of the Arilvax; yellow fever vaccine. *Expert Rev Vaccines* 3:413–420.

Dixit, A., Alam, S.I., Dhaked, R.K., Singh, L., 2006, Development of an immunodetection test for a botulinum-like neurotoxin produced by *Clostridium* sp. RKD. *Indian J Med Res* 124(3):355–362.

Fraser, C.M., Dando, M.R., 2001, Genomics and future biological weapons: the need for preventive action by the biomedical community. *Nature Genet* 29:253–256.

Friedlander, A.M., Evans, M.E., 1997, Tularemia. In: Sidell F.R., Takafuji, E.T., Franz, D.R., editors. Textbook of military medicine: Medical aspects of chemical and biological warfare, part 1. Washington, DC: *Walter Reed Army Medical Center*, 503–512.
Guillemin, J., 2005, Biological weapons and secrecy. *FASEB J* 19:1763–1765.
Henderson, D.A., Inglesby, T.V., Bartlett, J.G., Ascher, M.S., Eitzen, E., Jahrling, P.B., Hauer, J., Layton, M., McDade, J., Osterholm, M.T., O'Toole, T., Parker, G., Perl, T., Russell, P.K., Tonat, K., 1999, Smallpox as a biological weapon: Medical and public health management Working Group on civilian biodefense. *JAMA* 281:2127–2137.
Inglesby, T.V., Dennis, D.T., Henderson, D.A., Bartlett, J.G., Ascher, M.S., Eitzen, E., Fine, A.D., Friedlander, A.M., Hauer, J., Koerner, J.F., Layton, M., McDade, J., Osterholm, M.T., O'Toole, T., Parker, G., Perl, T.M., Russell, P.K., Schoch-Spana, M., Tonat, K., 2000, Plague as a biological weapon: Medical and public health management. Working Group on Civilian Biodefense. *JAMA* 283:2281–2290.
John, T.J., 1996, Emerging and re-emerging bacterial pathogens in India. *Indian J Med Res* 103:4–18.
Khushiramani, R., Tuteja, U., Shukla, J., Batra, H.V., 2009, Production and characterization of monoclonal antibodies against YopM effector protein of *Yersinia pestis*. *Hybridoma* (Larchmt). 2009 Feb 10. [Epub ahead of print]
Kumar, O., Lakshmana Rao, P.V., Pradhan, S., Jayaraj, R., Bhaskar, A.S., Nashikkar, A.B., Vijayaraghavan, R., 2007, Dose dependent effect of ricin on DNA damage and antioxidant enzymes in mice. *Cell Mol Biol* (Noisy-le-grand) 30;53(5):92–102.
Pradhan, S., Boopathi, M., Kumar, O., Baghel, A., Pandey, P., Mahato, T.H., Singh, B., Vijayaraghavan, R., 2009, Molecularly imprinted nanopatterns for the recognition of biological warfare agent ricin. *Biosens Bioelectron* 2009 Apr 5. [Epub ahead of print].
Prasad, GK., Singh, B., Ganesan, K., Batra, A., Kumeria, T., Gutch. P.K., Vijayaraghavan, R., 2009, Modified titania nanotubes for decontamination of sulphur mustard. *J Hazard Mater* 15:167 (1–3):1192–1197.
Polyak, C.S., Macy, J.T., Irizarry-De, La., Cruz, M., Lai, J.E., McAuliffe, J.F., Popovic, T., Pillai, S.P., Minitz, E.D., Emergency Operation Centre International Team, 2002, Bioterrorism-related anthrax: international response by the centers for disease control and prevention. *Emerg Infect Dis* 8:1056–1059.
Pomerantsev, A.P., Staritsin, N.A., Mockov, Yu.V., Marinin, L.I., 1997, Expression of cereolysine ab genes in *Bacillus anthracis* vaccine strain ensures protection against experimental hemolytic anthrax infection. *Vaccine* 15:1846–1850.
Rao, P.V., Jayaraj, R., Bhaskar, A.S., Kumar, O., Bhattacharya, R., Saxena, P., Dash, P.K., Vijayaraghavan, R., 2005, Mechanism of ricin-induced apoptosis in human cervical cancer cells. *Biochem Pharmacol* 1;69(5):855–865.
Sastry, K.S., Tuteja, U., Santhosh, P.K., Lalitha, M.K., Batra, H.V., 2003, Identification of Bacillus anthracis by a simple protective antigen-specific mAb dot-ELISA. *J Med Microbiol* 52:47–49.
Rotz, L.D., Khan, A.S., Lillibridge, S.R., Ostroff, S.M., Hughes, J.M., 2002, Public health assessment of potential biological terrorism agents. *Emerg Infect Dis* 8:225–230.
Schmoock, G., Ehricht, R., Melzer, F., Rassbach, A., Scholz, H.C., Neubauer, H., Sachse, K., Mota, R.A., Saqib, M., Elschner, M., 2009, DNA microarray-based detection and identification of *Burkholderia mallei*, *Burkholderia pseudomallei* and *Burkholderia* spp.. *Mol Cell Probes* 1:37–42.
Seely, T.D., Nowicke, J.W., Meselson, M., Guillemin, J., Akratanakkul, P., 1985, Yellow rain. *Sci Am* 253:128–137.
Török, T.J., Tauxe, R.V., Wise, R.P., Livengood, J.R., Sokolow, R., Mauvais, S., Birkness, K.A., Skeels, M.R., Horan, J.M., Foster, L.R., 1997, A large community outbreak of salmonellosis caused by intentional contamination of restaurant salad bars. *JAMA* 278:389–395.
Vijayaraghavan, R., Gautam, A., Kumar, O., Pant, S.C., Sharma, M., Singh, S., Kumar H.T., Singh, A.K., Nivsarkar, M., Kaushik, M.P., Sawhney, R.C., Chaurasia, O.P., Prasad G.B., 2006,

Protective effect of ethanolic and water extracts of sea buckthorn (Hippophae rhamnoides L.) against the toxic effects of mustard gas. *Indian J Exp Biol* 44(10):821–831.

Witkowski, J.A., Parish, L.C., 2002, The story of anthrax from antiquity to the present: a biological weapon of nature and humans. *Clin Dermatol* 20:336–342.

Wuthiekanun, V., Amornchai, P., Chierakul, W., Cheng, A.C., White, N.J., Peacock, S.J., Day, N.P., 2004, Evaluation of immunoglobulin M (IgM) and IgG rapid cassette test kits for diagnosis of melioidosis in an area of endemicity. *J Clin Microbiol* 42:3435–3437.

Chapter 30
Segments of Molecular Diagnostics – Market Place

Abstract The molecular diagnostics market is the fastest growing segment in the in vitro diagnostics and is being driven by multiple growth factors. This include the need for automated and easy-to-handle techniques which combine optimized sample preparation, analysis, data evaluation, and the growing availability of molecular diagnostic tests for monitoring the therapeutic efficacy of expensive drugs. Currently immunoassays account for approximately 25% of the global market while molecular diagnostics accounts for about 6%. However, things are about to change as molecular diagnostics is poised to take a larger share of the in vitro diagnostics in the years to come. Infectious diseases market is the largest segment of molecular diagnostics and the segment is expected to grow at an annual rate between 7 and 8% over the next few years and this is followed by cancer and cardiovascular segments. In terms of technological push, genomics and proteomics are the major drivers of the molecular diagnostics market. Nanobiotechnology and biochips are also expected to drive future growth. Introduction of new diagnostics tests, primarily in the infectious disease application area is likely to keep the momentum going for the molecular diagnostics business in the short-term.

World molecular diagnostics market is poised to deliver strong double-digit annual growth over the next several years. Global market size, as measured in terms of dollars, is expected to reach $8.00 billion by the year 2015. United States represents the largest market, accounting for a 42.75% share estimated in the year 2008. Europe represents the second largest market, distantly followed by Japan. Market for infectious disease testing in Europe is projected to reach $1.50 billion by the year 2012. Opportunities for growth exists in emerging segments of pharmacogenomics and cancer screening, both of which are forecast to grow at a CAGR of 11.05 and 10.71% respectively over the period 2011–2015.

Keywords Molecular diagnostics segments · Infectious diseases · Cancer · Cardiovascular markers · Market drivers · Point of care diagnostic testing · Market forecasts

M. Debnath et al., *Molecular Diagnostics: Promises and Possibilities*,
DOI 10.1007/978-90-481-3261-4_30,

30.1 Prologue

Molecular diagnostic tests typically analyze key DNA, RNA, or protein biomarkers to identify a disease, determine its course, evaluate response to therapy, or predict individual predisposition to a disease. The techniques applied in this category involve analysis of DNA sequences, DNA methylation patterns, gene expression profiles, proteins, protein expression, or combinations of these biomarkers. Such biomarkers provide direct information about genotypic and/or phenotypic changes associated with specific diseases or responses to treatment. Biomarker assays are thus increasingly used throughout the process of disease management. Biomarker analysis has also become an important tool in drug discovery, preclinical drug development, and patient monitoring during clinical trials. The technologies that constitute molecular diagnostics include first-generation tools such as amplification, DNA probes, fluorescent in situ hybridization (FISH), second-generation biochips and microfluidics, next-generation signal detection, biosensors, and nanotechnology and proteomic technologies. In the past few years, this rapidly evolving field has seen several fascinating developments.

Molecular diagnostics combined with therapeutics forms an important component of integrated healthcare. Initial applications of molecular diagnostics were mostly for infections but are now increasing in the areas of genetic disorders, preimplantation screening and cancer. Molecular diagnostic technologies are also involved in development of personalized medicine based on pharmacogenetics and pharmacogenomics. Genetic screening tests, despite some restrictions is a promising area for future expansion of in vitro diagnostics market. Currently, there has been a considerable interest in developing rapid biomarker detection methods for point-of-care and biowarfare agents such as anthrax.

30.2 Concept

The markets for molecular diagnostics technologies are difficult to estimate. Molecular diagnostics markets overlap with markets for non-molecular diagnostic technologies in the in vitro diagnostic market and are less well defined than those for pharmaceuticals. In vitro diagnostics, although occupies a small bite in overall health care market (Fig. 30.1) have proved themselves as modern tools of this era and will play an important role in practice of medicine, public health, pharmaceutical industry, forensics and biological warfare.

Immunoassays are the most 'visible' In Vitro Diagnostic (IVD) technologies on the market mostly due to their application in infectious disease diagnostics. The global IVD market for 2006 has been valued at approximately $32 billion and is expected to grow by an average of 7% annually. Immunoassays account for approximately 25% of the global market while molecular diagnostics accounts for about 6%. Although molecular diagnostics currently occupies a modest position in overall IVD market (Fig. 30.2), things are about to change and molecular diagnostics will emerge as a major player of the global IVD.

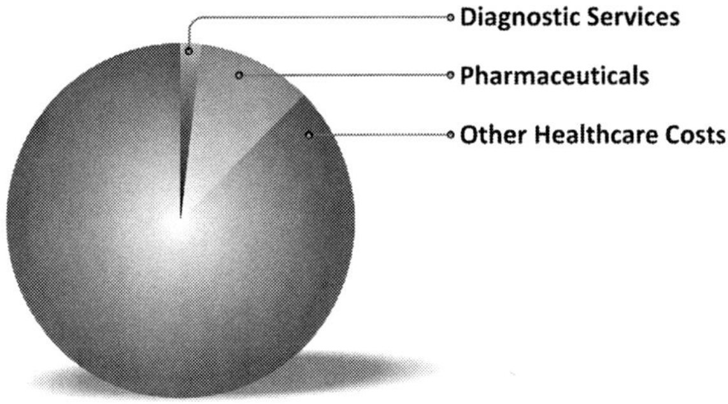

Fig. 30.1 Break down of world wide health care spending ($2.5 trillion)

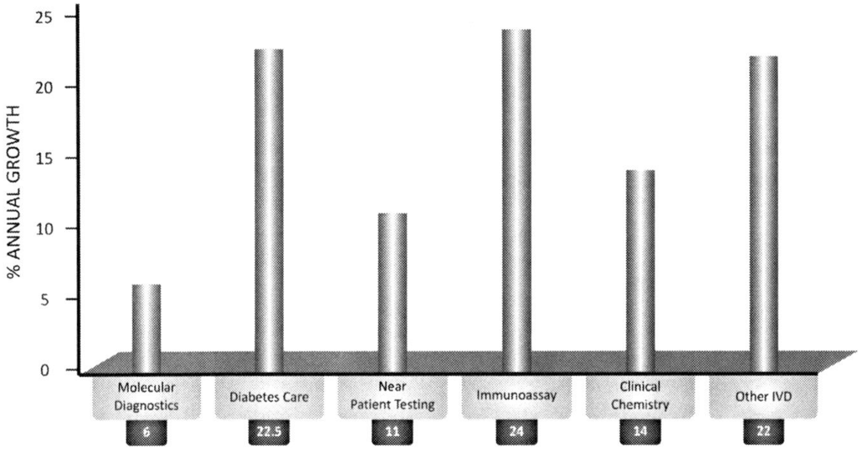

Fig. 30.2 In vitro diagnostic market segments (source: BBI report, 2005)

Molecular diagnostics is the fastest growing segment of the in vitro diagnostics market, with multiple opportunities for entry and growth, as well as competition and change. The current growth rate of Molecular Diagnostics is estimated in excess of 25% (Fig. 30.3) and is increasing at a rapid rate for various reasons. A major portion of the molecular diagnostic market can be attributed to advances in genomics and proteomics. It's a market that includes sales of reagents, instruments, and kits to clinical laboratories and research reagents that can be used by laboratories to develop their own in-house procedures. It also includes testing services by those clinical labs that have developed their own products, plus diagnostics companies that operate their own branded, certified testing services. The availability of diagnostic tools based on diverse approaches has rapidly been expanding since the advent of modern medicine akin to the rapid developments in the field of digital technology that occurred over the last decade.

The focus of this chapter is on market segments within the molecular diagnostics industry that are currently experiencing robust growth. This report focuses on

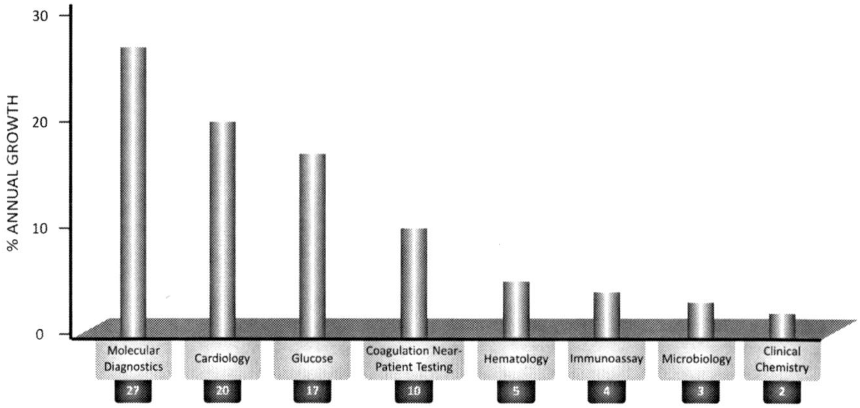

Fig. 30.3 Annual growth rate of various diagnostic segments (source: BBI report, 2005)

markets and business trends in the last 2–3 yrs and discusses the feasible market opportunities in different clinical diagnostic areas.

30.3 Market Drivers

As technical breakthroughs continue to drive the IVD market, the healthcare industry is becoming aware of their increasingly important impact on the way medicine is practiced. By focusing on predicting patient outcomes, targeted medicine, earlier disease detection and more advanced automation, diagnostics has secured considerable market growth potential. Molecular diagnostics currently is $2–$3 billion market and is growing at a 25% compounded annual rate (Fig. 30.3) and is predicted to grow to $3.5–$4 billion by 2010. Infectious disease and blood screening make up the majority (>70%) of this market. Other market segments include traditional genetics, personalized medicine, and cancer with 13, 9, and 8% of the U.S. clinical labs services market, respectively. Because infectious diseases are a mature segment, these other categories are growing more rapidly on a relative basis. Molecular diagnostics is also starting to emerge as an important tool for inherited disorders, cardiovascular disorders, and other disease areas. The driving force behind the rapid expansion of the diagnostic market can mostly be attributed to explosive developments in immunologic and DNA technologies propelling both new and established companies forward. Biochip and nanobiotechnology are expected to make a significant contribution to the growth of molecular diagnostics. Recent developments in the field of Pharmacogenomics are expected to make rapid developments in making disease diagnosis and therapy tailored to an individual's genetic makeup (often referred to as *personalized medicine*). Almost all of the Molecular Diagnostics tests today are based on PCR, but other technologies (most notably microarrays) are taking market share away.

30.4 Trends in Infectious Diseases Testing Market

Diseases caused by bacteria, viruses, fungi and other parasites are major causes of death, disability, and social and economic disruption for millions of people. Between 14 and 17 million people die each year due to infectious diseases – nearly all live in developing countries. In developing countries, lower respiratory infections, HIV/AIDS, diarrheal diseases, malaria and tuberculosis are among the top 10 causes of death. Early and sensitive detection of viral and bacterial infections is currently achieved by PCR based amplification technologies. Viral load testing has become standard for management of HIV, hepatitis C virus (HCV), and human cytomegalovirus infections, while nucleic acid probe-based testing for HIV and HCV is now in general use in specialized clinics and hospitals. The first molecular diagnostic tests to reach the market were infectious disease tests, and are one of the most rapidly growing segments of the in vitro diagnostics industry. This segment is dominated by tests for HIV, sexually transmitted diseases and hepatitis and this remains the largest segment of the molecular diagnostics market. Among the main driving forces behind the growth of this dynamic market is the spread of AIDS, which remains the world's major health threat and the key factor contributing to the rise in incidence of opportunistic infections. In 2004, the infectious disease diagnostics segment contributed up to 80.0% to the European molecular diagnostics markets, indicating high end-user acceptance of them. New analysis from Frost & Sullivan finds that the U.S. Infectious Disease Diagnostics Markets, as defined in the research service, earned revenues of $3.8 billion in 2005 and estimates this to reach $6.4 billion in 2012. The segment is expected to grow at an annual rate between 7 and 8% over the next few years and will present some excellent market opportunities for molecular diagnostic competitors.

30.5 Trends in Cancer Diagnostic Testing World Markets

Cancer is the second most common cause of death after cardiovascular disease in developed countries and cancer diagnostics is a growing market due to the ageing population. Cancer testing is one of the most important growth opportunities for the next three to five years in the diagnostics segment. Currently, cancer diagnostics market consists mainly of methods based on immunodiagnostics for detecting tumor markers. A number of companies are developing predisposition tests based on genomic variations associated with earlier stages of cancer. Co-development of molecular diagnostics and targeted therapeutics has already been proven to be a successful strategy in the development of novel anti-cancer drugs. Pharmacogenomic tests that permit tailoring chemotherapeutic therapy to the genetic characteristics of an individual patient and tumor have began to appear on the market. The unraveling of human genome is enabling biomedical scientists to workout molecular signatures of different cancers which may permit earliest possible pre-symptomatic diagnosis. A near-term market opportunity for cancer companion diagnostic tests exists in drug selection for cancer therapy. The National Cancer Institute estimates that about ten million Americans have or have had some form of cancer. Traditionally cancer has been treated with surgery, chemotherapy, hormones and radiation therapy, alone or in combination. Emerging technologies include photodynamic therapy, gene therapy, and immunotherapy and angiogenesis inhibitors. Pharmaceutical and biotechnology companies are investing billions of dollars to search out and develop weapons for the arsenal in the war against cancer. Overall costs of the disease are \$126 billion annually. Pharmaceutical companies are developing more than 300 new medicines for cancer, some of which are in development for more than one type of the disease, for a total of more than 500 ongoing R&D projects. Molecular oncology tests have been emerging as a second significant – and rapidly growing – market segment. One additional infectious agent, human papillomavirus, has joined the list of major tests. Cancer and pharmacogenomic diagnostics generated 10.0% of revenues in 2004.

30.6 Trends in Cardiac Marker Diagnostic Testing Markets

This immunoassay segment growth will result from tests based on recently validated biomarkers. One such growth opportunity example is in the cardiovascular disease drug therapy area where, currently, the relevant tests are only worth a few percent of the IVD market. Most tests relating to heart disease, e.g., cholesterol and triglycerides are done on high-throughput clinical chemistry analyzers using small volumes of inexpensive reagents. Thus the large numbers of cardiovascular tests performed annually worldwide do not translate into large markets.

However, sales of the newer cardiac markers, which are immunoassays, are growing strongly. European growth rates in 2003 exceeded 35%. These molecular diagnostics, used in risk stratification and therapy monitoring, illustrate the increasing role of specific biomarkers in selecting and guiding drug therapy.

30.7 Point of Care Diagnostic Testing World Markets

Point of care diagnostics is one of the most dynamic parts of the diagnostics industry and an area where several major competitors are battling for market share. Point-of-care testing (POC) is defined as diagnostic testing at or near the site of patient care. Over the years, the increasing introduction of portable and handheld instruments has resulted in the migration of POC testing from the hospital environment to a range of medical environments including the workplace, home, disaster care and most recently, convenience clinics. A recent market survey of US hospitals indicates significant growth in the number of hospitals performing point-of-care testing. Several established hospitals across the globe, now uses point-of-care testing for glucose, urinalysis, fecal occult blood, gastric occult blood, blood gases, provider microscopy, pregnancy, rapid cardiac markers, rapid strep and rapid influenza.

Growth in point-of-care testing will come about as more tests become available such as a new point-of-care technology for continuous non-invasive monitoring of blood gases and electrolytes during surgery. This system replaces repetitive testing by the central laboratory which can provide only intermittent results. In addition, the rapidly expanding market for home testing and patient self-monitoring will continue to drive growth.

The POC testing market is currently estimated to be worth US$11.3 billion in 2007 and is growing at 11% a year. POC testing accounts for approximately 34% of the US$33.6 billion global IVD testing market. Diagnostic testing accounts for only between 1and 2% of government healthcare expenditures worldwide, yet influences between 60 and 70% of healthcare decisions. Unsurprisingly, the POC market has also seen moves by large multinationals which previously had no presence in the IVD market. Siemens, for example, made major acquisitions to establish its diagnostics division, with Diagnostics Products Corporation and Bayer Diagnostics coming under the Siemens umbrella in 2006 at a cost of US$5.7 billion. The company has also just completed the acquisition of leading diagnostics company, Dade Behring, in a transaction valued at approximately US$7 billion.

It is believed that Over-the-Counter Diagnostic Products World Markets market segment has the potential for growth rates above 15% per year over the next five years in some of the less well-explored and well-developed areas.

30.8 Next Generation Molecular Diagnostics

The current molecular diagnostics products consist mainly of tests designed around a single biomarker associated with a disease state. Next-generation molecular diagnostics will employ multiplexing platforms, such as DNA or protein microarrays, to perform parallel biomarker analyses. Increasing automation and miniaturization, combined with new technologies that allow ultrasensitive quantitative multiplexed detection, is expected to increase sensitivity and reduce cost for many diagnostic applications. There are indications that the next advance will involve measuring several biomarker types simultaneously (for example DNA and protein). Such innovations relate to constructs, such as DNA/antibody microarrays capable of assaying proteins and NATs simultaneously, or to immunoassays using nucleic acid labels, where the results are amplified and read via PCR. Such tests would allow immunoassays and nucleic acid tests to be performed on a single instrument.

Also observed is a growing trend from low- to high-margin molecular diagnostics as developers have become more concentrated and have capitalized on patent-protected protein and gene-based biomarkers. This should make investment in molecular diagnostics more attractive to investors. Innovation in the molecular diagnostics area appears to be robust as demonstrated by an analysis of issued patents and published patent applications from January 2000 to September 2008.

30.9 The Rise of Companion Diagnostics

The term companion diagnostic means that the particular diagnostic test under evaluation is specifically linked to a known therapeutic drug. This linkage could be important in the therapeutic application and clinical outcome of a drug (personalized medicine), or an important component of the drug development process. Pharmacogenomics, though still developing, holds immense potential to develop into a lucrative market as the demand for personalized medicine increases.

Many drug companies are evaluating business models based on molecular diagnostics for pharmacogenomic testing as the driving force behind the shift toward personalized medicine. A number of hurdles confront the personalized medicine model, including regulatory policy, patient-provider cost coverage, and patient record privacy.

30.10 Market Considerations and Forecasts

The number of companies involved in molecular diagnostics has increased remarkably during the past few years. More than 500 companies have been identified to be involved in developing molecular diagnostics and 250 of these are profiled in the report along with tabulation of 573 collaborations.

The molecular diagnostic segments are projected to outperform the overall diagnostics market, growing from $13.8 billion in 2005 to $22.7 billion in 2010 (AGR of 10.4%), in contrast to the overall forecasted increase from $36.5 billion in 2005 to $53.6 billion in 2010 (AGR of 8.0%) for the entire IVD market. Although the overall laboratory-based immunoassay segment is a mature market, the molecular diagnostics segment of the immunoassay market of $12.8 billion in 2005, is expected to increase at an AGR of 8.9% to $19.5 billion in 2010. The sales of the newer cardiac markers, which are immunoassays, are growing strongly. European growth rates in 2003 exceeded 35%.

The first generation of marketed molecular diagnostics has been based on established immunoassays and nucleic acid testing (NAT) technologies. Although the NAT market continues to be dominated by Roche, Abbott Laboratories and Bayer, many biotechnology companies have recently acquired significant intellectual property rights in this area.

The company leaders in immunoassay and NAT identified in the D&MD Report include Roche Diagnostics, Abbott Laboratories, and the Ortho Clinical Diagnostics and Lifescan units of Johnson & Johnson, Beckman Coulter, Bayer Diagnostics, Becton Dickinson, Dade Behring, and bioMerieux. Important niche suppliers include Agendia, Bio-Rad, Biosite Diagnostics, Cepheid, Chembio Diagnostic Systems, Chiron (Novartis), Ciphergen Biosystems, Correlogic Systems, Cytyc, DiagnoSwiss, Diagnostic Products, Digene, DiaSorin, Epigenomics, EXACT Sciences, Gen-Probe, Genaissance Pharmaceuticals, HandyLab, Innogenetics,

Interleukin Genetics, Matritech, Myriad Genetics, Nanogen, Nuvelo, Nymox Pharmaceutical, OXIS International, Proteome Sciences, Sequenom, SomaLogic, Sysmex, and Third Wave Technologies.

Roche (40% market share), Gen-Probe, and Chiron lead the market with their dominant positions in infectious disease and blood-screening areas. The top nine players (Roche, Chiron, Gen-Probe, Abbott, Digene, Bayer, Myriad Genetics, Becton Dickinson, and bioMerieux) own approximately 75% of these markets.

Leon projects that the cancer segment is growing at about 20% a year, traditional genetics about 15% a year, and personalized medicine about 20% a year, compared to the 5–10% growth rate for infectious diseases. Growth also varies by individual tests within segments. For the Digene HPV test, the company reported a 38% increase in sales from fiscal 2005 to 2006, or $97.4 million to $134.4 million, then a 42% growth rate for the first nine months of fiscal 2007 compared with the same period in 2006. The company has since been acquired by Qiagen. For the Oncotype DX test, Genomic Health reported revenues of $4.8 million in 2005, $27 million in 2006, and $27.7 million for just the first six months of 2007.

In a recent web survey, over 250 participants ranked the segments poised for growth over the next two years and the next five years. Many of the survey respondents work in research areas such as biomarker discovery and validation, diagnostics research, diagnostics product development, discovery or preclinical research, and clinical research. The responses hence reflect more of the R&D phase rather than already commercialized products. Near term, respondents predicted significant growth in oncology at 33% and infectious disease at 30.1% with less growth seen for genetic testing at 14.5%, and pharmacogenomics at 12.5%.

Longer-term growth, as predicted by the survey respondents, remained the same categorically, but priorities changed. Increase in the oncology market was about the same at 34%, the pharmacogenomics prediction increased to 26.5% long term, infectious diseases growth over the long haul fell to 14.6%, and genetic testing was also forecasted to see less growth at just 12.3%.

In vitro molecular diagnostics have established themselves as effective tools for all aspects of disease management, especially in areas of unmet clinical need. Such tests have been developed for screening and prognosis as well as for applications, such as determination of genetic predisposition to disease, detection of presymptomatic disease, and prediction of individual drug response. Molecular diagnostics are the basis of pharmacogenomics, enabling the evolution of personalized medicine. Current molecular diagnostics are primarily single-analyte tests involving the detection of a single gene or protein. However, many disease-related processes are multifactorial, involving the abnormal expression of multiple genes or proteins. Second-generation molecular diagnostics are anticipated to utilize novel detection technologies and multiplexing platforms to allow the measurement of a large number of analytes simultaneously. These innovations will increasingly utilize multiplexing platforms such as DNA microarrays that perform parallel biomarker analyses.

Regulatory events of note during the last two years include FDA approval of the first DNA microarray instrumentation system for in vitro diagnostic use

(Affymetrix' GeneChip System 3000Dx) and the first highly multiplexed diagnostic microarrays: Roche's AmpliChip CYP450 test and two cystic fibrosis tests – Tm Bioscience's Tag-It and Osmetech's eSensor. Although not yet approved by the FDA, tests based on transcriptomic profiling have also debuted successfully on the market. There will be a steady increase in demand for new diagnostic testing services in the next five-year period, along with pressures to improve the quality of healthcare delivered in the clinic and also to lower its costs.

Molecular diagnostics is a rapidly growing and rapidly changing market. Our forecast annual growth rates for the leading application groups are: tumour markers 8%, speciality immunoassay markers 15%, NAT for infectious disease 21% (except genotyping applications, where we forecast 44%), cancer screening (NAT) 34%, genetic testing 10%, and predispositional/theranostic testing (NAT) 26%. Geographically, the markets for these tests overall will grow slowly in Japan, somewhat faster in the US, and relatively rapidly in Europe. With in a span of 2–3 yrs both the US and Europe will each have around 35% of the worldwide diagnostics market. US demand for in vitro diagnostic (IVD) products will grow 5.1% annually through 2011. Clinical chemistry and immunoassay methods will remain the top two IVD sciences, while nucleic acid testing products will continue to generate the fastest gains. Hospital labs will stay the largest market while home health care will grow the fastest. With the exception of blood screening, adoption of molecular diagnostics by the rest of world and Japan is limited to the infectious disease segment.

Faster growth rates and therefore superior commercial opportunities are expected in few developing regions. The major suppliers of established immunoassay systems have proved able to leverage their experience in dealing with clinical laboratories in order also to become leading suppliers of NAT products. Such companies include Roche, Abbott, and Bayer, which have at least 30% of the market between them.

The Molecular diagnostics segment is increasing at a rapid rate for two reasons viz., introduction of early-stage, new cutting edge diagnostic technologies by major market players and secondly due to rapid development of automated systems and throughput technologies. Studies indicate that molecular diagnostic companies are expecting continued strong growth. Many of today's molecular diagnostic tests are complex and low-volume compared to other traditional in vitro diagnostic tests and is only performed by 25% of the clinical laboratories that perform in vitro diagnostic tests. However, as test volumes increase and automation improves, molecular diagnostic tests will eventually be routinely performed by all segments of the clinical laboratory. As advances in molecular diagnostics continue, molecular diagnostic enterprises will become more attractive to venture capitalists, a group that previously stayed away from such investments. The wave of new information provided by molecular diagnostics has ushered in an era, evidenced by the growing importance of FDA guidelines on pharmacogenomics and the coupling of in vivo imaging techniques with molecular tests and molecular markers. Companies choosing to play in this space must understand its complexities and pitfalls.

References

The information presented in this report is derived from publicly available information sources such as company, government, and medical organization reports and internet resources such as Molecular and Nucleic Acid Testing Market Reports from web based Kalorama information services.

Avi Chaudhuri, 2008. Molecular Diagnostics: Future Potential for the Indian Marketplace, PAC Med Biotech Pvt. Ltd.

Blood: The Worldwide Market for Blood Products, Blood Testing, Blood Equipment, and Synthetic Blood Products. Nov 1, 2008 – *KLI1816262*

Diagnostic Test Service Commercialization in Multiplex and Esoteric Testing: A Roadmap to Diagnostics in the 21st Century. Aug 1, 2008 – *KLI1806317*

Drug Discovery & Development. *Magazine* Vol. 9, No. 9, September 2007, p. 68.

Glucose Testing Markets. Nov 1, 2008 – *KLI1911068*

Genetic engineering & Biotechnology News 2007, Vol. 27, No. 3

HIV: Markets for Diagnostics and therapeutics. Oct 1, 2008 – *KLI1646600*

Lab Automation Markets Worldwide (Next-Generation System, Key Companies, Forecasts and trends) Aug 1, 2008 – KLI1670001

Marena Subitra Rajeevan., 2006. Immunoassays versus Molecular Diagnostics: Who has the upper hand? *Research Analyst, Healthcare*. March issue.

Retail Clinics and the In-Store Healthcare Market 2009: Bright Spot in the Current Economy? Mar 1, 2009 – *KLI2041111*

The World Market for Molecular Diagnostics. Jan 1, 2009 – *KLI1914575*

The Worldwide Market for In Vitro Diagnostic (IVD) Tests, 6th edition. Dec 1, 2008 – *KLI1430266*

The Clinical Laboratory Services Market (Growth Opportunities, Competitive Analysis and Competitor Profiles) Nov 1, 2008 – *KLI1646590*

Worldwide In Vitro Diagnostics (IVD) Market Research Bundle (Worldwide Market for In Vitro Diagnostic Tests, SNP Genotyping Markets, Companion Diagnostics, Testing Services Markets) Oct 1, 2008 – *KLI1927625*

Index

A

Alzheimers, 201
Amplification, 7, 15, 38–39, 59–60, 130–143, 174–175, 217–218, 233–234, 238–241, 309–320, 340, 366, 387, 457, 460–461, 475–476
Amplification based test, 313
Analytes, 8, 20, 196, 211–212, 298, 385, 477, 513
Antemortem, 430
Anthrax, 6, 146, 175, 277, 390, 484–485, 487, 489, 492, 496–497, 506
Anthropometry, 431–432
Antibiogram, 231
Antibody, 19–20, 43–44, 51, 56, 65, 77, 137, 172–173, 175–177, 182, 184, 187–188, 190, 212–213, 216, 249, 276, 295, 301, 317–318, 320–321, 362, 387–388, 414–415, 473–474, 495, 511
Antibody-based microarray techniques, 175
Antibody based test, 19, 295, 306, 321, 473–474
Antigen, 20, 40, 42, 48, 62, 93, 103, 145, 172–176, 189, 229, 294, 298–299, 303, 318–319, 358–359, 387–388, 421–422, 503
Antigen-antibody interactions, 56, 173
Aptamers, 20, 55–65, 298
Autopsy, 427, 432

B

Bacterial infections, 130, 186, 241, 321–323, 509
Bacteriocin typing, 231
Biochips, 10, 222, 383–391, 505–506
Biodefence, 483
Biological weapons, 484–490, 495–500
Biomarkers, 13–14, 17–23, 26, 102, 166, 204–206, 251, 287–307, 357, 359, 362, 395, 408, 510–511
Biomedical research, 2, 8, 34, 85, 104–106, 206, 263, 347, 406, 417, 498
Biopharmaceuticals, 40–41, 48, 414–417, 419–423
Biopharming, 41–42
Bioreceptors, 212
Biosensor, 61, 209–223, 280–283, 387, 390, 469, 473, 477–478, 506
Bioterrorism, 5, 203, 217, 483–484, 497
Biotyping, 228, 230–231, 310
Bioweapons, 484, 490, 496–497
BRCA1, 4, 115, 144, 290, 294, 300, 343–344, 403, 406, 445

C

Cancer, 5, 90, 92, 102, 293, 297–302, 304, 306, 343, 356–360, 389, 419–420, 448–449, 509–510
Cancer biomarkers, 20, 293, 300, 389
Candidate markers, 104, 257, 295, 355
Cardiovascular markers, 505
CAT, 73, 75, 78, 80
CDNA cloning, 35
CDNA libraries, 42–43, 186, 190–192
CDNA sequences, 35, 141
CFTR, 45, 336–337
Chromosome, 1, 13, 38, 42, 50, 72, 87, 91–92, 116–117, 158–166, 332, 334–335, 339, 349, 353–354, 361, 367–369, 373–374, 389, 406, 429–430, 462
Chromosome painting, 159, 165
Clinical diagnostics, 2–3, 18, 81, 94, 147, 206, 512
Clinical forensic medicine, 432–433
Clinical medicine, 94–95, 113–114, 173, 222, 386–387, 426

M. Debnath et al., *Molecular Diagnostics: Promises and Possibilities*,
DOI 10.1007/978-90-481-3261-4, © Springer Science+Business Media B.V. 2010

Clinical testing, 4, 282, 335, 446
Clone, 35–36, 40, 43–44, 46–47, 58–59, 131, 139, 147, 183–184, 188–191, 201, 263, 295, 321, 325, 389
Cloning allergens, 190–191
Colorectal cancer, 90, 205, 303, 343, 348, 387, 440, 443
Combinatorial library, 57–58
Comparative genome hybridization, 197, 337, 343
Complex diseases, 23, 101, 256, 347–375, 400, 402
Computational tools, 16, 24, 92, 298
Cystic fibrosis, 4, 45, 88, 93, 115–116, 121, 144–145, 216, 222, 336–338, 341, 343, 350, 423, 514
Cystic Fibrosis Transmembrane Conductance Regulator gene, 331
Cytochrome P450, 273, 390, 405, 439–442
Cytogenetics, 153, 157, 159, 162–164, 334, 337, 348

D

Dendrimers, 269, 271
Detections, 477
Diabetes, 9, 23, 26, 34, 88–89, 100, 103, 341–342, 348, 363–364, 367, 406, 418
Diagnosis, 3–5, 18, 20, 28, 49–51, 86–87, 89, 97–106, 111–112, 144–146, 164–166, 177–178, 203–204, 213, 216–217, 228–229, 232, 257, 276, 291, 294, 310–323, 331–344, 348–350, 352–360, 386–388, 417, 470, 509–510
Diagnostics
  Diagnostic medicine, 3
  Diagnostic test, 2, 4–6, 9, 124, 174, 210, 229, 238–241, 302, 310–311, 319–321, 334, 337, 341–343, 366, 395, 449–450, 506, 509–512, 514
DNA
  aptamers, 56, 59–60, 62, 64–65
  biosensor, 209–223
  diagnosis, 331, 348
  diagnostics, 50, 209, 213, 417
  fingerprint, 45, 201, 233, 428–430, 457–458
  fingerprinting, 201, 233, 428–430, 458
  methylation, 431, 506
  profiling, 324, 455–456, 461
  selection, 55–65
  sequencing, 36, 46, 87, 92, 130, 145, 343, 437
  typing, 136, 455–459, 460, 463–464
Drug
  delivery, 262–263, 265, 267–270, 274–275, 280–283, 408
  design, 91, 184, 191, 394, 406
  discovery, 10–11, 20, 76–77, 98–99, 186–187, 190, 197, 202–204, 256, 288, 390, 396–399, 407–409, 438–439, 444–448, 506
  response, 16, 23, 29, 106, 200, 202–203, 374, 385, 398–399, 402, 404–405, 407, 409, 436–438, 441, 449, 451–452, 513
  screening, 81, 217, 282
Duchenne muscular dystrophy, 90, 336, 338

E

Electrochemical biosensors, 214, 216–218
ELISA, 73, 77–78, 173–178, 213, 306, 321, 323, 473–474, 487–488
End labeling, 156
Environment monitoring, 209
Environmentome, 103
Enzyme-linked immunosorbant assay, 173
Epitope Mapping, 185–186, 188
Ethical legal social issues, 393
Ethics, 111–127
Expressed sequence tags (ESTs), 124
Expression
  of genes, 44, 143, 161
  profile, 12, 14–17, 23, 28, 199–200, 204–206, 363, 390, 506
  signature, 11

F

Factor V (Leiden) mutation, 342
Fingerprint, 4, 19, 45, 118, 200–201, 232–233, 291, 324–325, 347, 428–430, 456–458, 463, 472
FISH, 10, 14, 74, 121, 156–160, 162–166, 249, 258, 299, 337, 470, 506
Fluorescence, 8, 10, 18, 64, 75, 81, 141, 145, 156–159, 161–162, 164–165, 174, 196, 210–211, 229, 249, 278, 299, 391, 476, 487
Food
  allergens, 469–470
  borne disease, 468, 470–471, 479
  diagnostics, 467–481
  safety, 219, 280, 467–469, 471–473, 479–481
Forensic, 2, 4, 10, 45, 143, 147, 217, 278, 365, 425–433, 455–464, 506
Forensic medicine, 4, 425–433, 462

Functional genomics, 5, 13, 86, 98, 101–102, 186, 194, 201, 393
Fungal infections, 323

G
GC-content, 133
Gene
  cloning, 36, 413–415
  expression, 8, 12–17, 24–25, 49, 62–63, 73–76, 79, 80–82, 103, 106, 139, 161, 166, 194–197, 200–206, 245–246, 262, 272, 275, 336, 371, 384, 386, 390, 403
  sequence, 14, 43, 86, 95, 123, 183, 279, 401, 414, 416, 474, 484
  technology, 49, 82, 391, 455–464
  therapy, 45, 50, 72, 81–82, 90, 106, 119–120, 125, 245, 270–272, 367, 398, 405, 414–415, 422, 510
  transcript, 14, 82
  transfer, 47, 51, 106, 125
  variation, 395, 437, 443
Genetically modified organism, 124, 468
Genetic assay, 3, 113, 257
Genetic disease, 9, 49–50, 85, 99, 116, 119, 126, 144, 201, 270, 282, 331, 333–334, 336–343, 348–349, 438
Genetic disorders, 49, 114, 120, 127, 130, 162, 331–344, 394, 421, 506
Genetic markers, 9, 93, 202, 288, 359, 364, 374, 400–401, 442, 456–459
Genetic test, 3, 6, 50, 94–95, 111–115, 120–124, 126, 209, 334–336, 343–344, 350, 357, 364–365, 374–375, 390, 400, 405, 407–409, 436, 442, 450, 513
Genetic testing, 3, 6, 95, 113–115, 121–123, 332, 334–336, 343–344, 357, 374–375, 407–409, 439, 450, 513
Genetic testing diagnostics, 3, 6, 113–114, 144, 334, 343, 374, 450
Genome
  cloning, 35–36
  DNA libraries, 42
  medicine, 95
  patents, 123–124
  technology, 95, 123–126
Genomics, 5, 12–15, 21, 23, 25–26, 86, 94, 98, 101–102, 104, 186, 194, 245–247, 254, 282, 288, 375, 390, 404, 406–407, 415–416, 436, 449, 499–500, 507
Genosensors, 214–215
Genotyping, 15, 24, 136, 145–146, 237, 266, 314, 323, 325–326, 343–344, 366–367, 398–399, 407–408, 445, 450–451, 472
Germ line therapy, 119
GFP, 73–74, 76–77, 81
Gleevec, 15, 300, 302, 398, 406
Glycosylations, 39–40, 49, 250
Glycotherapeutics, 421
GUS, 73, 79, 129, 214, 489–490

H
Haemophilia, 90, 336, 338, 365–367
Haemophilia A and B, 336
Hamster cell lines, 40, 48
Haplotype, 95, 288, 290, 339–340, 355, 371, 446
HapMap, 95
Health care, 4–6, 47, 85–86, 106, 112, 114, 120, 122, 254–257, 375, 404–405, 413–423, 439, 444, 449, 506
Hemochomatosis, 342
Herceptin, 300, 343, 398, 401, 405–406, 450
HGH, 48, 77, 238
HGP, 85–96, 112–114, 117–121, 126, 209
High throughput technology, 11
HUGO, 86
Human cell lines, 40–41, 49
Human genome project, 14, 17, 26, 85–87, 89, 92–95, 98–99, 112, 123–126, 143, 249–251, 347, 400, 428, 446, 500
Human Monoclonal Antibodies, 421
Humulin, 34
Hybridization, 7, 56, 142, 145–146, 153–166, 197, 201, 204, 213–218, 220, 237, 240, 276, 299, 301, 311–312, 319, 324–325
Hybridoma cell lines, 40

I
Immobilisation, 214
Immunoassays, 3, 171–178, 212–213, 417, 469, 474, 487, 495, 506, 510–512
Immunogenicity, 39, 60, 64, 175, 177, 270
Immunological method, 173
Immunomagnetic separation, 473–475
ImmunoRCA, 174
Immunosensor, 219
Immunostimulatory vaccines, 496
Individualised drug treatment, 91
Inducer, 74–75
Infections, 40–41, 51, 62, 130, 144, 177, 186, 233, 237, 274, 310–323, 394, 439, 470–471, 487, 506
Infectious disease, 4–5, 7–9, 41, 48, 50–51, 100, 104, 144–145, 203, 217, 228, 240, 309–311, 324–328, 484–485, 508–509, 513–514

Inherited disorder, 44–45, 85, 332, 508
Insect cell lines, 40
In situ hybridization, 10, 142, 145, 153–166, 299, 301, 337, 506

K
Kary Mullis, 129

L
Lac, 78, 200
LacZ, 74, 78–79
Luc, 73, 75–76

M
Marfan syndrome, 350
Market drivers, 508
Market Forecasts, 505
Mass spectrometry, 17–20, 22, 25, 101, 173, 216, 246, 249–252, 256–258, 301, 306, 464
Melting temperature, 133–134, 215
Metabolic profiling, 22
Metabolome, 12, 21–23, 25, 98
Metabolomics, 12, 21–23, 25–28, 99, 101–103, 194, 406
Microarray, 5, 8–9, 13, 16–17, 20, 24–25, 98, 166, 175, 193–206, 238, 282, 298, 301, 343–344, 384–386, 388–389, 391, 407, 437
Microbiological testing, 227–241, 479
Microsatellite, 146, 349, 365–366, 430, 446, 461
Mimetics and peptidomimetics, 421
Miniaturization, 10, 81, 95, 210, 219, 222, 282, 386–388
Mitochondrial DNA, 100, 430, 461–462
Mitochondrial DNA analysis, 430, 461
Molecular diagnostic industry, 6, 401–402
Molecular diagnostics, 4–10, 15, 75, 99, 114, 144–145, 204–205, 217–218, 228–230, 237, 239–241, 276, 310–311, 313, 317, 337, 342–343, 386, 390, 394, 407, 431, 451–452, 468, 506–507, 509, 512, 514
Molecular diagnostics segments, 505
Molecular epidemiology, 324–325
Molecular farming, 419–420, 422–423
Molecular imaging, 81, 275, 308
Molecular medicine, 9, 51–52, 119, 272–273, 360
Molecular methods, 164, 234–236, 314–316, 323–327, 372, 497
Molecular signature, 4, 6, 15, 204, 289, 297, 403, 509
Molecular techniques, 232, 236–237, 316, 319, 323, 326
Molecular tests, 4, 9, 228, 239, 241, 309–310, 323, 325–326, 514
Monogenic disorders, 332, 348, 400
MonoLEX, 60
MtDNA, 430, 459, 461–462
Multifactorial disorders, 332, 348–349
Multigenetic diseases, 104
Multiple endocrine neoplasia, 332, 341–342, 357
Multiple sclerosis, 348, 360–363
Mutations, 3, 24, 44–45, 49–50, 90, 93, 102, 104–105, 115, 136–137, 144–145, 215–216, 237–238, 290, 310, 314–315, 326, 332–333, 336–344, 347–351, 387–388, 400, 403, 406
Mycotoxins, 469, 474, 488–489, 491, 494, 496

N
Nanocarrier, 192, 264, 267–271
Nanochips, 337
Nanodiagnostics, 276–278
NanoDLSA, 175
Nanofabrication, 222
Nanoparticles, 214, 216, 262–268, 270, 274–275, 391, 408
Nanosensor, 216, 477
Nanospheres, 268, 271
Nanotechnology, 10, 216, 256, 261–276, 278, 280–283, 386–387, 406, 506
Nanotube, 214, 262, 265, 270, 273–274, 279, 281–282
Non PCR based method, 459
Nucleic acid hybridization, 50, 214, 217–218, 319
Nucleic acid probe, 43, 140, 142, 154, 212, 233, 311, 325, 509
Nutrigenomics, 26, 370

O
Obesity, 26, 95, 100–101, 104, 347–348, 367–370
Oligomers, 56
Oligonucleotide, 50, 56–60, 134, 137, 140, 154–156, 174, 195–197, 202, 204, 214–215, 217, 262, 271–272, 310, 324, 337, 384, 388
Oligonucleotide library, 57

Omics, 11–29, 101–102, 287, 406
Oncology, 4, 6, 16, 163–164, 173, 266, 343, 445, 510

P
Patent, 9, 49, 105, 117, 123–126, 239, 408–409, 417–418, 423, 511
Pathogens, 42, 61–62, 146, 165, 210, 217–219, 230, 232–234, 236–238, 257, 311–314, 468–478, 484, 497
Pathology, 1–3, 19, 21, 25–26, 50, 166, 199, 204, 288, 295, 307, 362, 398, 400, 426
PCR, 7–8, 44, 131–147, 238–241, 294, 313, 326, 337–340, 366, 429–430, 456–457, 459–463, 474–476
PCR based methods, 324, 459–463
Peptide nucleic acid, 161, 165, 215
Personalised medicine, 206, 393–410
Phage display, 17–18, 181–191, 274, 421
Pharmaceutical industry, 10, 52, 63, 104–105, 171, 266, 268, 278, 356, 394–396, 398, 413–416, 418, 423, 448–449
Pharmaceuticals, 27, 33–52, 64, 87, 122, 398, 414–418
Pharmacogenetics, 6, 91, 387, 394, 405, 441, 444–445, 447–450, 506
Pharmacogenomics, 6, 86, 102, 106, 202–203, 326, 390, 394–395, 399, 403–407, 435–453, 508
Pharmacogenomic tests, 401, 439, 442–444, 509
Phenylketonuria, 336, 339, 374
PNA, 10, 161, 165, 215
Point of Care Diagnostic Testing, 505
Polymerase chain reaction, 1, 7, 38, 44–45, 129–148, 228, 233–234, 310–311, 317, 319–320, 359, 366, 387–388, 429–430
Polymerisation, 138
Polymorphism, 116, 137, 141, 218, 233, 301, 338, 355, 359, 364–366, 386–387, 430, 452, 457, 461–462
Post genomic, 97–106, 246, 307, 375, 384
Post mortem, 373, 433
Post translational modifications, 41, 289, 362, 375
Predisposition, 2–3, 5, 9, 89, 115, 117, 201, 276, 325, 332–333, 349–351, 386, 395, 400, 506, 513
Primer Design, 133
Probe, 5, 7–8, 43–44, 147, 153–161, 201, 214–217, 233, 237, 279, 311–312, 315, 324–325, 339, 384, 391, 418, 430, 457
Profiling microarray, 8, 20, 25, 166, 200, 359, 386
Prognostic, 4, 6, 10, 16, 23, 28, 50, 100, 166, 204–206, 256, 343, 355, 399
Promoter, 16, 43, 45, 47, 51, 58, 71–75, 80–81, 194, 321, 469
Promoter assay, 80
Protein arrays, 20–21, 205, 253, 306, 385
Protein microarray, 17, 20, 197–199, 205, 256, 495, 511
Protein microchips, 177
Protein profiling, 18, 20, 25, 251, 375
Protein – protein interaction, 17, 19, 21, 25, 77, 81, 217, 249, 257–258, 289
Proteome, 5, 12, 17–21, 25, 27, 98, 102, 184, 205, 246, 250–253, 257–258, 375, 389
Proteomics, 5, 12, 17–21, 26–28, 99, 101–105, 246–259, 291, 297–298, 307

Q
QTL, 103, 351, 353, 367–369
Quantitative trait locus, 351
Quantum dots, 262, 265–266, 275, 280, 294, 408

R
Radioimmunoassay, 172
Real-Time PCR, 59, 140–141, 146–147, 236–238, 337
Recombinant
  DNA, 33–52, 124–125, 182, 413–415, 469, 499
  DNA pharmaceuticals, 33–52
  DNA technology, 34–38, 47–51, 413–414, 499
  protein, 34, 39, 41–42, 46, 48–49, 245, 393, 414–415, 422
  Reporter gene assay, 72, 74–75, 80–81
Reporter, 7, 71–82, 140–141, 175, 217, 221
Reporter gene construct, 73, 81
Resistogram, 231
Restriction fragment length polymorphism testing, 430
Retinoblastoma, 45, 163, 290, 331, 357
RFLP, 145, 147, 232, 315, 337–340, 365–366, 459–460, 462
Ribotyping, 232, 324, 497
RNA, 1, 3, 6–7, 12–13, 16–17, 36–38, 46, 58–60, 105, 135–136, 145, 154, 162, 196, 238, 290–291, 320, 359, 384, 415, 431, 468, 475, 477, 506
RNA aptamers, 59–65

RT-PCR, 16, 44, 130, 137, 139, 141–142, 145, 233, 235, 291, 319
Ruc, 75–77

S
Screening, 3, 6, 21, 43–44, 81, 102, 104, 115, 121, 130, 145, 184, 190–191, 198, 205, 218, 222, 230, 251, 255, 294, 300, 302, 314, 317, 334, 349, 358–359, 364, 366, 391, 394, 396, 399–400, 403, 406, 415, 417, 423
SELEX, 57–62, 64
Sensor technology, 215, 220–221, 476–478
Sequencing functional genomics, 5, 13, 86, 98, 101–102, 186, 194, 201, 393
Sex determination, 146, 161, 163, 429–430
Sexually transmitted diseases, 317, 418, 509
Signaling, 14, 17, 216, 249, 267, 348, 359, 372, 386
Simplex diseases, 350–351
Single gene disease, 88–90, 374
Single gene disorders, 89–90, 332
Single locus probe, 430
Single nucleotide polymorphism, 10, 24, 89, 116, 123, 137, 140, 201, 218, 266, 290, 301, 332, 365–366, 386, 389, 394–395, 436, 443, 445, 451–452, 462
Site specific mutation, 130
SLP, 315, 430
SNP, 10, 24, 89, 116, 123, 201, 266, 290, 301, 332, 337, 343, 349, 355, 366, 386, 389, 399–400, 407, 437, 443, 445, 452, 462
Somatic gene therapy, 119–120
STR, 430, 457, 461–462, 464
Systems biology, 11–13, 19, 25–27, 92, 217, 258, 307

T
Taq polymerase, 129, 132
Template, 16, 38, 42, 58, 130, 132–135, 137–138, 140–141, 147, 174, 273–274
Thalassaemia, 336, 339
Therapeutics, 2, 5–6, 10, 28, 61, 63–64, 99, 175, 184, 187, 262–263, 267–289, 288, 394, 398, 402, 405–408, 414, 441–442, 486, 506, 509
Thermus auquaticus, 129
T_m, 129, 133–134, 514
Toxicity, 1–2, 10, 22–23, 86, 106, 200, 202, 204, 212, 220, 256, 264, 268–269, 275, 326, 398–399, 405, 407, 440, 443, 449
Toxicogenomics, 25
Transcriptional factors, 193
Transcriptomics, 12, 15–17, 23–27, 99, 103, 105, 194
Tumor markers, 102, 257, 357–358, 386, 406, 509

V
Vaccines, 47–49, 51, 99, 184, 187–188, 398, 414–415, 417–422, 439, 485–486, 491, 496, 497, 499
Validation, 3, 98, 171, 205–206, 239–241, 291, 296–298, 301, 307, 315, 327, 441, 444, 463, 471, 513
Vector borne diseases, 309
Vectors, 36, 38, 42–43, 45, 47, 76, 79, 81–82, 106, 138, 143, 182, 186–187, 269–271, 389, 414, 422, 485
Viral infections, 313, 318–321
Virions, 183

X
X-gal, 74, 78–79
X-linked inheritance, 348

Y
Yeast two hybrid analysis, 17, 74

CPSIA information can be obtained at www.ICGtesting.com
Printed in the USA
LVOW070605030412

275894LV00006B/85/P

9 789048 132607